The Mathematics of Generalization

The Mathematics of Generalization

The Proceedings of the SFI/CNLS Workshop on Formal Approaches to Supervised Learning

Editor

David H. Wolpert
Santa Fe Institute
Santa Fe, NM 87505

Proceedings Volume XX

Santa Fe Institute
Studies in the Sciences of Complexity

CRC Press
Taylor & Francis Group
Boca Raton London New York

CRC Press is an imprint of the
Taylor & Francis Group, an **informa** business

A CHAPMAN & HALL BOOK

Director of Publications, Santa Fe Institute: *Ronda K. Butler-Villa*
Publications Assistant, Santa Fe Institute: *Della L. Ulibarri*

First published 1995 by Westview Press

Published 2018 by CRC Press
Taylor & Francis Group
6000 Broken Sound Parkway NW, Suite 300
Boca Raton, FL 33487-2742

Visit the Taylor & Francis Web site at
http://www.taylorandfrancis.com

and the CRC Press Web site at
http://www.crcpress.com

ISBN 13: 978-0-201-40983-3 (pbk)
ISBN 13: 978-0-201-40985-7 (hbk)

This volume was typeset using T_EXtures on a Macintosh II computer.

About the Santa Fe Institute

   The *Santa Fe Institute* (SFI) is a multidisciplinary graduate research and teaching institution formed to nurture research on complex systems and their simpler elements. A private, independent institution, SFI was founded in 1984. Its primary concern is to focus the tools of traditional scientific disciplines and emerging new computer resources on the problems and opportunities that are involved in the multidisciplinary study of complex systems—those fundamental processes that shape almost every aspect of human life. Understanding complex systems is critical to realizing the full potential of science, and may be expected to yield enormous intellectual and practical benefits.

All titles from the *Santa Fe Institute Studies in the Sciences of Complexity* series will carry this imprint which is based on a Mimbres pottery design (circa A.D. 950–1150), drawn by Betsy Jones. The design was selected because the radiating feathers are evocative of the outreach of the Santa Fe Institute Program to many disciplines and institutions.

Santa Fe Institute
Studies in the Sciences of Complexity

Lectures Volumes

Vol.	Editor	Title
I	D. L. Stein	Lectures in the Sciences of Complexity, 1989
II	E. Jen	1989 Lectures in Complex Systems, 1990
III	L. Nadel & D. L. Stein	1990 Lectures in Complex Systems, 1991
IV	L. Nadel & D. L. Stein	1991 Lectures in Complex Systems, 1992
V	L. Nadel & D. L. Stein	1992 Lectures in Complex Systems, 1993

Lecture Notes Volumes

Vol.	Author	Title
I	J. Hertz, A. Krogh, & R. Palmer	Introduction to the Theory of Neural Computation, 1990
II	G. Weisbuch	Complex Systems Dynamics, 1990
III	W. D. Stein & F. J. Varela	Thinking About Biology, 1993

Reference Volumes

Vol.	Author	Title
I	A. Wuensche & M. Lesser	The Global Dynamics of Cellular Automata: Attraction Fields of One-Dimensional Cellular Automata, 1992

Proceedings Volumes

Vol.	Editor	Title
I	D. Pines	Emerging Syntheses in Science, 1987
II	A. S. Perelson	Theoretical Immunology, Part One, 1988
III	A. S. Perelson	Theoretical Immunology, Part Two, 1988
IV	G. D. Doolen et al.	Lattice Gas Methods for Partial Differential Equations, 1989
V	P. W. Anderson, K. Arrow, D. Pines	The Economy as an Evolving Complex System, 1988
VI	C. G. Langton	Artificial Life: Proceedings of an Interdisciplinary Workshop on the Synthesis and Simulation of Living Systems, 1988
VII	G. I. Bell & T. G. Marr	Computers and DNA, 1989
VIII	W. H. Zurek	Complexity, Entropy, and the Physics of Information, 1990
IX	A. S. Perelson & S. A. Kauffman	Molecular Evolution on Rugged Landscapes: Proteins, RNA and the Immune System, 1990
X	C. G. Langton et al.	Artificial Life II, 1991
XI	J. A. Hawkins & M. Gell-Mann	The Evolution of Human Languages, 1992
XII	M. Casdagli & S. Eubank	Nonlinear Modeling and Forecasting, 1992
XIII	J. E. Mittenthal & A. B. Baskin	Principles of Organization in Organisms, 1992

Proceedings Volumes (con't.)

Contributors to This Volume

Hussein Almuallim, King Fahd University of Petroleum and Minerals
Ghulum Bakiri, University of Bahrain
Leo Breiman, University of California
Christopher C. J. Burges, AT&T Bell Laboratories
Richard Campbell, University of Wisconsin
Peter Cheeseman, NASA Ames Research Center
John S. Denker, AT&T Bell Laboratories
Thomas G. Dietterich, Oregon State University
Chong Gu, University of Wisconsin
David Haussler, University of California
Geoffrey E. Hinton, University of Toronto
Alan S. Lapedes, Los Alamos National Laboratory
Steven J. Nowlan, The Salk Institute
Naftali Tishby, Hebrew University
Grace Wahba, University of Wisconsin
Yuedong Wang, University of Wisconsin
Manfred Warmuth, University of California
David H. Wolpert, Santa Fe Institute

Contents

David H. Wolpert
Santa Fe Institute,1660 Old Pecos Trail, Suite A, Santa Fe, NM 87505
e-mail: dhw@santafe.edu

Preface

This book grew out of a workshop held under the auspices of the Center for Nonlinear Studies at Los Alamos and the Santa Fe Institute in the summer of 1992. The idea for the workshop arose from a perception that there were many different fields that address supervised learning, but by and large these fields were not communicating with one another. (Examples of such fields are neural nets, conventional Bayesian statistics, conventional sampling theory statistics, computational learning theory, AI, and machine learning.) In particular, there were many different mathematical frameworks for addressing supervised learning. All had their own jargon, their own concerns, and their own results. And for the most part they were not interacting.

This was clearly a less than optimal state of affairs; we all have much to learn from one another, not only in terms of raw mathematical results, but also (perhaps more importantly) in perceptions of what the crucial issues are and how they should be addressed. Unfortunately, although it seems that this problem is abating, the rate of improvement is quite small. It seems possible that a general lack of communication amongst its practitioners will characterize supervised learning theory for some time to come.

The purpose of the workshop was try to (begin to) rectify this situation. A small group of researchers from several of the different supervised learning fields was brought together and, in effect, forced to mingle. The format of the workshop was an intensive two-day session of talks and discussion. The participants were encouraged to have their talks be as informative as possible to practitioners of the other fields. Obviously such a workshop could not make all the participants instant experts in each other's fields. Rather the hope was that the participants would go home with a heightened sensitivity to the other fields. They could then follow up and explore those other fields "on their own time," as it were. In this the workshop appears to have been successful, to judge by the comments made by the participants subsequent to the workshop.

This volume is an attempt to try to replicate the success of the workshop in a broader context. Its purpose is to do for the reader what the workshop did for its participants: help a practitioner in one of the fields that make up supervised learning become acquainted with the relevant work by his or her colleagues in other fields.

Obviously (and unfortunately) it is not possible to duplicate in a reader of a book the experience of "an intensive two-day session...(of being) forced to mingle... (with) researchers from different fields." Given the different format, slightly different means are needed to achieve the same ends. Accordingly, it was decided that the chapters in this volume should not so much be a formal compendium of the talks presented at the workshop as an overview of the work being performed by the researchers who attended the workshop. Some of the work represented in these chapters had not even been completed at the time of the workshop. Some of the other chapters are reprints of work published shortly before or soon after the workshop. However, all of the chapters were chosen by their authors with the same goal in mind: to help those from other supervised learning fields get acquainted with the lay of those authors' lands. Moreover, the instructions to the authors were that they should not try to provide tutorials on their individual fields. (There are many other sources for such tutorials.) Rather they should present current cutting-edge perspectives and work that provide an intuitive understanding of what their field "is all about."

The first part of the book contains chapters that have an almost purely theoretical bent. The second part of the book contains chapters that are more directly concerned with real-world applications.

In "The Status of Supervised Learning Science circa 1994—The Search for a Consensus," I try to lay out the primary issues associated with integrating the approaches to supervised learning theory, i.e., with transforming supervised learning theory from a collection of techniques used by researchers with shared interests into a single unified field. In particular, I address the issue of what our goals should be and what questions we should ask if we wish to be successful at this task.

In "Reflections After Referring Papers for NIPS," Leo Breiman presents a global view of theoretical approaches to supervised learning. He does not so much concentrate on the specifics of the frameworks represented at the workshop as present lessons gleaned from his experience observing the rise and fall of various theoretical approaches within his own field of conventional statistics. He then draws lessons from those observations for the broad issue of what concerns theoretical approaches to supervised learning can/should be addressing.

In "The Probably Approximately Correct (PAC) and Other Learning Models," David Haussler and Manfred Warmuth review one of the newer and more vibrant forms of supervised learning theory, the PAC framework. At present this framework is the only one to incorporate the concerns of computer science theory into supervised learning. Haussler and Warmuth both recapitulate PAC and discuss extensions of it that take into account some of the concerns of other theoretical approaches to supervised learning.

In "Decision Theoretic Generalizations of the PAC Model for Neural Net and Other Learning Applications," David Haussler presents work specifically designed to bridge the gap between some of the theoretical approaches to supervised learning. The work presented in this chapter extends the PAC framework in a number of different respects. In particular, it presents extensions that incorporate some of the concerns of the Bayesian decision theory community.

In "The Relationship Between PAC, the Statistical Physics Framework, the Bayesian Framework, and the VC Framework," I present several of the more prominent frameworks represented at the workshop in terms of a single over-arching formalism. The hope is that such a presentation facilitates a comparison of the frameworks, and that it perhaps even facilitates their cross-fertilization. (As a possible example of such cross-fertilization I present a "Bayesian correction" to the familiar bias-plus-variance formula.)

In "Statistical Physics Models of Supervised Learning," Naftali Tishby and his co-authors present research that exemplifies the "statistical physics" approach to supervised learning. In particular, they use thermodynamic limits and the annealed approximation—concepts borrowed from statistical physics—to analyze the learning curves for some perceptron models.

In "On Exhaustive Learning," Alan Lapedes and myself analyze some of the properties of the "zero-temperature" version of one of the learning algorithms discussed in the article by Tishby et al. In particular, we analyze how much of the learning curve's behavior simply reflects memorization, and whether the input space is always effectively countable, due to the assumptions of the model.

In "A Study of Maximal-Coverage Learning Algorithms," Hussein Almuallim and Tom Dieterich extend Dieterich's previous research in which the supervised learning problem is viewed "in reverse" of the usual view found in the computational learning theory community: rather than start with a concept class and try to find an algorithm that is good at learning concepts from within that class, instead try to find algorithms that are good at learning classes that are as broad as possible.

In "On Bayesian Model Selection," Peter Cheeseman provides an overview of the Bayesian approach to all statistical analysis problems, of which supervised learning is a subset. He discusses how "overfitting" is avoided by the use of priors, their relationship with minimum message length principles, and the relative merits of choosing the most probable model versus taking a weighted average of models.

In "Soft Classification, a.k.a. Risk Estimation, via Penalized Log Likelihood and Smoothing Spline Analysis of Variance," Grace Wahba and her co-authors present work that is in many respects a canonical example of the conventional statistics approach to supervised learning. Involving splines and analysis of variance, this work is not simply a set of theoretical concerns. Rather Wahba et al. also use those concerns to derive real-world algorithms. This is what theory should do: directly guide practice.

In "Current Research," Leo Breiman gives an overview of some of his recent applied supervised learning work. He discusses "hinges and ramps," surface-fitting techniques that are extremely fast, and combining the estimates of different learning algorithms using the technique of stacking. He also discusses how to optimize tree structures, how to implement them on parallel machines, and "hypertrees," which do not restrict themselves to the axis-parallel splits common to conventional trees.

In their article "Simplifying Neural Networks by Soft Weight Sharing" and the accompanying preface for this volume, Steve Nowlan and Geoff Hinton reify some of the abstract notions of integrating formalisms. They present a successful system for supervised learning that can be gainfully viewed simply as a neural net technique (a variant of techniques like weight decay and weight sharing), or instead as a Bayesian technique, or as a minimum description length technique.

In "Error-Correcting Output Codes: A General Method for Improving Multi-class Inductive Learning Programs," Tom Dietterich and Ghulum Bakiri demonstrate the use of error-correcting codes to find good representations for output variables. They go on to present a series of tests showing how use of such codes can improve performance on the NETtalk data set.

Finally, in "Image Segmentation and Recognition," John Denker and Christopher Burges present a system that in many respects constitutes the "next step" in recognition (optical character, speech, etc.) algorithms. Although very much based on real-world considerations, this is a system that was designed to conform to a number of theoretically based criteria.

My primary thanks has to go to my co-organizer of the conference, Alan Lapedes. Without his encouragement, I frankly never would have thought that such a stellar group of researchers could be brought together. Other thanks go to the staff at the CNLS and the SFI that helped put together the workshop. In particular, thanks go to Barbara Rhodes of the CNLS and Andi Sutherland of the SFI. I would also like to thank Ronda Butler-Villa and Della Ulibarri of the SFI publications staff for helping me put together this book. Finally, I would like to thank all the people who were in some way connected with the workshop: David Campbell of the CNLS and Mike Simmons of the SFI, Kim Green of Los Alamos, Ginger Richardson and Patrisia Brunello of the SFI, and the staff of University House at Los Alamos.

Of course, thanks are also due to the agencies that support the CNLS and the SFI, and thereby supported the workshop. In particular, I would like to thank the DOE, the NSF, Citibank, and the MacArthur Foundation. I would also like to thank the NIH and TXN Inc., who provided additional support during the compilation of this volume.

July, 1994   David Wolpert
                                              Santa Fe

David H. Wolpert
Santa Fe Institute, 1399 Hyde Park Road, Santa Fe, NM 87501;
e-mail: dhw@santafe.edu

The Status of Supervised Learning Science Circa 1994: The Search for a Consensus

The field of "supervised learning" goes by many other names: inductive inference, statistical inference, (certain subfields of) machine learning, regression and classification, etc. For most purposes the problem can be formulated as follows: We have an *input space* X and an *output space* Y. There is an unknown relationship between X and Y which will be referred to as the *target* relationship. As an example, in much of supervised learning this relationship is assumed to take the form of function + noise—a Y value is given by a deterministic function of an X value, with noise added. In such a case the function involved is known as the *target function*.

One is given a set of m samples of the target relationship (the *training set* or *learning set*), created from that relationship according to a specified set of rules (the *likelihood function*). One is then given a value from the input space as a *question*. At this point, "learning" comes in: the problem, in its general form, is to use the training set to guess the X–Y relationship appropriate for the question. As an example, in the function+noise scenario, the problem might be to use the training set to guess what output space value on the target function corresponds to the given question. In general though, the precise meaning of "appropriate" can be quite complicated, and is determined through one or more loss functions.[21]

The Mathematics of Generalization, Ed. David Wolpert, SFI Studies in
the Sciences of Complexity, Proc. Vol XX, Addison-Wesley, 1995

1

Such a guessed relationship (function) from questions to outputs is known as a *hypothesis* relationship (function). An algorithm that produces a hypothesis relationship, basing the guess on the training set of $m(X \times Y)$ vectors generated by the target relationship (perhaps in conjunction with knowledge of the likelihood and/or some other prior knowledge), is called a *generalizer* or a *learning algorithm*.

To illustrate these definitions, note that if one uses cross validation to choose amongst a fixed set of generalizers, the resultant system is itself a generalizer. However, if we change the set of candidate generalizers, we change the behavior of the resultant cross-validating system. In this sense, if the set of candidate generalizers is not specified, cross validation does not constitute a generalizer.

Some examples of generalizers are back-propagated neural nets,[22] Holland's classifier system,[13] and some implementations of Rissanen's minimum description length principle[19,20] (which, along with many of the other schemes which attempt to exploit Occam's razor, is analyzed by Wolpert[29]). Other important examples are memory-based reasoning schemes,[23] regularization theory,[15,17] and similar schemes for overt surface fitting of a hypothesis function to the training set.[10,16,27] Conventional classifiers which work via Bayes' theorem, information theory, clustering analysis or the like (e.g., ID3,[18] Bayesian classifiers like Schlimmer's Stagger system described by Dietterich,[6] the systems described by Duda and Hart,[8] etc.) also serve as examples of generalizers. However, such classifiers usually can only guess a particular output value if that value occurs in the training set. This chapter assumes no such restriction on the guessing. Finally, it should also be noted that whole swathes of the fields of Bayesian and non-Bayesian statistics are concerned with the supervised learning problem. See, for example, Buntine and Weigend,[5] Geman,[11] Berger,[2] and Eubank,[9] and references therein.

Given this far-ranging nature of the supervised-learning problem, it should come as no surprise that theoretical investigations of supervised learning have been conducted from a number of different vantage points. In addition to the standard Bayesian and non-Bayesian approaches alluded to above, it has also been investigated using PAC,[3,25] Vapnik's "uniform convergence" framework,[1,26] and even statistical physics.[12,24,32] (See also the relevant articles in this proceedings.)

Such a multitude of approaches provides a certain richness. Unfortunately though, it also leads to much rediscovery of the wheel, and it sows much confusion. Lots of sound and fury, smoke and babble. Lots of playing around. In short, although the problem of supervised learning is so important to many disciplines, since those disciplines haven't gotten together for a good, long, heart-to-heart, there isn't yet really a *science* of supervised learning, in the sense of an agreed-upon, unified approach to the theory of the field.

How do we get from here to there? What's necessary to create a science of supervised learning? The first thing to do is create an inventory of what we already have. This means disentangling the various approaches to supervised learning and, in particular, answering questions like the following: What are the relative strengths of the different approaches to supervised learning? What are the relative weaknesses? How do the questions they are addressing differ? How do their assumptions differ?

How much do they duplicate one another? What fundamentally distinct ideas and insights do they collectively embody? What can be gained by combining techniques from the different approaches?

To answer these questions requires an over-arching theoretical framework.[1] However, as illuminating as such a framework and the answers it provides might be, it's important to realize that it's only a first step; such a framework does not a science make. In addition to such a framework there must be a shared culture, an explicit shared set of understandings as to what is important and what is not. There must be a shared set of understandings of how the science relates to other sciences, both in terms of its content (our technique can be used to solve your problem), and also in terms of its level of sophistication and maturity (we may be young compared to you guys, but we're enthusiastic). Without such a shared culture, even given a single shared theoretical superstructure, supervised learning is more a set of engineering techniques and tools for working with them than it is a bona fide science.

1. So what exactly goes into such a shared culture? In part, it's the answers to the following questions:

    a. Why do we care?

    b. Why should anyone else care?

Discussion: Currently, the answer one provides to (1a) often reflects what one's specialty was before one started conducting research in supervised learning, even to the extent that (for example) some physicists find supervised learning interesting simply in how it relates to physics (more precisely, statistical mechanics). Although such a broad range of reasons-for-caring is in many ways healthy, it also reflects a certain immaturity of the field. It means that the core of supervised learning, the reasons why it's interesting *purely in and of itself*, is only vaguely specified.

Discussion: Much of the current theory of supervised learning appears to have limited applicability to the real world. (For example, aside from a few exceptions (e.g., Drucker et al.[7]), PAC, the statistical physics approach, and the uniform convergence framework have not contributed to the field of applied supervised learning; see the chapter by Breiman[4] in this proceedings). Given this, much of current theory violates almost any conceivable answer to (1b). How should we formulate a theory of supervised learning so as to ensure automatic real-world relevance? Should we simply abandon those approaches (like PAC) which currently appear to have little such relevance? For that matter, what exactly are the aspects of an approach that seem to limit its relevance? Can we somehow excise those limiting aspects from the approaches that appear to have little real-world relevance?

---

[1] A preliminary attempt at such a framework can be found the article by Wolpert[33] appearing in this proceedings. That article also discusses why none of the existing approaches—including Bayesianism—is the holy grail, answering all we might wish to know.

Discussion: There is, of course, an entire field, much of which tries to perform theoretical supervised learning in a manner with real-world applicability: conventional Bayesian and non-Bayesian statistics. This raises the question of just how different the more recent, not-so-applicable approaches to supervised learning are from conventional statistics. Are these approaches reinventing the wheel? Or (perhaps worse) investigating issues long understood by others to be irrelevant dead ends? If either, are they going about things in a way which at least casts new light on these old wheels and ends? Or is there something fundamentally new underpinning them, something that simply hasn't worked its way out into the open yet?

Discussion: There are other ways in which the issue of comparing these alternative approaches to conventional statistics bears on our question. To a large degree, people care about conventional statistics because they have some data set and they want to see what it tells them. However, many of the other approaches to supervised learning, especially some of the more heuristic ones (like artificial intelligence, or Koza's genetic programming[14]), care for quite different reasons. These other approaches had their genesis in Big Issues. Like building an artificial intelligence. Or untangling how the natural sciences perform (or perhaps should perform?) the inference tasks involved in constructing a scientific theory. Even the straight data-manipulation problems associated with these other approaches to supervised learning (e.g., problems in visual pattern recognition) tend to be different from those of conventional statistics.

So is supervised learning simply conventional statistics, enlarged to where it's applied to problems other than those to which conventional statistics is usually applied, by people with relatively little statistical training, who come up with ad hoc (and potentially interesting) new techniques? If so, is there really any room for a theory of supervised learning which goes beyond that provided by conventional statistics?

Discussion: There is in fact a whole spectrum of issues potentially addressed by supervised learning. At the one end are practical, well-posed problems, concerning things like separating signal from noise. At the other end are profound, ill-posed problems, like a theory of how one should infer theories from experimental data, or how to build a true AI.

To what degree should we expect the techniques and concepts useful at one end of the spectrum to translate to the other? Might one fairly say that fields like statistics and function approximation are all the way over on the practical end of the spectrum, and that fields like neural nets are trying to explore the ground a little over from that end?

2. Given the answer to (1), what precise questions should we be addressing?

   a. What do we want to know? What do we want to know that we might be able to figure out?

b. As researchers in the field of theoretical supervised learning, what should our goals be? Where do we want to be five years from now? Ten years? Twenty?

Discussion: Much of supervised learning is tool driven; one takes concepts from other fields (e.g., in PAC, one takes the notion of computational complexity from computer science, or in the statistical physics approach, one takes the replica method from theoretical physics) and sees what questions they can answer for us. Although the results of such tool-driven work is useful, in a certain sense it is backwards—we should first determine the questions of interest, and only then try to answer them, inventing wholly new tools, if need be. This is the way of all natural sciences; we must answer (2a), and *then* poke around in our toolboxes to see what goodies they can provide us.

As one example, independent of what is easiest to do with the tools currently at our disposal, the question arises as whether we should be concentrating on i.i.d. (independent, identically distributed) statistics, as opposed to non-i.i.d., off-sample statistics (i.e., as opposed to concentrating on those questions lying outside of the training set). Which is of more interest in light of possible answers to (1)? I.i.d. statistics, the stuff of conventional statistics, is much easier to work with, and is usually of interest in the I-want-this-dataset-analyzed kinds of problems (although it turns out to be surprisingly common for testing scenarios to be governed by different distributions than those that created the training set, in which case the usual i.i.d. assumption must be modified[2]). However, if our goal is to build an artificial intelligence, the issue of how well it performs in situations it's never seen before looms large, and off-sample statistics—difficult as it is—becomes important.

Discussion: The tools that have been used in supervised learning usually come bundled, with some more or (all too often) less explicit assumptions. In this, essentially all the theoretical work done with supervised learning has been "assumptions driven." One makes some (hopefully "reasonable") assumptions about the physical universe, implicitly or otherwise, and then works out the statistical consequences. Obviously those starting assumptions can be wrong, and in fact in any real-world

[2] As an example of when the testing scenario is governed by different processes than those that created the training set, consider the problem of inferring protein secondary structure from primary structure (see Zhang et al.[35]). In this problem the training set is produced by evolution, and consists of a corpus of naturally occurring proteins whose primary and secondary structures are both known. Via natural selection, the process creating this corpus has strong coupling between output space values (i.e., secondary structure—the phenotype) and the viability of the associated input space values (i.e., primary structure—the genotype). This results in a strong statistical coupling between the distribution over input space values and the distribution taking inputs to outputs. In real-world testing however, often one is interested in predicting secondary structure for novel primary structures that either don't occur in nature or are taken out of their ecological context. In such a process there is essentially no coupling between the input space distribution and the distribution taking inputs to outputs (or more precisely, what coupling there is is determined by the research program rather than natural evolution). So training and testing are according to different processes.

application **all of them will always be at least partially wrong** (the universe being a real-number-valued place, with a truly ludicrous number of degrees of freedom, all of which are quantum mechanical). This is true even for as seemingly a "common sense" assumption as i.i.d. behavior, never mind heuristic assumptions such as those underlying the use of cross validation or Occam's razor. Do we really want our formalism to be predicated on at least partially incorrect assumptions, or is there some alternative?

3. Since everything is currently driven by (at least partially) incorrect assumptions, should we not think more about the general subjects of assumptions? Assumptions are the whole game, but very little work has been done on how best to deal with them.

     a. Is there a useful (!) formalism which explicitly takes into account the (perhaps unavoidable) incorrectness of our assumptions?

Discussion: The problem of inappropriate assumptions applies just as well to classical Bayesian analysis as to other formalisms. Most applied Bayesian analysis is just as guilty of not allowing for the possibility of any errors in its assumptions, and thereby imposing itself (i.e., its models and priors) on the universe, rather than letting the universe speak for itself.

Indeed, it is possible that the popular hierarchical Bayesian approach is by construction incapable of circumventing this issue. (In that approach, rather than directly manipulate the probabilities of the objects of interest, one manipulates probabilites of parameters of those objects, probabilities of those probabilities and of the associated parametriztions, etc.) Properly speaking, in such an approach one should average over all possible models (i.e., all possible paramertrizations) one might use to describe a particular problem. In practice, one settles for only considering a finite number of models. See the article by Cheeseman in this volume. But it's not clear that one can even define the space of all possible models, never mind put a measure on it. This means one can not claim an analysis using a restricted model space is in some sense an approximation to the "full calculation"[34] involving the entire space of all possible models. The practicing Bayesian not only tells God what priors to use, the Bayesian also tells God what space the prior should be over; (s)he tells God how to conceive of the problem. (Of course, this particular kind of hubris is by no means restricted to the hiearchical Bayesian approach.)

     b. Might it be possible to use the natural sciences—e.g., physics, psychology— somehow to *deduce* something of what our assumptions should be? In other words, can we use empirical knowledge concerning the rules by which our particular universe operates to tell us what assumptions to use? ("Universe" here meaning physical and/or cognitive universe.)

Discussion: In the extreme case, the answer would appear to be yes; in theory at least, one could use quantum mechanics in conjunction with the (currently unknown) spatio-temporal boundary conditions of the universe to *solve* for the prior probability that a human will investigate a problem generated according to some particular target function. The question before us is if there is a more tractable way of using the natural sciences to aid inductive inference.

Discussion: The empirical knowledge making up the natural sciences has been garnered via statistical inference, both in the small sense (error analysis of an experiment) and in the large sense (all science is statistical inference, the extrapolation from training sets ["empirical data"] to entire input-output relationships ["theories of how nature works"]). At a first cut, it makes sense to ignore this: to simply assume that science is "right," and to see if we can then use it to aid us in investigating the (relatively mundane) problems we supervised-learningtypes are usually concerned with. The more fascinating (and infinitely more difficult) course of action is to somehow take into account the inherently statistical nature of science when we use it to help us.

　　c.　If we must make assumptions without exploiting the natural sciences, what new *kinds* of assumptions should we be exploring?

Discussion: For example, does it make sense to spend more effort exploring an assumption concerning it like "It's better to have a generalizer whose behavior is invariant under application of the Euclidean symmetry groups to the training set together with the question" (an assumption that rules out linear model generalizers), rather than the rather hackneyed traditional assumptions concerning it, like "It's better to have hypothesis functions which are smooth"? Or might it make sense to use an assumption like the following: Rather than prefer hypothesis functions which are easily described "vertically," as patterns taking inputs to outputs, prefer hypothesis functions which are easily described "horizontally," as patterns relating the function at one part of the input space to the function at another part.[31] As another alternative, there is a close formal parallel between generalization (in the sense of using the training set to manipulate a set of candidate hypothesis functions) and meta-generalization (in the sense of using partitions of the training set to manipulate a set of candidate generalizers).[30] Given the empirically observed power of assumptions concerning meta-generalization (like cross validation), should we concentrate more on exploring variants of the common meta-generalization assumptions (e.g., "cross-validation works"), and less on exploring variants of conventional generalization assumptions (e.g., "favoring smoother input-output functions works")?

Discussion: Shouldn't we demand that our unified supervised learning theory, the starting point of our supervised learning science, at least be able to *address* the kinds of assumptions discussed above? Most formalisms cannot. Indeed, shouldn't we demand that our theory at least be able to address those assumptions that are the most innocuous, and are already widely used? For example, one might argue that the assumption "minimizing cross-validation error is good" is about as innocuous as they come. If so, shouldn't we require that we can integrate it into our theoretical structure? (This integration has never been satisfactorally carried out in many theoretical approaches to supervised learning, like PAC, Bayesianism, etc.)

Discussion: Alternatively, are there other more sophisticated ways of using the assumptions we already have? For example, rather than simply penalizing inferences according to certain generalization criteria and in this way choosing amongst those inferences, can we instead solve the inverse problem, use those criteria to specify the "optimal" inference, and then use this scheme for all our learning problems? It turns out that the answer is no if one of the "generalization criteria" is zero cross-validation error: we cannot combine the requirement of zero cross-validation error with other requirements (like perfect Euclidean invariance of the generalizer) and thereby specify a unique optimal generalizer.[3] But might it be possible to specify a unique "optimal" generalizer using other requirements, none of which involve cross validation? No one knows.

The fact that all of these issues are so open-ended suggests that as a science supervised learning is immature. Corroborating evidence for this view arises if one views supervised learning from the perspective of physics in particular and the natural sciences in general. For example, in mature versions of the natural sciences, experiment both comes up with phenomena for theory to explain and also answers questions posed by theory. In supervised learning, there's been essentially none of the latter process; there are no questions, having more than one possible answer, posed by theory for the experimentalists to answer (at least, there are no such questions beyond the ubiquitous "does this technique work well?").

Indeed, some approaches to supervised learning even appear immature when viewed from the perspective of other approaches. For example, most of the approaches lying outside of conventional statistics appear this way when compared to conventional statistics.

To summarize, we are missing

a. any kind of consensus on what framework to use;

b. any kind of consensus on where we want to go and what issues we want to answer;

c. any kind of consensus on what new approaches to the issue of assumptions we should be investigating.

For supervised learning to become a science, these missing consensus' must be found.

---

[3] For any such set of other requirements, either no generalizer satisfies all the other requirements, or multiple generalizers do. See paper by Wolpert.[28]

REFERENCES

1. Baum, E., and D. Haussler. "What Size Neural Net Gives Valid Generalization? " *Neural Comp.* **1** (1989): 151–160.
2. Berger, J. O. *Statistical Decision Theory and Bayesian Analysis.* Berlin: Springer-Verlag, 1985.
3. Blumer, A., A. Ehrenfeucht, D. Haussler, and M. Warmuth. "Learnability and Vapnik-Chervonenkis Dimension." *J. ACM* **36** (1989): 929–965.
4. Breiman, L. "Reflections After Referring Papers for NIPS." This volume.
5. Buntine, W., and A. Weigend. "Bayesian Back-Propagation." *Complex Systems* **5** (1991): 603–643.
6. Dietterich, T. "Machine Learning. " *Ann. Rev. Comp. Sci.* **4** (1990): 255–306.
7. Drucker, H. et al. "Improving Performance in Neural Networks Using a Boosting Algorithm." In *Neural Information Processing Systems 5*, edited by S. Hanson et al. San Mateo, CA: Morgan-Kaufman, 1993.
8. Duda, R., and P. Hart. *Pattern Classification and Scene Analysis.* New York: Wiley, 1973.
9. Eubank, R. *Spline Smoothing and Non-Parametric Regression.* New York: Marcel Dekker, 1988.
10. Farmer, J., and J. Sidorowich. "Exploiting Chaos to Predict the Future and Reduce Noise." Los Alamos Report LA-UR-88-901, Los Alamos, NM, 1988.
11. Geman, S., E. Bienenstock, and R. Doursaat. "Neural Networks and the Bias/Variance Dilemma." *Neural Comp.* **4** (1992): 1–58.
12. Hertz, J., A. Krogh, and R. G. Palmer, eds. *Introduction to the Theory of Neural Computation.* Santa Fe Institute Studies in the Sciences of Complexity. Lecture Vol. I. Redwood City, CA: Addison-Wesley, 1991.
13. Holland, J. *Adaptation in Natural and Artificial Systems.* Ann Arbor, MI: University of Michigan Press, 1975.
14. Koza, J. R. *Genetic Programming.* Cambridge, MA: MIT Press, 1992.
15. Morozov, V. *Methods for Solving Incorrectly Posed Problems.* New York: Springer-Verlag, 1984.
16. Omohundro, S. "Efficient Algorithms with Neural Network Behavior." Report UIUCSCS-R-87-1331. Computer Science Department. University of Illinois at Urbana-Champaign, Urbana, IL, 1987.
17. Poggio, T., and staff, MIT AI Lab. "Progress in Understanding Images. " In *Proceedings of the Image Understanding Workshop*, edited by L. Bauman. McLean, VA: MIT Press, 1988.
18. Quinlan, J. "Induction of Decision Trees. " *Machine Learning* **1** (1986): 81–106.
19. Rissanen, J. "A Universal Prior for Integers and Estimation by Minimum Description Length." *Ann. Stat.* **11** (1994): 416–431.
20. Rissanen, J. "Stochastic Complexity and Modeling. " *The Annals of Statistics* **14** (1986): 1080–1100.

21. Rosen, D. B. "Scoring the Forecaster by Mean Resulting Payoff of a Distribution of Decision Problems." Presented at The 13th Annual Workshop on Maximum Entropy and Bayesian Methods. Boston: Kluwer Academic, 1994.

22. Rumelhart, D., and J. McClelland. *Explorations in the Microstructure of Cognition*, Vol. I and II. Cambridge, MA: MIT Press, 1986.

23. Stanfill, C., and D. Waltz. "Toward Memory-Based Reasoning." *Comm. ACM* **29** (1986): 1213–1228.

24. Tishby, N., S. Solla, and E. Levin. "Consistent Inference of Probabilities in Layered Networks: Predictions and Generalization." In *IJCNN International Joint Conference on Neural Networks*, Vol. II, 403–409. New York: IEEE, 1989.

25. Valiant, L. "A Theory of the Learnable." *Comm. ACM* **27** (1984): 1134–1142.

26. Vapnik, V. *Estimation of Dependences Based on Empirical Data*. New York: Springer-Verlag, 1982.

27. Wolpert, D. "Constructing a Generalizer Superior to NETtalk via a Mathematical Theory of Generalization. " *Neural Nets*. **3** (1990): 445–452.

28. Wolpert, D. "A Mathematical Theory of Generalization: Part II." *Complex Systems* **4** (1990): 201–249.

29. Wolpert, D. "The Relationship Between Occam's Razor and Convergent Guessing." *Complex Systems* **4** (1990): 319–368.

30. Wolpert, D. "On the Connection Between In-Sample Testing and Generalization Error." *Complex Systems* **6** (1991): 47–94.

31. Wolpert, D. "Horizontal Generalization." Working Paper 92-07-033, Santa Fe Institute, Santa Fe, New Mexico, 1992.

32. Wolpert D., and A. Lapedes. "An Investigation of Exhaustive Learning." Working Paper 92-04-20, Santa Fe Institute , Santa Fe, New Mexico, 1992.

33. Wolpert, D. "The Relationship Between PAC, the Statistical Physics Framework, the Bayesian Framework, and the VC Framework." This volume.

34. Wolpert, D. "On the Bayesian 'Occam Factors' Argument for Occam's Razor." To appear in *Computational Learning Theory and Natural Learning Systems: Volume III Natural Learning Systems*, Edited by S. Hanson et al. Cambridge, MA: MIT Press: 1994

35. Zhang, X., et al. (1992). "Hybrid System for Protein Secondary Structure Prediction. " *J. Mol. Biol.* **225** (1992): 1049; erratum in *J. Mol. Biol.* **232** (1993): 1227.

Leo Breiman
Statistics Department, University of California, Berkeley, CA 94305;
e-mail: leo@stat.berkeley.edu

Reflections After Refereeing Papers for NIPS

In refereeing papers for NIPS I was struck by the growing emphasis on mathematical theory. Having lived for forty years in a field plagued with theory, and beginning life as a probability theorist, I thought I might make a few remarks that summarize what experience I have with theory as it may be relevant to machine learning.

1. WHAT IS THEORY?

This may be difficult to define. A rough definition is:

> THEORY = mathematical framework plus theorems plus proofs.
> "No theorems" implies "no theory."

One problem in the field of statistics has been that everyone wants to be a theorist. Part of this is envy—the real sciences are based on mathematical theory. In the universities for this century, the glamor and prestige has been in mathematical models and theorems, no matter how irrelevant.

As a result of the would-be mathematicians in statistics, it has been dominated by useless theory and fads.

- Decision Theory
- Asymptotics
- Robustness
- Nonparametric One and Two Sample Tests
- One-Dimensional Density Estimation
- Etc.

If statistics is an applied field and not a minor branch of mathematics, then more than 99% of the published papers are useless exercises. (The other colleagues in statistics I have spoken to say this is an exaggeration and peg the percentage at 95%. Either way it is significant). The result is a downgrading of sensibility and intelligence.

But among all of the trash, there are a few places where theory has been useful. To understand the potential usefulness of theory, I look at this a bit more.

2. USES OF THEORY

- **Comfort**: We knew it worked, but it's nice to have a proof.
- **Insight**: Aha! So that's why it works.
- **Innovation**: At last, a mathematically proven idea that applies to data.
- **Suggestion**: Something like this might work with data.

3. EXAMPLES: POST WORLD WAR II

3.1 COMFORT

Mainly asymptotics.

1. Gordon and Olshen proved CART consistent; i.e., as the sample size goes to infinity, the CART risk converges to the Bayes risk. The estimated sample size for the proof to take force is in the neighborhood of a million.
2. Cover proves that as the sample size becomes infinite, the nearest-neighbor risk becomes less than twice the Bayes risk (also provides insight).
3. Many theorems that say: given a specified class of basis functions, linear combinations of these are dense in some large class of smooth functions; i.e., any sufficiently smooth function can be arbitrarily closely approximated by a linear combination of the basis functions.

4. Charles Stone proves that if data $\{(y_n, \mathbf{x}_n), n = 1, \ldots, N\}$ where $\mathbf{x}_n$ is a vector in M dimensions, is sampled from $y = f(\mathbf{x}) + \varepsilon$, and some procedure used to estimate f, then there is always a continuous differentiable f such that the root mean squared error in the estimate is at least order of $N^{(-2/(2+M))}$. Thus, no procedure can hope to approximate all continuous differentiable functions in a high-dimensional space without the use of very large sample sizes.

Stone also proved later that if f is known to be in the class of continuous differentiable functions consisting of sums of functions of $J < M$ variables, then there is an approximation method such that the root mean squared error is uniformly of order $N^{(-2/(2+J))}$. These two results also provide insight.

But sometimes comfort results can be misleading. For instance, things that hold for nearly infinite sample size blow up for finite sample sizes. One example is the use of Akaike's penalty to select the dimension of a model. Its asymptotic properties simply do not hold for moderate sample sizes and its use gives poor results.

3.2 INSIGHT

1. Donoho's recent work on image processing under sparsity constraints.
2. Recent work on tomography.

Note that in both cases, physical laws made possible more precise modeling of data.

3.3 SUGGESTION

Here it is difficult to cite examples because the suggestive effect of some piece of theory on applied results in a different area are often undocumented. In my own work:

1. The pi-method, for approximating functions using noisy data, was suggested by results in mathematical approximation theory.
2. The ACE algorithm is a data implementation of results suggested by random variable inequalities.

3.4 INNOVATION

1. The theoretical work by Weiner and others on the spectral analysis of stationary time series penetrated statistics following Tukey's heuristic work on estimation of the spectrum. This opened up the field of time series analysis.
2. Shannon's work on information theory led to some important early work by statisticians, but the applications and further work has passed onto the engineering fields.

3. Efron's invention of the bootstrap and his early work on its asymptotic prop- erties established its credentials and it is now extensively used in applied work. But even so, the analytics of its performance are known only in simple cases.

The above list of useful theories was not meant to be inclusive, but even a more inclusive list would be very short. A possible reason is that it is difficult to formulate reasonable analytic models for complex data.

Notice also that none of the useful theory on the list were of the Grand Unifica- tion Type theory. Following WW II there was an effort to provide GUT theory for statistics in the form of decision theory, and works that hung off of this framework. In spite of intense activity, none of this work has had any effect on the day-to-day practice of statistics, or even on present-day theory. It slumbers in its own sanctified graveyard.

Mathematical theory is not critical to the development of machine learning.

But scientific inquiry is.

3.5 INQUIRY

INQUIRY = sensible and intelligent efforts to understand what is going on. For example:

- mathematical heuristics
- simplified analogies (like the Ising Model)
- simulations
- comparisons of methodologies
- devising new tools
- theorems where useful (rare!)
- shunning panaceas

Regarding this last point, every field gets frozen when a certain tool becomes a panacea (a.k.a. fad). For instance, for years in statistics, everything had to be robust. In machine learning, the current panacea is a sigmoid network fitted using backpropagation.

My colleague, Jerry Friedman, once told me an old folk saying: "Give a man a hammer and every problem looks like a nail." What is needed is not one hammer, but many different tools along with a sense of which ones to use.

For example, I have image data of healthy and diseased bones that was fit by use of neural networks at Mayo Clinic and produced 95% test set accuracy. Crude CART runs got 96% accuracy. I think that any reasonable classification method used on this data would produce comparable accuracy, possibly even linear discriminant analysis.

Our fields would be better off with far fewer theorems, less emphasis on faddish stuff, and much more scientific inquiry and engineering. But the latter requires real thinking.

For instance, there are many important questions regarding neural networks which are largely unanswered. There seem to be conflicting stories regarding the following issues:

- Why don't heavily parameterized neural networks overfit the data?
- What is the effective number of parameters?
- Why doesn't backpropagation head for a poor local minima?
- When should one stop the backpropagation and use the current parameters?

It makes research more interesting to know that there is no one universally best method. What is best is data dependent. Sometimes "least glamorous" methods such as nearest neighbor are best. We need to learn more about what works best where. But emphasis on theory often distracts us from doing good engineering and living with the data.

David Haussler and Manfred Warmuth
Baskin Center for Computer Engineering and Information Sciences, University of California,
Santa Cruz, CA 95064; e-mail: haussler@cse.ucsc.edu, manfred@cse.ucsc.edu

The Probably Approximately Correct (PAC) and Other Learning Models

Reprinted from *Proceedings of the 8th National Conference on Artificial Intelligence*, co-chaired by Tom Dietterich and Bill Swartout, 1101–1108. Cambridge, MA: MIT, 1990.

This paper surveys some recent theoretical results on the efficiency of machine learning algorithms. The main tool described is the notion of Probably Approximately Correct (PAC) learning, introduced by Valiant. We define this learning model and then look at some of the results obtained in it. We then consider some criticisms of the PAC model and the extensions proposed to address these criticisms. Finally, we look briefly at other models recently proposed in computational learning theory.

1. INTRODUCTION

It's a dangerous thing to try to formalize an enterprise as complex and varied as machine learning so that it can be subjected to rigorous mathematical analysis. To be tractable, a formal model must be simple. Thus, inevitably, most people will feel that important aspects of the activity have been left out of the theory. Of course, they will be right. Therefore, it is not advisable to present a theory of machine learning as having reduced the entire field to its bare essentials. All that can be hoped for is that some aspects of the phenomenon are brought more clearly into focus using the tools of mathematical analysis, and that perhaps a few new insights are gained. It is in this light that we wish to discuss the results obtained in the last few years in what is now called PAC (Probably Approximately Correct) learning theory.[7]

Valiant[53] introduced this theory in 1984 to get computer scientists, who study the computational efficiency of algorithms, to look at learning algorithms. By taking some simplified notions from statistical pattern recognition and decision theory, and combining them with approaches from computational complexity theory, he came up with a notion of learning problems that are feasible, in the sense that there is a polynomial time algorithm that "solves" them, in analogy with the class **P** of feasible problems in standard complexity theory. Valiant was successful in his efforts. Since 1984 many theoretical computer scientists and AI researchers have either obtained results in this theory, or complained about it and proposed modified theories, or both.

The field of research that includes the PAC theory and its many relatives has been called computational learning theory. It is far from being a monolithic mathematical edifice that sits at the base of machine learning; it's unclear whether such a theory is even possible or desirable. We argue, however, that insights have been gained from the varied work in computational learning theory. The purpose of this short monograph is to survey some of this work and reveal those insights.

2. DEFINITION OF PAC LEARNING

The intent of the PAC model is that successful learning of an unknown target concept should entail obtaining, with high probability, a hypothesis that is a good approximation of it. Hence the name Probably Approximately Correct. In the basic model, the instance space is assumed to be $\{0,1\}^n$, the set of all possible assignments to n Boolean variables (or *attributes*), and concepts and hypotheses are subsets of $\{0,1\}^n$. The notion of approximation is defined by assuming that there is some probability distribution D defined on the instance space $\{0,1\}^n$, giving the probability of each instance. We then let the *error* of a hypothesis h with respect to a fixed target concept c, denoted $error(h)$ when c is clear from the context, be defined by

$$error(h) = \sum_{x \in h \Delta c} D(x),$$

where Δ denotes the symmetric difference. Thus, $error(h)$ is the probability that h and c will disagree on an instance drawn randomly according to D. The hypothesis h is a good approximation of the target concept c if $error(h)$ is small.

How does one obtain a good hypothesis? In the simplest case, one does this by looking at independent random examples of the target concept c, each example consisting of an instance selected randomly according to D, and a label that is "+" if that instance is in the target concept c (*positive example*), otherwise "−" (*negative example*). Thus, training and testing use the same distribution, and there is no "noise" in either phase. A learning algorithm is then a computational procedure that takes a sample of the target concept c, consisting of a sequence of independent random examples of c, and returns a hypothesis.

For each $n \geq 1$, let C_n be a set of target concepts over the instance space $\{0,1\}^n$, and let $\mathbf{C} = \{C_n\}_{n \geq 1}$. Let H_n, for $n \geq 1$, and $\mathbf{H}$ be defined similarly. We can define PAC learnability as follows: The concept class $\mathbf{C}$ is PAC learnable by the hypothesis space $\mathbf{H}$ if there exists a polynomial–time learning algorithm A and a polynomial $p(\cdot, \cdot, \cdot)$ such that for all $n \geq 1$, all target concepts $c \in C_n$, all probability distributions D on the instance space $\{0,1\}^n$, and all ϵ and δ, where $0 < \epsilon, \delta < 1$; if the algorithm A is given at least $p(n, 1/\epsilon, 1/\delta)$ independent random examples of c drawn according to D, then with probability at least $1 - \delta$, A returns a hypothesis $h \in H_n$ with $error(h) \leq \epsilon$. The smallest such polynomial p is called the *sample complexity* of the learning algorithm A.

The intent of this definition is that the learning algorithm must process the examples in polynomial time, i.e., be computationally efficient, and must be able to produce a good approximation to the target concept with high probability using only a reasonable number of random training examples. The model is worst case in that it requires that the number of training examples needed be bounded by a single fixed polynomial for all target concepts in $\mathbf{C}$ and all distributions D in the instance space. It follows that if we fix the number of variables n in the instance space and the confidence parameter δ, and then invert the sample complexity function to plot the error ϵ as a function of training sample size, we do not get what is usually thought of as a learning curve for A (for this fixed confidence), but rather the upper envelope of all learning curves for A (for this fixed confidence), obtained by varying the target concept and distribution on the instance space. Needless to say, this is not a curve that can be observed experimentally. What is usually plotted experimentally is the error versus the training sample size for particular target concepts on instances chosen randomly according to a single fixed distribution on the instance space. Such a curve will lie below the curve obtained by inverting the sample complexity. We will return to this point later.

Another thing to notice about this definition is that target concepts in a concept class **C** may be learned by hypotheses in a different class **H**. This gives us some flexibility. Two cases are of interest. The first is that **C** = **H**; i.e., the target class and hypothesis space are the same. In this case we say that **C** is *properly* PAC learnable. Imposing the requirement that the hypothesis be from the class **C** may be necessary, e.g., if it is to be included in a specific knowledge base with a specific inference engine. However, as we will see, it can also make learning more difficult. The other case is when we don't care at all about the hypothesis space **H**, so long as the hypotheses in **H** can be evaluated efficiently. This occurs when our only goal is accurate and computationally efficient prediction of future examples. Being able to freely choose the hypothesis space may make learning easier. If **C** is a concept class and there exists some hypothesis space **H**, such that hypotheses in **H** can be evaluated on given instances in polynomial time, and such that **C** is PAC learnable by **H**, then we will say simply that **C** is *PAC learnable*.

There are many variants of the basic definition of PAC learnability. One important variant defines a notion of syntactic complexity of target concepts and, for each $n \geq 1$, further classifies each concept in C_n by its syntactic complexity. Usually the syntactic complexity of a concept c is taken to be the length of (number of symbols or bits in) the shortest representation of c in a fixed concept representation language. In this variant of PAC learnability, the number of training examples is also allowed to grow polynomially in the syntactic complexity of the target concept. This variant is used whenever the concept class is specified by a concept representation language that can represent any Boolean function, for example, when discussing the learnability of DNF (Disjunctive Normal Form) formulae or decision trees. Other variants of the model let the algorithm request examples, use separate distributions for drawing positive and negative examples, or use randomized (i.e., coin flipping) algorithms.[31] It can be shown that these latter variants are equivalent to the model described here, in that, modulo some minor technicalities, the concept classes that are PAC learnable in one model are also PAC learnable in the other.[26] Finally, the model can easily be extended to non-Boolean attribute-based instance spaces[20] and instance spaces for structural domains such as the blocks world.[22] Instances can also be defined as strings over a finite alphabet so that the learnability of finite automata, context-free grammars, etc. can be investigated.[44]

3. OUTLINE OF RESULTS FOR PROPER PAC LEARNABILITY: HARDNESS RESULTS BASED ON THE ASSUMPTION P ≠ NP

A number of fairly sharp results have been found for the notion of proper PAC learnability. The following summarizes some of these results. For precise definitions of the concept classes involved, the reader is referred to the literature cited. The negative results are based on the complexity theoretic assumption that $\mathbf{RP} \neq \mathbf{NP}$.[45]

1. Conjunctive concepts are properly PAC learnable,[53] but the class of concepts in the form of the disjunction of two conjunctions is not properly PAC learnable,[45] and neither is the class of existential conjunctive concepts on structural instance spaces with two objects.[22]
2. Linear threshold concepts (perceptrons) are properly PAC learnable on both Boolean and real-valued instance spaces,[10] but the class of concepts in the form of the conjunction of two linear threshold concepts is not properly PAC learnable.[12] The same holds for disjunctions and linear thresholds of linear thresholds (i.e., multilayer perceptrons with two hidden units). In addition, if the weights are restricted to 1 and 0 (but the threshold is arbitrary), then linear threshold concepts on Boolean instances spaces are not properly PAC learnable.[45]
3. The classes of k-DNF, k-CNF, and k-decision lists are properly PAC learnable for each fixed k,[47,54] but it is unknown whether the classes of all DNF functions, all CNF functions, or all decision trees are properly PAC learnable.

Most of the difficulties in proper PAC learning are due to the computational difficulty of finding a hypothesis in the particular form specified by the target class. For example, while Boolean threshold functions with 0-1 weights are not properly PAC learnable on Boolean instance spaces (unless $\mathbf{RP} = \mathbf{NP}$), they are PAC learnable by general Boolean threshold functions. Here we have a concrete case where enlarging the hypothesis space makes the computational problem of finding a good hypothesis easier. The class of all Boolean threshold functions is simply an easier space to search than the class of Boolean threshold functions with 0-1 weights. Similar extended hypothesis spaces can be found for the two classes mentioned in item 1 above that are not properly PAC learnable. Hence, it turns out that these classes are PAC learnable.[45] However, it is not known if any of the classes of DNF functions, CNF functions, decision trees, or multilayer perceptrons with two hidden units are PAC learnable.

4. METHODS FOR PROVING PAC LEARNABILITY; FORMAL-IZATION OF BIAS

All of the positive learnability results above are obtained by:

1. showing that there is an efficient algorithm that finds a hypothesis in a particular hypothesis space that is consistent with a given sample of any concept in the target class and
2. showing that the sample complexity of any such algorithm is polynomial.

By *consistent* we mean that the hypothesis agrees with every example in the training sample. An algorithm that always finds such a hypothesis (when one exists) is called a *consistent algorithm*.

As the size of the hypothesis space increases, it may become easier to find a consistent hypothesis, but it will require more random training examples to insure that this hypothesis is accurate with high probability. In the limit, when any subset of the instance space is allowed as a hypothesis, it becomes trivial to find a consistent hypothesis, but a sample size proportional to the size of the entire instance space will be required to insure that it is accurate. Hence, there is a fundamental trade-off between the computational complexity and the sample complexity of learning.

Restriction to particular hypothesis spaces of limited size is one form of *bias* that has been explored to facilitate learning.[41] In addition to the cardinality of the hypothesis space, a parameter known as the Vapnik-Chervonenkis (VC) dimension of the hypothesis space has been shown to be useful in quantifying the bias inherent in a restricted hypothesis space.[20] The VC dimension of a hypothesis space H, denoted $VCdim(H)$, is defined to be the maximum number d of instances that can be labeled as positive and negative examples in all 2^d possible ways, such that each labeling is consistent with some hypothesis in H.[15,55] Let $\mathbf{H} = \{H_n\}_{n\geq 1}$ be a hypothesis space and $\mathbf{C} = \{C_n\}_{n\geq 1}$ be a target class, where $C_n \subseteq H_n$ for $n \geq 1$. Then it can be shown[51] that any consistent algorithm for learning $\mathbf{C}$ by $\mathbf{H}$ will have sample complexity at most

$$\frac{1}{\epsilon(1 - \sqrt{\epsilon})} \left(2VCdim(H_n)\ln\frac{6}{\epsilon} + \ln\frac{2}{\delta} \right).$$

This improves on earlier bounds given by Blumer et al.,[10] but may still be a considerable overestimate. In terms of the cardinality of H_n, denoted $|H_n|$, it can be shown[9,42,55] that the sample complexity is at most

$$\frac{1}{\epsilon} \left(\ln|H_n| + \ln\frac{1}{\delta} \right).$$

For most hypothesis spaces on Boolean domains, the second bound gives the better bound. However, linear threshold functions are a notable exception, since the VC dimension of this class is linear in n, while the logarithm of its cardinality is quadratic in n.[10] Most hypothesis spaces on real-valued attributes are infinite, so only the first bound is applicable.

5. LEARNABILITY PRESERVING REDUCTIONS: A COMPLEXITY THEORETIC APPROACH

We now return to the computational difficulty of PAC learning. The theory of complexity classes and reducibilities (e.g., **NP**-completeness) has been particularly useful in providing evidence for the intractability of general computational problems. Pitt and Warmuth[46] developed a similar complexity theory for learnability.[1]

Following the approach taken in complexity theory, we would like to allow a learning algorithm to receive more training examples (and to spend more time) before achieving accurate learning, depending on the "complexity" of the hidden target concept to be learned. As mentioned in the section "Definition of PAC Learning," a reasonable measure of this complexity is the length (e.g., number of bits) of the representation of the concept in some given concept representation language. Thus the learnability will depend on what type of representations of the concepts we have chosen. For example, we may choose to represent regular languages by DFAs, NFAs, regular expressions, etc. We would like to ask the question "Are DFAs learnable?" rather than the question "Are regular languages learnable?"

This motivates the following definition: A *learning problem* consists of a concept representation language and a mapping from representations in this language to concepts. The concept class of a learning problem is the class of all concepts that can be represented in the given representation language. Whereas in previous sections we have referred to the PAC learnability of concept classes, here we will be more precise and speak of the PAC learnability of learning problems, emphasizing the fact that the difficulty of learning may depend on the concept representation language used to represent the concept class.

As in complexity theory, when faced with a new learning problem we can first attempt to reduce it to a problem that is known to be learnable, thus showing that the new problem is learnable as well. If we are unsuccessful, instead we can try to reduce a problem that is known or suspected to be unlearnable to the new problem, thus gathering evidence that the new problem is not learnable.

The type of reduction from one learning problem to another introduced by Pitt and Warmuth[46] is called a *polynomial-time, learning-preserving reduction.* This type of reduction generalizes those used previously.[22,31,35] A polynomial-time, learning-preserving reduction consists of two mappings: a polynomial-time computable function f that maps unlabeled examples of the first learning problem to unlabeled examples of the second learning problem, and a function g that maps representations of concepts used in the first problem to representations of concepts used in the second problem.[2]

---

[1]This was done with respect to a notion of "(polynomial) predictability" which is equivalent to "learnability" as defined in this chapter.[26]

[2]An interesting feature of the definition of reduction is that the mapping g need not be computable. It is only required that g be length-preserving within a polynomial.

For example, one such reduction shows that learning Boolean functions represented as k-CNFs reduces to learning Boolean functions represented as conjunctive concepts. Here k-CNFs are Boolean formulas in conjunctive normal form with at most k literals per clause, where k is a constant, and conjunctive concepts are Boolean conjunctions of literals (also called *monomials*). If the number of variables in the k-CNF problem is n, then the learning problem for conjunctive concepts will have $q(n, k)$ variables where $q(n, k)$ is the number of possible conjunctions of up to k literals. Clearly $q(n, k) = O(n^k)$. The function f maps assignments for the n original variables of the k-CNF learning problem into assignments for the $q(n, k)$ variables of the conjunctive learning problem in the obvious way: the ith variable in the set of $q(n, k)$ variables is assigned one if the ith conjunction with at most k literals evaluates to true on the assignment to the n original variables, and zero otherwise. The function g simply maps a k-CNF with l clauses to the corresponding l-literal conjunctive concept over the larger variable set.

In the Pitt and Warmuth paper,[46] learning problems are classified by the complexity of their evaluation problems. The *evaluation problem* for a given learning problem is defined as follows: given an unlabeled example and a representation of a concept in the concept class of the learning problem, is the example a positive or a negative example of the concept? The main results of the paper show that certain learning problems are *learning-complete* in the class of all learning problems whose evaluation problems are in a given complexity class; that is, it is shown that if any of these problems are learnable, then all learning problems whose evaluation problem is in that complexity class are learnable. In particular, it is shown that the learning problem for Deterministic Finite Automata (DFAs) is learning-complete for the complexity class deterministic logspace (**LOG**). This means that if one could find an efficient PAC learning algorithm for DFAs, then there would be an efficient learning algorithm for all concept classes with representations that can be evaluated in deterministic logspace. This gives evidence that there is no efficient PAC learning algorithm for DFAs.

In analogy with the case of Deterministic Finite Automata, it is shown by Pitt and Warmuth[46] that the learning problems for Nondeterministic Finite Automata (NFAs), Context Free Grammars (CFGs), and alternating DFAs are learning-complete for the complexity classes **NLOG**, **LOGCFL**, and **P**, respectively, and that the problem of learning Boolean formulas is learning-complete for the complexity class **NC**^{1}. Of particular interest are the learning problems that are learning-complete for **P**. The learnability of one such problem would imply the learnability of all learning problems with representations that can be evaluated in polynomial time. In addition to the alternating DFA learning problem, a number of such problems have been found:

- Convex Vertex Represented Polytope[40]: the concept is an unknown convex polytope represented by the list of its vertices; positive examples are points in the polytope and negative examples points outside of the polytope.

- Horn Clause Consistency[46]: the concept is an unknown conjunction of Horn clauses; positive examples are sets of facts that are consistent with the conjunction.
- Augmented CFG Emptiness[46]: the concept is an unknown context-free grammar; positive examples are sets of productions that, when added to the grammar, yield a grammar generating the empty language.

It is unlikely that problems that are learning-complete for **P** are learnable. As a matter of fact, as discussed in the next section, there is convincing evidence that the opposite is the case.

---

6. HARDNESS FOR PAC LEARNABILITY BASED ON CRYPTOGRAPHIC ASSUMPTIONS

It is a much stronger result to show that a learning problem is not PAC learnable than it is to show that it is not properly PAC learnable, since the former result implies that the problem is not PAC learnable by any reasonable hypothesis space. Indeed, it follows from the work of Goldreich, Goldwasser, and Micali[18] that problems that are learning-complete for **P** are not PAC learnable (even in an extremely weak sense) assuming the existence of any cryptographically secure pseudo-random bit generator, which is equivalent to the existence of a certain type of one-way function.[34] While such an assumption is stronger than the assumption that **RP** ≠ **NP**, there is still convincing evidence for its validity.

Simpler problems can also be shown not to be PAC learnable based on stronger cryptographic assumptions. In particular, Kearns and Valiant[33] show that a polynomial-time learning algorithm for DFAs can be used to invert certain cryptographic functions. This is done by first showing that learning arbitrary Boolean formulas is as hard as inverting the given cryptographic functions. Then, since it can be shown that learning Boolean formulas reduces to learning DFAs, it follows that DFAs are not polynomially learnable based on the same cryptographic assumptions.

Such hardness results are disheartening. However, note that all of these hardness results are worst case with respect to the distribution and target concept. Thus when faced with learning a problem that is learning-complete for a reasonably large complexity class, the practitioner might look for assumptions that can be made on the distribution of the examples that will make the problem easier on average in some suitable sense. Further one might assume that the target concept is drawn at random according to some reasonable distribution rather that assuming that the target concept is worst case. We discuss some ideas along these lines in the following section, but only from the perspective of sample complexity. To date there has been very little general work on the average-case computational complexity of machine learning.

7. CRITICISMS OF THE PAC MODEL

The two criticisms most often leveled at the PAC model by AI researchers interested in empirical machine learning are:

1. the worst-case emphasis in the model makes it unusable in practice,[14,50] and
2. the notions of target concepts and noise-free training data are too restrictive in practice.[1,13]

We take these in turn.

There are two aspects of the worst-case nature of the PAC model that are at issue. One is the use of the worst-case model to measure the computational complexity of the learning algorithm, and the other is the definition of the sample complexity as the worst-case number of random examples needed over all target concepts in the target class and all distributions on the instance space. Here we address only the latter issue.

As pointed out in the section "Definition of PAC Learning" above, the worst-case definition of sample complexity means that even if we could calculate the sample complexity of a given algorithm exactly, we would still expect it to overestimate the typical error of the hypothesis produced as a function of the training set size on any particular target concept and particular distribution on the instance space. This is compounded by the fact that we usually cannot calculate the sample complexity of a given algorithm exactly even when it is a relatively simple consistent algorithm. Instead we are forced to fall back on the upper bounds on the sample complexity that hold for any consistent algorithm, given in the previous section, which themselves may contain overblown constants.

The upshot of this is that the basic PAC theory is not good for predicting learning curves. Some variants of the PAC model come closer, however. One simple variant is to make it distribution specific; i.e., define and analyze the sample complexity of a learning algorithm for a specific distribution on the instance space, e.g., the uniform distribution on a Boolean space.[11,50] There are two potential problems with this. The first is finding distributions that are both analyzable and indicative of the distributions that arise in practice. The second is that the bounds obtained may. be very sensitive to the particular distribution analyzed, and not be very reliable if the actual distribution is slightly different.

A more refined, Bayesian extension of the PAC model is explored by Buntine.[14] Using the Bayesian approach involves assuming a prior distribution over possible target concepts as well as training instances. Given these distributions, the average error of the hypothesis as a function of training sample size, and even as a function of the particular training sample, can be defined. Also, $1 - \delta$ confidence intervals like those in the PAC model can be defined as well. Experiments with this model on small learning problems are encouraging, but further work needs to be done on sensitivity analysis, and on simplifying the calculations so that larger problems can be analysed. This work, and the other distribution specific learning work, provides an increasingly important counterpart to PAC theory.

Another variant of the PAC model designed to address these issues is the "probability of mistake" model explored by Hausser et al.[26,27] and Opper and Haussler.[43] This model is designed specifically to help understand some of the issues in incremental learning. Instead of looking at sample complexity as defined above, the measure of performance here is the probability that the learning algorithm incorrectly guesses the label of the tth training example in a sequence of t random examples. Of course, the algorithm is allowed to update its hypothesis after each new training example is processed, so as t grows, we expect the probability of a mistake on example t to decrease. For a fixed-target concept and a fixed distribution on the instance space, it is easy to see that the probability of a mistake on example t is the same as the average error of the hypothesis produced by the algorithm from $t - 1$ random training examples. Hence, the probability of mistake on example t is exactly what is plotted on empirical learning curves that plot error versus sample size and average several runs of the learning algorithm for each sample size.

In Haussler et al.'s paper[25] the focus is on the worst-case probability of mistake on the tth example, over all possible target concepts and distributions on the training examples. In related papers[27,43] the probability of mistake on the tth example is examined when the target concept is selected at random according to a prior distribution on the target class and the examples are drawn at random from a certain fixed distribution. This is a Bayesian approach. The former we will call the *worst-case probability of mistake* and the latter we will call the *average-case probability of mistake*. The results can be summarized as follows. Let $\mathbf{C} = \{C_n\}_{n \geq 1}$ be a concept class and $d_n = VCdim(C_n)$ for all $n \geq 1$.

First, for any concept class $\mathbf{C}$ and any consistent algorithm for $\mathbf{C}$ using hypothesis space $\mathbf{C}$, the worst-case probability of mistake on example t is at most $O((d_n/t)\ln(t/d_n))$, where $t > d_n$. Furthermore, there are particular consistent algorithms and concept classes where the worst-case probability of mistake on example t is at least $\Omega((d_n/t)\ln(t/d_n))$; hence, this is the best that can be said in general of arbitrary consistent algorithms.

Second, for any concept class $\mathbf{C}$ there exists a (universal) learning algorithm for $\mathbf{C}$ (not necessarily consistent or computationally efficient) with worst-case probability of mistake on example t at most d_n/t. On the other hand, any learning algorithm for $\mathbf{C}$ must have worst-case probability of mistake on example t at least $\Omega(d_n/t)$, so this universal algorithm is essentially optimal.

Third, if we focus on average-case behavior, then there is a different universal learning algorithm, which is called *Bayes optimal learning algorithm* (or the *weighted majority algorithm*[39]) and there is a closely related, more efficient algorithm called the *Gibbs* (or *randomized weighted majority*) algorithm that have average-case probability of mistake on example t at most d_n/t and $2d_n/t$, respectively. Furthermore, there are particular concept classes $\mathbf{C}$, particular prior probability distributions on the concepts in these classes, and particular distributions on the instance spaces of these classes, such that the average-case probability of mistake on example t is at least $\Omega(d_n/t)$ for any learning algorithm (with constant $\approx 1/2$). This indicates that the above general bounds are tight to within a small constant. Even better forms of these upper and lower bounds can be given for specific distributions on the examples, specific target concepts, and even specific sequences of examples.

These results show two interesting things. First, certain learning algorithms perform better than arbitrary consistent learning algorithms in the worst case and average case; therefore, even in this restricted setting there is definitely more to learning than just finding any consistent hypothesis in an appropriately biased hypothesis space. Second, the worst case is not always much worse than the average case. Some recent experiments in learning perceptrons and multilayer perceptrons have shown that in many cases d_n/t is a rather good predictor of actual (i.e., average case) learning curves for backpropagation on synthetic random data.[8,52] However, it is still often an overestimate on natural data,[49] and in other domains such as learning conjunctive concepts on a uniform distribution.[50] Here the distribution (and algorithm) specific aspects of the learning situation must also be taken into account. Thus, in general we concur that extensions of the PAC model are required to explain learning curves that occur in practice. However, no amount of experimentation or distribution-specific theory can replace the security provided by a distribution-independent bound.

The second criticism of the PAC model is that the assumptions of well-defined target concepts and noise-free training data are unrealistic in practice. This is certainly true. However, it should be pointed out that the computational hardness results for learning described above, having been established for the simple noise-free case, must also hold for the more general case. The PAC model has the advantage of allowing us to state these negative results simply and in their strongest form. Nevertheless, the positive learnability results have to be strengthened before they can be applicable in practice, and some extensions of the PAC model are needed for this purpose. Many have been proposed (see Angluin and Laird[5] and Kearns and Li[30]).

Since the definitions of target concepts, random examples, and hypothesis error in the PAC model are just simplified versions of standard definitions from statistical pattern recognition and decision theory, one reasonable thing to do is to go back to these well-established fields and use the more general definitions that they have developed. First, instead of using the probability of misclassification as the only measure of error, a general *loss function* can be defined that for every pair consisting of a guessed value and an actual value of the classification, gives a nonnegative real number indicating a "cost" charged for that particular guess given that particular actual value. Then the error of a hypothesis can be replaced by the average loss of the hypothesis on a random example. If the loss is 1 if the guess is wrong and 0 if it is right (*discrete loss*), we get the PAC notion of error as a special case. However, using a more general loss function, we can also choose to make false positives more expensive than false negatives or vice versa, which can be useful. The use of a loss function also allows us to handle cases where there are more than two possible values of the classification. This includes the problem of learning real-valued functions, where we might choose to use $|guess - actual|$ or $(guess - actual)^2$ as loss functions.

Second, instead of assuming that the examples are generated by selecting a target concept and then generating random instances with labels agreeing with this target concept, we might assume that, for each random instance, there is also some randomness in its label. Thus, each instance will have a particular probability of being drawn and, given that instance, each possible classification value will have a particular probability of occurring. This whole random process can be described as making independent random draws from a single joint probability distribution on the set of all possible labeled instances. Target concepts with attribute noise, classification noise, or both kinds of noise can be modeled in this way. The target concept, the noise, and the distribution on the instance space are all bundled into one joint probability measure on labeled examples. The goal of learning is then to find a hypothesis that minimizes the average loss when the examples are drawn at random according to this joint distribution.

The PAC model, disregarding computational complexity considerations, can be viewed as a special case of this set-up using the discrete loss function, but with the added twist that learning performance is measured with respect to the worst case over all joint distributions in which the entire probability measure is concentrated on a set of examples that are consistent with a single target concept of a particular type. Hence, in the PAC case it is possible to get arbitrarily close to zero loss by finding closer and closer approximations to this underlying target concept. This is not possible in the general case, but one can still ask how close the hypothesis produced by the learning algorithm comes to the performance of the best possible hypothesis in the hypothesis space. For an unbiased hypothesis space, the latter is known as Bayes optimal classifier.[16]

Some recent PAC research has used this more general framework. By using the quadratic loss function mentioned above in place of the discrete loss, Kearns and Shapire investigate the problem of efficiently learning a real-valued regression function that gives the probability of a "+" classification for each instance.[32] In Haussler's paper,[24] it is shown how the VC dimension and related tools, originally developed by Vapnik, Chervonenkis, and others for this type of analysis, can be applied to the study of learning in neural networks. Here no restrictions whatsoever are placed on the joint probability distribution governing the generation of examples; i.e., the notion of a target concept or target class is eliminated entirely. Using this method, specific sample complexity bounds are obtained for learning with feedforward neural networks under various loss functions.

8. OTHER THEORETICAL LEARNING MODELS

A number of other theoretical approaches to machine learning are flourishing in recent computational learning theory work. One of these is the *total mistake bound* model.[35] Here an arbitrary sequence of examples of an unknown target concept is fed to the learning algorithm, and after seeing each instance the algorithm must predict the label of that instance. This is an incremental learning model like the probability of mistake model described above; however, here it is not assumed that the instances are drawn at random, and the measure of learning performance is the *total* number of mistakes in prediction in the worst case over all sequences of training examples (arbitrarily long) of all target concepts in the target class. We will call this latter quantity the *(worst-case) mistake bound* of the learning algorithm. Of interest is the case when there exists a polynomial time learning algorithm for a concept class $\mathbf{C} = \{C_n\}_{n \geq 1}$ with a worst-case mistake bound for target concepts in C_n that is polynomial in n. As in the PAC model, mistake bounds can also be allowed to depend on the syntactic complexity of the target concept.

The perceptron algorithm for learning linear threshold functions in the Boolean domain is a good example of a learning algorithm with a worst-case mistake bound. This bound comes directly from the bound on the number of updates given in the perceptron convergence theorem (see Duda and Hart[16]). The worst-case mistake bound of the perceptron algorithm is polynomial (and at least linear) in the number n of Boolean attributes when the target concepts are conjunctions, disjunctions, or any concept expressible with 0–1 weights and an arbitrary threshold.[19] A variant of the perceptron learning algorithm with multiplicative instead of additive weight updates was developed that has a significantly improved mistake bound for target concepts with small syntactic complexity.[16] The performance of this algorithm has also been extensively analysed in the case when some of the examples may be mislabeled.[37]

It can be shown that if there is a polynomial time learning algorithm for a target class $\mathbf{C}$ with a polynomial worst-case mistake bound, then $\mathbf{C}$ is PAC learnable. General methods for converting a learning algorithm with a good worst-case mistake bound into a PAC learning algorithm with a low sample complexity are given by Littlestone.[36] Hence, the total mistake bound model is actually not unrelated to the PAC model.

Another fascinating transformation of learning algorithms is given by the *weighted majority method*.[39] This is a method of combining several incremental learning algorithms into a single incremental learning algorithm that is more powerful and more robust than any of the component algorithms. This method extends the Bayesian-style weighted majority algorithm mentioned in the previous section. The idea is simple. All the component learning algorithms are run in parallel on the same sequence of training examples. For each example, each algorithm makes a prediction and these predictions are combined by a weighted voting scheme to determine the overall prediction of the "master" algorithm. After receiving feedback on its prediction, the master algorithm adjusts the voting weights for each of the component algorithms, increasing the weights of those that made the correct prediction, and decreasing the weights of those that guessed wrong, in each case by a multiplicative factor. It can be shown that this method of combining learning algorithms is very robust with regard to mislabeled examples. More importantly, the method produces a master algorithm with a worst-case mistake bound that approaches the worst-case mistake bound of the best component learning algorithm.[39] Thus the performance of the master algorithm is almost as good as that of the best component algorithm. This is particularly useful when a good learning algorithm is known but a parameter of the algorithm has to be tuned for the particular application.[29] In this case the weighted majority method is applied to a pool of component algorithms, each of which is a version of the original learning algorithm with a different setting of the parameter. The master algorithm's performance approaches the performance of the component algorithm with the best setting of the parameter.

The weighted majority method can also be adapted to the case when the predictions of the component algorithms are continuous.[39] This leads to a method for designing a master algorithm whose worst-case loss approaches the worst-case loss of the best linear combination of the component learning algorithm.[38] Here instead of the total number of mistakes, the loss is the total squared prediction error. Finally, a version of the weighted majority method can also be used to obtain good mistake bounds in the case when the best component algorithm changes in various sections of the trial sequence. More general learning problems for "drifting" target concepts have been investigated as well.[28] This represents an interesting new direction in learning research.

Both the PAC and total mistake bound models can be extended significantly by allowing learning algorithms to perform experiments or make queries to a teacher during learning.[7] The simplest type of query is a *membership query*, in which the learning algorithm proposes an instance in the instance space and then is told whether or not this instance is a member of the target concept. The ability to make membership queries can greatly enhance the ability of an algorithm to efficiently learn the target concept in both the mistake bound and PAC models. It has been shown that there are polynomial time algorithms that make polynomially many membership queries and have polynomial worst-case mistake bounds for learning

1. monotone DNF concepts (Disjunctive Normal Form with no negated variables),[7]
2. μ-formulae (Boolean formulae in which each variable appears at most once),[3]
3. deterministic finite automata,[6] and
4. Horn sentences (propositional PROLOG programs).[2]

In addition, there is a general method for converting an efficient learning algorithm that makes membership queries and has a polynomial worst-case mistake bound into a PAC learning algorithm, as long as the PAC algorithm is also allowed to make membership queries. Hence, all of the concept classes listed above are PAC learnable when membership queries are allowed. This contrasts with the evidence from cryptographic assumptions that classes 2 and 3 above are not PAC learnable from random examples alone.[33]

Surprisingly, it can be shown, based on cryptographic assumptions, that slightly richer classes than those listed above list are not PAC learnable even with membership queries.[4] These include:

1. nondeterministic finite automata and
2. intersections of deterministic finite automata.

This is shown by generalizing the notion of polynomial-time learning-preserving reduction[46] (described in a previous section) to the case when membership queries are allowed, and then reducing known cryptographically secure problems to the above learning problems.

9. CONCLUSION

In this brief survey we were able to cover only a small fraction of the results that have been obtained recently in computational learning theory. For a glimpse at some of these further results we refer the reader to Haussler and Pitt[21]; Rivest et al.[48]; Fulk and Case[17]; and Valiant and Warmuth.[56] However, we hope that we have at least convinced the reader that the insights provided by this line of investigation, such as those about the difficulty of searching hypothesis spaces, the notion of bias and its effect on required training size, the effectiveness of majority voting methods, and the usefulness of actively making queries during learning, have made this effort worthwhile.

ACKNOWLEDGEMENTS

We gratefully acknowledge the support from ONR grants N00014-86-K-0454-P00002, N00014-86-K-0454-P00003, and N00014-91-J-1162. A preliminary version of this chapter appeared in *Proceedings of the 8th National Conference on Artificial Intelligence*.[23]

REFERENCES

1. Amsterdam, J. *The Valiant Learning Model: Extensions and Assessment.* Master's Thesis, Department of Electrical Engineering and Computer Science, Massachusetts Institute of Technology, January 1988.
2. Angluin, D., M. Frazier, and L. Pitt. "Learning Conjunctions of Horn Clauses." In *31th Annual IEEE Symposium on Foundations of Computer Science*, 186–192, 1990.
3. Angluin, D., L. Hellerstein, and M. Karpinski. "Learning Read-Once Formulas with Queries." *J. ACM* **40** (1993): 185–210.
4. Angluin, D., and M. Kharitonov. "Why Won't Membership Queries Help?" In *Mach. Learning* **9** (1992): 147–164.
5. Angluin, D., and P. Laird. "Learning From Noisy Examples." *Mach. Learning* **2(4)** (1988): 343–370.
6. Angluin, D. "Learning Regular Sets from Queries and Counterexamples." *Infor. & Comp.* **75** (1987): 87–106.
7. Angluin, D. "Queries and Concept Learning." *Mach. Learning* **2** (1988): 319–342.
8. Baum, E. "When are K-Nearest Neighbor and Back Propogation Accurate for Feasible Sized Sets of Examples." In *Snowbird Conference on Neural Networks for Computing.* Unpublished manuscript, 1990.
9. Blumer, A., A. Ehrenfeucht, D. Haussler, and M. K. Warmuth. "Occam's Razor." *Infor. Proc. Let.* **24** (1981): 377–380.
10. Blumer, A., A. Ehrenfeucht, D. Haussler, and M. K. Warmuth. "Learnability and the Vapnik-Chervonenkis Dimension." *J. Assoc. Comp. Mach.* **36(4)** (1989): 929–965.
11. Benedek, G. M., and A. Itai. "Learnability by Fixed Distributions." In *Proc. 1988 Workshop on Comp. Learning Theory*, 80–90. San Mateo, CA: Morgan Kaufmann, 1988.
12. Blum, A., and R. L. Rivest. "Training a Three-Neuron Neural Net is NP-Complete." *Proceedings of the 1988 Workshop on Computational Learning Theory*, 9–18. San Mateo, CA: Morgan Kaufmann, 1988.
13. Bergadano, F., and L. Saitta. "On the Error Probabilty of Boolean Concept Descriptions." In *Proceedings of the 1989 European Working Session on Learning*, 25–35, 1989.
14. Buntine, W. L. "A Theory of Learning Classification Rules." Ph.D. Thesis, University of Technology, Sydney, Australia, 1990.
15. Cover, T. M. "Geometrical and Statistical Properties of Systems of Linear Inequalities with Applications in Pattern Recognition." *IEEE Trans. on Electronic Computers* **EC-14** (1965): 326–334.
16. Duda, R. O., and P. E. Hart. *Pattern Classification and Scene Analysis.* New York: Wiley, 1973.

17. Fulk, M., and J. Case, eds. *Proceedings of the 1990 Workshop on Computational Learning Theory.* San Mateo, CA: Morgan Kaufmann, 1990.

18. Goldreich, O., S. Goldwasser, and S. Micali. "How to Construct Random Functions." *J. ACM* **33(4)** (1986): 792–807.

19. Hampson, S. E., and D. J. Volper. "Linear Function Neurons: Structure and Training." *Bio. Cyber.* **53** (1986): 203–217.

20. Haussler, D. "Quantifying Inductive Bias: AI Learning Algorithms and Valiant's Learning Framework." *Art. Intel.* **36** (1988): 177–221.

21. Haussler, D., and L. Pitt, eds. *Proceedings of the 1988 Workshop on Computational Learning Theory.* San Mateo, CA: Morgan Kaufmann, 1988.

22. Haussler, D. "Learning Conjunctive Concepts in Structural Domains." *Mach. Learning* **4** (1989): 7–40.

23. Haussler, D. "Probably Approximately Correct Learning." In *Proceedings of the 8th National Conference on Artificial Intelligence,* 1101–1108. San Mateo, CA: Morgan Kaufmann, 1990.

24. Haussler, D. "Decision Theoretic Generalizations of the PAC Model for Neural Net and Other Learning Applications." *Infor. & Comp.* **100(1)** (1992): 78–150.

25. Haussler, D., N. Littlestone, and M. K. Warmuth. "Predicting $\{0, 1\}$-Functions on Randomly Drawn Points." Technical Report UCSC-CRL-90-54, University of California Santa Cruz, Computer Research Laboratory, December 1990. To appear in *Infor. & Comp.*

26. Haussler, D., M. Kearns, N. Littlestone, and M. K. Warmuth. "Equivalence of Models for Polynomial Learnability." *Infor. & Comp.* **95** (1991): 129–161.

27. Haussler, D., M. Kearns, and R. Schapire. "Bounds on the Sample Complexity of Bayesian Learning Using Information Theory and the VC Dimension." In *Proceedings of the Fourth Workshop on Computational Learning Theory,* 61–74. San Mateo, CA: Morgan Kaufmann, 1991.

28. Helmbold, D., and P. Long. "Tracking Drifting Concepts Using Random Examples." In *Proceedings of the 1991 Workshop on Computational Learning Theory,* 13–23. San Mateo, CA: Morgan Kaufmann, August 1991.

29. Helmbold, D., R. Sloan, and M. K. Warmuth. "Learning Nested Differences of Intersection Closed Concept Classes." *Mach. Learning* **5** (1990): 165–196.

30. Kearns, M., and M. Li. "Learning in the Presence of Malicious Errors." In *Proceedings of the 20th ACM Symposium on Theory of Computing,* 267–279. Chicago: ACM, 1988.

31. Kearns, M., M. Li, L. Pitt, and L. Valiant. "On the Learnability of Boolean Formulae." In *Proceedings of the 19th ACM Symposium on Theory of Computing,* 285–295. New York: ACM Press, 1987.

32. Kearns, M. J., and R. E. Schapire. "Efficient Distribution-Free Learning of Probabilistic Concepts." In *Proceedings of the 31st Annual Symposium on Foundations of Computer Science,* 382–391. Los Alamitos, CA: IEEE Computer Society, 1990.

33. Kearns, M., and L. Valiant. "Cryptographic Limitations on Learning Boolean Formulae and Finite Automata." In *21st ACM Symposium on Theory of Computing*, 433–444. New York: ACM Press, 1989.
34. Levin, L. A. "One-Way Functions and Pseudorandom Generators." *Combinatorica* **7(4)** (1987): 357–363.
35. Littlestone, N. "Learning Quickly When Irrelevant Attributes Abound: A New Linear-Threshold Algorithm." *Mach. Learning* **2** (1988): 285–318.
36. Littlestone, N. "From On-Line to Batch Learning." In *Proceedings of the Second Annual Workshop on Computational Learning Theory*, 269–284. San Mateo, CA: Morgan Kaufmann, 1989.
37. Littlestone, N. "Mistake Bounds and Logarithmic Linear-Threshold Learning Algorithms." Ph.D. Thesis, University of California, Santa Cruz, 1989.
38. Littlestone, N., P. Long, and M.K. Warmuth. "On-Line Learning of Linear Functions." Technical Report UCSC-CRL-91-29, University of California Santa Cruz, October 1991. For an extended abstract see: *Proceedings of 23rd Annual ACM Symposium on Theory of Computing*, 465–475. New Orleans, LA: ACM Press, 1991.
39. Littlestone, N., and M. K. Warmuth. "The Weighted Majority Algorithm." Technical Report UCSC-CRL-91-28, University of California Santa Cruz, October 1991. A preliminary version appeared in the proceedings of the *30th Annual IEEE Symposium on Foundations of Computer Science*, 256–261. IEEE Computer Society, Los Alamitos, CA, October 1989, revised October 1992.
40. Long, P., and M. K. Warmuth. "Composite Geometric Concepts and Polynomial Learnability." To appear in *Infor. & Comp.*
41. Mitchell, T. M. "The Need for Biases in Learning Generalizations." Technical Report CBM-TR-117, Rutgers University, New Brunswick, NJ, 1980.
42. Natarajan, B. K. "On Learning Sets and Functions." *Mach. Learning* **4(1)** (1989).
43. Opper, M., and D. Haussler. "Calculation of the Learning Curve of Bayes Optimal Classification Algorithm for Learning a Perceptron with Noise." In *Computational Learning Theory: Proceedings of the Fourth Annual Workshop*, 75–87. San Mateo, CA: Morgan Kaufmann, 1991.
44. Pitt, L. "Inductive Inference, DFAs, and Computational Complexity." Technical Report UIUCDCS-R-89-1530, University of Illinois at Urbana-Champaign, 1989.
45. Pitt, L., and L. Valiant. "Computational Limitations on Learning from Examples." *J. ACM* **35(4)** (1988): 965–984.
46. Pitt, L., and M. K. Warmuth. "Prediction Preserving Reducibility." Special issue of the for the Third Annual Conference of Structure in Complexity Theory. *J. Comp. Sys. Sci.* **41(3)** (1990): 430–467.
47. Rivest, R. L. "Learning Decision Lists." *Mach. Learning* **2** (1987): 229–246.

48. Rivest, R., D. Haussler, and M. K. Warmuth, eds. *Proceedings of the 1989 Workshop on Computational Learning Theory.* San Mateo, CA: Morgan Kaufmann, 1989.

49. Rumelhart, D. Personal communication, 1990.

50. Sarrett, W., and M. Pazzani. "Average Case Analysis of Empirical and Explanation-Based Learning Algorithms." *Mach. Learning* **9(4)** (1992): 349–372.

51. Shawe-Taylor, S. T., M. Anthony, and N. L. Biggs. "Bounding Sample Size with the Vapnik-Chervonenkis Dimension." Technical Report CSD-TR-618, University of London, Surrey, England, 1989.

52. Tesauro, G., and D. Cohn. "Can Neural Networks Do Better than the Vapnik-Chervonenkis Bounds?" In *Advances in Neural Information Processing, Vol. 3*, edited by R. Lippmann, J. Moody, and D. Touretzky, 911–917. San Mateo, CA: Morgan Kaufmann, 1991.

53. Valiant, L. G. "A Theory of the Learnable." *Comm. ACM* **27(11)** (1984): 1134–1142.

54. Valiant, L. G. "Learning Disjunctions of Conjunctions." In *Proc. 9th IJCAI*, Vol. 1, 560–566. San Mateo, CA: Morgan Kaufmann, 1985.

55. Vapnik, V. N. *Estimation of Dependences Based on Empirical Data.* New York: Springer-Verlag, 1982.

56. Valiant, L. G. and M. Warmuth, eds. *Proceedings of the 1991 Workshop on Computational Learning Theory.* San Mateo, CA: Morgan Kaufmann, 1991.

David Haussler

Baskin Center for Computer Engineering and Information Sciences, University of California, Santa Cruz, CA 95064; e-mail: haussler@saturn.ucsc.edu.

Decision Theoretic Generalizations of the PAC Model for Neural Net and Other Learning Applications

This chapter, reprinted by permission, originally appeared in *Information and Computation* **100(1)** (1992): 78–150. Copyright © by Academic Press.

We describe a generalization of the PAC learning model that is based on statistical decision theory. In this model the learner receives randomly drawn examples, each example consisting of an instance $x \in X$ and an outcome $y \in Y$, and tries to find a decision rule $h : X \to A$, where $h \in \mathcal{H}$, that specifies the appropriate action $a \in A$ to take for each instance x, in order to minimize the expectation of a loss $l(y, a)$. Here X, Y, and A are arbitrary sets, l is a real-valued function, and examples are generated according to an arbitrary joint distribution on $X \times Y$. Special cases include the problem of learning a function from X into Y, the problem of learning the conditional probability distribution on Y given X (regression), and the problem of learning a distribution on X (density estimation).

We give theorems on the uniform convergence of empirical loss estimates to true expected loss rates for certain decision rule spaces $\mathcal{H}$, and show

how this implies learnability with bounded sample size, disregarding computational complexity. As an application, we give distribution-independent upper bounds on the sample size needed for learning with feedforward neural networks. Our theorems use a generalized notion of VC dimension that applies to classes of real-valued functions, adapted from Vapnik and Pollard's work, and a notion of *capacity* and *metric dimension* for classes of functions that map into a bounded metric space.

1. INTRODUCTION

The introduction of the Probably Approximately Correct (PAC) model[4,86] of learning from examples has done an admirable job of drawing together practitioners of machine learning with theoretically oriented computer scientists in the pursuit of a solid and useful mathematical foundation for applied machine learning work. These practitioners include both those in mainstream artificial intelligence and in neural net research. However, in attempting to address the issues that are relevant to this applied work in machine learning, a number of shortcomings of the model have cropped up repeatedly. Among these are the following:

1. The model is defined only for $\{0,1\}$-valued functions. Practitioners would like to learn functions on an instance space X that take values in an arbitrary set Y; e.g., multivalued discrete functions, real-valued functions, and vector-valued functions.

2. Some practitioners are wary of the assumption that the examples are generated from an underlying "target function," and are not satisfied with the noise models that have been proposed to weaken this assumption (see Angluin and Laird,[5] Sloan,[80] and Shackelford and Volper[78]). They would like to see more general regression models investigated in which the y component in a training example $(x, y) \in X \times Y$ is randomly specified according to a conditional distribution on Y, given x. Here the general goal is to approximate this conditional distribution for each instance $x \in X$. In the computational learning theory literature, a model of this type is investigated by Kearns and Schapire,[45] with $Y = \{0, 1\}$, and in a more general case by Yamanishi.[96]

3. Many learning problems are unsupervised; i.e., the learner has access only to randomly drawn, unlabeled examples from an instance space X. Here learning can often be viewed as some form of approximation of the distribution that is generating these examples. This is usually called *density estimation* when the instance space X is continuous and no specific parametric form for the underlying distribution on X is assumed. It is often called *parameter estimation* when specific parametric probability models are used. One example of this in the computational learning theory literature is the recent investigation of

Abe and Warmuth into the complexity of learning the parameters in a hidden Markov model.[1]

Our purpose here is twofold. First, we propose an extension of the PAC model, based on the work of Vapnik and Chervonenkis[89] and Pollard,[70,72] that addresses these and other issues. Second, we use this extension to obtain distribution-independent upper bounds on the size of the training set needed for learning with various kinds of feedforward neural networks,[69,76] a popular learning method which is not covered by the basic PAC model.

1.1 OVERVIEW OF THE PROPOSED FRAMEWORK

To extend the PAC model, we propose a more general framework based on statistical decision theory (see Ferguson,[34] Kiefer,[46] and Berger[14]). Valiant's original learnability proposal borrowed ideas from statistical decision theory only in a very limited fashion, which helped keep the model simple enough so that work could easily focus on computational issues. However, now that the field has matured, it is time to reexamine the statistical roots of the model and consider extensions that employ more of this general framework.

In this general framework, we assume the learner receives randomly drawn training examples, each example consisting of an instance $x \in X$ and an outcome $y \in Y$, where X and Y are arbitrary sets called *instance* and *outcome spaces*, respectively. These examples are generated according to a joint distribution on $X \times Y$, unknown to the learner. This distribution comes from a (known) class $\mathcal{P}$ of joint distributions on $X \times Y$, representing possible "states of nature." After training, the learner will receive further random examples drawn from this same joint distribution. For each example (x, y), the learner will be shown only the instance x. Then he will be asked to choose an action a from a set of possible actions A, called the *decision space*. Following this, the outcome y will be revealed to the learner. In the case that we examine here, the outcome y depends only on the instance x and not on the action a chosen by the learner. For each action a and outcome y, the learner will suffer a loss, which is measured by a fixed real-valued *loss function* l on $Y \times A$. We assume that the loss function is known to the learner. The learner tries to choose his actions so as to minimize his loss.

Here we look at the case in which, based on the training examples, the learner develops a deterministic strategy that specifies what he believes is the appropriate action a for each instance x in X. He then uses this strategy on all future examples. Thus we look at "batch" learning rather than "incremental" or "on-line" learning.[53] The learner's strategy, which is a function from the instance space X into the decision space A, will be called a *decision rule*. We assume that the decision rule is chosen from a fixed *decision rule space* $\mathcal{H}$ of functions from X into A. For example, instances in X may be encoded as inputs to a neural network, and outputs of the network may be interpreted as actions in A. In this case the network represents a decision rule, and the decision rule space $\mathcal{H}$ may be all functions represented by networks obtained by varing the parameters of a fixed underlying network. The goal of learning is to find a decision rule in $\mathcal{H}$ that minimizes the expected loss, when examples are drawn at random from the unknown joint distribution on $X \times Y$.

This learning framework can be applied in a variety of situations. We now give several illustrations. For further discussion, we refer the reader to the excellent surveys by White,[95] Barron,[9] Devroye,[24] and Vapnik,[89] to which we are greatly indebted.[1] We also recommend the text by Kiefer[46] for a general introduction to statistical inference and decision theory.

1.1.1 BETTING EXAMPLE. For our first example, consider the problem of learning to maximize profit (or minimize loss!) at the horse races. Here an instance x in X is a race, an action a in A consists of placing or not placing a certain bet, and an outcome y in Y is determined by the winner and the second and third place finishers. The loss $l(y, a)$ is the amount of money lost when bet a is placed and the outcome of the race is y. A negative loss is interpreted as gain. The joint distribution on $X \times Y$ represents the probability of various races and outcomes. The outcome y in Y depends on the race x through a conditional distribution on Y given x derived from this joint distribution. This joint distribution is unknown to the learner; he only has random examples $(x_1, y_1), \ldots, (x_m, y_m)$, each consisting of a race/outcome pair generated from this distribution. From these examples, the learner develops a deterministic betting strategy (decision rule). The best decision rule h is one that specifies a bet a for each race x that minimizes the expectation of the loss $l(y, a)$, when y is chosen randomly from the unknown conditional distribution on Y given x, which is determined by the underlying joint distribution on $X \times Y$. This (not necessarily unique) best decision rule minimizes the expected loss on a random example (x, y). It is known as *Bayes optimal decision rule*. The learner tries to approximate Bayes optimal decision rule as best he can using decision rules from a given decision rule space $\mathcal{H}$ (e.g., "simple" or "easy to compute" decision rules, or perhaps decision rules that can be represented by a particular kind of neural network).

1.1.2 CLASSIFICATION. As a second example, consider the problem of medical diagnosis. Here an instance x is a vector of measurements from medical tests conducted on the patient, an action a is a diagnosis of the patient's disease state, and an outcome y may be defined as the actual disease state of the patient. Here $A = Y$; i.e., the possible diagnoses are the same as the possible disease states. To specify the loss function l, we may stipulate that there is zero loss for the correct diagnosis $a = y$ but, for each pair (y, a) with diagnosis a differing from disease state y, there is some positive real loss $l(y, a)$, depending on the severity of the consequences of that

---

[1] For further discussion, see also Haussler.[40] However, there (as in previous versions of this paper) we assume that $Y = A$. In fact, we can always replace Y and A with $Y \cup A$ without loss of generality, but we now feel that the gain in economy of notation is more than offset by the confusion that results, and so have switched to the more standard notation used here.

particular misdiagnosis. Here a decision rule is a diagnostic method, and a Bayes optimal decision rule is the one that minimizes the expected loss from misdiagnosis when examples (x, y) of test results and associated disease states occur randomly according to some unknown "natural" joint distribution.

This medical diagnosis situation is a typical example of a *classification learning* problem in the field of pattern recognition (see Duda and Hart[25]). The problem of learning a Boolean function from noise-free examples, as investigated in the PAC model, is a special case of classification learning. Here the outcome space Y is $\{0, 1\}$ and only the instance x in an example (x, y) is drawn at random. The outcome y is $f(x)$ for some unknown Boolean *target function* f, rather than being determined stochastically. As above, the decision space A is the same as the outcome space Y, and the action a can be interpreted as a prediction of the outcome y. Hence, a decision rule h maps from the instance space X into the outcome space Y, just as the target function does. In much of AI, and in PAC learning in particular, it is common to refer to h as a *hypothesis* in this case, and to $\mathcal{H}$ as the *hypothesis space*.

This same setup, where the outcome y is a function of the instance x, can be applied to any function learning problem by letting X and Y be arbitrary sets. In the general function learning problem, the loss function $l(y, a)$ usually measures the distance between the prediction a and the actual value y in some metric. In the PAC model, l is the discrete metric: $l(y, a) = 0$ if $a = y$, else $l(y, a) = 1$. Thus the expected loss of the decision rule (or hypothesis) is just the probability that it predicts incorrectly, the usual PAC notion of the *error* of the hypothesis. In general, Y may be a set of strings, graphs, real vectors, etc., in which case other distance metrics or more general kinds of loss functions may be more appropriate.

For function learning, Bayes optimal decision rule is the decision rule (not necessarily in the decision rule space $\mathcal{H}$) that has minimal average distance from the target function, which is the target function itself whenever l is defined such that $a = y$ implies that $l(y, a) = 0$. The general function learning problem can also be considered as a special case of the classification learning problem if we allow infinite outcome (and decision) spaces in our formulation of the latter problem.

1.1.3 REGRESSION. The general problem of regression has a different character from that of classification learning, but can also be addressed in the decision theoretic learning framework. To illustrate this, as a third example consider a variant of the medical diagnosis situation in which the doctor provides an estimate of the probability that the patient has each of several diseases, rather than predicting that he has one specific disease or asserting that he is healthy. (Here we assume that the actual disease state includes at most one disease.) For example, the doctor may say "Given these test results x, I would say you have disease 1 with probability 55%, disease 2 with probability 5%, and no disease at all with probability 40%." Here the doctor is actually trying to estimate the conditional distribution on disease states Y given the test results x. Her action a entails providing a vector of parameters that determine that estimated distribution, e.g., $(0.55, 0.05, 0.4)$. The decision space A is the set of all such parameter vectors.

Now let Y be an arbitrary discrete outcome space. Keeping the instance x fixed, for each parameter vector a in A and outcome y in Y, let $\widehat{P}(y; a)$ denote the probability of outcome y with respect to the distribution on Y defined by the parameter vector a. Thus when we take action a on instance x, we are asserting that, given the instance x, we estimate the conditional probability of outcome y to be $\widehat{P}(y; a)$ for each outcome y in Y. Let $P(y)$ denote the actual conditional probability of outcome y, given the instance x, with respect to the unknown joint distribution on $X \times Y$. (The distributions P and $\widehat{P}$ can be replaced by densities when Y is continuous.) Let us define[2] the loss function l by setting $l(y, a) = -\log \widehat{P}(y; a)$. This is called the (negative) *log likelihood* loss function. If we define loss in this way, then the expected loss resulting from action a has a natural information theoretic interpretation[3]: it is the *Kullback-Leibler divergence*[49] (or *information gain*[74]) from the actual conditional probability distribution P to the estimated conditional distribution $\widehat{P}$, plus the entropy of P.

An appropriate loss function l to use for regression can be derived from the *maximum likelihood principle*. Let us fix the instance x. For each outcome y in Y, let $P(y)$ denote the actual conditional probability of outcome y, given test results x. For each vector a of parameters, let $\widehat{P}(y; a)$ denote the probability of outcome y with respect to the distribution on Y defined by the parameter vector a. When y is fixed and $\widehat{P}(y; a)$ is viewed as a function of a, it is called the *likelihood* of a. If a single outcome y is available, the maximum likelihood principle suggests choosing the action a that maximizes the likelihood $\widehat{P}(y; a)$; if several independent outcomes $y_1, \ldots, y_n$ are available, the principle suggests choosing the action a that maximizes $\prod_{i=1}^{n} \widehat{P}(y_i; a)$.

For a given x, the entropy of the true conditional distribution P is a constant, independent of the action a. Thus choosing the action a for each instance x that minimizes the expected log likelihood loss is equivalent to choosing the action a that gives the closest estimate $\widehat{P}$ to the true conditional distribution P over possible outcomes in Y as measured by the Kullback-Leibler divergence, given that instance x. It is well known that the Kullback-Leibler divergence is minimized when $\widehat{P} = P$. This is Bayes optimal decision rule in regression.

In the regression version of our medical diagnosis situation, the definition of the log likelihood loss function depends on the interpretation of the components of the parameter vector a. If there are k possible diseases and the patient can have at most one of these, then we might have $k + 1$ possible mutually exclusive disease

---

[2] We assume $\widehat{P}(y; a) > 0$ for all y in Y.

[3] The Kullback-Leibler divergence from P to $\widehat{P}$, denoted $I(P\|\widehat{P})$, is defined as $\sum_{y \in Y} P(y)(\log P(y)/\widehat{P}(y; a))$ for countable Y. The entropy of P, denoted $H(P)$, is $-\sum_{y \in Y} P(y) \log P(y)$. Thus $I(P\|\widehat{P}) + H(P) = -\sum_{y \in Y} P(y) \log \widehat{P}(y; a)$, which is the expectation of the (negative) log likelihood loss. Analogous results hold for densities when the relevant quantities are finite.[49]

states $y_1, \ldots, y_{k+1}$, where y_{k+1} means healthy. Hence $Y = \{y_1, \ldots, y_{k+1}\}$. Then we might specify that an action a takes the form

$$a = (a_1, \ldots, a_{k+1}),$$

where $a_i = \widehat{P}(y_i; a)$, the estimated probability of disease state y_i. Here the components of the vector a must be positive and sum to one. In this case the log likelihood loss would be $l(y_i, a) = -\log a_i = -\log \widehat{P}(y_i; a)$.

Often the constraints on the components of a are a nuisance, so other interpretations of a are used, e.g., that $a_i = \log \widehat{P}(y_i; a) - \log \widehat{P}(y_{k+1}; a)$ for each i, $1 \le i \le k+1$. In this case the $a_1, \ldots, a_k$ are arbitrary real numbers and $a_{k+1} = 0$, and hence can be ignored. Since $\widehat{P}(y_i; a) = e^{a_i} / \sum_{j=1}^{k+1} e^{a_j}$, the log likelihood loss is $l(y_i, a) = -a_i + \log \sum_{j=1}^{k+1} e^{a_j} = -a_i + \log(1 + \sum_{j=1}^{k} e^{a_j})$. This is known as the *logistic loss*.[9,56] A third interpretation would be to allow the possibility that the patient may have more than one disease, and assume, for the purposes of estimation, that diseases occur independently. Then the disease state y might be defined as a binary vector of length k, where the ith bit y_i is 1 if and only if the ith disease is present. Hence $Y = \{0, 1\}^k$. Similarly, the vector a would be a vector of independent probabilities $(a_1, \ldots, a_k)$, where a_i is the estimated probability of the patient having the ith disease. In this case

$$\widehat{P}(y; a) = \prod_{i=1}^{k} a_i^{y_i} (1 - a_i)^{(1 - y_i)}$$

and the log likelihood loss is

$$l(y, a) = -\sum_{i=1}^{k} (y_i \log a_i + (1 - y_i) \log(1 - a_i)),$$

which we will call the *cross entropy loss*.

In the medical diagnosis example, the outcome space Y is discrete. However, in most uses of regression, Y is real valued; i.e., the outcome y is the measurement of some real-valued quantity, and the instance x represents the experimental conditions under which this quantity was measured. In this case regression is usually defined as estimating the conditional expectation of Y given the instance x. Thus $A \subset \Re$, and the action $a \in A$ for a given instance x consists of an estimate of the mean of the various outcomes y that would typically be observed for that instance x. It is easy to show that by using the *quadratic* loss function $l(y, a) = (a - y)^2$, the expected loss is minimized when a is the true mean, and hence this version of regression also fits naturally[4] into the decision theoretic framework. An alternate approach is to use the L_1 loss function $l(y, a) = |a - y|$, in which case the expected loss is minimized when a is the median of the conditional distribution Y given the instance x. (See White[95] and Haussler.[40])

---

[4] In fact, the standard version of regression, defined as estimating the conditional mean of Y given instance x using the quadratic loss function, is actually a special case of the general version

1.1.4 DENSITY AND PARAMETER ESTIMATION. Finally, the problems of parameter estimation and density estimation can also be viewed as special cases of this decision theoretic framework. For parameter estimation, note that when the instance space X has only one element, then the particular instance x can be ignored entirely. Thus the regression problem reduces to the problem of estimating the parameters of a single distribution on the outcome space Y from a sample of random outcomes y from Y; i.e., to the simpler problem of parameter estimation. Here the decision rule is not a function but merely a single vector of parameters, and the decision rule space $\mathcal{H}$ is the same as the decision space A.

As above, if Y is discrete and a is a vector of parameters in the decision space A, let $\widehat{P}(y; a)$ denote the probability of outcome y for the distribution represented by the parameter vector a. When y is fixed and $\widehat{P}(y; a)$ is viewed as a function of a, it is called the *likelihood* of a. We have suggested that the (negative) log likelihood loss function, $l(y, a) = -\log \widehat{P}(y; a)$, is a natural choice for regression, and hence also for the special case of parameter estimation, since it has a nice information theoretic interpretation. However, the use of this loss function for parameter estimation is usually introduced as an application of the *maximum likelihood principle*. This principle suggests that in estimating the parameters of a distribution on the basis of given data, one should choose the parameter vector that maximizes the likelihood function. In our case the random outcomes from Y are the data. If a single outcome y is available, the maximum likelihood principle suggests choosing the action a that maximizes the likelihood $\widehat{P}(y; a)$; if several independent random outcomes $y_1, \ldots, y_m$ from Y are available, the principle suggests choosing the action a that maximizes $\prod_{i=1}^{n} \widehat{P}(y_i; a)$. This is equivalent to choosing the action a that minimizes $-1/n \sum i = 1^n \log \widehat{P}(y_i; a)$. This latter expression is the expectation of the loss $l(y, a)$ for the *empirical* distribution over Y generated by the independent random outcomes $y_1, \ldots, y_n$.

For density estimation, we can consider the dual case in which the outcome space Y has only one element and, hence, can be ignored. Thus examples are unlabeled instances x drawn randomly from some density $p(x)$ on X. Let the decision set A be the positive real numbers and each decision rule h in $\mathcal{H}$ be a density on X. Then, as above, information theoretic considerations suggest the loss function $l(y, a) = l(a) = -\log a$. Again, as above, the expected loss of h is minimized when

of regression defined above, where for continuous outcome spaces Y, the object is to estimate the parameters specifying the conditional density of Y given instance x, using the log likelihood loss function. To see this, assume that we represent the conditional density on Y with a Gaussian density $\hat{p}(y; \mu, \sigma) = (2\pi\sigma^2)^{-1/2} e^{-(\mu-y)^2/2\sigma^2}$, where μ is the mean and σ^2 the variance. Let the variance be fixed, independent of x, so that the estimate $\hat{p}(y; \mu, \sigma)$, of the conditional density on Y given x is completely determined by the mean μ. Thus the decision space $A \subset \Re$, and each action a in A is interpreted as specifying the mean of a Gaussian density. Substituting $\mu = a$ and evaluating $-\log \hat{p}(y; \mu, \sigma)$, the log likelihood loss is seen to be $l(y, a) = (1/2\sigma^2)(a - y)^2 + 1/2 \log(2\pi\sigma^2)$. For fixed variance σ^2, this is equivalent, for learning, to the quadratic loss $(a - y)^2$, since additive and multiplicative constants in the definition of l only rescale it without changing the value of a that minimizes its expectation.

h is the true density p. Further, if p is not a member of $\mathcal{H}$, then the best decision rule in $\mathcal{H}$, in terms of minimizing the expected loss, is the one with the smallest Kullback-Leibler divergence from the true density p.[49] Here, perhaps the simplest example is the representation of a mixture of Gaussian densities by a neural network with one hidden layer, as described Nowlan,[65] and Poggio and Girosi.[69]

When the instance space X is discrete, we are not estimating a density on X but rather a probability distribution. The same ideas as above carry over, except that we let the decision space $A = (0, 1)$ and each decision rule h in $\mathcal{H}$ represent a probability distribution on X. Here we can also use the same loss function, and it has the same properties.

These examples illustrate the diversity of the learning problems that can be cast in the proposed decision theoretic framework, even under the restrictive assumptions we make here; i.e., that the outcome y does not depend on the action a, and that the learner always observes both the outcome and the loss. By weakening these assumptions, we can model other types of learning as well, including *associative reinforcement learning*[11,35] and the theory of *learning automata* (with static environment).[58] However, we will not pursue this here.

1.2 SUMMARY AND DISCUSSION OF THE RESULTS PRESENTED HERE

There are three major practical issues in this decision theoretic view of learning. The first is the number of random examples needed in order to be able to produce a good decision rule in the decision rule space $\mathcal{H}$; i.e., a decision rule whose expected loss is near the minimum of all decision rules in $\mathcal{H}$. If too few examples are used, we run into the problem of *overfitting*, where the decision rule produced performs well on the training data, but not on further random examples drawn from the same joint distribution that generated this training data. The second is the adequacy of the decision rule space $\mathcal{H}$. If $\mathcal{H}$ does not contain any decision rule with expected loss close to that of Bayes optimal decision rule for the particular joint distribution we are dealing with, then we can never hope to achieve near optimal performance using this decision rule space. Choosing the right decision rule space often requires considerable insight into the particular problem domain. Note that in general, enlarging the decision rule space tends to increase the number of situations in which it is adequate, but also the number of examples that will be required to avoid overfitting. Hence, we must find a happy medium to address both these issues. Finally, the third practical problem is the computational complexity of the method we use to produce our decision rule from the training examples. This issue has been addressed extensively in the PAC literature, and is also addressed by Kearns and Schapire[45] and Abe and Warmuth.[1] The size of the decision rule space is not correlated with the computational complexity of finding a good decision rule. Enlarging it can make the problem easier or harder. Of these three important issues, here we examine only the first. This issue is referred to as the problem of estimating the "sample complexity" of the learning problem in the PAC literature.[31]

The number of random training examples needed to avoid overfitting depends critically on the nature of the decision rule space used. Different kinds of decision rule spaces are used in different areas of learning research, partly because different kinds of instance and outcome spaces are used. In pattern recognition and statistics, the instance space X is usually a finite-dimensional real vector space; i.e., each instance consists of a vector of real-valued measurements of some attributes. In density estimation, a decision rule represents a density on X, and many choices are possible. One common choice is a mixture of Gaussian densities (see Duda and Hart[25] and Nowlan[65]). In standard regression, the outcome and decision spaces Y and A are identical and real valued, and linear functions are most often used as decision rules.

From a statistical point of view, the goal is to find a linear function h from X into Y such that for each x in X, $h(x)$ is the mean of the best Gaussian estimate of the conditional density on Y given x, where the distance between the estimate and true density is measured by the information gain. Happily, even when the true density $p(y|x)$ is not Gaussian, the information gain is minimized when $h(x)$ is the mean of $p(y|x)$ (see White[95]). So we have a simpler, more common interpretation of linear regression as an attempt to find a linear function that gives a good estimate of the average value of the outcome Y for each instance x. (The average value of Y given x is often called the *regression function*. By minimizing the expectation of the quadratic loss over all linear functions, we obtain the linear function that is closest to the regression function in the L^2 norm. Hence, in this view, regression is like function approximation with noisy data.)

For more complex outcome spaces such as those in the medical diagnosis example given above, the decision rule space for regression is usually defined using a *generalized linear model*.[56] Similarly, in binary classification, where there are only two possible outcomes in Y as in the PAC model, linear threshold functions are most often used as decision rules, and there are straightforward generalizations for the case of k-ary classification (see Duda and Hart[25]). This "linear bias" in pattern recognition and statistics is in contrast to that in the PAC model and other AI areas, including work in neural networks, in which a rich variety of decision rule spaces are used (see Touretsky[84,85] and Haussler[38,39]). Our main goal here is to develop analytic tools to help understand the problem of overfitting in these more complex decision rule spaces.

In order to focus on the problem of overfitting, we take a simplified view of learning, in which the learner chooses a decision rule space $\mathcal{H}$, and then tries to find a decision rule in $\mathcal{H}$ with near minimal expected loss. To do this, the learner looks for a decision rule that minimizes the observed average loss on the training examples, which is called *empirical loss* or *empirical risk*. For example, in standard linear regression[5] the learning algorithm is the method of least squares; i.e., we find the linear function h that minimizes the average of $l(y, h(x)) = (h(x) - y)^2$

---

[5]For general regression with the negative log likelihood loss function, the principle of minimizing empirical loss is the same as the principle of *maximum likelihood*.[14,46]

over all examples (x, y) in our training set. As shown above, this can be viewed as an application of the principle of maximum likelihood. It is well known that if we have too few training examples, then we tend to overfit them, and the function we find does not come close to minimizing the actual expected quadratic loss, which would be obtained by integrating over all possible (mostly unseen) examples with respect to the unknown joint distribution on them. This same situation occurs with all nontrivial decision rule spaces, including the nonlinear regression models defined by feedforward neural nets.

Using certain measures of the "dimension" or "capacity" of the decision rule space $\mathcal{H}$ and classes derived from $\mathcal{H}$ (see below), we obtain general upper bounds on the number of random training examples needed so that with high probability, any decision rule in $\mathcal{H}$ that has small empirical loss on the training examples will have small actual expected loss; i.e., we get uniform convergence results for empirical estimates like those given by Vapnik,[88] Dudley,[27] and Pollard.[70,72] We show how these give upper bounds on sufficient training sample size like those derived by Blumer et al.,[16] and elsewhere using the notion of the VC dimension, and generalize those results.

As an application, we give specific bounds on the number of training examples needed to avoid overfitting when learning with the decision rule space of feedforward neural nets,[76] extending previous work by Baum and Haussler[12] and White[94] (see also related work by Anthony and Shawe-Taylor[7]). These are the nets most widely used in current neural net learning research. Our model for feedforward neural nets is quite general in that it allows many types of units in the nets, including quasi-linear units,[76] radial basis units,[69] and product units.[29]

In our general setting, successful learning means finding a decision rule with average loss close to minimal over all decision rules in the given decision rule space, rather than loss close to zero as in the PAC model. In addition to using an additive model as did Lineal et al.,[51] we also define "close to" using a measure of relative difference (the d_ν metric) similar to the standard multiplicative measure of approximation used in combinatorial optimization. This allows us to state the relevant uniform convergence bounds as generalized "Chernoff-style" bounds,[6] as in results by Pollard[71] and Breiman et al. (Chapter 12),[17] rather than "Hoeffding-style" bounds (as in Pollard's results[70]), giving better bounds on sufficient training sample size in some important cases. These two types of bounds are analogous to the two types of bounds that Vapnik gives in his book[88] in that one uses a measure of absolute difference and the other a measure of relative difference. However, both of our bounds are "two-sided"; i.e., they bound deviations both above and below the mean.

We give these upper bounds on required sample size only to give some indication of the order-of-magnitude dependence of sample size on certain critical parameters of the learning problem, and to illustrate the theory. They are still too crude to be used directly in practice, e.g., as explicit formulae for choosing an appropriate sample size. Cross-validation techniques, in which some of the training examples are held in reserve and used instead to test the performance of the decision rules produced by the learning algorithm, are likely to perform better for this task in practice. Nevertheless, cross validation is only a means of estimating the amount of overfitting in the learning method in particular cases; i.e., it is only an engineering trick and provides no scientific explanation of the phenomenon. Our goal is to understand and explain overfitting in general decision rule spaces, from a scientific rather than an engineering viewpoint.

Finally, we should note that in practice, many learning algorithms do more than just search for a decision rule in a fixed decision rule space that minimizes empirical loss. For example, it is common to let the decision rule space depend on the number of training examples available, using richer and richer decision rule spaces as more examples become available (see White[94] and Blumer et al.[16]). This can allow the learning algorithm to produce a sequence of decision rules with expected losses that approach the loss of Bayes optimal decision rule in the limit of infinite training sample size for a large class of possible joint distributions. The results given here can be used to estimate the appropriate rate at which the decision rule space should grow relative to the sample size to avoid overfitting. Other approaches, e.g., the method of *structural risk minimization* introduced by Vapnik,[88] and the *Bayesian*[14,19,54] and *minimum description length* (MDL) approaches,[10,75] try to find a decision rule that minimizes some function of empirical loss and decision rule complexity. These can also achieve expected loss approaching that of Bayes optimal decision rule in the limit, and may be more effective in practice. Although uniform convergence results such as those we develop here are also used in the analysis of such methods[88] (and in the analysis of cross-validation methods[63]), the full treatment of such approaches is beyond the scope of the present paper. It should also be noted that Bayesian methods and structural risk minimization can be applied even when the decision rule space includes only neural networks of a fixed size. An example is the recent work using weight penalty functions in neural net training.[19,50,54,66,90] Such approaches may significantly reduce the training sample size needed to avoid overfitting in practice.

1.3 OVERVIEW OF METHODS USED

We now briefly discuss the methodology and previous work used in obtaining our results. Our work builds directly on the work of Vapnik and Chervonenkis, Pollard, and Dudley on the uniform convergence of empirical estimates[27,70,88] and its application to pattern recognition.[24,88,89] It also builds on the work of Benedek and Itai on PAC learnability with respect to specific probability distributions,[13] and is related to the work of Natarajan and Tadepalli on extensions of the VC dimension to multivalued functions[60,62] and PAC learnability with respect to classes of probability distributions.[59,61] In addition, Quiroz and Kulkarni have each independently generalized the PAC model in a related manner.[48,73]

One of the key ideas we use is the notion of an ε-cover of a metric space[13,27,61,70,73] and the associated idea of *metric dimension*[47] (also called the *fractal* dimension[32]). This notion of dimension has played an important role in the now very active study of fractals in nature,[55] especially in connection with chaos in dynamical systems.[32,33] Here we build further on the beautiful results of Vapnik and Chervonenkis,[88] Dudley,[26] and Pollard,[70] which relate a type of generalized VC dimension for a decision rule space to the number of balls of radius ε required to cover the space, with respect to certain metrics. The sizes of the smallest such covers determine the *metric dimension* of the space. Our treatment closely parallels the approach given by Pollard.[72] It is interesting to note that related results connecting ε-covers with the VC dimension have also been independently developed by Benedek and Itai[13] and in recent computational geometry work.[6] This work seems to lead to a potentially rich area of investigation that combines elements of combinatorics, topology and geometry, and probability and measure in a novel framework. We feel that this area is not only fascinating from a purely mathematical standpoint, but also potentially very useful in machine learning and other applied fields.

1.4 ORGANIZATION OF THE PAPER

The remainder of the paper is organized as follows. The learning framework we have described above in Section 1.3 is defined more formally in Section 2. There we also look at the question of evaluating the performance of learning algorithms in terms of the number of training examples they use. This question is also formalized from a decision theory perspective. We then provide a lemma (Lemma 1) that can be used to evaluate the performance of learning algorithms that work by minimizing empirical loss. To use this lemma, we need bounds on the rate of uniform convergence of empirical loss estimates to true expected losses. These are given in Section 3. The key bound is given in Theorem 2 in Section 3, and in a more general version in Theorem 3.

To use the bound from Theorem 2, we need bounds on the "random covering numbers" associated with the decision rule space $\mathcal{H}$, the loss function l, and the distribution P. These are related to the idea of an ε-cover described above. In Section 4 we introduce Pollard's notion of the pseudo dimension as a means of bounding the random covering numbers. Applications of this method to several learning problems are described in Section 5.

The techniques of Sections 4 and 5 only apply to the case when the action set A is real valued. Tools for bounding the random covering numbers that apply in more general cases are developed in Section 6. Here we introduce the notion of the capacity of the decision rule space $\mathcal{H}$ (for a particular loss function l), and the related notion of the metric dimension of $\mathcal{H}$. In Section 7 we use these notions to obtain bounds on the performance (in terms of the number of training examples

---

[6] See Welzl.[92] Specifically, Lemma 7.13 by Dudley[26] is nearly equivalent to Lemma 4.1 by Welzl[92] (using the primal space instead of the dual). This result also gives a stronger version of Theorem 4, part(3) of Benedek and Itai.[13] We give a still stronger version of this result in Theorem 6 below.

used) of learning algorithms that use multilayer feedforward neural networks, and work by minimizing empirical loss (Corollary 3). Finally, some further discussion of our results is given in the conclusion, Section 8.

Many of the more technical proofs and definitions have been moved into the appendix to make the paper more readable. The appendix has several sections. Section 9.1 contains a brief overview of the theory of metric spaces, ε-covers, and metric dimension. Notation from this section is used in several places in the paper. Section 9.2 deals with certain technical measurability requirements. Section 9.3 gives an analogue of Chernoff and Hoeffding bounds using the d_ν metric. Section 9.4 contains the proof of Theorem 2. Finally, Section 9.5 contains a result on feedforward neural networks of linear threshold functions that is similar to that given by Baum and Haussler,[12] and provides a counterpart to Corollary 3 in Section 7.

1.5 NOTATIONAL CONVENTIONS

We denote the real numbers by $\Re$ and the nonnegative real numbers by $\Re^+$. By log and ln we denote the logarithm base 2 and the natural logarithm, respectively. We use $\mathbf{E}(\cdot)$ to denote the expectation of a random variable, and $\mathbf{Var}(\cdot)$ to denote the variance of a random variable. When the probability space is defined implicitly from the context, we use $\mathbf{Pr}(\cdot)$ to denote the probability of a set. However, usually the measure on the underlying probability space will be defined explicitly using the symbol P.

Here, P will usually denote a probability measure on some appropriate[7] σ-algebra over the set $Z = X \times Y$, where X is the instance space and Y is the outcome space. We use P^m to denote the m-fold product measure on Z^m. Functions on Z and subsets of Z mentioned in what follows will be assumed to be measurable without explicit reference. Alternately, we will also view X and Y as random variables on some other, unspecified, probability space, e.g., when they are viewed as real-valued measurements. In this case P is viewed as a joint distribution on X and Y. In either case, the probability of a set $T \subset Z$ is defined by

$$P(T) = \int_T dP(z)$$

(where $z = (x, y)$ with $x \in X$ and $y \in Y$), and the expectation of function f on Z is denoted by

$$\mathbf{E}(f) = \int_Z f(z)dP(z).$$

When Z is countable, we will, with some abuse of notation, also use P for the probability mass function; i.e., for $z \in Z$, $P(z)$ denotes $P(\{z\})$. Hence $P(T) =$

---

[7]If Z is countable, then we assume this σ-algebra contains all subsets of Z; otherwise we assume that Z is a complete, separable metric space (see Section 9.1) and that this σ-algebra is the smallest σ-algebra that contains the open sets of Z (i.e., the σ-algebra of Borel sets).

$\sum_{z \in T} P(z)$ and $\mathbf{E}(f) = \sum_{z \in Z} f(z)P(z)$ in this case. When Z is continuous, a density associated with P (if it exists) is denoted by p.

TABLE 1

$X, Y, A, \mathcal{H}$, and l	Sections 1.1 and 2.1
$\mathcal{P}$	Section 2.1
$\mathbf{r}_{h,l}(P)$, $\mathbf{r}_h(P)$ (true risk)	Section 2.2
$\mathbf{r}_l^*(P)$, $\mathbf{r}^*(P)$ (optimal risk)	Section 2.2
$\hat{\mathbf{r}}_h(\vec{z})$ (empirical risk)	Section 2.2
$\hat{\mathbf{r}}^*(\vec{z})$ (optimal empirical risk)	Section 2.2
d_ν	Section 2.2
L, L_ε, $L_{\alpha,\nu}$ (regret functions)	Section 2.3
R (big "L" risk)	Section 2.3
$m(\varepsilon, \delta)$, $m(\alpha, \nu, \delta)$ (sample complexity)	Section 2.4
$\mathcal{N}$ (covering number)	Sections 10.1 and 3.2
$\mathcal{M}$ (packing number)	Section 10.1
dim (metric dimension)	Section 10.1
dimp (pseudo dimension)	Section 4
$\mathcal{C}$ (capacity)	Section 6
ρ_l	Section 6
$l_\mathcal{H}$	Section 3
$\mathbf{F}_{|\vec{z}}$	Section 3
$\widehat{\mathbf{E}}$ (empirical expectation)	Section 3
d_{L^1} (L^1 distance for vectors)	Section 3.2
$d_{L^1(P)}$ (L^1 distance for functions)	Section 4
$d_{L^1(P,\rho)}$ (L^1 distance for functions)	Section 6

When Z is countable, we use $P(y|x)$ to denote the probability that $Y = y$ given that $X = x$ (viewing X and Y as random variables) and similarly for $P(x|y)$. Hence $P(\cdot|x)$ denotes the conditional distribution on Y, given $X = x$. The marginal distribution in X is defined by[8] $P_{|X}(x) = \sum_{y \in Y} P(x, y)$. Here and elsewhere, we abbreviate $P((x, y))$ by $P(x, y)$.

Finally, in Table 1 we list some other notation that is used several places in the text, indicating which section it is defined in.

[8] When Z is uncountable, the marginal and conditional distributions are defined so that

$$\int_Z f(x, y) dP(x, y) = \int_X \left(\int_Y f(x, y) dP(y|x) \right) dP_{|X}(x)$$

for every bounded measurable function f (see Dudley,[27] Lemma 1.2.1).

2. LEARNING AND OPTIMIZATION

We now further formalize the basic problem of learning, as introduced in Section 1.1. We will introduce a formal notion of a learning algorithm, and a higher-level loss function, which we will call a *regret function*, that measures how well the learning algorithm performs. The regret function will be defined in terms of the low-level loss function l discussed in the previous section. Finally, we will show how an algorithm can solve the learning problem by solving a related optimization problem.

2.1 THE BASIC COMPONENTS $X, Y, A, \mathcal{H}, \mathcal{P}$, AND l

We first review and further formalize the six components of the basic learning problem introduced in the previous section: X, Y, A, $\mathcal{H}$, $\mathcal{P}$, and l. The first four components are the instance, outcome, decision, and decision rule spaces, respectively. The first three of these are arbitrary sets, and the fourth, $\mathcal{H}$, is a family of functions from X into A. These have been discussed extensively in the previous section.

The fifth component $\mathcal{P}$ is a family of joint probability distributions on $X \times Y$. These represent the possible "states of nature" that might be governing the generation of examples.[9] The set $Z = X \times Y$ will be called the *sample space*. We assume that examples are drawn independently at random according to some probability distribution $P \in \mathcal{P}$ on the sample space Z. A sequence of examples will be called a *sample*. In what follows, we will usually assume that $\mathcal{P}$ includes all probability distributions on Z. Hence our results will be distribution independent.[10]

The last component, the loss function l, is a mapping from $Y \times A$ into $\Re$. More general forms of loss function are possible without essential changes to the theoretical results given below; e.g., l may also depend explicitly on the instance x and on the particular decision rule h used to determine the action a[89,95]; we have merely adopted this convention to keep the notation simple. In this chapter we will assume that l is bounded and nonnegative; i.e., $0 \leq l \leq M$ for some real

---

[9] Thus $(Z, \mathcal{A}, P)$ forms a probability space for some appropriate σ-algebra $\mathcal{A}$ on Z for all $P \in \mathcal{P}$. If Z is finite or countably infinite, then we assume $\mathcal{A}$ includes all subsets of Z. Otherwise we assume that X and Y are complete, separable metric spaces and let $\mathcal{A}$ be the family of Borel sets with respect to the induced topology on Z; i.e., the smallest σ-algebra containing the open sets of Z.

[10] It is, however, possible and, in fact, common to assume that $\mathcal{P}$ is a very specific class of probability distributions on Z. For example, let $X = \Re^n$. Then if we are doing classification learning and Y is discrete, we may assume that y is selected according to an arbitrary distribution on Y, and for each y, $P(x|y)$ is a multivariate Gaussian distribution on X.[25] On the other hand, if we are doing linear regression, then Y is real valued and we might assume that x is selected according to an arbitrary distribution on X, and y is a linear function of x with additive Gaussian noise. In PAC learning theory, we have a discrete analog of the latter case. Here we usually have $X = \{0,1\}^n$, $Y = \{0,1\}$, and y a Boolean function of x of a particular type (e.g., defined by a small disjunctive normal form formula), possibly plus random noise.

M. When Y and A are finite, it is always possible to enforce this condition by simply adding a constant to l, which doesn't change the learning problem in any essential way. When either Y or A is infinite, the learning problem sometimes needs to be restricted to meet this condition. For example, in regression we might restrict the possible parameter vectors in A and/or the possible outcomes in Y such that for every $y \in Y$ and $a \in A$, $\widehat{P}(y; a) \geq b$ for some constant b.[11] We can then take $M = -\log b$. In density estimation, the same thing can be accomplished by restricting the instance space X to a bounded subset of $\Re^n$ on which all densities in $\mathcal{H}$ have values uniformly greater than b and less than B for constants $0 < b < B$. We can then add $\log B$ to the loss function to make it positive. The same method works for estimating distributions on discrete spaces: we restrict ourselves to a finite instance space X and demand that for all $x \in X$ and all probability distributions $h \in \mathcal{H}$, $h(x) \geq b > 0$ (see Abe and Warmuth[1] and Yamanishi[96]). Here, $h(x) \leq 1$, so there is no problem with negative loss. These restrictions are often reasonable in practice; e.g., most measurements naturally have bounded ranges, but they can be annoying (see Vapnik[89] and Pollard[70,72] for alternative approaches for unbounded loss functions).

2.2 MEASURING DISTANCE FROM OPTIMALITY WITH THE d_ν METRIC

For a given decision rule $h \in \mathcal{H}$ and distribution P on the sample space Z, the expected loss of h is the average value of $l(y, h(x))$, when the example (x, y) is drawn at random according to P. It is defined by

$$\mathbf{r}_{h,l}(P) = \mathbf{r}_h(P) = \mathbf{E}(l(y, h(x))) = \int_Z l(y, h(x)) dP(x, y)$$

(the subscript l will be omitted when the loss function is clear from the context). Since l is bounded, this expectation is finite for every distribution P. In decision theory the expected loss $\mathbf{r}_h(P)$ is called the *risk* of h when P is the true underlying distribution. This quantity generalizes the notion of the *error* of h used in computational learning theory.

In Section 1.1 we stated the goal of learning quite informally: Given examples chosen independently at random from some unknown probability distribution $P \in \mathcal{P}$, find a decision rule $\widehat{h}$ in $\mathcal{H}$ that comes "close to" minimizing the risk $\mathbf{r}_h(P)$ over all $h \in \mathcal{H}$. Let $\mathbf{r}_l^*(P)$ (or $\mathbf{r}_l^*(P)$ when l is clear from the context) denote the infimum of $\mathbf{r}_h(P)$ over all h in the decision rule space $\mathcal{H}$. To formalize our notion of a basic learning problem, we first need to say what we mean that $\mathbf{r}_{\widehat{h}}(P)$ is "close to" $\mathbf{r}^*(P)$.

Let $r = \mathbf{r}_h(P)$ and $s = \mathbf{r}^*(P)$. One natural interpretation is to demand that $|r - s| \leq \varepsilon$ for some small $\varepsilon > 0$. However, we will see in Section 3.1 that sometimes

---

[11] Note that to get bounded loss in linear regression, X must a bounded subset of $\Re^n$ as well, since we cannot bound Y without bounding X. The coefficients of the functions in $\mathcal{H}$ must also be bounded.

it is better to use a relative measure of distance. For any real $\nu > 0$, let d_ν be the function defined by

$$d_\nu(r, s) = \frac{|r - s|}{\nu + r + s}$$

for any nonnegative reals r and s. It is straightforward but tedious to verify that d_ν is a metric on $\Re^+$. The d_ν metric is similar to the standard function

$$\frac{|r - s|}{s}$$

used to measure the difference between the quality r of a given solution and the quality s of an optimal solution in combinatorial optimization. However, our measure has been modified to be well behaved when one or both of its arguments are zero, and to be symmetric in its arguments (so that it is a metric). Three other properties of d_ν are also useful.

1. For all nonnegative reals r and s, $0 \leq d_\nu(r, s) < 1$.
2. For all nonnegative $r \leq s \leq t$, $d_\nu(r, s) \leq d_\nu(r, t)$ and $d_\nu(s, t) \leq d_\nu(r, t)$.
3. For $0 \leq r, s \leq M$, $|r - s|/(\nu + 2M) \leq d_\nu(r, s) \leq |r - s|/\nu$.
4. And in particular, if $\varepsilon > 0$, $\nu = 2M$ and $\alpha = \varepsilon/4M$, then $d_\nu(r, s) \leq \alpha \Rightarrow |r - s| \leq \varepsilon$.

We will refer to the second property by saying that d_ν is *compatible with the ordering on the reals*. This second property also clearly holds for the absolute difference metric. The key motivation for the d_ν metric is to capture the idea that sometimes the magnitude of the difference between the positive numbers r and s should be measured relative to sizes of these numbers when they are large with respect to some normative value ν. This is a common intuition for natural numbers, for example. People tend to see much more significance in the difference between 1 and 2 than in the difference between $100,001$ and $100,002$, even though the absolute difference is the same (as is the usual bounded variant of the absolute difference $d(r, s) = |r - s|/(\nu + |r - s|)$). Indeed the difference *is* more significant relative to the normative value $\nu = 1$. This is reflected in the d_ν metric: $d_\nu(1, 2) = 0.25$ and $d_\nu(100,001 \ , \ 100,002) \approx 0.000005$ for $\nu = 1$. This distinction is especially important in the the present context, because how close we can get to s will depend on the size of the random training sample we have. It turns out that, assuming an arbitrary distribution P and a bounded nonnegative loss function, we need a much smaller sample size to find (with high probability) a decision rule with risk r within 1 of the optimum s when $s = 1$ than when $s = 100,000$. We discuss this further in the following section.

2.3 THE REGRET FUNCTION L AND THE BIG "L" RISK R

Once we have specified how we measure closeness to optimality, we still need to specify our criteria for a successful learning algorithm. Do we need to have the risk of the decision rule found close to the optimum $r_l^*(P)$ with high probability, or should its average distance from $r_l^*(P)$ be small? Do we measure success in terms of the performance of the algorithm on the worst-case distribution in $\mathcal{P}$, or do we use some average-case analysis over distributions in $\mathcal{P}$? These questions lead us right back to decision theory again, but this time at a higher level in the analysis of learning.

To see this, consider the structure of a learning algorithm $\mathcal{A}$. For any sample size m, the algorithm $\mathcal{A}$ may be given a sample $\vec{z} = ((x_1, y_1), \ldots, (x_m, y_m))$ drawn at random from Z^m according an unknown product distribution P^m, where $P \in \mathcal{P}$. For any such $\vec{z}$ it will choose a decision rule $\mathcal{A}(\vec{z}) \in \mathcal{H}$. Thus abstractly, the algorithm defines a function $\mathcal{A}$ from the set of all samples over Z into $\mathcal{H}$; i.e., $\mathcal{A} : \bigcup_{m \geq 1} Z^m \to \mathcal{H}$. Since we are not requiring computability here, we will call such $\mathcal{A}$ a *learning method*. When $P \in \mathcal{P}$ is the actual "state of nature" governing the generation of examples, and the algorithm produces the decision rule $h \in \mathcal{H}$, let us say that we suffer a nonnegative real-valued *regret* $L(P, h)$. Thus, formally $L : \mathcal{P} \times \mathcal{H} \to \Re^+$. In our treatment here, the regret function L will be derived from the loss function l, and will measure the extent to which we have failed to produce a near optimal decision rule, assuming P is the true state of nature (i.e., the amount of "regret" we feel for not having produced the optimal decision rule). Finally, for each possible state of nature P, the average regret suffered by the algorithm, over all possible training samples $\vec{z} \in Z^m$, is the *big "L" risk* of that algorithm under P for sample size m. This big "L" risk is defined formally by

$$R_{L, \mathcal{A}, m}(P) = \int_{\vec{z} \in Z^m} L(P, \mathcal{A}(\vec{z})) dP^m(\vec{z}).$$

The goal of learning is to minimize big "L" risk.

We illustrate these definitions with a few examples. First suppose we want to capture the notion of successful learning that is used in the PAC model. Then one possibility is to introduce an *accuracy parameter* $\varepsilon > 0$ and define the regret function $L = L_\varepsilon$ by letting $L_\varepsilon(P, h) = 1$ if $r_{h,l} - r_l^*(P) > \varepsilon$, and $L_\varepsilon(P, h) = 0$ otherwise. Thus we suffer regret only when the decision rule h produced by the learning algorithm has risk that is more than ε from optimal, measured by the absolute difference metric. For this definition of regret, the big "L" risk $R_{L_\varepsilon, \mathcal{A}, m}(P)$ measures the probability that the decision rule produced by $\mathcal{A}$ has risk more than ε from optimal, when $\mathcal{A}$ is given m random training examples drawn according to P. We then demand that this big "L" risk be small; i.e., smaller than some given *confidence parameter* $\delta > 0$.

In the PAC model, it is commonly assumed that the examples given to the algorithm $\mathcal{A}$ are noise-free examples of some underlying target function $f \in \mathcal{H}$.

In this case the risk $\mathbf{r}^*(P)$ of the optimal decision rule in $\mathcal{H}$ is zero and, hence, $L_\varepsilon(P, h) = 1 \leftrightarrow \mathbf{r}_{h,l}(P) > \varepsilon$. Hence, demanding big "L" risk at most δ gives the usual PAC criterion that the risk (or "error") of the decision rule (or "hypothesis") produced by $\mathcal{A}$ be greater than ε with probability at most δ.

The regret function L can also be defined similarly, but using the d_ν metric to measure distance from optimality, instead of the absolute difference. Specifically, for every $\nu > 0$ and $0 < \alpha < 1$, we can define the regret function $L_{\alpha,\nu}$ by letting $L_{\alpha,\nu}(P, h) = 1$ if $d_\nu(\mathbf{r}_{h,l}(P), \mathbf{r}_l^*(P)) > \alpha$, and $L_{\alpha,\nu}(P, h) = 0$ otherwise. In this case the big "L" risk $R_{L_{\alpha,\nu},\mathcal{A},m}(P)$ measures the probability that the risk of the decision rule produced by the algorithm $\mathcal{A}$ has distance more than α from optimal in the d_ν metric, when the algorithm is given m random training examples drawn according to P. We see in Sections 2.4 and 3.1 why this sometimes gives a more useful and flexible definition of regret. In fact, in this chapter, we will give our main results in terms of the family $\{L_{\alpha,\nu} : \nu > 0 \text{ and } 0 < \alpha < 1\}$ of regret functions, and show how corresponding results may be derived as corollaries for the family $\{L_\varepsilon : \varepsilon > 0\}$ of regret functions.

Other regret functions are also possible and lead to different learning criteria. For example, another, perhaps simpler, way to define regret is to let $L(P, h) = \mathbf{r}_{h,l}(P) - \mathbf{r}_l^*(P)$. When $\mathbf{r}_l^*(P) = 0$, as it does in the standard noise-free PAC model, this definition makes the regret L equal to the risk $\mathbf{r}_{h,l}(P)$, i.e., the expectation of the underlying loss l. In this case the big "L" risk $R_{L,\mathcal{A},m}(P)$ measures the expectation of the loss incurred by the learning algorithm $\mathcal{A}$ when it is given m random training examples drawn according to P, forms a decision rule h, and then uses h to determine the action on one further independent random example drawn according to P. This gives a generalization of the learning criterion studied by Haussler et al.[41] When $\mathbf{r}_l^*(P) \neq 0$, then the big "L" risk gives the expectation of the amount of such loss above and beyond the expected loss that would be suffered if the optimal decision rule were used. In particular, in density estimation, where P and h are both densities on the instance space X, if $P \in \mathcal{H}$, then defining the regret by $L(P, h) = \mathbf{r}_{h,l}(P) - \mathbf{r}_l^*(P)$ makes it equal to the Kullback-Leibler divergence from P to h. Hence, the big "L" risk is the expected Kullback-Leibler divergence of the decision rule h returned by the algorithm from the true density (see Sections 1.1.3 and 1.1.4).

It is also possible to define the regret function L directly, without using an underlying loss function l. For example, in density estimation it is possible to use other measures of the distance between two densities, e.g., the Hellinger distance or the total variational distance, as given by Barron and Cover[10] and Yamanishi.[96] The criterion from Kearns and Schapire[45] for inferring a good *model of probability* can also be defined using an appropriate regret function, without defining an underlying loss l.

2.4 FULL FORMALIZATION OF THE BASIC LEARNING PROBLEM

Having defined the regret function, and thereby the big "L" risk function, we still face one last issue: do we want to minimize big "L" risk in the worst case over all possible states of nature P in $\mathcal{P}$, or do we want to assume a *prior distribution* on possible distributions in $\mathcal{P}$, so that we can define a notion of "average case" big "L" risk to be minimized. The former goal is know as *minimax optimality*, and has been used in the PAC model. The later is the *Bayesian* notion of optimality,[14,46] and has been used in several approaches to learning in neural nets based on statistical mechanics.[23,36,67,68,81,83] Unfortunately this last question has no clear cut answer, and leads us directly into a longstanding unresolved debate in statistics (see Lindley[52] and following discussion). Since we have set out to generalize the PAC model, and since our results are best illustrated in the minimax setting, we will formalize the notion of a basic learning problem using the minimax criterion. However, as with our choice of regret function, this choice should not be interpreted as a dogmatic statement that this is the only "correct" model. In subsequent work we hope to further explore this Bayesian setting. (For recent work in Bayesian approaches to neural network learning, see MacKay[54] and Buntine and Weigend,[19] and for Bayesian versions of the PAC model see Haussler et al.[43] and Buntine.[18])

We can now define exactly what we mean by a basic learning problem, and what it means for a learning method to solve this problem in this minimax setting.

DEFINITION 1. A basic learning problem is defined by six components $X, Y, \mathcal{A}, \mathcal{H}$, $\mathcal{P}$, and $\mathcal{L}$, where the first five components are as defined in Section 2.1, and the last component, $\mathcal{L}$, is a family of regret functions as defined in Section 2.3 (e.g., $\mathcal{L} = \{L_{\alpha,\nu} : \nu > 0 \text{ and } 0 < \alpha < 1\}$, or $\mathcal{L} = \{L_\varepsilon : \varepsilon > 0\}$ for some loss function l). Let $\mathcal{A}$ be a learning method as defined in Section 2.3. We say that $\mathcal{A}$ solves the basic learning problem if, for all $L \in \mathcal{L}$ and all $0 < \delta < 1$, there exists a finite sample size $m = m(L, \delta)$ such that

$$\forall P \in \mathcal{P}, \quad R_{L,\mathcal{A},m}(P) \leq \delta.$$

The *sample complexity* of the learning method $\mathcal{A}$ is the smallest such integer-valued function $m(L, \delta)$. When $\mathcal{L} = \{L_{\alpha,\nu} : \nu > 0 \text{ and } 0 < \alpha < 1\}$, we will denote $m(L_{\alpha,\nu}, \delta)$ by $m(\alpha, \nu, \delta)$ and, when $\mathcal{L} = \{L_\varepsilon : \varepsilon > 0\}$, we will denote $m(L_\varepsilon, \delta)$ by $m(\varepsilon, \delta)$.

As discussed above, this definition generalizes the PAC criterion, and several others as well. In fact, this definition is quite generous, in that sample size needed to get the big "L" risk less than δ is only required to be finite for each $\delta > 0$. In particular, using property (3) of the d_ν metric from Section 2.2, when the underlying loss function l is bounded, as we assume here, any algorithm $\mathcal{A}$ solves the basic learning problem using the $L_{\alpha,\nu}$ class of regret functions if and only if it solves it using the L_ε class. Thus it doesn't matter which of these two classes of regret functions we use. However, in practice it is the sample complexity of $\mathcal{A}$ that is critical, and this will depend on which class of regret functions are used.

The nature of this dependence is seen more clearly when we expand the condition

$$R_{L,\mathcal{A},m}(P) \leq \delta$$

for $L = L_\varepsilon$ and $L = L_{\alpha,\nu}$. When $L = L_\varepsilon$, this condition means that given m random training examples drawn according to P, with probability at least $1-\delta$, the decision rule $\hat{h}$ produced by the algorithm $\mathcal{A}$ satisfies

$$\mathbf{r}_{\hat{h},l}(P) \leq \mathbf{r}_l^*(P) + \varepsilon;$$

i.e., the risk of $\hat{h}$ is at most ε greater than that of the optimal decision rule in $\mathcal{H}$. When $L = L_{\alpha,\nu}$, this condition is the same, except that we require

$$\mathbf{r}_{\hat{h},l}(P) \leq \frac{1+\alpha}{1-\alpha}\mathbf{r}_l^*(P) + \frac{\alpha\nu}{1-\alpha}.$$

Thus in the former case, the sample complexity is defined in terms of small additive deviations from optimality, and in the latter, we allow both additive and multiplicative deviations. These deviations are controlled by the parameters α and ν.

For example, when $\mathbf{r}_l^*(P) = 0$ as in the standard PAC model, then setting $\alpha = 1/2$ and $\nu = \varepsilon$ makes the L_ε and $L_{\alpha,\nu}$ conditions equivalent; each reduces to the PAC condition

$$\mathbf{r}_{\hat{h},l}(P) \leq \varepsilon.$$

When $\mathbf{r}_l^*(P) > 0$, then $L_{\alpha,\nu}$ condition approximates the L_ε condition when α is small and $\nu \approx \varepsilon/\alpha$. In particular, since we are assuming that the underlying loss function l is bounded between 0 and M, we have $0 \leq \mathbf{r}_{\hat{h},l}(P), \mathbf{r}_l^*(P) \leq M$ and property 3 of the d_ν metric shows that the $L_{\alpha,\nu}$ condition with $\nu = 2M$ and $\alpha = \varepsilon/4M$ implies the L_ε condition. This shows how the two parameter $L_{\alpha,\nu}$ condition is generally more flexible than the single parameter L_ε condition.

2.5 RELATION BETWEEN LEARNING AND OPTIMIZATION

Let us assume that the underlying loss function l is fixed, and we are using either the L_ε or $L_{\alpha,\nu}$ regret functions derived from l. In order to solve a basic learning problem, we must find, with high probability, a decision rule $\hat{h}$ with risk close to optimal. As the true distribution P is unknown, to do this we must rely on estimates of $\mathbf{r}_h(P)$ for the various $h \in \mathcal{H}$ which are derived from the given random training sample. For a given $h \in \mathcal{H}$ and training sample $\vec{z} = (z_1, \ldots, z_m)$, where $z_i = (x_i, y_i) \in Z$, let $\hat{\mathbf{r}}_h(\vec{z})$ denote the empirical risk on $\vec{z}$, i.e., $\hat{\mathbf{r}}_h(\vec{z}) = (1/m)\sum_{i=1}^m l(y_i, h(x_i))$. Let $\hat{\mathbf{r}}^*(\vec{z}) = \inf\{\hat{\mathbf{r}}_h(\vec{z}) : h \in \mathcal{H}\}$. We can then define a natural optimization problem associated with the basic learning problem: given the training sample $\vec{z}$, find a decision rule $\hat{h} \in \mathcal{H}$ such that $\hat{\mathbf{r}}_{\hat{h}}(\vec{z})$ is close to $\hat{\mathbf{r}}^*(\vec{z})$; i.e., a decision rule whose empirical risk on the training sample is close to minimal.

Solving the optimization problem does not automatically solve the learning problem. We need to have good empirical risk estimates as well. Since l is bounded, for every $h \in \mathcal{H}$, as the sample size $m \to \infty$, $\hat{\mathbf{r}}_h(\vec{z}) \to \mathbf{r}_h(P)$ with probability 1. We will say that the empirical risk estimates of decision rules in $\mathcal{H}$ converge *uniformly* to the true risk if, for all ε and $\delta > 0$, there exists a sample size m such that when the $z_i \in \vec{z}$, $1 \le i \le m$, are drawn independently at random from Z according to the distribution P, with probability at least $1 - \delta$, we have $\rho(\hat{\mathbf{r}}_h(\vec{z}), \mathbf{r}_h(P)) \le \varepsilon$ for all $h \in \mathcal{H}$. Here ρ is some metric on $\Re^+$; e.g., either the absolute difference or the d_ν metric.

The following result shows that uniform convergence of the empirical risk estimates, along with a learning method $\mathcal{A}$ that gives a randomized solution to the optimization problem on the estimates, gives a solution to the basic learning problem. We state it for the d_ν metric, but the same argument works also for the absolute difference metric.

LEMMA 1. Let $\nu > 0$ and $0 < \alpha, \delta < 1$. Suppose the sample size $m = m(\alpha, \nu, \delta)$ is such that for all probability distributions $P \in \mathcal{P}$

$$\mathbf{Pr}(\exists h \in \mathcal{H} : d_\nu(\hat{\mathbf{r}}_h(\vec{z}), \mathbf{r}_h(P)) > \alpha/3) \le \delta/2,$$

where the $z_i \in \vec{z}$, $1 \le i \le m$, are drawn independently at random from Z according to the distribution P. Suppose also that the algorithm $\mathcal{A}$ is such that for all $P \in \mathcal{P}$

$$\mathbf{Pr}(d_\nu(\hat{\mathbf{r}}_{\mathcal{A}(\vec{z})}(\vec{z}), \hat{\mathbf{r}}^*(\vec{z})) > \alpha/3) \le \delta/2,$$

where $\vec{z}$ is drawn randomly by P as above. Then for all $P \in \mathcal{P}$

$$\mathbf{Pr}(d_\nu(\mathbf{r}_{\mathcal{A}(\vec{z})}(P), \mathbf{r}^*(P)) > \alpha) \le \delta,$$

i.e., $\mathcal{A}$ solves the basic learning problem for the family of $L_{\alpha,\nu}$ regret functions and has sample complexity at most $m(\alpha, \nu, \delta)$.

PROOF. By the triangle inequality for d_ν, if

1. $d_\nu(\mathbf{r}_{\mathcal{A}(\vec{z})}(P), \hat{\mathbf{r}}_{\mathcal{A}(\vec{z})}(\vec{z})) \leq \alpha/3$,
2. $d_\nu(\hat{\mathbf{r}}_{\mathcal{A}(\vec{z})}(\vec{z}), \hat{\mathbf{r}}^*(\vec{z})) \leq \alpha/3$, and
3. $d_\nu(\hat{\mathbf{r}}^*(\vec{z}), \mathbf{r}^*(P)) \leq \alpha/3$,

then

$$d_\nu(\mathbf{r}_{\mathcal{A}(\vec{z})}(P), \mathbf{r}^*(P)) \leq \alpha.$$

The second assumption of the lemma states that item 2 holds with probability at least $1 - \delta/2$. The first assumption implies that both items 1 and 3 hold with probability at least $1 - \delta/2$. (If item 3 fails then we can find a decision rule $h \in \mathcal{H}$ such that $d_\nu(\hat{\mathbf{r}}_h(\vec{z}), \mathbf{r}_h(P)) > \alpha/3$. Here we use the compatibility of d_ν with the ordering on the reals.) Hence with probability at least $1 - \delta$ all of 1–3 hold. The result follows. $\square$

In statistics, this type of result is called a *consistency theorem* about the "statistic" (i.e., the decision rule) computed by the learning method $\mathcal{A}$. This use of the term "consistency" differs sharply from that common in PAC learning research.

3. UNIFORMLY GOOD EMPIRICAL ESTIMATES OF MEANS

In this section, we concentrate on the problem of bounding the number of random examples needed to get good empirical estimates of the risk of each of the decision rules in a decision rule space $\mathcal{H}$. For each decision rule $h \in \mathcal{H}$ and example $z = (x, y) \in Z$, let $l_h(z) = l(y, h(x))$. As in the previous section, we assume that l is a nonnegative bounded loss function taking values in the interval $[0, M]$; thus, for each decision rule h, l_h defines a random variable taking values in $[0, M]$. The value of l_h on an example (x, y) is the loss incured when you use h to determine the action to take for instance x, and the outcome is y. The risk of h is just the expectation of l_h; i.e.,

$$\mathbf{r}_h(P) = \mathbf{E}(l_h) = \int_Z l_h(z) dP(z).$$

Furthermore, if $\vec{z} = (z_1, \ldots, z_m)$ is a sequence of examples from Z, then the empirical risk of h on $\vec{z}$ is the empirical estimate of the mean of l_h based on the sample $\vec{z}$, which we denote by $\widehat{\mathbf{E}}_{\vec{z}}(l_h)$; i.e.,

$$\hat{\mathbf{r}}_h(\vec{z}) = \widehat{\mathbf{E}}_{\vec{z}}(l_h) = \frac{1}{m} \sum_{i=1}^{m} l_h(z_i).$$

Let $l_{\mathcal{H}} = \{l_h : h \in \mathcal{H}\}$. We need to draw enough random examples to get a uniformly good empirical estimate of the expectation of every random variable in $l_{\mathcal{H}}$.

The general problem of obtaining a uniformly good estimate of the expectation of every function in a class $\mathbf{F}$ of real-valued functions has been widely studied (see Vapnik,[88] Pollard,[70] and Dudley,[27] and their references). If no assumptions at all are made about the functions in $\mathbf{F}$, we immediately run into the problem that some functions in $\mathbf{F}$ could take on arbitrarily large values with arbitrarily small probabilities, making it impossible to obtain uniformly good empirical estimates of all expectations with any finite sample size. This problem can be avoided by making assumptions about the moments of the functions in $\mathbf{F}$, as given by Vapnik,[88] or by assuming that there exists a single nonnegative function with a finite expectation (called an *envelope*) that lies above the absolute value of every function in $\mathbf{F}$, as given by Pollard[70] and Dudley.[27] In our case, when the loss takes only values in $[0, M]$, then the constant function M serves as an envelope. This case is especially nice since this same envelope works for all distributions on the domain Z of the functions in $\mathbf{F}$.

The usual measure of deviation of empirical estimates from true means is simply the absolute value of the difference. Thus we would say that the empirical estimates for the expectations of the functions in $\mathbf{F}$ converge uniformly to the true expectations if as the size of the random sample $\vec{z}$ grows,

$$\mathbf{Pr}\left(\exists f \in \mathbf{F} : |\widehat{\mathbf{E}}_{\vec{z}}(f) - \mathbf{E}(f)| > \varepsilon\right)$$

goes to zero for any $\varepsilon > 0$. (This is called *(uniform) convergence in probability;* see e.g., Billingsly.[15]) Vapnik, Dudley, Pollard, and others have obtained general bounds on the sample size needed so that

$$\mathbf{Pr}\left(\exists f \in \mathbf{F} : |\widehat{\mathbf{E}}_{\vec{z}}(f) - \mathbf{E}(f)| > \varepsilon\right) < \delta$$

for $\varepsilon, \delta > 0$.[27,70,88] Vapnik also obtains better bounds in some important cases by considering the relative deviation of empirical estimates from true expectations. He looks at bounds on the sample size needed so that

$$\mathbf{Pr}\left(\exists f \in \mathbf{F} : \frac{\mathbf{E}(f) - \widehat{\mathbf{E}}_{\vec{z}}(f)}{\mathbf{E}(f)} > \varepsilon\right) < \delta,$$

and also bounds on the sample size needed so that

$$\mathbf{Pr}\left(\exists f \in \mathbf{F} : \frac{\mathbf{E}(f) - \widehat{\mathbf{E}}_{\vec{z}}(f)}{\sqrt{\mathbf{E}(f)}} > \varepsilon\right) < \delta.$$

(Anthony and Shawe-Taylor also obtain bounds of the latter form.[7]) Note that these are one-sided bounds, in that they only bound the probability that the empirical mean is significantly smaller than the true mean. While extremely useful, as we mentioned in the previous section, these measures of deviation suffer from a discontinuity at $\mathbf{E}(f) = 0$, and a lack of convenient metric properties. Like Pollard,[71] we will give bounds on the sample size needed so that

$$\mathbf{Pr}\left(\exists f \in \mathbf{F} : d_\nu(\widehat{\mathbf{E}}_{\vec{z}}(f), \mathbf{E}(f)) > \alpha\right)$$

$$= \mathbf{Pr}\left(\exists f \in \mathbf{F} : \frac{|\widehat{\mathbf{E}}_{\vec{z}}(f) - \mathbf{E}(f)|}{\nu + \widehat{\mathbf{E}}_{\vec{z}}(f) + \mathbf{E}(f)} > \alpha\right) < \delta;$$

i.e., the deviation measured using the d_ν metric.[12] By setting ν and α appropriately, we obtain results similar to those given by Pollard[70] and Vapnik[88] as special cases of our main theorem. However, our results are restricted to the case that all functions in $\mathbf{F}$ are positive and uniformly bounded.

3.1 THE CASE OF FINITE F

Before considering the general case, it is useful to see what bounds we can get in the case that $\mathbf{F}$ is a finite set of functions. Here we can easily prove the following.

THEOREM 1. Let $\mathbf{F}$ be a finite set of functions on Z with $0 \leq f(z) \leq M$ for all $f \in \mathbf{F}$ and $z \in Z$. Let $\vec{z} = (z_1, \ldots, z_m)$ be a sequence of m examples drawn independently from Z according to any distribution on Z, and let $\varepsilon > 0$. Then

$$\mathbf{Pr}\left(\exists f \in \mathbf{F} : |\widehat{\mathbf{E}}_{\vec{z}}(f) - \mathbf{E}(f)| > \varepsilon\right) \leq 2|\mathbf{F}|e^{-2\varepsilon^2 m/M^2}.$$

For $0 < \delta \leq 1$ and sample size

$$m \geq \frac{M^2}{2\varepsilon^2}\left(\ln |\mathbf{F}| + \ln \frac{2}{\delta}\right),$$

[12] Pollard[71] also gives results that can be used to bound the sample size needed so that

$$\mathbf{Pr}\left(\exists f \in \mathbf{F} : \frac{|\widehat{\mathbf{E}}_{\vec{z}}(f) - \mathbf{E}(f)|}{\nu + \sqrt{\widehat{\mathbf{E}}_{\vec{z}}(f)} + \sqrt{\mathbf{E}(f)}} > \alpha\right) < \delta,$$

in analogy with the second type of bound given by Vapnik, except that these bounds are two-sided. We do not pursue these further here.

this probability is at most δ. Further, for any $\nu > 0$ and $0 < \alpha < 1$,

$$\mathbf{Pr}\left(\exists f \in \mathbf{F} : d_\nu(\widehat{\mathbf{E}}_{\bar{z}}(f), \mathbf{E}(f)) > \alpha\right) \leq 2|\mathbf{F}|e^{-\alpha^2 \nu m/M}.$$

For $0 < \delta \leq 1$ and sample size

$$m \geq \frac{M}{\alpha^2 \nu}\left(\ln|\mathbf{F}| + \ln\frac{2}{\delta}\right),$$

this probability is at most δ.

PROOF. For the second part of the theorem, using Bernstein's inequality (see Pollard[70]), it is easy to show that for any single function f with $0 \leq f \leq M$,

$$\mathbf{Pr}\left(d_\nu(\widehat{\mathbf{E}}_{\bar{z}}(f), \mathbf{E}(f)) > \alpha\right) < 2e^{-\alpha^2 \nu m/M}.$$

Details are given in Lemma 9, part (2) in the appendix. It follows that the probability that there is any $f \in \mathbf{F}$ with $d_\nu(\widehat{\mathbf{E}}_{\bar{z}}(f), \mathbf{E}(f)) > \alpha$ is at most $2|\mathbf{F}|e^{-\alpha^2 \nu m/M}$. Setting this bound to δ and solving for m gives the result on the sample size. The proof of the first part of the lemma is similar, except we use Hoeffding's inequality (see Pollard[70]), which implies that for any single f,

$$\mathbf{Pr}\left(|\widehat{\mathbf{E}}_{\bar{z}}(f) - \mathbf{E}(f)| > \varepsilon\right) \leq 2e^{-2\varepsilon^2 m/M^2}.$$

□

By letting $\mathbf{F} = l_{\mathcal{H}}$, this theorem can be used in conjunction with Lemma 1 from the previous section to obtain bounds on the sample complexity of learning algorithms that minimize empirical risk. Here we can use either the L_ε or $L_{\alpha,\nu}$ family of regret functions. In the former case we get a sample complexity

$$m(\varepsilon, \delta) = O\left(\frac{M^2}{\varepsilon^2}\left(\log|l_{\mathcal{H}}| + \log\frac{1}{\delta}\right)\right). \tag{1}$$

In the latter case we get a sample complexity

$$m(\alpha, \nu, \delta) = O\left(\frac{M}{\alpha^2 \nu}\left(\log|l_{\mathcal{H}}| + \log\frac{1}{\delta}\right)\right). \tag{2}$$

As shown in the previous section, a generalization of the PAC learning model can be obtained by using either the L_ε or $L_{\alpha,\nu}$ regret functions, in the latter case by setting $\alpha = 1/2$ and $\nu = \varepsilon$. Note that plugging this latter setting into Eq. (2) gives a sample complexity

$$m(\varepsilon,\delta) = O\left(\frac{M}{\varepsilon}\left(\log|l_{\mathcal{H}}| + \log\frac{1}{\delta}\right)\right), \tag{3}$$

a significant improvement over Eq. (1), which is quadratic in M/ε. Thus the generalization of the PAC model using the d_ν metric to measure distance from optimality, and the resulting $L_{\alpha,\nu}$ family of regret functions, offers new insight in this regard. (Vapnik's[88] use of the relative difference between empirical estimates and true expectations also has this advantage; see Anthony and Shawe-Taylor[7] and also the appendix of Blumer et al.[16])

3.2 THE GENERAL CASE

The main task of this section is to generalize Theorem 1 to infinite collections of uniformly bounded functions. The basic idea is simple: we replace the infinite class $\mathbf{F}$ of functions with a finite class $\mathbf{F_0}$ that "approximates" it, in the sense that each function in $\mathbf{F}$ is close to some function in $\mathbf{F_0}$, and argue that some type of uniform convergence of empirical estimates for $\mathbf{F_0}$ implies uniform convergence for $\mathbf{F}$. In the simplest version of this technique, the choice of $\mathbf{F_0}$ depends only on $\mathbf{F}$ and the distribution P, as in the "direct method" discussed in section II.2 of Pollard[70] (see also Vapnik,[88] Section 6.6; Dudley,[27] Chapter 6; Benedek and Itai,[13] and White[94]). However, more general results (apart from certain measurability constraints) are obtained by allowing $\mathbf{F_0}$ to depend on the particular random sample $\vec{z}$ (see Pollard,[70] Chapter 2). Here $\mathbf{F_0}$ is called a "random cover," and its size is called a "random covering number." It is this type of result that we derive here.

We will need a few preliminary definitions to introduce the notion of ε-covers and metric dimension. A more general treatment of these ideas is given in the appendix, Section 9.1. This more general treatment will be used later, but the following definitions suffice for this section.

For any real vectors $\vec{x} = (x_1,\ldots,x_m)$ and $\vec{y} = (y_1,\ldots,y_m)$ in $\Re^m$, let $d_{L^1}(\vec{x},\vec{y}) = (1/m)\sum_{i=1}^m |x_i - y_i|$. Thus d_{L^1} is the L^1 distance metric. Let T be a set of points that lie in a bounded region of $\Re^m$. For any $\varepsilon > 0$, an ε-cover for T is a finite set $N \subset \Re^m$ (not necessarily contained in T) such that for all $\vec{x} \in T$ there is a $\vec{y} \in N$ with $d_{L^1}(\vec{x},\vec{y}) \le \varepsilon$. The function $\mathcal{N}(\varepsilon,T)$ denotes the size of the smallest ε-cover for T. We refer to $\mathcal{N}(\varepsilon,T)$ as a *covering number*.

Following Kolmogorov and Tihomirov,[47] we define the *upper metric dimension* of the set T of points by

$$\overline{\dim}(T) = \limsup_{\varepsilon\to 0}\frac{\log\mathcal{N}(\varepsilon,T)}{\log(1/\varepsilon)}.$$

The *lower metric dimension*, denoted by <u>dim</u>, is defined similarly using **liminf**. When $\overline{\dim}(T) = \underline{\dim}(T)$, then this quantity is denoted $\dim(T)$, and referred to simply as the *metric dimension* of T. Note that if $\mathcal{N}(\varepsilon, T) = (g(\varepsilon)/\varepsilon)^n$, where $g(\varepsilon)$ is polylogarithmic in $1/\varepsilon$, then $\dim(T) = n$. Hence the metric dimension essentially picks out the exponent in the rate of growth of the covering number as a function of $1/\varepsilon$.

Assume all functions in $\mathbf{F}$ map from Z into $[0, M]$. For any sample $\vec{z} = (z_1, \ldots, z_m)$, with $z_i \in Z$, let

$$\mathbf{F}_{|\vec{z}} = \{(f(z_1), \ldots, f(z_m)) : f \in \mathbf{F}\}.$$

We call $\mathbf{F}_{|\vec{z}}$ the *restriction* of $\mathbf{F}$ to $\vec{z}$. Note that $\mathbf{F}_{|\vec{z}}$ is a set of points in the m-cube $[0, M]^m$. We can consider the size of the covering number $\mathcal{N}(\varepsilon, \mathbf{F}_{|\vec{z}})$ as giving some indication of the "richness at scale $\approx \varepsilon$" of the class $\mathbf{F}$ of functions, restricted to the domain $z_1, \ldots, z_m$. The metric dimension of $\mathbf{F}_{|\vec{z}}$ gives some indication of the "number of essential degrees of freedom" in this restriction of $\mathbf{F}$.

When $z_1, \ldots, z_m$ are drawn independently at random from Z, the *random covering number* $\mathbf{E}(\mathcal{N}(\varepsilon, \mathbf{F}_{|\vec{z}}))$ gives some indication of the "richness" of $\mathbf{F}$ on a "typical" set of m points in the domain Z. Note that for finite $\mathbf{F}$, we have $\mathcal{N}(\varepsilon, \mathbf{F}_{|\vec{z}}) \leq |\mathbf{F}|$ for all ε and all samples $\vec{z}$ and, hence, the random covering number $\mathbf{E}(\mathcal{N}(\varepsilon, \mathbf{F}_{|\vec{z}})) \leq |\mathbf{F}|$ for all ε, all sample sizes m, and all distributions on Z. The main result about uniform empirical estimates for infinite classes of functions is similar to Theorem 1 except that the random covering numbers are used in place of $|\mathbf{F}|$.

THEOREM 2. (See Pollard.[71]) Let $\mathbf{F}$ be a permissible[13] set of functions on Z with $0 \leq f(z) \leq M$ for all $f \in \mathbf{F}$ and $z \in Z$. Let $\vec{z} = (z_1, \ldots, z_m)$ be a sequence of m examples drawn independently from Z according to any distribution on Z. Then for any $\nu > 0$ and $0 < \alpha < 1$,

$$\mathbf{Pr}\left(\exists f \in \mathbf{F} : d_\nu(\widehat{\mathbf{E}}_{\vec{z}}(f), \mathbf{E}(f)) > \alpha\right) \leq 4\mathbf{E}\left(\mathcal{N}(\alpha\nu/8, \mathbf{F}_{|\vec{z}})\right) e^{-\alpha^2\nu m/16M}.$$

COROLLARY 1. (See Pollard.[70]) Under the same assumptions as above, for all $\varepsilon > 0$,

$$\mathbf{Pr}\left(\exists f \in \mathbf{F} : |\widehat{\mathbf{E}}_{\vec{z}}(f) - \mathbf{E}(f)| > \varepsilon\right) \leq 4\mathbf{E}\left(\mathcal{N}(\varepsilon/16, \mathbf{F}_{|\vec{z}})\right) e^{-\varepsilon^2 m/128M^2}.$$

---

[13] This is a measurability condition defined by Pollard[70] which need not concern us in practice. Further details are given in Section 9.2 of the appendix.

Proof of Corollary 1. This follows directly from the above result by setting $\nu = 2M$, and $\alpha = \varepsilon/4M$. To see this, note that property (3) of the d_ν metric (Section 2.2) implies that $|r - s| \le \varepsilon$ whenever $d_\nu(r, s) \le \alpha$ for all $0 \le r, s \le M$ when this setting of ν and α is used. □

The constants in these results are only crude estimates. No serious attempt has been made to minimize them. (See the recent results of Talagrand[82] for much better constants for Corollary 1.)

The bound in this latter result depends critically on the relative magnitudes of the negative exponent in $e^{-\varepsilon^2 m/128M^2}$ and the exponent in the expectation of the covering number $\mathcal{N}(\varepsilon/16, \mathbf{F}_{|\bar{z}})$, which reflects the extent to which $\mathbf{F}_{|\bar{z}}$ "fills up" the m-cube $[0, M]^m$. For example, if $\mathbf{F}_{|\bar{z}}$ has metric dimension at most n for all m and all $\bar{z}$, then there is a constant c_0 such that for any $\eta > 0$, $\mathcal{N}(\varepsilon/16, \mathbf{F}_{|\bar{z}}) \le (c_0 M/\varepsilon)^{n+\eta}$ for suitably small ε. In this case the negative exponential term eventually dominates the expected covering number, and beyond a critical sample size

$$m_0 = O\left(\frac{nM^2}{\varepsilon^2} \log \frac{M}{\varepsilon}\right),$$

the bound goes to zero exponentially fast. Hence, beyond sample size m_0, the probability that any empirical estimate is very far from the true expectation drops off exponentially fast. We will see examples of this in the following section, where we give bounds on the metric dimension of $\mathbf{F}_{|\bar{z}}$ in terms of a combinatorial parameter called the pseudo dimension of $\mathbf{F}$. The theorem actually shows that this exponential drop off occurs even if this metric dimension bound holds only for "most" $\bar{z}$.

On the other hand, if with high probability $\mathbf{F}_{|\bar{z}}$ "fills up" the m-cube $[0, M]^m$ to the extent that $\mathcal{N}(\varepsilon/16, \mathbf{F}_{|\bar{z}}) \approx (c_0/\varepsilon)^m$, which is as large as possible, then the covering number dominates, and the bound is trivial. Results given by Vapnik (Theorem A.2, p. 220)[88] indicate that uniform convergence does not take place in this case. Similar remarks apply to the bound given in Theorem 2, which uses the d_ν metric.

The proof of Theorem 2 follows the proof of Pollard's Theorem 24 (see Pollard,[70] p. 25) in general outline. However, the use of the d_ν metric necessitates a number of substantial modifications. The approach taken here is different from that taken (independently, but prior to this work) by Pollard.[71] Still different, and more involved, techniques are used in the more general theory of weighted empirical processes developed by Alexander.[2,3] While the proof of Theorem 2 we give is simpler, it is still somewhat lengthy, so it is given in Appendix 9.4.

Actually, we can prove a slightly stronger result than Theorem 2. This result is obtained by bounding the probability of uniform convergence on a sample of length m in terms of the expected covering numbers associated with a sample of length $2m$, and by expanding the expectation to include the negative exponential term with a "truncation" at 1. It turns out that this saves us a factor of $1/2$ in the negative exponential term. We also include special bounds for the case that $\mathbf{F}_{|\vec{z}}$ is always finite. This case comes up, for example, when $\mathbf{F} = l_{\mathcal{H}}$ and we use the discrete loss function l, as in the PAC learning model.

THEOREM 3. Let $\mathbf{F}$ be a permissible set of functions on Z with $0 \le f(z) \le M$ for all $f \in \mathbf{F}$ and $z \in Z$. Assume $\nu > 0$, $0 < \alpha < 1$, and $m \ge 1$. Suppose that $\vec{z}$ is generated by m independent random draws according to any probability measure on Z. Let

$$ p(\alpha, \nu, m) = \mathbf{Pr}\left\{ \vec{z} \in Z^m : \exists f \in \mathbf{F} \text{ with } d_\nu(\widehat{\mathbf{E}}_{\vec{z}}(f), \mathbf{E}(f)) > \alpha \right\}. $$

Then

$$ p(\alpha, \nu, m) \le 2\mathbf{E}(\min(2\mathcal{N}(\alpha\nu/8, \mathbf{F}_{|\vec{z}}, d_{L^1})e^{-\alpha^2 \nu m/8M}, 1)), $$

where the expectation is over $\vec{z}$ drawn randomly from Z^{2m}. If in addition $\mathbf{F}_{|\vec{z}}$ is finite for all $\vec{z} \in Z^{2m}$, then

$$ p(\alpha, \nu, m) \le 2\mathbf{E}(\min(2|\mathbf{F}_{|\vec{z}}|e^{-\alpha^2 \nu m/2M}, 1)). $$

Theorem 2 is obtained as a corollary of this result by substituting $m/2$ for m and not taking the minimum with 1 in the left hand side of the first bound for $p(\alpha, \nu, m)$. We will use Theorem 3 to obtain slightly better constants in some of the results in the sequel.

4. PSEUDO DIMENSION OF CLASSES OF REAL-VALUED FUNCTIONS

In this section we will look at one way that bounds on the covering numbers appearing in Theorem 2 can be obtained. This technique, due to Pollard,[70] who extended methods from Dudley,[26] is based on certain intuitions from combinatorial geometry. It generalizes the techniques based on the Vapnik-Chervonenkis dimension used by Blumer et al.,[16] which apply only to $\{0, 1\}$-valued functions. We begin by establishing some basic notation.

DEFINITION 2. For $x \in \Re$, let $\text{sign}(x) = 1$ if $x > 0$, else $\text{sign}(x) = 0$. For $\vec{x} = (x_1, \ldots, x_d) \in \Re^d$, let $\text{sign}(\vec{x}) = (\text{sign}(x_1), \ldots, \text{sign}(x_d))$ and, for $T \subset \Re^d$, let $\text{sign}(T) = \{\text{sign}(\vec{x}) : \vec{x} \in T\}$. For any Boolean vector $\vec{b} = (b_1, \ldots, b_d)$, $\{\vec{x} \in \Re^d : \text{sign}(\vec{x}) = \vec{b}\}$ is called the $\vec{b}$-*orthant* of $\Re^d$, where we have, somewhat arbitrarily, included points with value zero for a particular coordinate in the associated lower orthant. Thus $\text{sign}(T)$ denotes the set of orthants intersected by T. For any $T \subset \Re^d$, and $\vec{x} \in \Re^d$, let $T + \vec{x} = \{\vec{y} + \vec{x} : \vec{y} \in T\}$, i.e., the translation of T obtained by adding the vector $\vec{x}$. We say that T is *full* if there exists $\vec{x} \in \Re^d$ such that $\text{sign}(T + \vec{x}) = \{0,1\}^d$, i.e., if there exists some translation of T that intersects all 2^d orthants of $\Re^d$.

The following result is well known and can be proved in a variety of ways. For example, it follows easily from well-known bounds on the number of cells in arrangements of hyperplanes (see, e.g., Edelsbrunner[30]). We give an elementary proof using a technique from Dudley.[26]

LEMMA 2. No hyperplane in $\Re^d$ intersects all orthants of $\Re^d$.

PROOF. Let T be a hyperplane in $\Re^d$. Choose a vector $\vec{x} \in \Re^d$ as follows. If T includes the origin, then let $\vec{x}$ be any vector that is orthogonal to T and has at least one strictly negative coordinate. (For any nonzero orthogonal vector $\vec{x}$, if $\vec{x}$ doesn't have a negative coordinate, then $-\vec{x}$ does.) Otherwise, let $\vec{x}$ be the (nonzero) vector in T on the line perpendicular to T that passes through the origin. To complete the proof, we show that for all $\vec{y} \in T$, $\text{sign}(\vec{y}) \neq \vec{1} - \text{sign}(\vec{x})$, where $\vec{1}$ denotes the all 1's vector.

Suppose to the contrary that $\text{sign}(\vec{y}) = \vec{1} - \text{sign}(\vec{x})$ for some $\vec{y} \in T$. This implies that the inner product $\sum_{i=1}^{d} x_i y_i$ is nonpositive, and is in fact strictly negative if either $\vec{x}$ or $\vec{y}$ contain a strictly negative coordinate. However, by our choice of $\vec{x}$, either $\vec{x}$ is orthogonal to $\vec{y}$ and contains a strictly negative coordinate, giving an immediate contradiction, or $\vec{x}$ is nonzero and $\vec{x}$ is orthogonal to $\vec{y} - \vec{x}$. In this last case,

$$\sum_{i=1}^{d} x_i y_i = \sum_{i=1}^{d} x_i^2,$$

which is again a contradiction, since the left side is nonpositive while the right side is strictly positive. $\square$

It follows from this lemma that if T is contained in a hyperplane of $\Re^d$, then T is not full.

DEFINITION 3. Let $\mathbf{F}$ be a family of functions from a set Z into $\Re$. For any sequence $\vec{z} = (z_1, \ldots, z_d)$ of points in Z, let $\mathbf{F}_{|\vec{z}} = \{(f(z_1), \ldots, f(z_d)) : f \in \mathbf{F}\}$. If $\mathbf{F}_{|\vec{z}}$ is full, then we say that $\vec{z}$ is *shattered* by $\mathbf{F}$. The *pseudo dimension* of $\mathbf{F}$, denoted $\mathbf{dim_P}(\mathbf{F})$, is the largest d such that there exists a sequence of d points in Z that is shattered by $\mathbf{F}$. If arbitrarily long finite sequences are shattered, then $\mathbf{dim_P}(\mathbf{F})$ is infinite.

It is clear that when $\mathbf{F}$ is a set of $\{0, 1\}$-valued functions, then for any sequence $\vec{z}$ of d points in Z, $\mathbf{F}_{|\vec{z}}$ is full if and only if $\mathbf{F}_{|\vec{z}} = \{0, 1\}^d$. Thus in this case $\mathbf{dim_P}(\mathbf{F})$ is the length d of the longest sequence of points $\vec{z}$ such that $\mathbf{F}_{|\vec{z}} = \{0, 1\}^d$. This is the definition of the Vapnik-Chervonenkis dimension of a class $\mathbf{F}$ of $\{0, 1\}$-valued functions.[16,37,88] Thus the pseudo dimension generalizes the Vapnik-Chervonenkis dimension to arbitrary classes of real-valued functions.

The pseudo dimension also generalizes the algebraic notion of the dimension of a vector space of real-valued functions.

THEOREM 4. (Dudley[26]) Let $\mathbf{F}$ be a d-dimensional vector space of functions from a set Z into $\Re$. Then $\mathbf{dim_P}(\mathbf{F}) = d$.

PROOF. Fix any sequence $\vec{z} = (z_1, \ldots, z_{d+1})$ of points in Z. For any $f \in \mathbf{F}$ let $\Psi(f) = (f(z_1), \ldots, f(z_{d+1}))$. Then Ψ is a linear mapping from $\mathbf{F}$ into $\Re^{d+1}$, and the image of Ψ is $\mathbf{F}_{|\vec{z}}$. Since $\mathbf{F}$ is a vector space of dimension d, this implies that $\mathbf{F}_{|\vec{z}}$ is a subspace of $\Re^{d+1}$ of dimension at most d. Hence by Lemma 2, $\mathbf{F}_{|\vec{z}}$ is not full. This implies $\mathbf{dim_P}(\mathbf{F}) \leq d$. On the other hand, if $\mathbf{F}$ is a d-dimensional vector space of real-valued functions on Z, then there exists a sequence $\vec{z}$ of d points in Z such that $\mathbf{F}_{|\vec{z}} = \Re^d$. Hence $\vec{z}$ is shattered, implying that $\mathbf{dim_P}(\mathbf{F}) \geq d$. $\square$

There are many other ways that the VC dimension can be generalized to real-valued functions.[28,61,62,70,89] Dudley[28] compares several such generalizations, albeit in a different context. The generalization we have proposed here, the pseudo dimension, is a minor variant of the notion used by Pollard[70] to define classes of real-valued functions of polynomial discrimination, called VC-subgraph classes by Dudley.[28] The pseudo dimension will be used in the form defined above in Pollard's new book.[72]

The pseudo dimension has a few invariance properties that are useful (see Pollard[72] for further results of this type).

THEOREM 5. Let $\mathbf{F}$ be a family of functions from Z into $\Re$. Fix any function g from Z into $\Re$ and let $\mathbf{G} = \{g + f : f \in \mathbf{F}\}$. Let I be a real interval (possibly all of $\Re$) such that every function in $\mathbf{F}$ takes values only in I. Fix any nondecreasing (resp. nonincreasing) function $h : I \to \Re$ and let $\mathbf{H} = \{h \circ f : f \in \mathbf{F}\}$, where $\circ$ indicates function composition. Then

1. $\dim_\mathbf{P}(\mathbf{G}) = \dim_\mathbf{P}(\mathbf{F})$[93] and
2. $\dim_\mathbf{P}(\mathbf{H}) \leq \dim_\mathbf{P}(\mathbf{F})$, with equality if h is continuous and strictly increasing (resp. continuous and strictly decreasing).[28,63]

PROOF. Part (1) follows directly from the fact that the notion of a set of points being full is invariant under translation. For part (2) it suffices to prove the results for h nondecreasing and h continuous and strictly increasing. Let $\vec{z} = (z_1, \ldots, z_d)$ be such that $\mathbf{H}_{|\vec{z}}$ is full; i.e., such that $\mathbf{H}_{|\vec{z}} - \vec{x}$ intersects all 2^d orthants of $\Re^d$ for some vector $\vec{x} = (x_1, \ldots, x_d)$ in $\Re^d$. Then for every Boolean vector $\vec{b} \in \{0,1\}^d$ there exists a function $f_{\vec{b}} \in \mathbf{F}$ such that for every i, $1 \leq i \leq d$, we have $h \circ f_{\vec{b}}(z_i) > x_i$ if and only if the ith bit of $\vec{b}$ is 1. For each i, $1 \leq i \leq d$, let

$$u_i = \min\{f_{\vec{b}}(z_i) : \text{the } i\text{th bit of } \vec{b} \text{ is } 1\}$$

and

$$l_i = \max\{f_{\vec{b}}(z_i) : \text{the } i\text{th bit of } \vec{b} \text{ is } 0\}.$$

Since h is nondecreasing, we have $u_i > l_i$ for each i. Let $r_i = (u_i + l_i)/2$ for each i and $\vec{r} = (r_1, \ldots, r_d)$. Let $T = \{f_{\vec{b}} : \vec{b} \in \{0,1\}^d\}$. Then clearly $T - \vec{r}$ intersects every orthant of $\Re^d$, so $T_{|\vec{z}}$ is full. Since $T \subset \mathbf{F}$, this implies that $\mathbf{F}_{|\vec{z}}$ is full, and hence $\dim_\mathbf{P}(\mathbf{H}) \leq \dim_\mathbf{P}(\mathbf{F})$. Equality follows when h is continuous and strictly increasing since we obtain the class $\mathbf{F}$ from $\mathbf{H}$ by composing with h^{-1}. □

By putting a probability measure on Z, we can view a class $\mathbf{F}$ of real-valued functions on Z as a pseudo metric space. As always, we assume that the functions in $\mathbf{F}$ are measurable. The distance between two functions is the integral of the absolute value of their difference, i.e., the L^1 distance, relative to the given measure. To make this work, we need to make some assumptions about the integrability of the functions in $\mathbf{F}$ under the given measure. Since we will be concerned only with families of functions taking values in a bounded range in this chapter, this will cause no problems for us. For convenience, we choose this range to be $[0, M]$. For a more general treatment, see Pollard[70] and Dudley.[27]

DEFINITION 4. Let $\mathbf{F}$ be a class of functions from Z into $[0, M]$, where $M > 0$, and P be a probability measure on Z. Then $d_{L^1(P)}$ is the pseudo metric on $\mathbf{F}$ defined by

$$d_{L^1(P)}(f, g) = \mathbf{E}(|f - g|) = \int_Z |f(z) - g(z)| dP(z) \text{ for all } f, g \in \mathbf{F}.$$

The notions of ε-cover and metric dimension used in the previous section can be generalized to arbitrary pseudo metric spaces. This generalization is given in Section 9.1 of the appendix. In the remainder of the chapter we will use the concepts and notation given there without further special reference.

Using techniques that go back to Dudley,[26] Pollard has obtained a beautiful theorem bounding the metric dimension of $(\mathbf{F}, d_{L^1(P)})$ by $\mathbf{dim_P}(\mathbf{F})$ for any probability measure P on Z. Actually this result is much stronger in that it gives explicit bounds on the packing numbers for $\mathbf{F}$ using $d_{L^1(P)}$ balls of radius ε. Since packing numbers are closely related to covering numbers (Theorem 12 in Section 9.1), these bounds can then be used with Theorem 2 to obtain uniform convergence results for empirical estimates of functions in $\mathbf{F}$. We now state and prove a version of Pollard's result (see Pollard,[70] Lemma 25, p. 27) for the special case when $\mathbf{F}$ is a class of functions taking values in the interval $[0, M]$ with somewhat better bounds on the packing numbers.

THEOREM 6. (Pollard) Let $\mathbf{F}$ be a class of functions from a set Z into $[0, M]$, where $\mathbf{dim_P}(\mathbf{F}) = d$ for some $1 \leq d < \infty$. Let P be a probability measure on Z. Then for all $0 < \varepsilon \leq M$,

$$\mathcal{M}(\varepsilon, \mathbf{F}, d_{L^1(P)}) < 2 \left(\frac{2eM}{\varepsilon} \ln \frac{2eM}{\varepsilon} \right)^d.$$

The proof we give uses essentially the same techniques as Pollard's, with some minor modifications. It relies on a few lemmas, which we give now. The first, which we give without proof, was discovered independently by a number of people (see Assouad[8]), including Vapnik and Chervonenkis,[87] but is most often attributed to Sauer[77] in the computer science literature.

LEMMA 3. (Sauer) Let $\mathbf{F}$ be a class of functions from $S = \{1, 2, \ldots, m\}$ into $\{0, 1\}$ with $|\mathbf{F}| > 1$ and let d be the length of the longest sequence of points $\vec{z}$ from S such that $\mathbf{F}_{|\vec{z}} = \{0, 1\}^d$. Then[15]

$$|\mathbf{F}| \leq \sum_{i=0}^{d} \binom{m}{i} \leq (em/d)^d,$$

where e is the base of the natural logarithm.

In the next lemma we bound the packing numbers of $(\mathbf{F}, d_{L^1(P)})$ in terms of the expected number of orthants intersected by a random translation of a random restriction of $\mathbf{F}$. This is the key lemma of the proof.

LEMMA 4. Let $\mathbf{F}$ be a family of functions from a set Z into $[0, M]$ and let P be a probability measure on Z. Let $\vec{r} = (r_1, \ldots, r_m)$ be a random vector in $[0, M]^m$ where each r_i is drawn independently at random from the uniform distribution on $[0, M]$. Let $\vec{z} = (z_1, \ldots, z_m)$ be a random vector in Z^m where each z_i is drawn independently at random from P. Then for all $\varepsilon > 0$,

$$\mathbf{E}(|\mathrm{sign}(\mathbf{F}_{|\vec{z}} - \vec{r})|) \geq \mathcal{M}(\varepsilon, \mathbf{F}, d_{L^1(P)}) \left(1 - \mathcal{M}(\varepsilon, \mathbf{F}, d_{L^1(P)}) e^{-\varepsilon m/M}\right).$$

PROOF. For all $f \in \mathbf{F}$ we will denote $(f(z_1), \ldots, f(z_m))$ by $f_{|\vec{z}}$. Choose $\varepsilon > 0$. Let $\mathbf{G}$ be an ε-separated subset of $\mathbf{F}$ (with respect to $d_{L^1(P)}$), with $|\mathbf{G}| = \mathcal{M}(\varepsilon, \mathbf{F}, d_{L^1(P)})$. Then

$$
\begin{aligned}
&\mathbf{E}(|\mathrm{sign}(\mathbf{F}_{|\vec{z}} - \vec{r})|) \\
&\geq \mathbf{E}(|\mathrm{sign}(\mathbf{G}_{|\vec{z}} - \vec{r})|) \\
&\geq \mathbf{E}(|\{f \in \mathbf{G} : \mathrm{sign}(f_{|\vec{z}} - \vec{r}) \neq \mathrm{sign}(g_{|\vec{z}} - \vec{r}) \text{ for all } g \in \mathbf{G}, g \neq f\}|) \\
&= \sum_{f \in \mathbf{G}} \mathbf{Pr}\left(\mathrm{sign}(f_{|\vec{z}} - \vec{r}) \neq \mathrm{sign}(g_{|\vec{z}} - \vec{r}) \text{ for all } g \in \mathbf{G}, g \neq f\right) \\
&= \sum_{f \in \mathbf{G}} \left(1 - \mathbf{Pr}\left(\exists g \in \mathbf{G}, g \neq f : \mathrm{sign}(f_{|\vec{z}} - \vec{r}) = \mathrm{sign}(g_{|\vec{z}} - \vec{r})\right)\right) \\
&\geq \sum_{f \in \mathbf{G}} \left(1 - |\mathbf{G}| \max_{g \in \mathbf{G}, g \neq f} \mathbf{Pr}\left(\mathrm{sign}(f_{|\vec{z}} - \vec{r}) = \mathrm{sign}(g_{|\vec{z}} - \vec{r})\right)\right).
\end{aligned}
$$

---

[15]See Dudley,[27] Prop. 2.2.9; or Blumer et al.,[16] Appendix, for a proof of the second inequality. Note also that Vapnik and Chervonenkis[87] actually contains a slightly weaker result.

Let f and g be distinct functions in $\mathbf{G}$. Since $\mathbf{G}$ is ε-separated,

$$\int_Z |f(z) - g(z)| dP(z) > \varepsilon.$$

In addition, the range of f and g is $[0, M]$. Hence if z_i is drawn at random from P and r_i drawn at random from the uniform distribution on $[0, M]$, then the probability that r_i lies between $f(z_i)$ and $g(z_i)$ is at least ε/M. And $\operatorname{sign}(f_{|\vec{z}} - \vec{r}) = \operatorname{sign}(g_{|\vec{z}} - \vec{r})$ only if this fails to occur for each i, $1 \le i \le m$. Thus

$$\mathbf{Pr}\left(\operatorname{sign}(f_{|\vec{z}} - \vec{r}) = \operatorname{sign}(g_{|\vec{z}} - \vec{r})\right) \le \left(1 - \frac{\varepsilon}{M}\right)^m \le e^{-\varepsilon m/M}.$$

Since this holds for every distinct f and g in $\mathbf{G}$, combining this with the inequality above, we have

$$\mathbf{E}(|\operatorname{sign}(\mathbf{F}_{|\vec{z}} - \vec{r})|) \ge |\mathbf{G}|(1 - |\mathbf{G}|e^{-\varepsilon m/M}).$$

Since $|\mathbf{G}| = \mathcal{M}(\varepsilon, \mathbf{F}, d_{L^1(P)})$, this gives the result. $\square$

Proof of Theorem 6. Since $\dim_P(\mathbf{F}) = d$, it follows from Sauer's lemma that $|\operatorname{sign}(\mathbf{F}_{|\vec{z}} - \vec{r})| \le (em/d)^d$ for all $m \ge d$, $\vec{z} \in Z^m$, and $\vec{r} \in [0, M]^m$. Hence, the above lemma implies that

$$\left(\frac{em}{d}\right)^d \ge \mathcal{M}(\varepsilon, \mathbf{F}, d_{L^1(P)}) \left(1 - \mathcal{M}(\varepsilon, \mathbf{F}, d_{L^1(P)})e^{-\varepsilon m/M}\right) \qquad (4)$$

for all probability measures P on Z and $m \ge d$. It is easily verified that if

$$\frac{M}{\varepsilon} \ln(2\mathcal{M}(\varepsilon, \mathbf{F}, d_{L^1(P)})) < d,$$

then the upper bound given in Theorem 6 follows trivially using the fact that $\varepsilon \le M$. Thus we may assume that $(M/\varepsilon) \ln(2\mathcal{M}(\varepsilon, \mathbf{F}, d_{L^1(P)})) \ge d$. Hence, if $m \ge (M/\varepsilon) \ln(2\mathcal{M}(\varepsilon, \mathbf{F}, d_{L^1(P)}))$, then $m \ge d$ and

$$(1 - \mathcal{M}(\varepsilon, \mathbf{F}, d_{L^1(P)})e^{-\varepsilon m/M}) \ge \frac{1}{2}.$$

Since $\varepsilon \le M/2d$, we certainly have $m \ge d$ in this case. Thus from Eq. (4) we obtain

$$\left(\frac{eM \ln(2\mathcal{M}(\varepsilon, \mathbf{F}, d_{L^1(P)}))}{\varepsilon d}\right)^d \ge \frac{1}{2}\mathcal{M}(\varepsilon, \mathbf{F}, d_{L^1(P)}).$$

With some simple calculations, this gives the bound of Theorem 6. (These calculations are given in Lemma 14 in the appendix of Haussler.[40] To apply that lemma, let $a = \mathcal{M}(\varepsilon, \mathbf{F}, d_{L^1(P)})$ and $b = eM/\varepsilon \geq 2e > 5$.) $\square$

Since

$$\overline{\dim}(\mathbf{F}) = \overline{\lim_{\varepsilon \to 0}} \frac{\log \mathcal{M}(\varepsilon, \mathbf{F}, d_{L^1(P)})}{\log(1/\varepsilon)},$$

it follows easily from Theorem 6 that the (upper) metric dimension of $(\mathbf{F}, d_{L^1(P)})$ is at most $\dim_P(\mathbf{F})$ for any probability measure P on Z.

Using our results on uniform convergence from Sections 3.2 and 9.4, we can now show the following.

THEOREM 7. Let $\mathbf{F}$ be a permissible family of functions from a set Z into $[0, M]$ with $\dim_P(\mathbf{F}) = d$ for some $1 \leq d < \infty$. Assume $m \geq 1$, $0 < \nu \leq 8M$, and $0 < \alpha < 1$. Let $\vec{z}$ be generated by m independent draws according to any distribution on Z. Then

$$\mathbf{Pr}\left(\exists f \in \mathbf{F} : d_\nu(\widehat{\mathbf{E}}_{\vec{z}}(f), \mathbf{E}(f)) > \alpha\right) \leq 8\left(\frac{16eM}{\alpha\nu}\ln\frac{16eM}{\alpha\nu}\right)^d e^{-\alpha^2 \nu m/8M}.$$

Moreover, for $m \geq (8M/\alpha^2\nu)(2d\ln 8eM/\alpha\nu + \ln(8/\delta))$ this probability is at most δ.

PROOF. Let $\varepsilon = \alpha\nu/8$. Since $\alpha < 1$ and $\nu \leq 8M$, $\varepsilon \leq M$. For any sequence $\vec{z}$ of points in Z there is a trivial isometry between $(\mathbf{F}_{|\vec{z}}, d_{L^1})$ and $(\mathbf{F}, d_{L^1(P_{\vec{z}})})$, where $P_{\vec{z}}$ is the empirical measure induced by $\vec{z}$, in which each set has measure equal to the fraction of the points in $\vec{z}$ it contains. Thus by Theorem 12 of Section 9.1 and Theorem 6, we have

$$\mathcal{N}(\varepsilon, \mathbf{F}_{|\vec{z}}, d_{L^1}) \leq \mathcal{M}(\varepsilon, \mathbf{F}_{|\vec{z}}, d_{L^1}) \leq 2\left(\frac{2eM}{\varepsilon}\ln\frac{2eM}{\varepsilon}\right)^d,$$

for all $\vec{z} \in Z^{2m}$. Hence the given probability is at most

$$8\left(\frac{2eM}{\varepsilon}\ln\frac{2eM}{\varepsilon}\right)^d e^{-\alpha^2 \nu m/8M} = 8\left(\frac{16eM}{\alpha\nu}\ln\frac{16eM}{\alpha\nu}\right)^d e^{-\alpha^2 \nu m/8M}$$

by Theorem 3.

For the second result, setting the bound above equal to δ and solving for m gives

$$m \geq \frac{8M}{\alpha^2\nu}\left(d\ln\left(\frac{16eM}{\alpha\nu}\ln\frac{16eM}{\alpha\nu}\right) + \ln\frac{8}{\delta}\right).$$

It is easily verified that $\ln(a \ln a) < 2\ln(a/2)$ when $a \geq 5$, and from this the bound given in the second result follows. $\square$

COROLLARY 2. Under the same assumptions as above, for all $0 < \varepsilon \leq M$,

$$\mathbf{Pr}\left(\exists f \in \mathbf{F} : |\widehat{\mathbf{E}}_{\bar{z}}(f) - \mathbf{E}(f)| > \varepsilon\right) \leq 8 \left(\frac{32eM}{\varepsilon} \ln \frac{32eM}{\varepsilon}\right)^d e^{-\varepsilon^2 m / 64M^2}.$$

Moreover, for $m \geq (64M^2/\varepsilon^2)\,(2d\ln(16eM/\varepsilon) + \ln(8/\delta))$ this probability is at most δ.

PROOF. This follows directly from the above result by setting $\nu = 2M$, $\alpha = \varepsilon/4M$, and using property (3) of the d_ν metric, as in the proof of Corollary 1 in Section 3.2. $\square$

5. SOME APPLICATIONS OF PSEUDO DIMENSION IN LEARNING

We now look at how the theoretical results obtained in the previous two sections can be applied to certain types of learning problems. Suppose that we have a basic learning problem defined by X, Y, A, $\mathcal{H}$, $\mathcal{P}$, and $\mathcal{L}$, where $\mathcal{L}$ is the family of $L_{\alpha,\nu}$ regret functions for an underlying loss function l, as in Section 2. As before, let $Z = X \times Y$, $l_h : Z \to [0, M]$ be defined by $l_h(x, y) = l(y, h(x))$ for all $h \in \mathcal{H}$, and $l_{\mathcal{H}} = \{l_h : h \in \mathcal{H}\}$. In this section we will show how to obtain sample complexity bounds on algorithms for this basic learning problem using Theorem 7 above.

To obtain these bounds, we will need bounds on $\mathbf{dim}_\mathcal{P}(l_{\mathcal{H}})$. In this section we will look at some useful tricks for computing $\mathbf{dim}_\mathcal{P}(l_{\mathcal{H}})$ in the important case $A \subset \Re$, i.e., when each decision is represented by a real number. In the following section we discuss more general decision spaces.

When $A \subset \Re$, the functions in $\mathcal{H}$ are themselves real-valued, so we can talk about the pseudo dimension of $\mathcal{H}$ itself, without reference to any particular loss function. What makes this useful is that in many important cases the pseudo dimension of $\mathcal{H}$ is the same as the pseudo dimension of $l_{\mathcal{H}}$. Thus we can factor out the effects of the loss function in deriving our sample size bounds, and concentrate on the pseudo dimension of the decision rule space $\mathcal{H}$.

DEFINITION 5. Let $l : Y \times A \to \Re$ be a loss function, where $A \subset \Re$ and Y is an arbitrary set. For each $y \in Y$ define the function $f_y : A \to \Re$ by letting $f_y(a) = l(y, a)$ for each $a \in A$; i.e., f_y is the restriction of l obtained by fixing its first argument to y. We say that l is *monotone over* A if for every $y \in Y$, either f_y is strictly increasing on A, or f_y is strictly decreasing on A. Thus f_y may be increasing for some $y \in Y$ and decreasing for others.

LEMMA 5. If $A \subset \Re$ and l is a loss function on $A \times Y$ that is monotone over A, then $\dim_P(l_{\mathcal{H}}) = \dim_P(\mathcal{H})$.

PROOF. Suppose that $\vec{x} = (x_1, \ldots, x_d)$ is shattered by $\mathcal{H}$, where $x_i \in X$, $1 \le i \le d$. Then there is some real vector $\vec{r} = (r_1, \ldots, r_d)$ such that $\mathcal{H}_{|\vec{x}} - \vec{r}$ intersects all 2^d orthants of $\Re^d$. Hence, for every Boolean vector $\vec{b} \in \{0, 1\}^d$, there exists a function $h_{\vec{b}} \in \mathcal{H}$ such that, for every i, $1 \le i \le d$, we have $h_{\vec{b}}(x_i) > r_i$ if and only if the ith bit of $\vec{b}$ is 1. Fix an outcome $y \in Y$. Let $\vec{z} = (z_1, \ldots, z_d)$, where $z_i = (x_i, y)$ for all i, $1 \le i \le d$. Note that if f_y is strictly increasing, then for any $h \in \mathcal{H}$ and $1 \le i \le d$, $h(x_i) > r_i \leftrightarrow l_h(z_i) = l(y, h(x_i)) = f_y(h(x_i)) > f_y(r_i)$. Hence, for every Boolean vector $\vec{b} \in \{0, 1\}^d$ there exists a function $h_{\vec{b}} \in \mathcal{H}$ such that for every i, $1 \le i \le d$, we have $l_{h, \vec{b}}(z_i) > f_y(r_i)$ if and only if the ith bit of $\vec{b}$ is 1. A similar result holds if f_y is strictly decreasing. Thus $\vec{z}$ is shattered by $l_{\mathcal{H}}$. It follows that $\dim_P(l_{\mathcal{H}}) \ge \dim_P(\mathcal{H})$.

For the other direction, assume $\vec{z} = (z_1, \ldots, z_d)$ is shattered by $l_{\mathcal{H}}$, where $z_i = (x_i, y_i)$ for all i, $1 \le i \le d$. We will show that $\vec{x} = (x_1, \ldots, x_d)$ is shattered by $\mathcal{H}$. Since $\vec{z}$ is shattered by $l_{\mathcal{H}}$, there is some real vector $\vec{r} = (r_1, \ldots, r_d)$ such that for every Boolean vector $\vec{b} \in \{0, 1\}^d$ there exists a function $h_{\vec{b}} \in \mathcal{H}$ such that for every i, $1 \le i \le d$, we have $l_{h, \vec{b}}(z_i) > r_i$ if and only if the ith bit of $\vec{b}$ is 1. Let $A_0 = \{h_{\vec{b}}(x_i) : 1 \le i \le d \text{ and } \vec{b} \in \{0, 1\}^d\}$. For each outcome $y \in Y$ define the function $g_y : \Re \to A_0$ as follows. If f_y is increasing then $g_y(r) = \max\{a \in A_0 : f_y(a) \le r\}$ and if f_y is decreasing then $g_y(r) = \max\{a \in A_0 : f_y(a) > r\}$. Then for each i, $1 \le i \le d$, we either have

1. for all $h \in \mathcal{H}$, $l_h(z_i) = f_{y_i}(h(x_i)) > r_i \leftrightarrow h(x_i) > g_{y_i}(r_i)$ or
2. for all $h \in \mathcal{H}$, $l_h(z_i) = f_{y_i}(h(x_i)) > r_i \leftrightarrow h(x_i) \le g_{y_i}(r_i)$.

Hence, for every Boolean vector $\vec{b} \in \{0, 1\}^d$ there exists a function $h'_{\vec{b}} \in \mathcal{H}$ such that for every i, $1 \le i \le d$, we have $h'_{\vec{b}}(x_i) > g_{y_i}(r_i)$ if and only if the ith bit of $\vec{b}$ is 1. (To see this, let $\vec{c}$ be the Boolean vector derived from $\vec{b}$ by complementing the bit in each position i for which f_{y_i} is decreasing, and then let $h'_{\vec{b}} = h_{\vec{c}}$.) Thus $\vec{x} = (x_1, \ldots, x_d)$ is shattered by $\mathcal{H}$. It follows that $\dim_P(\mathcal{H}) \ge \dim_P(l_{\mathcal{H}})$, and combined with the above inequality, this gives the result. $\square$

Combined with Theorem 7, this gives the following result on the uniform convergence of empirical risk estimates for the basic learning problem. Here and below we use the notation $\hat{r}_{h,l}(\vec{z})$ and $r_{h,l}(P)$ introduced in Section 2 for the empirical risk estimate and true risk of a decision rule h, respectively.

THEOREM 8. Assume the decision space $A \subset \Re$, the loss function l is monotone over A and bounded between 0 and M, the decision rule space $\mathcal{H}$ is such that $l_{\mathcal{H}}$ is permissible, and $1 \leq d = \mathbf{dimp}(\mathcal{H}) < \infty$. Assume $m \geq 1$, $0 < \nu < 8M$ and $0 < \alpha < 1$. Let P be any probability distribution on Z. Let $\vec{z}$ be generated by m independent draws from Z according to P. Then

$$\mathbf{Pr}\left(\exists h \in \mathcal{H} : d_\nu(\hat{r}_{h,l}(\vec{z}), r_{h,l}(P)) > \alpha\right) \leq 8 \left(\frac{16eM}{\alpha\nu} \ln \frac{16eM}{\alpha\nu}\right)^d e^{-\alpha^2 \nu m / 8M}.$$

Moreover, for $m \geq (8M/\alpha^2\nu)(2d\ln(8eM/\alpha\nu) + \ln(8/\delta))$ this probability is at most δ.

PROOF. By Lemma 5, $\mathbf{dimp}(l_{\mathcal{H}}) = \mathbf{dimp}(\mathcal{H})$ in this case. Thus, since $\hat{r}_{h,l}(\vec{z}) = \hat{\mathbf{E}}_{\vec{z}}(l_h)$ and $r_{h,l}(P) = \mathbf{E}(l_h)$, the result follows directly from Theorem 7. $\square$

When its conditions are satisfied, this theorem, combined with Lemma 1 from Section 2.5, gives us a means of bounding in terms of $\mathbf{dimp}(\mathcal{H})$ the sample complexity $m(\alpha, \nu, \delta)$ of any algorithm that solves the basic learning problem by returning (with high probability) a decision rule with near minimal empirical risk on the training sample. The resulting bound is

$$m(\alpha, \nu, \delta) = O\left(\frac{M}{\alpha^2\nu}\left(\mathbf{dimp}(\mathcal{H})\log\frac{M}{\alpha\nu} + \log\frac{1}{\delta}\right)\right). \tag{5}$$

This is similar to the sample complexity

$$m(\alpha, \nu, \delta) = O\left(\frac{M}{\alpha^2\nu}\left(\log|\mathcal{H}| + \log\frac{1}{\delta}\right)\right)$$

that can be obtained by using Theorem 1, when $\mathcal{H}$ (and hence $l_{\mathcal{H}}$) is finite. The term $\mathbf{dimp}(\mathcal{H})\log(M/\alpha\nu)$ replaces the term $\log|\mathcal{H}|$. In particular, for the "PAC settings" $\alpha = \frac{1}{2}$ and $\nu = \varepsilon$ we get the sample complexity

$$m(\varepsilon, \delta) = O\left(\frac{M}{\varepsilon}\left(\mathbf{dimp}(\mathcal{H})\log\frac{M}{\varepsilon} + \log\frac{1}{\delta}\right)\right), \tag{6}$$

in place of the sample complexity

$$m(\varepsilon, \delta) = O\left(\frac{M}{\varepsilon}\left(\log|\mathcal{H}| + \log\frac{1}{\delta}\right)\right)$$

derived from Theorem 1. Moreover, these bounds are distribution-independent, so $\mathcal{P}$ can be taken to be the class of all probability distributions on Z.

We now give several examples to illustrate the use of this theorem. First, consider the standard PAC model in which the outcome space Y and the decision space A are both $\{0, 1\}$. In this case any loss function such that $l(0, 1) \neq l(0, 0)$ and $l(1, 1) \neq l(1, 0)$ is monotone over A, and in particular, the standard discrete loss function, $l(y, a) = 1$ if $y = a$, else $l(y, a) = 0$, is monotone. Clearly $M = 1$ in this case. Thus from Eq. 6 above we get a sample complexity bound of

$$O\left(\frac{1}{\varepsilon}\left(\mathbf{dim}_{\mathbf{P}}(\mathcal{H})\log\frac{1}{\varepsilon} + \log\frac{1}{\delta}\right)\right).$$

As mentioned in the previous section, $\mathbf{dim}_{\mathbf{P}}(\mathcal{H})$ is the same as the Vapnik-Chervonenkis dimension of **H** in this case; hence, this bound is, up to constants, the same as that given in Theorem 2.1 of Blumer et al.[16] A number of applications of this result are outlined by Blumer et al.[16] Further applications, specifically for learning problems that have been studied recently in the mainstream artificial intelligence work, are given by Haussler.[38,39]

For our second example, consider the case that the outcome space Y is $\{0, 1\}$, but the decision space A is $[0, 1]$. Assume that the loss function is $l(y, a) = |a - y|^q$ for some $q > 0$. This case was examined with $q = 2$ by Kearns and Schapire[45] in their investigation into the learnability of p-concepts. They showed that $\mathbf{dim}_{\mathbf{P}}(l_\mathcal{H}) = \mathbf{dim}_{\mathbf{P}}(\mathcal{H})$ in this case. Since the loss l is monotone in A for all $q > 0$, Lemma 5 shows that this result holds for other values of q as well. Hence the conditions of Theorem 8 are met. Some applications are given by Kearns and Schapire.[45] (It should be noted that it is important that $A = [0, 1]$ in this case. The result does not hold in general for larger A.)

For our third example, consider the problem of logistic regression, as described in Section 1.1.3. In the simplest case the outcome space again has only two values, denoted y_1 and y_2, where y_1 indicates that some event has taken place and y_2 indicates that it has not, and an action a represents an estimate of the log odds ratio $\ln(P(y_1)/P(y_2))$, where the probability P is conditioned on the observed instance x. Here $A = \Re$ and the log likelihood loss function is the logistic loss function, defined by $l(y_1, a) = \ln(1 + e^a) - a$ and $l(y_2, a) = \ln(1 + e^a)$. Again, it is easily verified that l is monotone in A. In logistic regression the standard assumption is that the instance space X is contained in $\Re^n$ for some $n \geq 1$ and $\mathcal{H}$ is contained in the family of all linear functions on X (see McCullagh and Nelder[56]). In this case,

$\dim_P(\mathcal{H}) \leq n+1$ by Theorem 4. By restricting A to a bounded range, we can then apply Theorem 8 to obtain sample complexity bounds that are linear in n.

Several other regression methods in the class of regression methods that use what are known as *generalized linear models*[56] can be handled this way as well.

As a last example, consider the problem of density estimation, as described in Section 1.1.4. Here there is only one outcome in Y, $A \subset \Re^+$, and $l(y,a) = l(a) = -\log a$. Thus clearly l is monotone in A. Thus we can apply Theorem 8 to the problem of denisity estimation as well, whenever the family of densities $\mathcal{H}$ is uniformly bounded (away from zero) and has finite pseudo dimension.

---

6. CAPACITY AND METRIC DIMENSION OF FUNCTION CLASSES

In Sections 4 and 5 we showed how the pseudo dimension can be used to obtain distribution independent bounds on the random covering numbers needed for Theorem 2, thereby obtaining bounds on the sample size needed for uniform convergence and learning results. In this section we develop an alternate way of obtaining distribution independent bounds on random covering numbers. This method can sometimes be used in conjunction with the method given in the previous sections to extend that method to cover cases where the decision space A is not contained in $\Re$. We will demonstrate this in our analysis of the sample size needed for learning in feedforward neural networks in the following section.

The key idea is to introduce a pseudo metric (see Section 9.1) on the decision space A. The distance between two actions is the maximum difference in loss for these actions, over all possible outcomes.

DEFINITION 6. For every loss function $l : Y \times A \rightarrow [0, M]$, by ρ_l we denote the pseudo metric on A defined by $\rho_l(a, b) = \sup_{y \in Y} |l(y, a) - l(y, b)|$ for all $a, b \in A$.

Note that (A, ρ_l) is a bounded pseudo metric space: no two actions in A are more than M apart.

The notions of ε-cover and metric dimension used in the previous section can be generalized to arbitrary pseudo metric spaces. This generalization is given in Section 9.1 of the appendix. In the remainder of the chapter we will use the concepts and notation given there without further special reference.

Since decision rules in $\mathcal{H}$ map from the instance space X into A, the pseudo metric ρ_l on A can be used to induce a pseudo metric on $\mathcal{H}$ in which two decision rules differ only to the extent that the actions that they proscribe differ with respect to ρ_l. There are several ways to do this. The easiest is to use an L^∞ function distance on $\mathcal{H}$, defining the distance between decision rules f and g as the supremum of $\rho_l(f(x), g(x))$ over all $x \in X$. This works, and is a useful method of obtaining uniform convergence and learning results (see related techniques used by White[94]). However, as we will see, the crucial issue is the size of the smallest ε-cover of the resulting pseudo metric space $\mathcal{H}$. In some cases we can get smaller covers, and hence better results, by using an L^1 function distance instead. Since the L^1 distance is never more than the L^∞ distance, the results are never worse. Thus we present this more powerful method here.

DEFINITION 7. Let $\mathcal{H}$ be a family of functions from a set X into a bounded pseudo metric space (A, ρ). Let P be a probability measure on X. Then $d_{L^1(P,\rho)}$ is the pseudo metric on $\mathcal{H}$ defined by

$$d_{L^1(P,\rho)}(f, g) = \mathbf{E}(\rho(f(x), g(x))) = \int_X \rho(f(x), g(x))dP(x)$$

for all $f, g \in \mathcal{H}$. For every $\varepsilon > 0$, let

$$\mathcal{C}(\varepsilon, \mathcal{H}, \rho) = \sup\{\mathcal{N}(\varepsilon, \mathcal{H}, d_{L^1(P,\rho)})\} \text{ over all probability measures } P \text{ on } X.$$

If $\mathcal{N}(\varepsilon, \mathcal{H}, d_{L^1(P,\rho)})$ is infinite for some measure P, or if the set in this supremum is unbounded, then $\mathcal{C}(\varepsilon, \mathcal{H}, \rho) = \infty$. (The third argument to $\mathcal{C}$ will be dropped when the pseudo metric ρ is clear from the context.) We call[15] $\mathcal{C}(\varepsilon, \mathcal{H}, \rho)$ the *capacity* of $\mathcal{H}$. In analogy with the definition of metric dimension, we define the *upper metric dimension* of $\mathcal{H}$ by

$$\overline{\dim}(\mathcal{H}) = \limsup_{\varepsilon \to 0} \frac{\log \mathcal{C}(\varepsilon, \mathcal{H}, \rho)}{\log(1/\varepsilon)},$$

and the *lower metric dimension*, denoted by $\underline{\dim}(\mathcal{H})$, is defined similarly using **liminf**. When $\overline{\dim}(\mathcal{H}) = \underline{\dim}(\mathcal{H})$, then this quantity is denoted $\dim(\mathcal{H})$, and referred to simply as the *metric dimension* of $\mathcal{H}$. If $\mathcal{C}(\varepsilon, \mathcal{H}, \rho) = \infty$ for some $\varepsilon > 0$, then $\dim(\mathcal{H}) = \infty$.

We now show how bounds on the capacity of $\mathcal{H}$ lead to distribution-independent bounds on the rate of uniform convergence of empirical risk estimates for functions in $\mathcal{H}$ with respect to the loss function l. As before, let $Z = X \times Y$, let P be a probability distribution on Z, and let $l_{\mathcal{H}}$ be the family of functions on Z defined by $l_{\mathcal{H}} = \{l_h : h \in \mathcal{H}\}$, where $l_h(x, y) = l(y, h(x))$. Let $P_{|X}$ be the marginal on X of the joint distribution P on $X \times Y$ (see Section 1.5).

---

[15]The term *metric entropy* is often used for the quantities $\log \mathcal{N}(\varepsilon, \mathcal{H}, d_{L^1(P,\rho)})$ and $\log \mathcal{C}(\varepsilon, \mathcal{H}, \rho)$.[28,73] It is also used for an analogous, but fundamentally distinct, concept in the dynamical systems literature (see Farmer[32]). The term *capacity* has also been used with many other related meanings.[12,33,47,55,88] Our usage here is taken from Dudley.[27]

LEMMA 6. For all $\varepsilon > 0$,

$$\mathcal{N}(\varepsilon, l_{\mathcal{H}}, d_{L^1(P)}) \leq \mathcal{N}(\varepsilon, \mathcal{H}, d_{L^1(P_{|X, \rho_l})}).$$

PROOF. For every $h \in \mathcal{H}$ let $\psi(h) = l_h$. Hence ψ maps from $\mathcal{H}$ onto $l_{\mathcal{H}}$. It suffices to show that ψ is a contraction, i.e., that

$$\text{for all } f, g \in \mathcal{H}, \; d_{L^1(P)}(\psi(f), \psi(g)) \leq d_{L^1(P_{|X, \rho_l})}(f, g).$$

Let f and g be any two functions in $\mathcal{H}$. Then

$$
\begin{aligned}
d_{L^1(P)}(\psi(f), \psi(g)) &= \int_Z |l(y, f(x)) - l(y, g(x))| dP(x, y) \\
&\leq \int_Z \rho_l(f(x), g(x)) dP(x, y) \\
&= \int_X \rho_l(f(x), g(x)) dP_{|X}(x) \\
&= d_{L^1(P_{|X, \rho_l})}(f, g).
\end{aligned}
$$

□

This gives the following theorem about distribution-independent uniform convergence of risk estimates for learning.

THEOREM 9. Assume that the decision rule space $\mathcal{H}$ and the loss function l are such that $l_{\mathcal{H}}$ is permissible. Let P be any probability distribution on $Z = X \times Y$. Assume $m \geq 1$, $\nu > 0$ and $0 < \alpha < 1$. Let $\vec{z}$ be generated by m independent draws from Z according to P. Then

$$\mathbf{Pr}\left(\exists h \in \mathcal{H} : d_\nu(\hat{r}_{h,l}(\vec{z}), r_{h,l}(P)) > \alpha\right) \leq 4\mathcal{C}(\alpha\nu/8, \mathcal{H}, \rho_l) e^{-\alpha^2 \nu m / 8M}.$$

PROOF. Let $\mathbf{F} = l_{\mathcal{H}}$. For any sequence $\vec{z}$ of points in Z there is a trivial isometry between $(\mathbf{F}_{|\vec{z}}, d_{L^1})$ and $(\mathbf{F}, d_{L^1(P_{\vec{z}})})$, where $P_{\vec{z}}$ is the empirical measure induced by $\vec{z}$, in which each set has measure equal to the fraction of the points in $\vec{z}$ it contains. Thus by Lemma 6 above, we have

$$\mathcal{N}(\varepsilon, \mathbf{F}_{|\vec{z}}, d_{L^1}) = \mathcal{N}(\varepsilon, \mathbf{F}, d_{L^1(P_{\vec{z}})}) \leq \mathcal{N}(\varepsilon, \mathcal{H}, d_{L^1(P_{\vec{z}|X}, \rho_l)}) \leq \mathcal{C}(\varepsilon, \mathcal{H}, \rho_l)$$

for all $\vec{z} \in Z^{2m}$. Hence, setting $\varepsilon = \alpha\nu/8$, the given probability is at most $4\mathcal{C}(\alpha\nu/8, \mathcal{H}, \rho_l)e^{-\alpha^2\nu m/8M}$ by Theorem 3. □

In order to apply the above theorem, we need tools for bounding the capacity of various decision rule spaces. Along these lines, we close this section by proving two basic lemmas, one about the capacity of the free product of a set of function classes, and the other about the capacity of compositions of functions classes.

DEFINITION 8. Let $(A_1, \rho_1), \ldots, (A_k, \rho_k)$ be bounded metric spaces. Let $A = A_1 \times \cdots \times A_k$ and ρ be the metric on A defined by

$$\rho(\vec{u}, \vec{v}) = \frac{1}{k} \sum_{j=1}^{k} \rho_j(u_j, v_j)$$

for any $\vec{u} = (u_1, \ldots, u_k)$ and $\vec{v} = (v_1, \ldots, v_k) \in A$. For each j, $1 \leq j \leq k$, let $\mathcal{H}_j$ be a family of functions from X into A_j. The *free product* of $\mathcal{H}_1$ through $\mathcal{H}_k$ is the class of functions

$$\mathcal{H} = \{(f_1, \ldots, f_k) : f_j \in \mathcal{H}_j, 1 \leq j \leq k\},$$

where $(f_1, \ldots, f_k) : X \to A$ is the function defined by

$$(f_1, \ldots, f_k)(x) = (f_1(x), \ldots, f_k(x)).$$

LEMMA 7. If $\mathcal{H}, \mathcal{H}_1, \ldots, \mathcal{H}k$ are defined as above, then

1. for any probability measure P on X and $\varepsilon > 0$,

$$\prod_{j=1}^{k} \mathcal{N}(2k\varepsilon, \mathcal{H}_j, d_{L^1(P, \rho_j)}) \leq \mathcal{N}(\varepsilon, \mathcal{H}, d_{L^1(P, \rho)}) \leq \prod_{j=1}^{k} \mathcal{N}(\varepsilon, \mathcal{H}_j, d_{L^1(P, \rho_j)}),$$

2. $\overline{\dim}(\mathcal{H}) = \sum_{j=1}^{k} \overline{\dim}(\mathcal{H}j)$, and similarly for $\underline{\dim}$ and $\dim$, when the latter is defined.

PROOF. We begin with the second inequality of part (1). For each $1 \leq j \leq k$, let U_j be an ε-cover for $\mathcal{H}_j$. Let

$$U = \prod_{j=1}^{k} U_j = \{(f_1, \ldots, f_k) : f_j \in U_j, 1 \leq j \leq k\}.$$

It suffices to show that U is an ε-cover for $\mathcal{H}$. Let $g = (g_1, \ldots, g_k)$ be any function in $\mathcal{H}$. For each j, $1 \leq j \leq k$, find $f_j \in U_j$ such that $d_{L^1(P,\rho_j)}(f_j, g_j) \leq \varepsilon$. Let $f = (f_1, \ldots, f_k)$. Then

$$d_{L^1(P,\rho)}(f,g) = \int_X \frac{1}{k} \sum_{j=1}^{k} \rho_j(f_j(x), g_j(x)) dP(x)$$

$$= \frac{1}{k} \sum_{j=1}^{k} \int_X \rho_j(f_j(x), g_j(x)) dP(x)$$

$$= \frac{1}{k} \sum_{j=1}^{k} d_{L^1(P,\rho_j)}(f_j, g_j)$$

$$\leq \varepsilon.$$

Hence U is an ε-cover for $\mathcal{H}$.

The first inequality of part (1) is verified similarly. For each $1 \leq j \leq k$ let V_j be an $k\varepsilon$-separated subset of $\mathcal{H}_j$. Let $V = \prod_{j=1}^{k} V_j$. Let $f = (f_1, \ldots, f_k)$ and $g = (g_1, \ldots, g_k)$ be distinct functions in V. Then

$$d_{L^1(P,\rho)}(f,g) = \frac{1}{k} \sum_{j=1}^{k} d_{L^1(P,\rho_j)}(f_j, g_j) > \varepsilon.$$

Hence V is an ε-separated subset of $\mathcal{H}$. It follows that

$$\prod_{j=1}^{k} \mathcal{M}(k\varepsilon, \mathcal{H}_j, d_{L^1(P,\rho_j)}) \leq \mathcal{M}(\varepsilon, \mathcal{H}, d_{L^1(P,\rho)}).$$

The first inequality of part (1) then follows using Theorem 12. From part (1) we have

$$\prod_{j=1}^{k} \mathcal{C}(2k\varepsilon, \mathcal{H}_j, \rho_j) \leq \mathcal{C}(\varepsilon, \mathcal{H}, \rho) \leq \prod_{j=1}^{k} \mathcal{C}(\varepsilon, \mathcal{H}j, \rho_j).$$

Part 2 follows easily from this. $\square$

DEFINITION 9. Let P be a probability measure on X and f be a measurable function from X into Y. Then P_f denotes the probability measure on Y induced by f; i.e.,

$$P_f(S) = P(f^{-1}(S)) \text{ for all measurable } S \subset Y.$$

DEFINITION 10. Let f be a function from a metric space (X, ρ) into a metric space (Y, σ). A *Lipschitz bound* on f is a real number $b > 0$ such that for all $x, y \in X$, $\sigma(f(x), f(y)) \le b \, \rho(x, y)$. The Lipschitz bound on f is the smallest such b. If $\mathcal{F}$ is a class of functions from (X, ρ) into (Y, σ) then b is a *uniform Lipschitz bound* on $\mathcal{F}$ if b is a Lipschitz bound on f for all $f \in \mathcal{F}$.

LEMMA 8. Let $(X_1, \rho_1), \dots, (X_{k+1}, \rho_{k+1})$ be metric spaces, where (X_j, ρ_j) is bounded, $2 \le j \le k$, and $\mathcal{H}_j$ be a class of functions with $f : X_j \to X_{j+1}$ for all $f \in \mathcal{H}_j$, $1 \le j \le k$. Let b_j be a uniform Lipschitz bound on $\mathcal{H}_j$ for all $2 \le j \le k$. Let $\mathcal{H}$ denote the class of all functions from X_1 into X_{k+1} defined by compositions of functions in the $\mathcal{H}j$'s; i.e.,

$$\mathcal{H} = \{f_k \circ f_{k-1} \circ \cdots \circ f_1 : f_j \in \mathcal{H}_j, 1 \le j \le k\}.$$

1. For any $\varepsilon, \varepsilon_1, \dots, \varepsilon_k > 0$ such that

$$\varepsilon = \sum_{j=1}^{k} \left(\prod_{l=j+1}^{k} b_l \right) \varepsilon_j,$$

   we have

$$\mathcal{C}(\varepsilon, \mathcal{H}, \rho_{k+1}) \le \prod_{j=1}^{k} \mathcal{C}(\varepsilon_j, \mathcal{H}_j, \rho_{j+1}).$$

2. $\overline{\dim}(\mathcal{H}) \le \sum_{j=1}^{k} \overline{\dim}(\mathcal{H}_j)$, and similarly for $\underline{\dim}$ and $\dim$, when the latter is defined.

PROOF. Fix a probability measure P on X_1. We define a tree-structured family of covers for the $\mathcal{H}_j$'s by induction as follows. For the basis case, let U be a minimum-sized ε_1-cover for $\mathcal{H}_1$ with respect to the measure P on X_1; i.e., $|U| = \mathcal{N}(\varepsilon_1, \mathcal{H}_1, d_{L^1(P, \rho_2)})$ and every function in $\mathcal{H}_1$ is $L^1(P, \rho_2)$-approximated to within ε_1 by some function in U. Now for each j, $2 \leq j \leq k$, and for each sequence of functions $f_1, \ldots, f_{j-1}$ where $f_1 \in U$, $f_2 \in U_{f_1}$, $f_3 \in U_{f_1, f_2}, \ldots$, $f_{j-1} \in U_{f_1, \ldots, f_{j-2}}$, let $U_{f_1, \ldots, f_{j-1}}$ be a minimum-sized ε_j cover for $\mathcal{H}_j$ with respect to the L^1 metric for the measure $P_{f_{j-1} \circ f_{j-2} \cdots \circ f_1}$ on X_j and the metric ρ_{j+1} on X_{j+1}.

Next we define a cover V for $\mathcal{H}$ by composing functions in the covers for the $\mathcal{H}_j$'s. If $k = 1$, then $V = U$. Otherwise

$$V = \{f_k \circ f_{k-1} \circ \cdots \circ f_1 : f_1 \in U, f_2 \in U_{f_1}, f_3 \in U_{f_1, f_2}, \cdots, \text{ and } f_k \in U_{f_1, \ldots, f_{k-1}}\}.$$

Since $\mathcal{N}(\varepsilon_j, \mathcal{H}_j, d_{L^1(P_{f_{j-1} \circ f_{j-2} \circ \cdots \circ f_1}, \rho_{j+1})}) \leq \mathcal{C}(\varepsilon_j, \mathcal{H}_j, \rho_{j+1})$ for all $1 \leq j \leq k$ and all $f_1, \ldots, f_{j-1}$, it is clear that $|V| \leq \prod_{j=1}^{k} \mathcal{C}(\varepsilon_j, \mathcal{H}_j, \rho_{j+1})$. Hence it remains to show that V is an ε-cover for $\mathcal{H}$.

Suppose that $g = g_k \circ g_{k-1} \circ \cdots \circ g_1 \in \mathcal{H}$. Find

- $f_1 \in U$ such that $d_{L^1(P, \rho_2)}(f_1, g_1) \leq \varepsilon_1$,
- $f_2 \in U_{f_1}$ such that $d_{L^1(P_{f_1}, \rho_3)}(f_2, g_2) \leq \varepsilon_2, \ldots$, and
- $f_k \in U_{f_1, \ldots, f_{k-1}}$ such that $d_{L^1(P_{f_{k-1} \circ f_{k-2} \circ \cdots \circ f_1}, \rho_{k+1})}(f_k, g_k) \leq \varepsilon_k$.

Let $f = f_k \circ f_{k-1} \circ \cdots \circ f_1 \in V$. It suffices to show that $d_{L^1(P, \rho_{k+1})}(f, g) \leq \varepsilon$. We prove that for all h, $1 \leq h \leq k$,

$$d_{L^1(P, \rho_{h+1})}(f_h \circ \cdots \circ f_1, g_h \circ \cdots \circ g_1) \leq \sum_{j=1}^{h} \left(\prod_{l=j+1}^{h} b_l \right) \varepsilon_j.$$

Since

$$\varepsilon = \sum_{j=1}^{k} \left(\prod_{l=j+1}^{k} b_l \right) \varepsilon_j,$$

part (1) of the result follows.

If $h = 1$, then the result follows directly from our definition of f. Otherwise

$$d_{L^1(P,\rho_{h+1})}(f_h \circ \cdots \circ f_1, g_h \circ \cdots \circ g_1)$$

$$= \int_{X_1} \rho_{h+1}(f_h \circ \cdots \circ f_1(x), g_h \circ \cdots \circ g_1(x))dP(x)$$

$$\leq \int_{X_1} \rho_{h+1}(g_h \circ f_{h-1} \circ \cdots \circ f_1(x), g_h \circ g_{h-1} \circ \cdots \circ g_1(x))dP(x)$$

$$+ \int_{X_1} \rho_{h+1}(f_h \circ f_{h-1} \circ \cdots \circ f_1(x), g_h \circ f_{h-1} \circ \cdots \circ f_1(x))dP(x)$$

$$\leq b_h \int_{X_1} \rho_h(f_{h-1} \circ \cdots \circ f_1(x), g_{h-1} \circ \cdots \circ g_1(x))dP(x)$$

$$+ \int_{X_h} \rho_{h+1}(f_h(y), g_h(y))dP_{f_{h-1},\ldots,f_1}(y) \text{ (by the Lipschitz assumption)}$$

$$\leq b_h \sum_{j=1}^{h-1} \left(\prod_{l=j+1}^{h-1} b_l \right) \varepsilon_j + \varepsilon_h \text{ (by inductive hypothesis and definition of } f_h)$$

$$= \sum_{j=1}^{h} \left(\prod_{l=j+1}^{h} b_l \right) \varepsilon_j.$$

To prove part (2), let $a_j = \prod_{l=j+1}^{k} b_l$ and set $\varepsilon_j = \varepsilon/ka_j$ for $1 \leq j \leq k$. By part (1), $\mathcal{C}(\varepsilon, \mathcal{H}, \rho_{k+1}) \leq \prod_{j=1}^{k} \mathcal{C}(\varepsilon_j, \mathcal{H}_j, \rho_{j+1})$. Thus

$$\overline{\dim}(\mathcal{H}) = \limsup_{\varepsilon \to 0} \left(\frac{\log(\mathcal{C}(\varepsilon, \mathcal{H}, \rho_{k+1}))}{\log(1/\varepsilon)} \right)$$

$$\leq \limsup_{\varepsilon \to 0} \left(\frac{\log(\prod_{j=1}^{k} \mathcal{C}(\varepsilon/ka_j, \mathcal{H}_j, \rho_{j+1}))}{\log(1/\varepsilon)} \right)$$

$$= \sum_{j=1}^{k} \limsup_{\varepsilon \to 0} \left(\frac{\log(\mathcal{C}(\varepsilon/ka_j, \mathcal{H}_j, \rho_{j+1}))}{\log(1/\varepsilon)} \right)$$

$$= \sum_{j=1}^{k} \limsup_{\varepsilon_j \to 0} \left(\frac{\log(\mathcal{C}(\varepsilon_j, \mathcal{H}_j, \rho_{j+1}))}{\log(1/\varepsilon_j) + \log(1/ka_j)} \right)$$

$$= \sum_{j=1}^{k} \limsup_{\varepsilon_j \to 0} \left(\frac{\log(\mathcal{C}(\varepsilon_j, \mathcal{H}_j, \rho_{j+1}))}{\log(1/\varepsilon_j)} \right)$$

$$= \sum_{j=1}^{k} \overline{\dim}(\mathcal{H}_j).$$

□

7. SAMPLE SIZE BOUNDS FOR LEARNING WITH MULTI-LAYER NEURAL NETS

We now present some applications of the results of the previous section to learning with feedforward neural nets (see Rumelhart and McClelland[69,76]). The decision rule space $\mathcal{H}$ represented by a feedforward neural net consists of a family of functions from an instance space $X \subset \Re^n$ into a decision space $A \subset \Re^k$ for some $k, n \geq 1$. To apply Theorem 9 of the previous section, we will need to obtain an upper bound on the capacity $\mathcal{C}(\varepsilon, \mathcal{H}, \rho_l)$ of such decision rule spaces for various loss functions l.

For many loss functions, the metric ρ_l on $A \subset \Re^k$ can be bounded in terms of the d_{L^1} metric, i.e., we can find a constant c_l such that for all $\vec{a} = (a_1, \ldots, a_k)$ and $\vec{b} = (b_1, \ldots, b_k)$ in A, $\rho_l(\vec{a}, \vec{b}) \leq c_l d_{L^1}(\vec{a}, \vec{b}) = (c_l/k) \sum_{i=1}^k |a_i - b_i|$. In this case it is clear that $\mathcal{C}(\varepsilon, \mathcal{H}, \rho_l) \leq \mathcal{C}(\varepsilon/c_l, \mathcal{H}, d_{L^1})$. Thus our problem is reduced to obtaining an upper bound on the capacity $\mathcal{C}(\varepsilon/c_l, \mathcal{H}, d_{L^1})$.

We now give a few examples to illustrate this reduction. First consider the common case in which the outcome space Y is also contained in $\Re^k$; e.g., we receive explicit feedback on each coordinate of our action $\vec{a} \in A \subset \Re^k$. This occurs when each coordinate a_i of the action $\vec{a}$ is a prediction of the corresponding coordinate of the outcome $\vec{y}$. Here the loss function l may itself be a metric on $\Re^k$ which measures the distance between the predicted vector and the actual outcome vector. When l is a metric, we have for any actions $\vec{a}, \vec{b} \in A$

$$\rho_l(\vec{a}, \vec{b}) = \sup_{\vec{y} \in Y} |l(\vec{y}, \vec{a}) - l(\vec{y}, \vec{b})| \leq l(\vec{a}, \vec{b})$$

by the triangle inequality for l. Thus if the metric l is bounded with respect to d_{L^1} metric, i.e., $l(\vec{a}, \vec{b}) \leq c_l d_{L^1}(\vec{a}, \vec{b})$ for all $\vec{a}, \vec{b} \in A$, then we have $\rho_l(\vec{a}, \vec{b}) \leq c_l d_{L^1}(\vec{a}, \vec{b})$. For example, if

$$l(\vec{y}, \vec{a}) = d_{L^2}(\vec{y}, \vec{a}) = \frac{1}{k} \left(\sum_{i=1}^k (y_i - a_i)^2 \right)^{\frac{1}{2}},$$

then we may take $c_l = 1$, and similarly for the other d_{L^q} metrics, for $q > 1$.

Note that the above trick does not apply to the mean squared loss

$$l(\vec{y}, \vec{a}) = \frac{1}{k} \sum_{i=1}^k (y_i - a_i)^2$$

since this loss does not satisfy the triangle inequality. However, in this case it is easy to show by direct calculation that if the outcome space Y is bounded, e.g., $Y \subset [0, M]^k$, then $\rho_l(\vec{a}, \vec{b}) \leq 2M d_{L^1}(\vec{a}, \vec{b})$, and hence we may take $c_l = 2M$.

For our final example, consider the case when $Y = \{0,1\}^k$, $A \subset [0,1]^k$, and l is the cross entropy loss

$$l(\vec{y}, \vec{a}) = -\sum_{i=1}^{k} \left(y_i \ln a_i + (1 - y_i) \ln(1 - a_i) \right).$$

As discussed in Section 1.1.3, this is the log likelihood loss for the regression problem in which the action $\vec{a}$ represents a vector of probabilities for independent Bernoulli variables, and the outcome $\vec{y}$ gives the observed values of these variables. This loss is bounded if we restrict the probabilities in $\vec{a}$ to be between B and $1 - B$ for some $0 < B \leq 1/2$. In this case

$$\rho_l(\vec{a}, \vec{b}) = \sup_{\vec{y} \in Y} \left| \sum_{i=1}^{k} \left(y_i \ln \frac{b_i}{a_i} + (1 - y_i) \ln \frac{(1 - b_i)}{(1 - a_i)} \right) \right|$$

$$\leq \sum_{i=1}^{k} \left| \ln \frac{b_i}{a_i} \right| + \sum_{i=1}^{k} \left| \ln \frac{(1 - b_i)}{(1 - a_i)} \right|$$

$$\leq \frac{2}{B} \sum_{i=1}^{k} |a_i - b_i|.$$

The latter inequality follows from the fact that for $x, y > 0$,

$$\left| \ln \frac{x}{y} \right| = \ln \frac{\max(x,y)}{\min(x,y)} \leq \frac{\max(x,y)}{\min(x,y)} - 1 = \frac{|x - y|}{\min(x,y)}.$$

Thus in this case we may take $c_l = 2k/B$.

We now turn to the task of obtaining an upper bound on the capacity $C(\varepsilon, \mathcal{H}, d_{L^1})$ when the decision rules in $\mathcal{H}$ map into a decision space $A \subset \Re^k$, and in particular, when these decision rules are represented by neural networks. When $k = 1$, i.e., the neural net has only one output, the decision rule space $\mathcal{H}$ is a family of real-valued functions and $d_{L^1}(a, b) = |a - b|$ for $a, b \in A$. In this case we can apply the methods and results of Section 4. We must first find an upper bound on $\mathbf{dim_P}(\mathcal{H})$, the pseudo dimension of $\mathcal{H}$. Then, when A is bounded, from the bound on $\mathbf{dim_P}(\mathcal{H})$ we get a bound on the capacity $C(\varepsilon, \mathcal{H}, d_{L^1})$ using Theorem 6. This also gives a bound on the metric dimension of $\mathcal{H}$.

THEOREM 10. Let $\mathcal{H}$ be a family of functions from X into $A = [0, M]$. Assume $\dim_P(\mathcal{H}) = d$ for some $1 \leq d < \infty$.

1. For all $0 < \varepsilon \leq M$,

$$C(\varepsilon, \mathcal{H}, d_{L^1}) < 2 \left(\frac{2eM}{\varepsilon} \ln \frac{2eM}{\varepsilon} \right)^d.$$

2. $\overline{\dim}(\mathcal{H}) \leq \dim_P(\mathcal{H})$.

PROOF. Let P be any probability measure on X. Then by Theorems 6 and 12 in Sections 4 and 9.1,

$$\mathcal{N}(\varepsilon, \mathcal{H}, d_{L^1(P)}) \leq \mathcal{M}(\varepsilon, \mathcal{H}, d_{L^1(P)}) < 2 \left(\frac{2eM}{\varepsilon} \ln \frac{2eM}{\varepsilon} \right)^d.$$

(Theorem 6 is applied with $Z = X$ and $\mathbf{F} = \mathcal{H}$.) This gives part (1), and part (2) follows easily from part (1). $\square$

In the general case, where $A \subset \mathfrak{R}^k$ for $k > 1$, we can apply the methods from the previous section, in addition to the pseudo dimension methods from Section 4, to obtain bounds on $C(\varepsilon, \mathcal{H}, d_{L^1})$. We illustrate this for the case when $\mathcal{H}$ is the class of decision rules represented by a feedforward neural network.

A feedforward neural network is defined as a directed acyclic graph in which the incoming edges to each node (or *unit*) are ordered and each incoming edge can carry a real number representing the *activation* on that edge. We will assume that all activations are restricted to the interval $[c_0, c_1]$ for some constants $c_0 < c_1$. The units are divided into *input units*, which have no incoming edges from other units and serve as input ports for the network (their activations are determined by these external inputs), and *computation units*, which have incoming edges from other units and compute an activation based on the activations on these incoming edges. After an activation has been determined, this activation is placed on the outgoing edges of the unit. Computation units with no outgoing edges are called *output units* and serve as output ports for the network. Computation units that are not output units are called *hidden units*. The network as a whole computes a function that maps from vectors of activation values in its input units to vectors of activation values in its output units by composing the functions computed by its computation units in the obvious way.

The action of a computation unit with n incoming edges can be specified by a function f from $[c_0, c_1]^n$ into $[c_0, c_1]$, where $f(\vec{x})$ is the resulting activation of the unit when the activations of its incoming edges are given be the vector $\vec{x} \in [c_0, c_1]^n$. In the nets we consider, the function f is defined by

$$ f(\vec{x}) = \sigma \left(\mu(\phi_1(\vec{x}), \ldots, \phi_k(\vec{x})) + \theta + \sum_{j=1}^{k} w_j \phi_j(\vec{x}) \right), $$

where the w_j's are adjustable real *weights*, θ is an adjustable real *bias*, $\phi_1, \ldots, \phi_k$ are fixed real-valued functions which we call the *input transformers*, $\mu : \Re^k \to \Re$ is a fixed function which we call the *global modifier*, and $\sigma : \Re \to [c_0, c_1]$ is a fixed non-increasing or non-decreasing function which we call the *squashing function*. Different units can have different modifiers, transformers and squashing functions. We say that the function f computed by a given computation unit with n incoming edges has *Lipschitz bound* b if for any $\vec{x}, \vec{y} \in [c_0, c_1]^n$, $|f(\vec{x}) - f(\vec{y})| \leq b d_{L^1}(\vec{x}, \vec{y})$.

We give a few examples to illustrate the flexibility of this model at the level of the individual computation unit. First assume that $k = n$, i.e., the number of input transformers is the same as the number of inputs, and that each input transformer simply extracts a component of the input, i.e., $\phi_j(\vec{x}) = x_j$, $1 \leq j \leq n$. In this case, which is the standard case for most neural net research, the overall input transformation is just the identity map and can be ignored. In this standard case, if the global modifier $\mu = 0$, we get what is known as a quasi-linear unit[76]:

$$ f(\vec{x}) = \sigma \left(\theta + \sum_{j=1}^{k} w_j x_j \right). $$

In the standard case, if $\mu(\vec{x}) = \sum_{j=1}^{n} x_j^2$, we get a unit that computes a function of the form

$$ f(\vec{x}) = \sigma \left(\theta' + \sum_{j=1}^{n} (x_j - a_j)^2 \right), $$

where $a_j = -w_j/2$ and $\theta' = \theta - \sum_{j=1}^{n} a_j^2$. This is similar to what is called a *radial basis* unit in the neural net literature.[57,69]

Now assume that $k = n$ but the input transformers take logs of the components of the inputs, i.e., $\phi_j(\vec{x}) = \log x_j$. (Here we assume $c_0 > 0$.) Let $\mu = 0$ and change the squashing function σ to σ', where $\sigma'(x) = \sigma(e^x)$. Then

$$ f(\vec{x}) = \sigma' \left(\theta + \sum_{j=1}^{n} w_j \log x_j \right) = \sigma \left(e^\theta \prod_{j=1}^{n} x_j^{w_j} \right), $$

giving what is commonly known as a *product unit*.[29]

A *feedforward neural net* is a directed acyclic graph with adjustable real-valued *weights* on each edge and an adjustable real-valued *bias* in each node. Each source (*input*) node is associated with one component of a real vector input, and each sink (*output*) node with one component of a real vector output. Noninput nodes are called *computation nodes*. For each computation node with k incoming edges, there is associated a function $f : \Re^k \to [0,1]$. (This range is chosen merely for convenience.) The *Lipschitz bound* for the node is the Lipschitz bound for f using the L^1 metric. If the node only receives inputs from a subset T of $\Re^k$, then this Lipschitz bound is calculated for f restricted to T. This occurs when the node only receives inputs from other computation nodes, in which case $T \subset [0,1]^k$.

In the nets we consider, the function f computed by a given node with k input edges is defined by

$$f(x_1,\ldots,x_k) = \sigma\left(\mu(x_1,\ldots,x_k) + \theta + \sum_{j=1}^{k} w_j x_j\right),$$

where the w_j's are the adjustable weights on the incoming edges, θ is the adjustable bias, $\mu : \Re^k \to \Re$ is a fixed function which we call the *modifier*, and $\sigma : \Re \to [0,1]$ is an arbitrary nonincreasing or nondecreasing function which we call the *activation function*. As shown in Section 5, this includes both the "standard" node found in neural networks, which sums its inputs, adds a threshold, and then applies a sigmoid activation function, and also nodes in radial basis networks. If the net has n input nodes and t output nodes, then it computes a function from $\Re^n$ into $[0,1]^t$ in the obvious way, by simply composing the functions computed by the nodes.

We define a feedforward *architecture* as a feedforward net with unspecified weights and biases; i.e., each computation unit has a fixed global modifier, a fixed squashing function, and fixed input transformers, but it has variable weights and a variable bias. We will be interested only in *layered* feedforward nets and architectures. In this case the nodes of the net can be decomposed into $h+2$ disjoint layers for some $h \geq 0$ such that layer 0 is the input layer and each node in this layer has an outgoing edge to every node in layer 1, layer $h+1$ is the output layer and each node in this layer has an incoming edge from every node in layer h, and such that each node in layer j has incoming edges from every node in layer $j-1$ and outgoing edges to every node in layer $j+1$, $1 \leq j \leq h$. Layers 1 through h are commonly called *hidden layers*, and nodes within them *hidden nodes*. We say a unit is *at depth* j in an architecture if the longest (directed) path from an input unit to that unit has j edges. Thus all input units are at depth 0, all computation units that have incoming edges only from input units are at depth 1, all computation units that have incoming edges only from input units and computation units at depth 1 are at depth 2, etc. The *depth of the architecture* is the depth of the deepest unit in it.

We can bound the capacity of the decision rule space represented by a feedforward architecture as follows.

THEOREM 11. Let $\mathcal{A}$ be a feedforward architecture as above with $n \geq 1$ input units, $k \geq 1$ output units, and depth $d \geq 1$. Let W be the total number of adjustable weights and biases in $\mathcal{A}$. Assume $b_j \geq 1$ for $2 \leq j \leq d$, and let $\mathcal{H}$ be all functions from $[c_0, c_1]^n$ into $[c_0, c_1]^k$ representable on $\mathcal{A}$ by setting the adjustable weights and biases such that for all j, $2 \leq j \leq d$, the average of the Lipschitz bounds of the functions computed by computation units at depth j is at most b_j. Then for all $0 < \varepsilon \leq c_2 - c_1$,

$$C(\varepsilon, \mathcal{H}, d_{L^1}) \leq \left(\frac{2e(c_2 - c_1)d \prod_{l=2}^{d} b_l}{\varepsilon} \right)^{2W}.$$

PROOF. For each j, $0 \leq j \leq d$, let n_j be the number of units at depth j in the architecture $\mathcal{A}$. For each j, $0 \leq j \leq d-1$, let $l_j = \sum_{i=0}^{j} n_j$, and let $l_d = n_d = k$. For each j, $1 \leq j \leq d+1$, let $X_j = [c_1, c_2]^{l_{j-1}}$. Then for each j, $1 \leq j \leq d$, we can define the family $\mathcal{H}_j$ of functions from X_j into X_{j+1} in the following manner.

First assume $j < d$. Let $u_1, \ldots, u_{n_j}$ be an enumeration of the computation units at depth j and $f_1, \ldots, f_{n_j}$ be functions such that f_i can be represented by u_i, $1 \leq i \leq n_j$, and the average Lipschitz bound on the f_is is at most b_j. Let h_j be the free product of $f_1, \ldots, f_{n_j}$ and l_{j-1} copies of the identity function on $[c_1, c_2]$. Thus $h_j : X_j \to X_{j+1}$. The function h_j represents a mapping from the sequence of all activations of units at depth at most $j-1$ to the sequence of all activations of units at depth at most j, where the activations at depth at most $j-1$ are unaltered, and the new activations; i.e., those at depth j, are calculated by $f_1, \ldots, f_{n_j}$. The family $\mathcal{H}_j$ consists of all functions h_j obtained in this manner, by varying the weights and biases in the units $u_1, \ldots, u_{n_j}$ at depth j in such a manner that the Lipschitz constraint is satisfied.

When $j = d$, no subsequent calculations will be performed so we no longer need to preserve the activations of shallower units. Hence, we omit the identity function components in each $h_d \in \mathcal{H}_d$. Otherwise the definition of $\mathcal{H}_d$ is the same as that for $\mathcal{H}_j$, where $j < d$.

It is clear that the class $\mathcal{H}$ in the statement of the theorem can be represented as the class of compositions of functions from classes $\mathcal{H}_1, \ldots, \mathcal{H}_d$. Since the identity function has Lipshitz bound $1 \leq b_j$, the average Lipschitz bound on the components of each function $h_j \in \mathcal{H}_j$ is at most b_j. It is easily verified that a free product function is Lipschitz bounded by the average of the Lipschitz bounds on its component functions. Hence by assumption, b_j is a uniform Lipschitz bound on $\mathcal{H}_j$, $2 \leq j \leq d$. For each j, $1 \leq j \leq d$, let $a_j = \prod_{l=j+1}^{d} b_l$ and $\varepsilon_j = \varepsilon/da_j$. Since $\varepsilon < c_2 - c_1$ and $a_j \geq 1$, $\varepsilon_j \leq c_2 - c_1$. Let ρ_j be the d_{L^1} metric on X_j, $1 \leq j \leq d+1$. Then by Lemma 8, part (1),

$$\mathcal{C}(\varepsilon, \mathcal{H}, d_{L^1}) \leq \prod_{j=1}^{d} \mathcal{C}(\varepsilon_j, \mathcal{H}_j, \rho_{j+1}).$$

For each j, $\mathcal{H}_j$ is contained in the free product of l_j function classes. Each class $\mathcal{F}$ in this product is either the trivial class containing only the identity function, or is a finite-dimensional vector space of real-valued functions, summed with a fixed modifier and then composed with a nonincreasing or nondecreasing squashing function. In the latter case, the dimension N of this vector space is the number of free parameters associated with the corresponding computation unit, i.e., the number of weights plus one (for the adjustable bias). Hence by Theorems 4 and 5 in Section 4, the pseudo dimension $\mathbf{dim_P}(\mathcal{F}) \leq N$. Thus by Theorem 10 above,

$$\mathcal{C}(\varepsilon_j, \mathcal{F}, d_{L^1}) \leq 2 \left(\frac{2e(c_2 - c_1)}{\varepsilon_j} \ln \frac{2e(c_2 - c_1)}{\varepsilon_j} \right)^N \leq \left(\frac{2e(c_2 - c_1)}{\varepsilon_j} \right)^{2N},$$

since $2 \ln x < x$ and $N \geq 1$. Since the capacity of a class with only one function is 1, it follows from Lemma 7, part (1) that

$$\mathcal{C}(\varepsilon_j, \mathcal{H}_j, \rho_{j+1}) \leq \left(\frac{2e(c_2 - c_1)}{\varepsilon_j} \right)^{2W_j},$$

where W_j is the total number of weights and biases of all computation nodes at depth j. Multiplying these bounds over all j, it follows that

$$\mathcal{C}(\varepsilon, \mathcal{H}, d_{L^1}) \leq \prod_{j=1}^{d} \left(\frac{2e(c_2 - c_1)}{\varepsilon_j} \right)^{2W_j}$$

$$= \prod_{j=1}^{d} \left(\frac{2e(c_2 - c_1)d \prod_{l=j+1}^{d} b_l}{\varepsilon} \right)^{2W_j}$$

$$\leq \left(\frac{2e(c_2 - c_1)d \prod_{l=2}^{d} b_l}{\varepsilon} \right)^{2W}.$$

$\square$

COROLLARY 3. Let n, k, $\mathcal{H}$, W, d, and $b_2, \ldots, b_d$ be as in the previous theorem. Let X be the instance space $[c_1, c_2]^n$, A be the decision space $[c_1, c_2]^k$, and Y be any outcome space. Let $l : Y \times A \to [0, M]$ be a loss function and c_l be a constant such that $\rho_l(\vec{a}, \vec{b}) \leq c_l d_{L^1}(\vec{a}, \vec{b})$ for all $\vec{a}, \vec{b} \in A$. Let $m \geq 1$, $0 < \nu \leq 8(c_2 - c_1)$, $0 < \alpha < 1$, and P be any probability distribution on $Z = X \times Y$. Let $\vec{z}$ be generated by m independent random draws from Z according to P. Then

1.
$$\mathbf{Pr}\left(\exists f \in \mathcal{H} : d_\nu(\hat{\mathbf{r}}_{f,l}(\vec{z})\hat{\mathbf{r}}_{f,l}(\vec{z})) > \alpha\right)$$
$$\leq 4\left(\frac{c_l 16 e(c_2 - c_1)d \prod_{l=2}^{d} b_l}{\alpha\nu}\right)^{2W} e^{-\alpha^2 \nu m/8M}.$$

2. Assume that for each computation unit at depth 2 and above the number of weights is at most $W_{\max}$, no weight is allowed to have absolute value greater than β, the input transformers are identity functions, the global modifier has Lipschitz bound at most r, and the squashing function has Lipschitz bound at most s, where $s(\beta W_{\max} + r) \geq 1$. Then for any $0 < \delta < 1$, the probability in part (1) is less than δ for sample size
$$m = O\left(\frac{M}{\alpha^2 \nu}\left(W\left(\log \frac{c_l(c_2 - c_1)}{\alpha\nu} + d\log(s(\beta W_{\max} + r))\right) + \log \frac{1}{\delta}\right)\right).$$

PROOF. Let $\varepsilon = \alpha\nu/8 \leq c_2 - c_1$. It can be verified that $l_{\mathcal{H}}$ is permissible for the decision rule space $\mathcal{H}$. Hence, using Theorem 9 and Theorem 11,

$$\mathbf{Pr}\left(\exists f \in \mathcal{H} : d_\nu(\hat{\mathbf{r}}_{f,l}(\vec{z}), \hat{\mathbf{r}}_{f,l}(P)) > \alpha\right)$$
$$\leq 4\mathcal{C}(\alpha\nu/8, \mathcal{H}, \rho_l)e^{-\alpha^2 \nu m/8M}$$
$$\leq 4\mathcal{C}(\alpha\nu/8c_l, \mathcal{H}, d_{L^1})e^{-\alpha^2 \nu m/8M}$$
$$\leq 4\left(\frac{c_l 16 e(c_2 - c_1)d \prod_{l=2}^{d} b_l}{\alpha\nu}\right)^{2W} e^{-\alpha^2 \nu m/8M}.$$

For the second bound, it is readily verified that the L^1 Lipschitz bound for a linear function defined by $W_{\max}$ weights and a bias is no more than $W_{\max}$ times the largest absolute value of any weight. Furthermore, the Lipschitz bound for the sum of two functions is no more than the sum of the Lipschitz bounds on the individual functions, and the Lipschitz bound for the composition of two functions is no more than the product of the Lipschitz bounds on the individual functions. Thus, if the input transformers are identity functions, the global modifier and squashing function have Lipschitz bounds r and s respectively and no weight is allowed to have absolute value greater than β, then the Lipschitz bound for a computation unit is at most $s(\beta W_{\max} + r)$. If this holds for all units at depth 2 and above, may take $b_j = s(\beta W_{\max} + r)$ for all $j \geq 2$ in the first bound. Solving for m, this gives the order-of-magnitude estimate of the second bound. $\square$

We give the constants in the upper bound of part (1) the above theorem only to show that they are not outlandishly large. We do not mean to suggest that the bound is tight. At present we cannot even verify that the asymptotic bound of part (2) is tight. In particular, we cannot show that the dependence on the Lipschitz bounds is necessary. Evidence that it may not be necessary comes from the analysis of the case where the squashing function σ is a sharp threshold function; i.e., $\sigma(x) = \mathrm{sign}(x)$. Corollary 3 does not apply in this case, because the jump in σ prevents us from obtaining a Lipschitz bound on the computation units. As we let a smooth σ approach the sign function, its slope increases without limit, and thus the bound given in Corollary 3 degenerates. Nevertheless, using the techniques given by Baum and Haussler,[12] it can be shown that results similar to Corollary 3 hold in this case, except that no Lipschitz bounds are required, and a bound on the sample size is

$$O\left(\frac{1}{\alpha^2 \nu}\left(W \log \frac{N}{\alpha \nu} + \log \frac{1}{\delta}\right)\right),$$

where N is the total number of computation units in the net. Details are given in Theorem 13 in the appendix.

Despite the uncertainty about the need for the Lipschitz bounds, the result does give some indication of the maximum training sample size that will be needed for many popular network configurations. For example, if the squashing function is chosen as $\sigma(x) = 1/(1 + e^{-x/T})$ for some *temperature* $T > 0$, then it can be shown that the Lipschitz bound s for σ is $1/4T$. When the modifier $\mu \equiv 0$, then $r = 0$. Thus in this case the term $d \log(s(\beta W_{\max} + r))$ in the bound of Corollary 3 becomes $d \log(\beta W_{\max}/T)$. If the maximum weight β, the temperature T and the depth d are constants, along with $c_2 - c_1$, M, and c_l, then the asymptotic bound of the theorem becomes

$$O\left(\frac{1}{\alpha^2 \nu}\left(W \log \frac{W_{\max}}{\alpha \nu} + \log \frac{1}{\delta}\right)\right),$$

which is similar to the bound obtained by Baum and Haussler.[12]

It should also be noted that Corollary 3 does have the feature that no Lipschitz bounds are required on the computation units at depth one. Thus if all computation units are at depth one; i.e., there are no hidden units, then no Lipschitz units are required at all. If the architecture has only one layer of hidden units at depth one and a single output unit at depth two, as is quite common, then Lipschitz bounds are required only on the output unit. This means that the weights and biases associated with the hidden units do not need to be bounded in order to get the rates of uniform convergence given by Corollary 3, as they would, for example, if the methods given by White[94] were used to obtain a result of this type.

For an example of the above, consider networks that implement generalized radial basis[16] functions, as described by Poggio and Girosi.[69] These networks have one layer of hidden units at depth 1 and one output unit at depth 2. The structure of the hidden units is as described in the example above: the input transformers are identity functions, the modifier is $\sum_{i=1}^{n} x_i^2$ and the squashing function is usually a smooth decreasing function. The output unit simply computes a weighted sum, so for this unit the modifier is the 0 function and the squashing function is the identity. Since this is the only unit at depth 2 and above, we require a Lipschitz bound only for this unit. If β is a bound on the maximum weight coming into the output unit, and $W_{\max}$ is the number of units in the hidden layer, then the term $d\log(s(\beta W_{\max} + r))$ in the above bound becomes $\log(\beta W_{\max})$. Again, fixing $\beta, c_2 - c_1, M$, and c_l gives the same sample size bound,

$$ O\left(\frac{1}{\alpha^2 \nu} \left(W \log \frac{W_{\max}}{\alpha \nu} + \log \frac{1}{\delta} \right) \right), $$

similar to that obtained by Baum and Haussler.[12]

Since W appears to be the dominant factor in these bounds, apart from the accuracy parameters α and ν, these bounds support the conventional wisdom that the training set size should be primarily related to the number of adjustable parameters in the net. They also support the notion that this relationship between appropriate training size and the number of parameters is nearly linear, at least in the worst case. Further work is needed to sharpen these relationships (see, e.g., the lower bounds obtained by Baum and Haussler[12]).

---

[16] The computation units in the network of radial basis functions described here are quite primitive in that they have no adjustable multiplicative parameter included in their basic radial distance calculation. Such parameters would be needed to do any reasonable type of kernel based density estimation (see Duda and Hart[25]). These parameters can be simulated by inserting another layer of computation units between the inputs and the layer described here. Alternately, the analysis can also be done directly for adjustable kernel units. This cleaner approach is detailed by Pollard.[71]

8. CONCLUSION

We have extended the PAC learning model to a more general decision-theoretic framework so that it addresses many of the concerns raised by machine learning practitioners, and also introduced a number of new theoretical tools. Here we concentrate on applications of the extended model to the problem of obtaining upper bounds on sufficient training sample size. Further work will be required to obtain lower bounds on sample size needed, and to determine the computational complexity of finding decision rules with near minimal empirical risk. Some promising results along these lines are given by Kearns and Schapire.[45] However, even granting that such results can be obtained, the extended model still has a number of shortcomings in its present form. Some of these can be easily remedied, others may be more problematic.

First, we define the model only for a fixed decision rule space $\mathcal{H}$. The model should be extended to learning problems on a sequence of decision rule spaces $\{\mathcal{H}_n : n \geq 1\}$, where $\mathcal{H}_n$ is a decision rule space on an n-attribute domain X_n (e.g., $[0,1]^n$), and to families of decision rule spaces of different "complexities" on a fixed domain,[16,42,44] so that trade-offs between decision rule complexity and empirical risk can be addressed. The former extension is easy, the latter more involved. One approach to the latter problem is via Vapnik's principle of structural risk minimization.[89] (See also Devroye.[24]) Other approaches include the MDL (see Barron and Cover[10]), regularization (see Poggio and Girosi[69]), and more general Bayesian methods (see Berger[14]).

Second, the constants in the upper bounds are still too large to give sample size estimates that are useful in practice. It may be difficult to improve them to the point where the results are directly usable in applied work. Thus even with matching asymptotic lower bounds, practitioners may still need to rely at least in part on empirically derived sample size bounds. It is possible that the Bayesian viewpoint may yield better tools for calculating sample complexities. Support for this belief is given by Clarke and Barron,[20,21] Haussler et al.,[43] and Opper and Haussler.[67] However, necessary sample size estimates for decision rule spaces as general as those studied from the minimax perspective using uniform convergence have not yet been tackled from the Bayesian perspective.

Finally, many other issues would need to be considered in a complete treatment of the problem of overfitting, including distribution specific bounds on sample complexity (Theorem 2 is actually distribution specific, since the random covering numbers are distribution specific, yet we only apply it here in a distribution independent setting), decision rule spaces with infinite pseudo and metric dimensions (these include various classes of "smooth" functions and their relatives; see Dudley,[27] and Quiroz[73]) and non-iid sources of examples (see White[94] and Nobel and Dembo[64]).

Despite these shortcomings, we feel that the theory we give here provides useful insights into the nature of the problem of overfitting in learning, and because of its generality will be a useful starting point for further research in this area.

ACKNOWLEDGEMENTS

I would like to thank Dana Angluin, David Pollard and Phil Long for their careful criticisms of an earlier draft of this chapter, and their numerous suggestions for improvements. I also thank Naoki Abe, Anselm Blumer, Richard Dudley, and Michael Kearns for helpful comments on earlier drafts. I would also like to thank Ron Rivest, David Rumelhart, Andrzej Ehrenfeucht, and Nick Littlestone for stimulating discussions on these topics.

9. APPENDIX

9.1 METRIC SPACES, COVERING NUMBERS, AND METRIC DIMENSION

A *pseudo metric* on a set S is a function ρ from $S \times S$ into $\Re^+$ such that for all $x, y, z \in S$, $x = y \Rightarrow \rho(x,y) = 0$, $\rho(x,y) = \rho(y,x)$ (symmetry), and $\rho(x,z) \leq \rho(x,y) + \rho(y,z)$ (triangle inequality). If in addition $\rho(x,y) = 0 \Rightarrow x = y$, then ρ is a *metric*. (S, ρ) is a *(pseudo) metric space*. (S, ρ) is *complete* if every Cauchy sequence of points in S converges to a point in S; (S, ρ) is *separable* if it contains a countable dense subset, i.e., a countable subset A such that for every $x \in X$ and $\varepsilon > 0$ there exists $a \in A$ with $\rho(x, a) < \varepsilon$. If $\rho(x,y) = 1 \leftrightarrow x \neq y$, then ρ is called the *discrete metric*.

The *diameter* of a set $T \subseteq S$ is $\sup\{\rho(x,y) : x, y \in T\}$. If the diameter of T is finite, then we say that T is *bounded*. For any $\varepsilon > 0$, an ε-*cover* for T is a finite set $N \subseteq S$ (not necessarily contained in T) such that for all $x \in T$ there is a $y \in N$ with $\rho(x,y) \leq \varepsilon$. If T has a (finite) ε-cover for all $\varepsilon > 0$, then T is *totally bounded*. (Note that this implies that (T, ρ) is separable and bounded.) In this case the function $\mathcal{N}(\varepsilon, T, \rho)$ denotes the size of the smallest ε-cover for T (with respect to the space S and the (pseudo) metric ρ). We refer to $\mathcal{N}(\varepsilon, T, \rho)$ as a *covering number*. A set $R \subseteq T$ is ε-*separated* if for all distinct $x, y \in R$, $\rho(x,y) > \varepsilon$. We denote by $\mathcal{M}(\varepsilon, T, \rho)$ the size of the largest ε-separated subset of T. We refer to $\mathcal{M}(\varepsilon, T, \rho)$ as a *packing number*. The third argument to $\mathcal{N}$ and $\mathcal{M}$ will be omitted when the metric ρ is clear from the context.

The following inequalities are easily verified (see Kolmogorov and Tihomirov[47]):

THEOREM 12. If T is a totally bounded subset of the (pseudo) metric space (S, ρ), then for any $\varepsilon > 0$,

$$\mathcal{M}(2\varepsilon, T, \rho) \leq \mathcal{N}(\varepsilon, T, \rho) \leq \mathcal{M}(\varepsilon, T, \rho).$$

Hence both these measures of boundedness, by covering number and by packing number, are equivalent to within a factor of 2 of ε. Following Kolmogorov and Tihomirov,[47] we define the *upper metric dimension* of a (pseudo) metric space (S, ρ) by

$$\overline{\dim}(S) = \limsup_{\varepsilon \to 0} \frac{\log \mathcal{N}(\varepsilon, S, \rho)}{\log(1/\varepsilon)}.$$

The *lower metric dimension*, denoted by $\underline{\dim}$, of a (pseudo) metric space (S, ρ) is defined similarly using **liminf**. When $\overline{\dim}(S) = \underline{\dim}(S)$, then this quantity is denoted $\dim(S)$, and referred to simply as the *metric dimension* of (S, ρ). This quantity has also been called the *fractal dimension*[32] and the *capacity dimension*.[33] A very lucid and intuitive treatment is given by Mandelbrot.[55]

9.2 PERMISSIBLE CLASSES OF FUNCTIONS

In order to obtain the uniform convergence results given in Theorem 2, certain measurability assumptions have to be made concerning the class of functions **F** when this class is uncountable. These we have indicated by saying that **F** must be a *permissible* class.[70] Here we give a definition of permissible that is a special case of that given by Pollard. This definition will be suitable for our purposes; we refer the reader to Pollard[70] and Dudley[27] for a more general treatment. See Pollard,[70] Exercise 10, p. 39, for an indication of the kind of problems that can come up with nonpermissible classes.

Throughout the paper we have assumed that **F** is a class of real-valued functions on a set Z, and that P is a measure defined on some σ-algebra $\mathcal{A}$ of subsets of Z such that each function in **F** is measurable. We will need further conditions on **F** when it is uncountable. Let us say that the class **F** is *indexed* by the set T if

$$\mathbf{F} = \{f(\cdot, t) : t \in T\},$$

where f is a real-valued function on $Z \times T$ and $f(\cdot, t)$ denotes the real-valued function on Z obtained from f by fixing the second parameter to t. We will say that the function f *indexes* **F** *by* T.

We will need some structure on T as well. If T is contained in a topological space $\overline{T}$, then we will say that T is a *Borel subspace* if T is a Borel set with respect to the topology on $\overline{T}$; i.e., if T is in the smallest σ-algebra on $\overline{T}$ containing the open sets. The σ-algebra $\mathcal{B}(T)$ of Borel sets on T is then the restriction to T of the σ-algebra $\mathcal{B}$ of Borel sets on $\overline{T}$, i.e.,

$$\mathcal{B}(T) = \{B \cap T : B \in \mathcal{B}\}.$$

Finally, if A is any σ-algebra on Z and $\mathcal{C}$ is any σ-algebra on T, then $\mathcal{A} \times \mathcal{C}$ denotes the smallest σ algebra on $Z \times T$ that contains $\{A \times C : A \in \mathcal{A}, C \in \mathcal{C}\}$. We are now ready for our main definition.

DEFINITION 11. We say that the class **F** is *permissible* if it can be indexed by a set T such that

1. T is a Borel subspace of a compact metric space $\overline{T}$ and
2. the function $f : Z \times T \to \Re$ that indexes **F** by T is measurable with respect to the σ-algebra $\mathcal{A} \times \mathcal{B}(T)$.

   Most uncountable classes of functions that come up in practice can be indexed by a finite number of real parameters (i.e., with $T = \Re^n$ for some $n \geq 1$) in such a way that condition 2 is satisfied. Condition 1 is satisfied as well in this case, since we can take $\overline{T}$ to be the one-point compactification of T, obtained by adding a *point at infinity* to T (see Simmons[79]). (Let $\bar{x}$ be the point $(1, 0, \dots, 0) \in \Re^{n+1}$. The one-point compactification of $\Re^n$ can be obtained by mapping $\Re^n$ onto the unit sphere in $\Re^{n+1}$ minus the point $\bar{x}$ by projection through $\bar{x}$, followed by the completion of the space by the addition of the point $\bar{x}$.)

   Results given by Pollard (Appendix C)[70] imply that the sets used in Lemmas 12 and 13 are measurable when **F** is permissible. He also shows that the packing numbers $\mathcal{M}(\varepsilon, \mathbf{F}_{|\bar{z}}, d_{L^1})$ are measurable functions of $\bar{z} \in Z^m$ for any $m \geq 1$. Since Theorem 12 relates these closely to the covering numbers $\mathcal{N}(\varepsilon, \mathbf{F}_{|\bar{z}}, d_{L^1})$, this allows us to further formalize our usage of random covering numbers. A more formal treatment would either replace the covering numbers with the packing numbers in our upper bounds, or reword probabilistic bounds on the covering numbers to use outer measure arguments.

9.3 MEASURING THE ACCURACY OF EMPIRICAL ESTIMATES WITH THE d_ν METRIC

In this section, we give two bounds on the probability of large deviation of empirical estimates from true means, as measured by the d_ν metric. One is derived from Chebyshev's inequality and the other from Bernstein's inequality. The first bound is better for estimates obtained from small samples, the latter for estimates obtained from larger samples.

LEMMA 9. Let $Z_1, \ldots, Z_n$ be iid random variables with range $0 \leq Z_i \leq M$ and $\mathbf{E}(Z_i) = \mu$, $1 \leq i \leq n$. Assume $\nu > 0$ and $0 < \alpha < 1$. Then

1. $\mathbf{Pr}\left(d_\nu\left(\dfrac{1}{n}\sum\limits_{i=1}^{n} Z_i, \mu\right) > \alpha\right) \leq \dfrac{M^2}{4\alpha^2\nu n(M+\nu)} < \dfrac{M}{4\alpha^2\nu n}.$

2. $\mathbf{Pr}\left(d_\nu\left(\dfrac{1}{n}\sum\limits_{i=1}^{n} Z_i, \mu\right) > \alpha\right)$

   $\leq 2e^{-18\alpha^2\nu n/(3+\alpha)^2 M} < 2e^{-(9/8)\alpha^2\nu n/M} < 2e^{-\alpha^2\nu n/M}.$

PROOF. Let $Y_i = Z_i - \mu$, $1 \leq i \leq n$. Then

$$\mathbf{Pr}\left(d_\nu\left(\frac{1}{n}\sum_{i=1}^{n} Z_i, \mu\right) > \alpha\right) = \mathbf{Pr}\left(\frac{|(\sum_{i=1}^{n} Z_i) - \mu n|}{\nu n + \sum_{i=1}^{n} Z_i + \mu n} > \alpha\right)$$

$$\leq \mathbf{Pr}\left(\left|\sum_{i=1}^{n} Y_i\right| > \alpha n(\nu + \mu)\right),$$

since $\sum_{i=1}^{n} Z_i \geq 0$.

To obtain the first bound, note that by Chebyshev's inequality,

$$\mathbf{Pr}\left(|\sum_{i=1}^{n} Y_i| > \alpha n(\nu + \mu)\right) \leq \frac{\sigma^2}{(\alpha n(\nu + \mu))^2}, \text{ where } \sigma^2 = n\mathbf{Var}(Y_i).$$

Let $\beta = \mu/M$. Thus $\mu = \beta M$ and $0 \leq \beta \leq 1$. This implies that $\mathbf{Var}(Y_i) \leq (1 - \beta)\beta M^2$. Hence

$$\mathbf{Pr}\left(\left|\sum_{i=1}^{n} Y_i\right| > \alpha n(\nu + \mu)\right) \leq \frac{M^2}{\alpha^2 n} \max\left\{\frac{(1 - \beta)\beta}{(\nu + \beta M)^2} : 0 \leq \beta \leq 1\right\}.$$

This maximum occurs at $\beta = \nu/(M+2\nu)$, giving a maximum value of $1/4\nu(M+\nu)$. Thus

$$\mathbf{Pr}\left(d_\nu\left(\frac{1}{n}\sum_{i=1}^{n} Z_i, \mu\right) > \alpha\right) \leq \frac{M^2}{4\alpha^2\nu n(M+\nu)}.$$

Since $0 \leq Z_i \leq M$, $\mathbf{Var}(Z_i) = \mathbf{Var}(Y_i) \leq \mu(M - \mu)$. Hence

$$\mathbf{Pr}\left(\left|\sum_{i=1}^{n} Y_i\right| > \alpha n(\nu + \mu)\right) \leq \frac{\mu(M - \mu)}{\alpha^2 n(\nu + \mu)^2}.$$

It is easily verified that the maximum value of this expression occurs at $\mu = (\nu M)/(2\nu + M)$, and that this gives an upper bound of

$$\frac{M^2}{4\alpha^2 \nu n(M+\nu)}.$$

To obtain the second bound, we apply Bernstein's inequality (see Pollard,[70] p. 92), which states that

$$\mathbf{Pr}\left(\left|\sum_{i=1}^{n} Y_i\right| > \eta\right) \leq 2e^{-\eta^2/2(n\mathbf{Var}(Y_i)+\frac{1}{3}B\eta)}$$

for any zero mean iid random variables $Y_1, \ldots, Y_n$ bounded in absolute value by B. Substituting $\eta = \alpha n(\nu + \mu)$ and upper bounds $B \leq M$ and $\mathbf{Var}(Y_i) \leq \mu M$, this gives a bound of

$$2e^{-\alpha^2 n^2(\nu+\mu)^2/2(n\mu M+(1/3)M\alpha n(\nu+\mu))} = 2e^{-3\alpha^2 n(\nu+\mu)^2/2M(\alpha\nu+(3+\alpha)\mu)}.$$

Since $(\nu+\mu)^2/(\alpha\nu+(3+\alpha)\mu)$ is minimized at $\mu = (3-\alpha)/(3+\alpha)\nu$, the latter expression is bounded by substituting this value of μ. This gives the first bound of part (2). The second bound follows from the fact that $\alpha < 1$. □

9.4 PROOF OF MAIN THEOREM ON UNIFORM CONVERGENCE OF EMPIRICAL ESTIMATES

In this section we prove Theorem 3 from Section 3. Actually, we will state and prove a slightly stronger result. This result is obtained by bounding the probability of uniform convergence on a sample of length m in terms of the expected covering numbers associated with a sample of length $2m$, and by expanding the expectation to include the negative exponential term with a "truncation" at 1. It turns out that this saves us a factor of $1/2$ in the negative exponential term. We also include special bounds for the case that $\mathbf{F}_{|\vec{z}}$ is always finite. This case comes up, for example, when $\mathbf{F} = l_{\mathcal{H}}$ and we use the discrete loss function l, as in the PAC learning model.

The proof is given in a series of lemmas. We first extend the metric d_ν to a pseudo-metric on vectors in $(\Re^+)^{2m}$ for $m \geq 1$. We will do so in a somewhat unusual manner that will be useful in what follows. In fact, this extension can be defined for any metric, so we will state it in its general form.

DEFINITION 12. For each integer $m \geq 1$, let Γ_{2m} denote the set of all permutations σ of $\{1, \ldots, 2m\}$ such that for all i, $1 \leq i \leq m$, either $\sigma(i) = m+i$ and $\sigma(m+i) = i$, or $\sigma(i) = i$ and $\sigma(m+i) = m+i$. Thus the permutations in Γ_{2m} swap selected indices in the first half of the sequence $\{1, \ldots, 2m\}$ with corresponding indices in the second half. For any $\vec{x} = (x_1, \ldots, x_{2m}) \in (\Re^+)^{2m}$ and $\sigma \in \Gamma_{2m}$, let $\mu_1(\vec{x}, \sigma) = (1/m) \sum_{i=1}^{m} x_{\sigma(i)}$ and $\mu_2(\vec{x}, \sigma) = (1/m) \sum_{i=m+1}^{2m} x_{\sigma(i)}$. For any metric d on $\Re^+$ and $\vec{x}, \vec{y} \in (\Re^+)^{2m}$, let $\vec{d}(\vec{x}, \vec{y}) = \max\{d(\mu_1(\vec{x}, \sigma), \mu_1(\vec{y}, \sigma)) + d(\mu_2(\vec{x}, \sigma), \mu_2(\vec{y}, \sigma)) : \sigma \in \Gamma_{2m}\}$.

It is easily verified that $\vec{d}$ is a pseudo-metric on $(\Re^+)^{2m}$. Symmetry is obvious, and the triangle inequality follows easily from the triangle inequality for d on $\Re^+$:

$$
\begin{aligned}
\vec{d}(\vec{x}, \vec{y}) + \vec{d}(\vec{y}, \vec{z}) &= \max\{d(\mu_1(\vec{x}, \sigma), \mu_1(\vec{y}, \sigma)) + d(\mu_2(\vec{x}, \sigma), \mu_2(\vec{y}, \sigma)) : \sigma \in \Gamma_{2m}\} \\
&\quad + \max\{d(\mu_1(\vec{y}, \sigma), \mu_1(\vec{z}, \sigma)) + d(\mu_2(\vec{y}, \sigma), \mu_2(\vec{z}, \sigma)) : \sigma \in \Gamma_{2m}\} \\
&\geq \max\{d(\mu_1(\vec{x}, \sigma), \mu_1(\vec{y}, \sigma)) + d(\mu_2(\vec{x}, \sigma), \mu_2(\vec{y}, \sigma)) \\
&\quad + d(\mu_1(\vec{y}, \sigma), \mu_1(\vec{z}, \sigma)) + d(\mu_2(\vec{y}, \sigma), \mu_2(\vec{z}, \sigma)) : \sigma \in \Gamma_{2m}\} \\
&\geq \max\{d(\mu_1(\vec{x}, \sigma), \mu_1(\vec{z}, \sigma)) + d(\mu_2(\vec{x}, \sigma), \mu_2(\vec{z}, \sigma)) : \sigma \in \Gamma_{2m}\} \\
&= \vec{d}(\vec{x}, \vec{z})
\end{aligned}
$$

We note the following additional property of this extension.

LEMMA 10. For all $\vec{x}, \vec{y} \in (\Re^+)^{2m}$ and $\sigma \in \Gamma_{2m}$, $d(\mu_1(\vec{x}, \sigma), \mu_2(\vec{x}, \sigma)) \leq d(\mu_1(\vec{y}, \sigma), \mu_2(\vec{y}, \sigma)) + \vec{d}(\vec{x}, \vec{y})$.

PROOF. We have

$$
\begin{aligned}
d(\mu_1&(\vec{x}, \sigma), \mu_2(\vec{x}, \sigma)) \\
&\leq d(\mu_1(\vec{y}, \sigma), \mu_2(\vec{y}, \sigma)) + d(\mu_1(\vec{x}, \sigma), \mu_1(\vec{y}, \sigma)) + d(\mu_2(\vec{x}, \sigma), \mu_2(\vec{y}, \sigma))
\end{aligned}
$$

by the triangle inequality on d. The last two terms of this sum combined are at most $\vec{d}(\vec{x}, \vec{y})$ by definition. $\square$

We now restrict ourselves to the case that the metric d is the metric d_ν for some $\nu > 0$. The following lemma will play a key role in establishing our basic exponential inequality.

LEMMA 11. Let $\vec{x} = (x_1, \ldots, x_{2m})$ be a sequence of reals such that $0 \leq x_i \leq M$, $1 \leq i \leq 2m$. Assume $\nu > 0$ and $0 < \alpha < 1$. Then if a permutation $\sigma \in \Gamma_{2m}$ is chosen uniformly at random,

$$\mathbf{Pr}\left(d_\nu(\mu_1(\vec{x}, \sigma), \mu_2(\vec{x}, \sigma)) > \alpha\right) \leq 2e^{-2\alpha^2 \nu m/M}.$$

PROOF. For each i, $1 \leq i \leq m$, let Y_i be an independent random variable such that $Y_i = x_i - x_{m+i}$ with probability $1/2$ and $Y_i = x_{m+i} - x_i$ with probability $1/2$. Note that for any $\sigma \in \Gamma_{2m}$,

$$
\begin{aligned}
d_\nu(\mu_1(\vec{x}, \sigma), \mu_2(\vec{x}, \sigma)) &= \frac{|(1/m)\sum_{i=1}^m x_{\sigma(i)} - (1/m)\sum_{i=m+1}^{2m} x_{\sigma(i)}|}{\nu + (1/m)\sum_{i=1}^{2m} x_{\sigma(i)}} \\
&= \frac{|\sum_{i=1}^m (x_{\sigma(i)} - x_{\sigma(m+i)})|}{\nu m + \sum_{i=1}^{2m} x_i}.
\end{aligned}
$$

Hence

$$
\begin{aligned}
&\mathbf{Pr}\left(d_\nu(\mu_1(\vec{x}, \sigma), \mu_2(\vec{x}, \sigma)) > \alpha\right) \\
&= \mathbf{Pr}\left(\left|\sum_{i=1}^m (x_{\sigma(i)} - x_{\sigma(m+i)})\right| > \alpha\left(\nu m + \sum_{i=1}^{2m} x_i\right)\right) \\
&= \mathbf{Pr}\left(\left|\sum_{i=1}^m Y_i\right| > \alpha\left(\nu m + \sum_{i=1}^{2m} x_i\right)\right),
\end{aligned}
$$

because each swap in a randomly chosen $\sigma \in \Gamma_{2m}$ is independent. Since $\mathbf{E}(Y_i) = 0$ and $-|x_i - x_{m+i}| \leq Y_i \leq |x_i - x_{m+i}|$, we can apply Hoeffding's inequality (see Pollard[70]) to bound the latter probability by

$$2e^{-\alpha^2\left(\nu m + \sum_{i=1}^{2m} x_i\right)^2 / 2\sum_{i=1}^m (x_i - x_{m+i})^2}.$$

Let $\beta = \sum_{i=1}^{2m} x_i$. Since $0 \leq x_i \leq M$,

$$\sum_{i=1}^m (x_i - x_{m+i})^2 \leq \sum_{i=1}^m M|x_i - x_{m+i}| \leq \beta M.$$

Hence we have

$$
\begin{aligned}
2e^{-\alpha^2\left(\nu m + \sum_{i=1}^{2m} x_i\right)^2 / 2\sum_{i=1}^m (x_i - x_{m+i})^2} &\leq 2e^{-\alpha^2(\nu m + \beta)^2 / 2\beta M} \\
&= 2e^{-(\alpha^2/2M)((\nu m + \beta)^2/\beta)}.
\end{aligned}
$$

The expression in parentheses is minimized and, therefore, the whole expression is maximized, by setting $\beta = \nu m$, giving a value of $4\nu m$. Hence

$$\mathbf{Pr}\left(d_\nu(\mu_1(\vec{x},\sigma),\mu_2(\vec{x},\sigma)) > \alpha\right) \leq 2e^{-2\alpha^2\nu m/M}.$$

$\square$

For our next lemma we will need some notation to refer to the separate empirical estimates based on the first and second halves of an even length sample.

DEFINITION 13. For all $m \geq 1$ and $\vec{z} \in Z^{2m}$, we let $\widehat{\mathbf{E}}'_{\vec{z}}(f) = (1/m)\sum_{i=1}^{m} f(z_i)$ and $\widehat{\mathbf{E}}''_{\vec{z}}(f) = (1/m)\sum_{i=m+1}^{2m} f(z_i)$.

LEMMA 12. Let $\mathbf{F}$ be a permissible set of functions on Z with $0 \leq f(z) \leq M$ for all $f \in \mathbf{F}$ and $z \in Z$. Assume $\nu > 0$, $0 < \alpha < 1$ and $m \geq 2M/(\alpha^2\nu)$. Then

$$\mathbf{Pr}\left\{\vec{z} \in Z^m : \exists f \in \mathbf{F} \text{ with } d_\nu(\widehat{\mathbf{E}}_{\vec{z}}(f), \mathbf{E}(f)) > \alpha\right\}$$
$$\leq 2\mathbf{Pr}\left\{\vec{z} \in Z^{2m} : \exists f \in \mathbf{F} \text{ with } d_\nu(\widehat{\mathbf{E}}'_{\vec{z}}(f), \widehat{\mathbf{E}}''_{\vec{z}}(f)) > \frac{\alpha}{2}\right\}.$$

PROOF. If Z and $\mathbf{F}$ are uncountable, the assumption of permissibility guarantees that these probabilities are well defined (see Section 9.2 and Pollard[70]). From Chebyshev's inequality (see Lemma 9, Part 1 in Section 9.3), for each individual $f \in \mathbf{F}$,

$$\mathbf{Pr}\left\{\vec{z} \in Z^m : d_\nu(\widehat{\mathbf{E}}_{\vec{z}}(f), \mathbf{E}(f)) > \frac{\alpha}{2}\right\} \leq \frac{M}{(\alpha^2\nu m)}.$$

Since $m \geq 2M/(\alpha^2\nu)$, this probability is at most $1/2$. Now consider any $f \in \mathbf{F}$ and sample $\vec{z}' \in Z^m$ such that $d_\nu(\widehat{\mathbf{E}}_{\vec{z}'}(f), \mathbf{E}(f)) > \alpha$. If we draw an independent random sample $\vec{z}'' \in Z^m$, then with probability at least $1/2$, $d_\nu(\widehat{\mathbf{E}}_{\vec{z}''}(f), \mathbf{E}(f)) \leq \alpha/2$. Whenever this happens we have $d_\nu(\widehat{\mathbf{E}}_{\vec{z}'}(f), \widehat{\mathbf{E}}_{\vec{z}''}(f)) > \alpha/2$ by the triangle inequality for d_ν. Thus

$$\mathbf{Pr}\left\{\vec{z} \in Z^{2m} : \exists f \in \mathbf{F} \text{ with } d_\nu(\widehat{\mathbf{E}}'_{\vec{z}}(f), \widehat{\mathbf{E}}''_{\vec{z}}(f)) > \frac{\alpha}{2}\right\}$$
$$\geq \mathbf{Pr}\{\vec{z}'\vec{z}'' \in Z^{2m} : \exists f \in \mathbf{F} \text{ with } d_\nu(\widehat{\mathbf{E}}_{\vec{z}'}(f), \mathbf{E}(f)) > \alpha$$
$$\text{and } d_\nu(\widehat{\mathbf{E}}_{\vec{z}''}(f), \mathbf{E}(f)) \leq \frac{\alpha}{2}\}$$
$$\geq \frac{1}{2}\mathbf{Pr}\left\{\vec{z}' \in Z^m : \exists f \in \mathbf{F} \text{ with } d_\nu(\widehat{\mathbf{E}}_{\vec{z}'}(f), \mathbf{E}(f)) > \alpha\right\}.$$

Again, when Z and $\mathbf{F}$ are uncountable, permissibility guarantees that the implied use of Fubini's theorem in obtaining the above inequalities is justified. ☐

We are now in a position to prove the following version of our theorem, using the extended metric $\vec{d_\nu}$ in place of the L^1 metric to measure covering numbers.

LEMMA 13. Let $\mathbf{F}$ be a permissible well-behaved set of functions on Z with $0 \leq f(z) \leq M$ for all $f \in \mathbf{F}$ and $z \in Z$. Assume $\nu > 0$, $0 < \alpha < 1$ and $m \geq 1$. Let

$$p(\alpha, \nu, m) = \mathbf{Pr}\left\{\vec{z} \in Z^m : \exists f \in \mathbf{F} \text{ with } d_\nu(\widehat{\mathbf{E}}_{\vec{z}}(f), \mathbf{E}(f)) > \alpha\right\}.$$

Then

$$p(\alpha, \nu, m) \leq 2\mathbf{E}(\min(2\mathcal{N}(\frac{\alpha}{4}, \mathbf{F}_{|\vec{z}}, \vec{d_\nu})e^{-\alpha^2\nu m/8M}, 1)),$$

and if in addition $\mathbf{F}_{|\vec{z}}$ is finite for all $\vec{z} \in Z^{2m}$, then

$$p(\alpha, \nu, m) \leq 2\mathbf{E}(\min(2|\mathbf{F}_{|\vec{z}}|e^{-\alpha^2\nu m/2M}, 1)),$$

where the expectations are over $\vec{z}$ drawn randomly from Z^{2m}.

PROOF. First note that both bounds are trivial if $m < 2M/(\alpha^2\nu)$, so we may assume $m \geq 2M/(\alpha^2\nu)$. Hence, by Lemma 12,

$$p(\alpha, \nu, m) \leq 2\mathbf{Pr}\left\{\vec{z} \in Z^{2m} : \exists f \in \mathbf{F} \text{ with } d_\nu(\widehat{\mathbf{E}}'_{\vec{z}}(f), \widehat{\mathbf{E}}''_{\vec{z}}(f)) > \frac{\alpha}{2}\right\}.$$

Thus it suffices to obtain bounds for the latter quantity.
We begin with the second bound, for the case when $\mathbf{F}_{|\vec{z}}$ is always finite. For any sample $\vec{z} = (z_1, \ldots, z_{2m}) \in Z^{2m}$ and $\sigma \in \Gamma_{2m}$, let $\sigma(\vec{z}) = (z_{\sigma(1)}, \ldots, z_{\sigma(2m)})$. For any fixed function $f \in \mathbf{F}$ and fixed $\vec{z} \in Z^{2m}$, if we select a permutation $\sigma \in \Gamma_{2m}$ uniformly at random,

$$\mathbf{Pr}\left(d_\nu(\widehat{\mathbf{E}}'_{\sigma(\vec{z})}(f), \widehat{\mathbf{E}}''_{\sigma(\vec{z})}(f)) > \frac{\alpha}{2}\right) \leq 2e^{-\alpha^2\nu m/2M} \tag{7}$$

by Lemma 11. Hence for any fixed $\vec{z} \in Z^{2m}$, if we select a permutation $\sigma \in \Gamma_{2m}$ uniformly at random,

$$\mathbf{Pr}\left(\exists f \in \mathbf{F} \text{ with } d_\nu(\widehat{\mathbf{E}}'_{\sigma(\vec{z})}(f), \widehat{\mathbf{E}}''_{\sigma(\vec{z})}(f)) > \frac{\alpha}{2}\right) \leq \min(2|\mathbf{F}_{|\vec{z}}|e^{-\alpha^2\nu m/2M}, 1). \tag{8}$$

Thus if we draw $\vec{z}$ at random from Z^{2m} and independently select a permutation $\sigma \in \Gamma_{2m}$ uniformly at random,

$$\mathbf{Pr}\left(\exists f \in \mathbf{F} \text{ with } d_\nu(\widehat{\mathbf{E}}'_{\sigma(\vec{z})}(f), \widehat{\mathbf{E}}''_{\sigma(\vec{z})}(f)) > \frac{\alpha}{2}\right) \leq \mathbf{E}(\min(2|\mathbf{F}_{|\vec{z}}|e^{-\alpha^2 \nu m/2M}, 1)).$$

However, since each of the $2m$ observations in $\vec{z}$ are independent, each of the samples $\sigma(\vec{z})$ for $\sigma \in \Gamma_{2m}$ are equally likely. Hence

$$\mathbf{Pr}\left(\exists f \in \mathbf{F} \text{ with } d_\nu(\widehat{\mathbf{E}}'_{\sigma(\vec{z})}(f), \widehat{\mathbf{E}}''_{\sigma(\vec{z})}(f)) > \frac{\alpha}{2}\right),$$

where both $\vec{z}$ and σ are chosen at random, is the same as

$$\mathbf{Pr}\left(\exists f \in \mathbf{F} \text{ with } d_\nu(\widehat{\mathbf{E}}'_{\vec{z}}(f), \widehat{\mathbf{E}}''_{\vec{z}}(f)) > \frac{\alpha}{2}\right),$$

where only the sample $\vec{z}$ is chosen at random. The second bound follows.

The proof of the first bound is similar, except for steps 7 and 8. From Lemma 9, using the extension $\vec{d}_\nu$ of the metric d_ν to even-length sequences of reals given in Definition 12 above, if $f, f^* \in \mathbf{F}$ are such that

$$\vec{d}_\nu((f(z_1), \dots, f(z_{2m})), (f^*(z_1), \dots, f^*(z_{2m}))) \leq \frac{\alpha}{4},$$

then, for any $\sigma \in \Gamma_{2m}$ such that

$$d_\nu(\widehat{\mathbf{E}}'_{\sigma(\vec{z})}(f), \widehat{\mathbf{E}}''_{\sigma(\vec{z})}(f)) > \frac{\alpha}{2},$$

we must have

$$d_\nu(\widehat{\mathbf{E}}'_{\sigma(\vec{z})}(f^*), \widehat{\mathbf{E}}''_{\sigma(\vec{z})}(f^*)) > \frac{\alpha}{4}.$$

Thus, if N is an $\alpha/4$-cover for $\mathbf{F}_{|\vec{z}}$ with respect to the metric $\vec{d}_\nu$, then whenever there exists $f \in \mathbf{F}$ with

$$d_\nu(\widehat{\mathbf{E}}'_{\sigma(\vec{z})}(f), \widehat{\mathbf{E}}''_{\sigma(\vec{z})}(f)) > \frac{\alpha}{2},$$

there exists $f^* \in N$ with

$$d_\nu(\widehat{\mathbf{E}}'_{\sigma(\vec{z})}(f^*), \widehat{\mathbf{E}}''_{\sigma(\vec{z})}(f^*)) > \frac{\alpha}{4}.$$

For fixed f^* and random σ, the probability of the latter event is at most $2e^{-\alpha^2 \nu m/8M}$ by Lemma 10. Hence for any fixed $\vec{z} \in Z^{2m}$, if we select a permutation $\sigma \in \Gamma_{2m}$ uniformly at random,

$$\mathbf{Pr}\left(\exists f \in \mathbf{F} \text{ with } d_\nu(\widehat{\mathbf{E}}'_{\sigma(\vec{z})}(f), \widehat{\mathbf{E}}''_{\sigma(\vec{z})}(f)) > \frac{\alpha}{2}\right)$$

$$\leq \min(2\mathcal{N}(\frac{\alpha}{4}, \mathbf{F}_{|\vec{z}}, \vec{d}_\nu)e^{-\alpha^2 \nu m/8M}, 1)).$$

The remainder of the proof is as above. $\square$

We need only one more lemma to complete our proof; one that can be used to relate the $\vec{d}_\nu$ covering numbers to the d_{L^1} covering numbers.

LEMMA 14. For any $m \geq 1$, $\vec{x}, \vec{y} \in (\Re^+)^{2m}$ and $\nu > 0$, $\vec{d}_\nu(\vec{x}, \vec{y}) \leq (2/\nu)d_{L^1}(\vec{x}, \vec{y})$.

PROOF. For any $\sigma \in \Gamma_{2m}$

$$d_\nu(\mu_1(\vec{x}, \sigma), \mu_1(\vec{y}, \sigma)) + d_\nu(\mu_2(\vec{x}, \sigma), \mu_2(\vec{y}, \sigma))$$

$$= \frac{|\sum_{i=1}^m (x_{\sigma(i)} - y_{\sigma(i)})|}{\nu m + \sum_{i=1}^m (x_{\sigma(i)} + y_{\sigma(i)})}$$

$$+ \frac{|\sum_{i=m+1}^{2m} (x_{\sigma(i)} - y_{\sigma(i)})|}{\nu m + \sum_{i=m+1}^{2m} (x_{\sigma(i)} + y_{\sigma(i)})}$$

$$\leq \frac{\sum_{i=1}^{2m} |x_{\sigma(i)} - y_{\sigma(i)}|}{\nu m} = \frac{2}{\nu} d_{L^1}(\vec{x}, \vec{y}).$$

The result follows. □

The theorem follows easily from the last two lemmas.

9.5 BOUNDS ON SAMPLE SIZE FOR LEARNING IN FEEDFORWARD NETS WITH SHARP THRESHOLDS

In this section we give the bounds for uniform convergence of empirical estimates in neural networks with sharp threshold functions claimed in Section 7.

THEOREM 13. Let $\mathcal{A}$ be a feedforward architecture as defined in Section 7 with $n \geq 1$ inputs, one output, $N \geq 2$ computation units, and a total of W weights and biases. Let $X = \Re^n$, $Y = A = \{0, 1\}$, and l be the discrete loss function. Assume that the squashing function for each computation unit has the form $\text{sign} \circ \sigma$, where σ is nondecreasing or nonincreasing. Different σ's can be used for different units. Let $\mathcal{H}$ be all functions from X into A representable on $\mathcal{A}$ by varying the weights and biases. Let $\nu > 0$, $0 < \alpha < 1$, and $m \geq 1$. Suppose that $\vec{z}$ is generated by m independent random draws from a probability measure P on $Z = X \times Y$. Then

$$\mathbf{Pr}\left(\exists h \in \mathcal{H} : d_\nu(\hat{\mathbf{r}}_{h,l}(\vec{z}), \mathbf{r}_{h,l}(P)) > \alpha\right) \leq 4(2eNm/W)^W e^{-\alpha^2 \nu m/2}.$$

This probability is at most δ for a sample size m that is

$$O\left(\frac{1}{\alpha^2 \nu}\left(W \log \frac{N}{\alpha \nu} + \log \frac{1}{\delta}\right)\right).$$

PROOF. Each computation unit in the network with k weights is associated with a class of $\{0, 1\}$-valued functions of the form

$$f(\vec{x}) = \text{sign}(\sigma(\phi_1(\vec{x}), \dots, \phi_k(\vec{x})) + \theta + \sum_{j=1}^{k} w_j \phi_j(\vec{x}))$$

where $\theta, w_1, \dots, w_k$ are adjustable real-valued parameters and $\phi_1, \dots, \phi_k$, μ, and σ are fixed functions, the latter monotone. By Theorems 4 and 5 this class of functions has pseudo-dimension at most $k+1$. Since the pseudo-dimension is the same as the Vapnik-Chervonenkis dimension for classes of indicator functions, this implies that the class has Vapnik-Chervonenkis dimension at most $k + 1$. Now let d be the sum of all the Vapnik-Chervonenkis dimensions of all the classes of functions associated with the computation units of the architecture $\mathcal{A}$. It follows that $d \leq W$, the total number of weights and biases in the network. For each $h \in \mathcal{H}$, let l_h be the loss function associated with h for the discrete loss l; i.e., $l_h(x, y) = 1$ if $y \neq h(x)$, $l_h(x, y) = 0$ if $y = h(x)$. Let $\mathbf{F} = l_{\mathcal{H}} = \{l_h : h \in \mathcal{H}\}$. Let $\vec{z} = ((x_1, y_1), \dots, (x_m, y_m))$ be any fixed sample and $\vec{x} = (x_1, \dots, x_m)$. It is easily verified that $|\mathbf{F}_{|\vec{z}|}| = |\mathcal{H}_{|\vec{x}|}|$. It is shown by Baum and Haussler,[12] Theorem 1 (and is also implied directly from results given by Cover[22]), that for any class $\mathcal{H}$ as above, $|\mathcal{H}_{|\vec{x}|}| \leq (Nem/d)^d$ for all $\vec{x} = (x_1, \dots, x_m)$, where d and N are as above. This implies that

$$|\mathbf{F}_{|\vec{z}|}| \leq (Nem/d)^d \leq (Nem/W)^W$$

for all samples $\vec{z}$ of length m. Since each $l_h \in \mathbf{F}$ is a random variable that is bounded between 0 and 1, Theorem 3 (second part) shows that

$$\mathbf{Pr}\left(\exists h \in \mathbf{F} : d_\nu(\hat{\mathbf{E}}_{\vec{z}}(h), \mathbf{E}(h)) > \alpha\right) \leq 4(2eNm/W)^W e^{-\alpha^2 \nu m/2}.$$

This gives the first bound.
For the second bound, it can be shown that for sample size

$$m = \frac{4}{\alpha^2 \nu}\left(2W \ln \frac{16N}{\alpha^2 \nu} + \ln \frac{4}{\delta}\right),$$

we have

$$4(2eNm/W)^W e^{-\alpha^2 \nu m/2} \leq \delta.$$

To see this, first note that by rearranging, we get

$$\alpha^2 \nu m/2 \geq W \ln(2eNm/W) + \ln(4/\delta).$$

thus it suffices to show

$$\alpha^2 \nu m/4 \geq W \ln(2eNm/W) \text{ and } \alpha^2 \nu m/4 \geq \ln(4/\delta).$$

The latter inequality is assured by the last term in the formula for m. For the former inequality, let us take m equal to the first term only; i.e.,

$$m = \frac{8W}{\alpha^2 \nu} \ln \frac{16N}{\alpha^2 \nu}.$$

Substituting this into the former inequality and simplifying, we get

$$2 \ln \frac{16N}{\alpha^2 \nu} \geq \ln \left(\frac{16eN}{\alpha^2 \nu} \ln \frac{16N}{\alpha^2 \nu} \right).$$

This further simplifies to

$$\frac{16N}{\alpha^2 \nu} \geq e \ln \frac{16N}{\alpha^2 \nu},$$

which holds, since $x \geq e \ln x$ for any x. Finally, since this inequality holds for the given m, it is easy to see that it will also hold for larger m. $\square$

REFERENCES

1. Abe, N., and M. Warmuth. "On the Computational Complexity of Approximating Distributions by Probabilistic Automata." In *Proceedings of the 3rd Workshop on Computational Learning Theory*, 52–66. San Mateo, CA: Morgan Kaufmann, 1990.
2. Alexander, K. "Rates of Growth for Weighted Empirical Processes." In *Proceedings of Berkeley Conference in Honor of Jerzy Neyman and Jack Kiefer*, Vol. 2, 475–493, 1985.
3. Alexander, K. "Rates of Growth and Sample Moduli for Weighted Empirical Processes Indexed by Sets." *Prob. Theory & Related Fields* **75** (1987): 379–423.
4. Angluin, D. "Queries and Concept Learning." *Mach. Learning* **2** (1988): 319–342.
5. Angluin, D., and P. Laird. "Learning from Noisy Examples." *Mach. Learning*, **2(4)** (1988): 343–370.

6. Angluin, D., and L. G. Valiant. "Fast Probabilistic Algorithms for Hamiltonian Circuits and Matchings." *J. Comp. & System Sci.* **18(2)** (1979): 155–193.

7. Anthony, M., and J. Shawe-Taylor. "A Result of Vapnik with Applications." Technical Report CSD-TR-628, University of London, 1990.

8. Assouad, P. "Densité et Dimension." *Ann. de l'Institut Fourier* **33(3)** (1983): 233–282.

9. Barron, A. "Statistical Properties of Artificial Neural Networks." In *28th Conference on Decision and Control*, 280–285, 1989.

10. Barron, A., and T. Cover. "Minumum Complexity Density Estimation." *IEEE Trans. Infor. Theory* **37(4)** (1990): 1034–1054.

11. Barto, A. G., and P. Anandan. "Pattern Recognizing Stochastic Learning Automata." *IEEE Trans. Sys., Man & Cybernetics* **15** (1985): 360–374.

12. Baum, E., and D. Haussler. "What Size Net Gives Valid Generalization?" *Neural Comp.* **1(1)** (1989): 151–160.

13. Benedek, G. M., and A. Itai. "Learnability by Fixed Distributions." In *Proceedings 1988 Workshop on Comp. Learning Theory*, 80–90. San Mateo, CA: Morgan Kaufmann, 1988.

14. Berger, J. *Statistical Decision Theory and Bayesian Analysis*. New York: Springer-Verlag, 1985.

15. Billingsley, P. *Probability and Measure*. New York: Wiley, 1986.

16. Blumer, A., A. Ehrenfeucht, D. Haussler, and M. K. Warmuth. "Learnability and the Vapnik-Chervonenkis Dimension." *J. Assoc. Comp. Mach.* **36(4)** (1989): 929–965.

17. Breiman, L., J. H. Friedman, R. A. Olshen, and C. J. Stone. *Classification and Regression Trees*. Montery, CA: Wadsworth International Group, 1984.

18. Buntine, W. L. "A Theory of Learning Classification Rules." Ph.D. Thesis, University of Technology, Sydney, 1990.

19. Buntine, W., and A. Weigend. "Bayesian Back Propagation." *Complex Systems* **5** (1991): 603–643.

20. Clarke, B. and A. Barron. "Information-Theoretic Asymptotics of Bayes Methods." *IEEE Trans. Infor. Theory* **36(3)** (1990): 453–471.

21. Clarke, B., and A. Barron. "Entropy, Risk and the Bayesian Central Limit Theorem." Manuscript, 1990.

22. Cover, T. M. "Capacity Problems for Linear Machines." In *Pattern Recognition*, edited by L. Kanal, 283–289. Springfield, IL: Thompson Books, 1968.

23. Denker, J., D. Schwartz, B. Wittner, S. Solla, R. Howard, L. Jackel, and J. Hopfield. "Automatic Learning, Rule Extraction, and Generalization." *Complex Systems* **1** (1987): 877–922.

24. Devroye, L. "Automatic Pattern Recognition: A Study of the Probability of Error." *IEEE Trans. Pattern Anal. & Mach. Int.* **10(4)** (1988): 530–543.

25. Duda, R. O., and P. E. Hart. *Pattern Classification and Scene Analysis*. New York: Wiley, 1973.

26. Dudley, R. M. "Central Limit Theorems for Empirical Measures." *Ann. Prob.*, **6(6)** (1978): 899–929.

27. Dudley, R. M. *A Course on Empirical Processes*. Lecture Notes in Mathematics, Vol. 1097, 2–142. New York: Springer-Verlag, 1984.

28. Dudley, R. M. "Universal Donsker Classes and Metric Entropy." *Ann. Prob.* **15(4)** (1987): 1306–1326.

29. Durbin, R., and D. E. Rumelhart. "Product units: A Computationaly Powerful and Biologically Plausible Extension to Backpropagation Networks." *Neural Comp.* **1(1)** (1989): 133–142.

30. Edelsbrunner, H. *Algorithms in Combinatorial Geometry*. New York: Springer-Verlag, 1987.

31. Ehrenfeucht, A., D. Haussler, M. Kearns, and L. Valiant. "A General Lower Bound on the Number of Examples Needed for Learning." *Infor. & Comp.* **82**: 247–261.

32. Farmer, J. Doyne. "Information Dimension and the Probabilistic Structure of Chaos." *Z. Naturforsch. A* **37** (1982): 1304–1325.

33. Farmer, J. Doyne, E. Ott, and J. A. Yorke. "The Dimension of Chaotic Attractors." *Physica D* **7** (1983): 153–180.

34. Ferguson, T. *Mathematical Statistics: A Decision Theoretic Approach*. New York: Academic Press, 1967.

35. Gullapalli, V. "A Stochastic Reinforcement Algortihm for Learning Real-Valued Functions." *Neural Networks* **3(6)** (1990): 671–692.

36. Gyorgyi, G., and N. Tishby. "Statistical Theory of Learning a Rule." In *Neural Networks and Spin Glasses*, edited by K. Theumann and R. Koeberle. Singapore: World Scientific, 1990.

37. Haussler, D., and E. Welzl. "Epsilon Nets and Simplex Range Queries." *Disc. Comp. Geometry* **2** (1987): 127–151.

38. Haussler, D. "Quantifying Inductive Bias: AI Learning Algorithms and Valiant's Learning Framework." *Art. Int.* **36** (1988): 177–221.

39. Haussler, D. "Learning Conjunctive Concepts in Structural Domains." *Mach. Learning* **4** (1989): 7–40.

40. Haussler, D. "Decision Theoretic Generalizations of the PAC Learning Model." In *Proc. First Workshop on Algorithmic Learning Theory*, 21–41. Tokyo, Japan: Japanese Society for Artificial Intelligence, 1990.

41. Haussler, D., N. Littlestone, and M. K. Warmuth. "Predicting $\{0,1\}$-Functions on Randomly Drawn Points." Technical Report UCSC-CRL-90-54, Computer

Research Laboratory, University of California Santa Cruz, December 1990. To appear in *Infor. & Comp.*

42. Haussler, D., M. Kearns, N. Littlestone, and M. K. Warmuth. "Equivalence of Models for Polynomial Learnability." *Infor. & Comp.* **95** (1991): 129–161.

43. Haussler, D., M. Kearns, and R. Schapire. "Bounds on the Sample Complexity of Bayesian Learning Using Information Theory and the VC Dimension." In *Proceedings of the Fourth Workshop on Computational Learning Theory*, 61–74. San Mateo, CA: Morgan Kaufman, 1991.

44. Kearns, M., M. Li, L. Pitt, and L. Valiant. " On the Learnability of Boolean Formulae." In *Proceedings of the Nineteenth Annual ACM Symposium on Theory of Computing*, 285–295. New York: ACM Press, May 1987.

45. Kearns, M., and R. E. Schapire. "Efficient Distribution-Free Learning of Probabilistic Concepts." In *31st Annual Symposium on Foundations of Computer Science*, 382–391. IEEE Computer Society Press, 1990.

46. Kiefer, J. *Introduction to Statistical Inference.* New York: Springer-Verlag, 1987.

47. Kolmogorov, A. N., and V. M. Tihomirov. "ε-Entropy and ε-Capacity of Sets in Functional Spaces." *Am. Math. Soc. Translations (Ser. 2)* **17** (1961): 277–364.

48. Kulkarni, S. "On Metric Entropy, Vapnik-Chervonenkis Dimension, and Learnability for a Class of Distributions." Technical Report LIDS-P-1910, Center for Intelligent Control Systems, Massachusetts Institute of Technology, 1989.

49. Kullback, S. *Information Theory and Statistics.* New York: Wiley, 1959.

50. LeCun, Y., J. Denker, and S. Solla. "Optimal Brain Damage." In *Advances in Neural Information Processing Systems*, edited by D. Touretsky, Vol. 2, 589. San Mateo, CA: Morgan Kaufmann, 1990.

51. Lineal, N., Y. Mansour, and R. Rivest. "Results on Learnability and the Vapnik-Chervonenkis Dimension." In *Proceedings of the 1988 Workshop on Computational Learning Theory*, 56–68. San Mateo, CA: Morgan Kaufmann, 1988.

52. Lindley, D. V. "The Present Position in Bayesian Statistics." *Stat. Sci.* **5(1)** (1990): 44–89.

53. Littlestone, N. "Learning Quickly when Irrelevant Attributes Abound: A New Linear-Threshold Algorithm." *Mach. Learning* **2** (1988): 285–318.

54. MacKay, D. "Bayesian Methods for Adaptive Models." Ph.D. Thesis, California Institute of Technology, 1992.

55. Mandelbrot, B. B. *The Fractal Geometry of Nature.* San Francisco: W. H. Freeman, 1982.

56. McCullagh, P., and J. A. Nelder. *Generalized Linear Models.* London: Chapman and Hall, 1989.

57. Mood, J., and C. Darken. "Fast Learning in Networks of Locally-Tuned Processing Units." *Neural Comp.* **1(2)** (1989): 281–294.

58. Narendra, K. S. and M. A. L. Thathachar. *Learning Automata—An Introduction.* Englewood Cliffs, NJ: Prentice-Hall, 1989.

59. Natarajan, B. K. "Learning over Classes of Distributions." In *Proceedings of the 1988 Workshop on Computational Learning Theory,* pages 408–409. San Mateo, CA: Morgan Kaufmann, 1988.

60. Natarajan, B. K., and P. Tadepalli. "Two New Frameworks for Learning." In *Proceedings of the 5th International Conference on Machine Learning,* 402–415. San Mateo, CA: Morgan Kaufmann, 1988.

61. Natarajan, B. K. "Probably-Approximate Learning over Classes of Distributions." Technical Report HPL-SAL-89-29, Hewlett Packard Labs, Palo Alto, CA, 1989.

62. Natarajan, B. K. "Some Results on Learning." Technical Report CMU-RI-TR-89-6, Carnegie-Mellon University, 1989.

63. Nolan, D., and D. Pollard. "U-Processes: Rates of Convergence." *Ann. Stat.* **15(2)** (1987): 780–799.

64. Nobel, A., and A. Dembo. "On Uniform Convergence for Dependent Processes." Technical Report 74, Department of Statistics, Stanford University, 1990.

65. Nowlan, S. "Maximum Likelihood Competitive Learning." In *Advances in Neural Information Processing Systems,* edited by D. Touretsky, Vol. 2, 574–582. San Mateo, CA: Morgan Kaufmann, 1990.

66. Nowlan, S., and G. Hinton. "Soft Weight-Sharing." Technical Report, Department of Computer Science, University of Toronto, 1991.

67. Opper, M., and D. Haussler. "Calculation of the Learning Curve of Bayes Optimal Classification Algorithm for Learning a Perceptron with Noise." In *Computational Learning Theory: Proceedings of the Fourth Annual Workshop,* 75–87. San Mateo, CA: Morgan Kaufmann, 1991.

68. Opper, M., and D. Haussler. "Generalization Performance of Bayes Optimal Classification Algorithm for Learning a Perceptron." *Phys. Rev. Lett.* **66(20)** (1991): 2677–2680.

69. Poggio, T., and F. Girosi. "A Theory of Networks for Approximation and Learning." A.I. Memo No. 1140, Massachusetts Institute of Technology, 1989.

70. Pollard, D. *Convergence of Stochastic Processes.* New York: Springer-Verlag, 1984.

71. Pollard, D. "Rates of Uniform Almost-Sure Convergence for Empirical Processes Indexed by Unbounded Classes of Functions." Manuscript, 1986.

72. Pollard, D. "Empirical Processes: Theory and Applications." In *NSF-CBMS Regional Conference Series in Probability and Statistics*, Vol. 2. Institute of Math. Stat. and Am. Stat. Assoc., 1990.
73. Quiroz, A. J. "Metric Entropy and Learnability." Unpublished manuscript. Universidad Simón Bolívar, Caracas, Venezuela, 1989.
74. Renyi, A. *Probability Theory*. Amsterdam: North Holland, 1970.
75. Rissanen, J. "Stochastic Complexity and Modeling." *Ann. Stat.* **14(3)**: 1080–1100.
76. Rumelhart, D. E., and J. L. McClelland. *Parallel Distributed Processing: Explorations in the Microstructure of Cognition. Volume 1: Foundations*. Cambridge, MA: MIT Press, 1986.
77. Sauer, N. "On the Density of Families of Sets." *J. Comb. Theory (Ser. A)* **13** (1972): 145–147.
78. Shackelford, G., and D. Volper. "Learning k-DNF with Noise in the Attributes." In *Proceedings of the 1988 Workshop on Computational Learning Theory*, 97–103. San Mateo, CA: Morgan Kaufmann, 1988.
79. Simmons, G. F. *Introduction to Topology and Modern Analysis*. New York: McGraw-Hill, 1963.
80. Sloan, R. "Types of Noise in Data for Concept Learning." In *Proceedings 1988 Workshop on Comp. Learning Theory*, 91–96 San Mateo, CA: Morgan Kaufmann, 1988.
81. Sompolinsky, H., N. Tishby, and H. S. Seung. "Learning from Examples in Large Neural Networks." *Phys. Rev. Lett.* **65** (1990): 1683–1686.
82. Talagrand, M. "Sharper Bounds for Gaussian and Empirical Processes." *Ann. Prob.* (to appear).
83. Tishby, N., E. Levin, and S. Solla. "Consistent Inference of Probabilities in Layered Networks: Predictions and Generalizations." In *IJCNN International Joint Conference on Neural Networks*, Vol. II, 403–409. Wash., DC: IEEE, 1989.
84. Touretsky, D. *Advances in Neural Information Processing Systems*, Vol. 1. San Mateo, CA: Morgan Kaufmann, 1989.
85. Touretsky, D. *Advances in Neural Information Processing Systems*, Vol. 2. San Mateo, CA: Morgan Kaufmann, 1990.
86. Valiant, L. G. "A Theory of the Learnable." *Comm. ACM* **27(11)** (1984): 1134–1142.
87. Vapnik, V. N., and A. Ya. Chervonenkis. "On the Uniform Convergence of Relative Frequencies of Events to Their Probabilities." *Theory Prob.* **XVI(2)** (1971): 264–280.
88. Vapnik, V. N. *Estimation of Dependences Based on Empirical Data*. New York: Springer-Verlag, 1982.

89. Vapnik, V. N. "Inductive Principles of the Search for Empirical Dependences (Methods Based on Weak Convergence of Probability Measures)." In *Proceedings of the 2nd Workshop on Computational Learning Theory*. San Mateo, CA: Morgan Kaufmann, 1989.

90. Weigend, A., B. Huberman, and D. Rumelhart. "Predicting the Future: A Connectionist Approach." *Intl. J. Neural Sys.* **1** (1990): 193–209.

91. Weiss, S., and C. Kulikowski. *Computer Systems that Learn*. San Mateo, CA: Morgan Kaufmann, 1991.

92. Welzl, E. "Partition Trees for Triangle Counting and Other Range Search Problems." In *4th ACM Symposium on Computer Geometry*, 23–33. Urbana, IL: ACM Press, 1988.

93. Wenocur, R. S., and R. M. Dudley. "Some Special Vapnik-Chervonenkis Classes." *Discrete Math.* **33** (1981): 313–318.

94. White, H. "Connectionist Nonparametric Regression: Multilayer Feedforward Networks Can Learn Arbitrary Mappings." *Neural Networks* **3** (1990): 535–549.

95. White, H. "Learning in Artificial Neural Networks: A Statistical Perspective." *Neural Comp.* **1(4)** (1990): 425–464.

96. Yamanishi, K. "A Learning Criterion for Stochastic Rules." In *Proceedings of the 3nd Workshop on Computational Learning Theory*, 67–81. San Mateo, CA: Morgan Kaufmann, 1990.

David H. Wolpert
The Santa Fe Institute, 1399 Hyde Park Road, Santa Fe, NM 87501;
e-mail: dhw@santafe.edu

The Relationship Between PAC, the Statistical Physics Framework, the Bayesian Framework, and the VC Framework

This chapter discusses the intimate relationships between the supervised learning frameworks mentioned in the title. In particular, it shows how all those frameworks can be viewed as particular instances of a single overarching formalism. In doing this, many commonly misunderstood aspects of those frameworks are explored. In addition the strengths and weaknesses of those frameworks are compared, and some novel frameworks are suggested (resulting, for example, in a "correction" to the familiar bias-plus-variance formula).

"Give a man a hammer, and every problem looks like a nail."

—Paraphrase of comment by J. Friedman

"Have toolbox will travel."

—Motto for The Handy Hardware Company (now out of business).

"Careful with that axe, Eugene."

—R. Waters

The Mathematics of Generalization, Ed. David Wolpert, SFI Studies in
the Sciences of Complexity, Proc. Vol XX, Addison-Wesley, 1995

---

1. INTRODUCTION
1.1 BACKGROUND

In the last decade several theoretical frameworks for addressing supervised learning
have been discussed in the neural net community. Some of them are represented in
the other papers in this proceedings. This chapter is primarily concerned with four
of those frameworks:

1. PAC,[2,5,6,14,15,23,24,32,38,49]
2. the "statistical physics of supervised learning,"[27,34,35,41,42,46,47,50,63]
3. Bayesian supervised learning,[4,8,10,17,22,26,28,30,33,38,61,58,62] and
4. the VC ("uniform convergence") framework.[2,3,13,14,22,24,27,32,38,51,52]

These four frameworks have very different ancestries. PAC explicitly betrays
its origins in computer science, and SP its origins in condensed matter physics. On
the other hand, the VC framework grew out of probability theory, and Bayesian
analysis has been a part of conventional statistics for centuries. This mixture of
backgrounds means that the frameworks all appear to differ markedly from one
another. The result has been a "tower of Babel" aspect to the theoretical analyses
one finds in the neural net community.

This chapter demonstrates how, with the help of an overarching supervised
learning formalism, it's possible to reconcile these four frameworks, and thereby
depart the tower of Babel. This reconciliation shows that the four frameworks are
just four different instantiations of the extended Bayesian formalism (the aforementioned "overarching supervised learning formalism"). This allows us to see how the
four frameworks are related to one another and clarifies those fundamental features
that do distinguish them. (Such features go far beyond relatively superficial issues
like worse-case vs. average-case analysis.) All of this can then be used to compare
the strengths and weaknesses of the frameworks on a foundational level. The reconciliation also suggests novel frameworks, i.e., novel "instantiations" of the extended
Bayesian formalism (EBF).

THE FOUR DEADLY SINS

One advantage of the EBF is that its structure makes it difficult to indulge in any of
the following (distressingly common) scientific sins found in current presentations
of the four frameworks:

1.2.1 NOT DECLARING WHAT EVENTS ONE'S PROBABILITIES ARE CONDITIONED UPON. I.e., just writing "Pr(event)" rather than "Pr(event | condition)." In its benign form, the presence of this sin in a framework means that the practitioners of that framework are aware of such conditioning events but interested "outsiders" are not. In its more dangerous form, even seasoned practitioners of the framework aren't fully aware of the conditioning events. By forcing all conditions to be explicit, the EBF avoids this sin, and allows one to see what a framework's results really mean, and therefore how (if at all) they can be used in the real world.

1.2.2 CONFUSING RANDOM VARIABLES. For example, confusing whether a distribution being assumed concerns "true," "target" events, or whether it concerns the guesses made by one's learning algorithm. The presence of this sin in a framework makes the precise mathematics of that framework hard to pin down. It also allows one to confuse assumptions that are tautologically true (e.g., "assumptions" concerning one's learning algorithm) with those that are problematic (e.g., assumptions concerning target events).

1.2.3 NOT MAKING ALL ASSUMPTIONS EXPLICIT. For example, the frameworks are often accompanied by the assumption that the distribution over the input space (the "sampling distribution") is statistically independent from the distribution governing the mapping between inputs and outputs (the target distribution). This assumption often simplifies the analysis substantially. However, it is almost never explicitly stated. Moreover, it often does not hold in the real world (see Section 2). Since it requires all distributions to be explicit, the EBF makes it difficult to engage in this kind of sin.

1.2.4 PUTTING THE TECHNIQUE-CART BEFORE THE ISSUE-HORSE. To one degree or another, the frameworks are built about what issues their practitioners can answer—i.e., what techniques their parent fields can bring to bear—rather than about what issues are most important. (See Wolpert[64] and the quotes at the beginning of this chapter.) Often researchers are not explicitly aware of this problem. Especially in combination with the other sins, the result can be a disjunction between the language a researcher uses when talking about their theoretical results and what the associated mathematics actually says. (See the discussion in Ripley's paper[37] for comments on this problem.)

This sin also means that some important supervised learning issues get ignored. As an example, consider the issue of off-training-set behavior. Many introductory supervised learning texts take the view that "the overall objective...is to learn from samples and to generalize to new, *as yet unseen* cases" (italics mine—see Weiss and Kulikowski,[54] for example). Similarly, it is common practice to try to avoid fitting the training set exactly, i.e., to try to avoid "overtraining." One of the major rationales given for this is that if one overtrains, "the resulting (system) is unlikely to classify *additional points* (in the input space) correctly" (italics mine—see Dietterich[15]).

Such language implies—correctly—that one of the major topics of interest in supervised learning is behavior off the training set. There are many reasons for concerning oneself with such behavior:

i. In the low-noise regime, optimal behavior on the training set is determined by look-up table memorization, and the only interesting issues concern off-training-set behavior.

ii. In particular, in that low-noise regime, if one uses a memorizing learning algorithm, then for test sets overlapping with training sets the upper bound on test set error shrinks as the training set grows. If one does not correct for this when comparing behavior for different different training sets (as when investigating learning curves), one is comparing apples and oranges. When there is no noise, correcting for this effect by renormalizing the range of possible errors is equivalent to requiring that test sets and training sets be distinct. (See Section 2.)

iii. In artificial intelligence—one of the primary fields concerned with supervised learning—the emphasis is often exclusively on generalizing to as-yet-unseen examples.

iv. Very often the process generating the training set is not the same as that governing testing. In such scenarios, the usual justification for testing with the same process that generated the training set (and with it the possibility that test sets overlap with training sets) does not apply. An example is provided by the footnote on secondary structure prediction in Wolpert.[64] Say we wish to learn tertiary protein structure from primary structure and then use that to aid drug design. We *already know* what tertiary structure corresponds to the primary structures in the training set. So we will never have those structures in the "test set" (i.e., in the set of nucleotide sequences whose tertiary structure we wish to predict by using the training data). So we will only be interested in off-training-set error.

v. Distinguishing the regime where test examples coincide with the training set from the one where there is no overlap amounts to splitting supervised learning along its natural "cleavage plane." Since behavior can be radically different in the two regimes, it is hard to see why one would not want to distinguish them.

None of this means that one should never test with the same process that generated the training set. Rather it means that off-training-set testing is an issue of major importance which warrants scrutiny. However, the four frameworks usually require that the test set can overlap with the training set. (They do this by having testing governed by an independent identically distributed (iid) rerunning of the process which generated the training set.) Therefore they mix together behavior on already-seen examples with that for not-already-seen examples, and accordingly cannot distinguish the two kinds of behavior.[1]

The primary reason that the four frameworks allow the test set to overlap with the training set is that much of their research has been tool-driven rather than issue-driven, and the tools that have been used are ill-suited for investigating off-training-set behavior. In fact, often the four frameworks use language that implies that their goal is understanding off-training-set behavior, even when they use a test set that can overlap with the training set. For example, in a paper by Blumer et al.,[5] in the context of noise-free supervised learning, we read that "the real value of a scientific explanation lies not in its ability to explain [what one has already seen], but in predicting events that have yet to occur," despite the fact that the subsequent analysis allows test sets to overlap with training sets.

It should not be too surprising that these four sins are common. They reflect the fact that the frameworks are, to a degree, applications to supervised learning of paradigms *developed for different fields*. As such, they carry with them the entire cultural baggage—implicit assumptions and all—of the fields in which they originated. Such assumptions tend to be "background" in those originating fields, so it is natural that they are also background—though perhaps no longer so innocuous— when practitioners of those fields cross over to supervised learning.

As an aside, these four sins concerning theoretical learning often have applied learning analogues. An important example is the manipulation of parametrizations of the object of interest with little concern for how those manipulations affect the object of interest itself. For example, many researchers impose priors on neural net weights, implement a bias in favor of fewer hidden neurons, etc. Often, for lack of an alternative, they do this without taking into account the ultimate effect on the direct object of interest, the input-output functions parametrized by those weights. (See Wolpert[58] and Neal.[33]) One advantage of the EBF is that it is nonparametric and, therefore, helps focus attention directly on the object of interest rather than on parametrizations of that object.

1.3 OVERVIEW OF THIS CHAPTER

In Section 2 of this chapter I present the EBF. In Section 3 I present the "no-free-lunch theorems" of the EBF (phrase due to D. Haussler). These are theorems that bound how much one can infer concerning the generalization error probability distribution before one has to make relatively strong assumptions concerning the real world. They serve as a broad context in which one should view the claims of any supervised learning framework. In Section 4 I present a cursory overview

[1]There is a popular heuristic argument that iid testing error and off-training-set testing error must become identical when the size of the input space grows very large. If true, this would mean that one can ignore the distinction between the two kinds of error. Unfortunately, this heuristic is often wrong. See, for example, the discussion below on the SP framework.

of the reconciliation of the four frameworks. Although not coming at the end of this chapter, in a certain sense this section can be viewed as the "conclusion" of the chapter. In Sections 5 through 8 I go into more detail, describing each of the four frameworks in turn in terms of the EBF. For expository simplicity, only "vanilla" versions of the frameworks are considered. (The purpose of this chapter is to illuminate the foundational structure of the four frameworks, not to provide an exhaustive overview of them.) In Section 9 I discuss "filling in the gaps" between the frameworks; i.e., I discuss some novel frameworks that, from the perspective of the EBF, are natural extensions of the four frameworks discussed in this chapter. Some interesting results arising in such extensions are presented (e.g., a "correction" to the familiar bias-plus-variance formula). Finally, in Section 10 I discuss the relative strengths and weaknesses of the four frameworks from the perspective of trying to use them to help real-world supervised learning.

A reader wishing to simply skim the results of this chapter should read part (i) of Section 2, and then Sections 3 and 4. A deeper skim might add in Sections 9 and 10. However, to fully understand the discussion in those last sections it is probably necessary to have at least glanced through Sections 5 through 8.

2. THE EXTENDED BAYESIAN FORMALISM

2.1 INTRODUCTION

Intuitively, the EBF is just the conventional Bayesian supervised learning framework, extended to add one extra random variable. In addition to "costs" or "generalization errors" c, and "target" input-output relationships f, from which are produced m-element "training sets" d, there are also hypotheses h. These are the outputs of one's learning algorithm, made in response to d. (Loosely speaking, h can be viewed as the algorithm's guess for f.) The EBF[56,58] is conventional probability theory applied to the space of quadruples $\{h, f, d, c\}$.

It is the inclusion of h in the space that allows the EBF to go beyond the conventional Bayesian supervised learning framework; without h, the EBF could not encompass the non-Bayesian frameworks of PAC, SP, and the VC framework. To understand this, note that one's learning algorithm (or "generalizer") is given by the conditional probability distribution $P(h|d)$. There is no direct analogue to $P(h|d)$ in the conventional Bayesian framework. In particular, the "likelihood function" of the Bayesian framework, which gives the data generation process, is $P(d|f)$; the "posterior distribution" referred to in that framework usually means $P(f|d)$; and the "prior" referred to in that framework usually means $P(f)$. (More generally, the terms "prior" and "posterior" mean not-conditioned and data-conditioned, respectively.) Viewed another way, the conventional Bayesian framework has $P(h|d)$ pre-fixed, to be the "Bayes-optimal" $P(h|d)$ associated with $P(f|d)$ (see Section 5).

No allowance is made for $P(h|d)$'s like those considered in the other frameworks (and found in the real world) that are not Bayes-optimal for any $P(f|d)$.

The EBF itself imposes no restrictions on h and f; all such restrictions are imposed by $P(f, h, d, c)$. For example, if one's algorithm always makes guesses lying in a class H, then $P(h) = 0$ for all $h \notin H$. Similarly, if one assumes that the truths come from some class F, then $P(f) = 0$ for all $f \notin F$. $P(h|d)$ can be "deterministic"—i.e., always guess the same h for the same d (as in a nearest neighbor algorithm)—or "stochastic"—i.e., potentially guess different h's even when d is fixed (as in backpropagation with a random initial weight).

The rest of this section presents a formal definition of the EBF. Those whose eyes turn glassy at such formal text are encouraged to skip to Section 3, referring back to the figure at the end of this section as needed.

2.2 FORMAL DEFINITION OF THE EBF

- In this chapter, random variables are denoted by capital letters, and instances of random variable with corresponding lower-case letters. For the purposes of this chapter, there is no reason to be concerned with quasi-philosophical distinctions between random variables and "parameters." (See the discussion in Wolpert[61] on prior information.)

Whenever possible, "P" notation will be used: the arguments of a "P" indicates if it is a probability or a density or a mixture of the two and, if it involves densities, what random variables they are over. When more precision is required, "$\Pr(A)$" will be used to indicate the probability of the event A, and lowercase "p" will be used to indicate a probability density. (So $P(z) = p_Z(z)$ is the probability density of random variable Z, evaluated at value z.) The notation "$E(Z|a)$" is defined to mean the expectation value $E(Z|A = a) = \int dz\, z\, P(z|a)$ (the integral being replaced by a sum if that is appropriate).

- The input space X has n elements, and the output space Y has r elements, where both n and r are countable (though perhaps infinite). Such discreteness of the spaces does not amount to an undue restriction, since it always holds in the real world, where measuring devices have finite precision and where the computers used to emulate learning algorithms are finite state machines. (Note also that our training and test sets will always be finite, and often we can restrict X to be just the input space values found in those sets.)

- The training set d consists of m ordered pairs of inputs and output values, $\{d_X(i), d_Y(i)\} : 1 \leq i \leq m$. The number of distinct values in d_X is indicated by m'.

- Let "f" be a function giving the probability of $y \in Y$ conditioned on $x \in X$. This is indicated by writing $P(y|x, f) = f_{x,y}$. (Note that f is a vector of real numbers, with components $f_{x,y}$.) When extra precision is required, "F" will indicate the random variable of which "f" is an instantiation. So, for example, $p_{Y|X,F}(y|x, f) = f_{x,y}$.

The variable "f" labels the "true" or "target" conditional distribution of y given x, in that training set output values are generated according to f. For the purposes of this chapter, this means that $P(d_Y|d_X, f) = \Pi_{i=1}^m f_{d_X(i),d_Y(i)}$. (Note that this equation need not fix anything concerning test set generation.)

- Let "g" be a function giving the probability of $x \in X$. This is indicated by writing $P(x|g) = g_x$. (Note that g is a vector of real numbers, with components g_x.) Training set input values are generated according to g: $P(d_X|g) = \Pi_i g_{d_X(i)}$. (This means that repeats are allowed in d_X.) In this chapter it is stipulated that $P(d|f, g)$ (which equals $P(d_Y|d_X, f, g)P(d_X|f, g)$) is equivalent to $P(d_Y|d_X, f)P(d_X|g)$. This can be viewed as part of the definition of f and g.

In addition to such definitional requirements, here it will be convenient if certain other properties of g are assumed. In particular, it is assumed that g and f are statistically independent. This means that

$$P(d|f) = \int dg\, P(d|f, g)P(g|f)$$

$$= \int dg\, P(d|f, g)P(g)$$

$$= P(d_Y|d_X, f) \int dg\, P(d_X|g)P(g)$$

$$= P(d_Y|d_X, f) \int dg\, \Pi_i g_{d_X(i)} P(g).$$

It is also assumed that $P(g)$ is a delta function about some "sampling" distribution $\pi(x)$. These assumptions mean that $P(d|f) = \Pi_i \pi(d_X(i)) f_{d_X(i),d_Y(i)}$, and the variable g can henceforth be ignored.

These assumptions concerning g are often unrealistic. Usually there is *some* coupling between f and g, and sometimes there is a lot. (See, for example, the footnote in Wolpert[64] concerning the secondary structure prediction problem.) Moreover, any techniques that try to use unsupervised learning to aid supervised learning (e.g., decision-directed learning—see Duda and Hart[17]) implicitly assume that g and f are coupled. However, aside from the VC framework, all four of the (vanilla versions of the) frameworks discussed in this chapter implicitly assume that g and f are independent. That is why that assumption is adopted here.[2]

- The fact that they are themselves distributions does not forbid either f or g from being arguments of probability distributions. For example, it is perfectly meaningful to write $P(f|d) = P(d|f)P(f)/P(d)$. In this way new data can update our estimation of what distribution generated that data.

- Let $h_{x,y}$ be the x-conditioned probability distribution over values y which is produced by our learning algorithm in response to d. Sometimes our algorithm's

---

[2]See Wolpert[59] for a discussion of the ramifications of the assumption that g is independent of f. For the views of conventional statistics on this issue, see also Titterington[48] and, in particular, the Dawid references therein concerning the "predictive" paradigm and the "diagnostic" paradigm.

"output" is a quantity based on h (e.g., our algorithm might produce a decision of some sort based on h), and sometimes h itself is the output of our algorithm. $P(h|d)$ is the rule for how hypotheses are produced from training sets and is known as a "generalizer." Examples are back-propagation applied to neural nets and memory-based (i.e., nearest neighbor) reasoners.

If our algorithm's output is a guessed function from X to Y (rather than a guessed distribution), then we can view that output as an h where $h_{x,y}$ is of the form $\delta(y, \eta(x))$ for some function $\eta(.)$ parametrized by d ($\delta(.,.)$ being the Kronecker delta function). Such an h is "single valued." As examples, schemes like nearest-neighbor classifiers, and "Bayes-optimal" classifiers are deterministic and produce single-valued h. This should be contrasted with schemes like softmax applied to neural nets,[7] in which the net gives a mapping from X to a distribution over Y, rather than from X to Y. (See Appendix 2 in Wolpert.[58]) If a generalizer produces a single-valued h that goes through all the elements of d, that generalizer is said to "reproduce" d.

- Unless explicitly stated otherwise, it will be assumed in this chapter that our algorithm produces single-valued h's. (However, it is *not* necessarily assumed that the algorithm is deterministic. See the section below on the SP framework.) It is similarly assumed that any f is of the form $\delta(y, \phi(x))$ for some single-valued function $\phi(x)$ from X to Y. (Or equivalently, it is assumed that $P(f)$ equals zero for any other f.) Note that due to the form of $P(d|f)$, this assumption is equivalent to requiring that the training set is generated without any noise. These two assumptions mean that, as restricted in this chapter, the EBF is not capable of addressing problems where the "target" is a (non-delta function) distribution, and you wish to guess that target, so h, too, is a distribution.

As with the assumptions concerning g, these assumptions concerning f and h are made to comply with the implicit assumptions in the vanilla versions of the four frameworks examined in this chapter. It is straightforward to use the EBF when none of these assumptions hold. However doing so introduces extra mathematics which obscures the underlying issues.

To simplify notation, without loss of generality, from now on I will use the symbols "h" and "f" to refer to single-valued functions (i.e., I will rewrite $\eta(x)$ as $h(x)$, and $\phi(x)$ as $f(x)$). It will be convenient to write $P(d_Y|d_X, f)$ as $\delta(d \subset f)$, i.e., $\delta(d \subset f) = 1$ if d lies on f, 0 otherwise. The notation is motivated by viewing f as a set of X–Y pairs, just like d. (Note though that repeats are allowed in d but not in f. So $\delta(d \subset f)$ can equal 1 even if the set d is not, formally speaking, contained in the set f.)

- One crucial stipulation—again, one that is adhered to in all four frameworks—is that the guess the learning algorithm makes depends only on d. This means that $P(h|f, d) = P(h|d)$; if d is held fixed but f changes, the learning algorithm behaves the same. (Note that the learning algorithm can be based on assumptions concerning f. But those are embodied in $P(h|d)$, and do not change if f changes.) An immediate corollary is that $P(f|h, d) = P(f|d)$. Note the symmetry between h and f.

• In this chapter only algorithms that work exclusively with full input-output (i-o) pairs are considered. As an alternative, one could have an algorithm that can try to learn even if some of the data is unlabeled or label-only (i.e., if we have a $d_X(i)$ without a corresponding $d_Y(i)$ or vice versa). For such cases, one must introduce a new random variable that specifies which of the elements of the data set consist of a full i-o pair, just an input, or just an output. Without such a variable there is ambiguity in the notation: $P(h|d_X)$ could either mean the distribution over h's when the generalizer tries to learn from the inputs-only values in d_X, or it could mean the average over d_Y of the distribution when the generalizer tries to generalize from a set of input-output pairs, $\{d_X, d_Y\}$ (i.e., $\sum_{d_Y} P(d_Y|d_X)P(h|d_X, d_Y)$).

• Now define a real-world random variable C which represents the "loss" (or what in some circumstances is called "cost" or "utility" or "value") associated with a particular f and h. Intuitively, C represents the real-world implications of a particular use of a learning algorithm. Formally, its meaning is set by the distribution $P(c|h, f, d)$. That distribution reflects how "test sets" are generated from f (and in particular whether it is in the same manner that training sets are generated), how big test sets are, how h is mapped (stochastically or otherwise) to a "decision" or an "action," how such a decision is combined with a test set to generate a real-world loss, etc. For current purposes, all of these kinds of details are irrelevant—only the final distribution $P(c|h, f, d)$ matters. This is similar to the fact that the only aspect of the learning algorithm that matters is the distribution $P(h|d)$. (The way a particular $P(h|d)$ is implemented—through a gradient descent, stochastic sampling, nearest neighbor rule, or whatever—is irrelevant.)

The "generalization error function" much discussed in the four frameworks is the expectation value $E(C|h, f, d)$. For example, when one is interested in "average misclassification rate error," one might have $E(C|h, f, d) = E(C|h, f) = \sum_x \pi(x)[1 - \delta(f(x), h(x))]$. This is the average (according to $\pi(x)$) number of times across X that h and f differ.

In this chapter it is assumed that we are interested in misclassification rates, and that $P(c|h, f, d)$ takes one of two forms. Formally, with $C = Er(F, H, D)$ (so that $P(c|f, h, d) = \delta[c, Er(f, h, d)]$), either C is independent of D and is given by

$$C = Er(F, H, D) = \sum_x \pi(x)[1 - \delta(F(x), H(x))] \text{ ("iid error")},$$

or C depends on D and is given by

$$C = Er(F, H, D) = \frac{\sum_{X \notin D_X} \pi(x)[1 - \delta(F(x), H(x))]}{\sum_{X \notin D_X} \pi(x)} \text{ ("off-training-set error")}.$$

Since $C = Er(F, H, D), E(C|f, h, d) = E(C|F = f, H = h, D = d) = Er(f, h, d)$, and $P(c \text{ obeys } property|stuff) = P(f, h, d \text{ such that } Er(f, h, d) \text{ obeys } property| stuff)$, $Er(f, h, d)$ will be called the "error function." Note that for any generalizer that reproduces the training set the off-training-set C is simply the iid C, renormalized so that the maximal value (over all f and h such that both $P(f|d)$ and $P(h|d)$ are nonzero) is always 1.

One should not confuse the error function with the "error surface" found in techniques like backpropagation. In the standard Bayesian formulation of backpropagation,[10,58] the error surface is (the log of) $P(\mathbf{w}|d)$, where $\mathbf{w}$ is the weight vector parametrizing f. So, for example, the term in that error surface that equals the squared error on the training set simply reflects the assumption that $P(d|f)$ is created with Gaussian noise. That squared error need have nothing to do with $Er(f, h, d)$, even if $Er(f, h, d)$ is quadratic in $(f - h)$ (unlike the misclassification rate error functions analyzed in this chapter).

- Although this chapter restricts itself to the noise-free scenario, it is worth briefly pointing out some of the subtleties involved with noise. Usually the best way to allow for noise is to have f be a non-delta-function distribution over Y. However, it is common to instead adopt a "function + noise" scenario. In this scenario, f is still a single-valued function, but now $P(d|f)$ reflects the process of adding noise to f to create d (i.e., $P(d|f)$ can be nonzero even if d does not lie on f). If one uses this scenario, care should be exercised in the choice of the error function. In particular, often we are interested in whether h agrees with a sample of f where that sample is created *with the noise process*. (This is usually the case when we measure performance with a "test set," for example.) In general, this differs from whether h agrees with a noise-free sample of f, and the definitions given above for C should be adjusted accordingly.

- All that is necessary for the EBF to be an appropriate formalism for a particular problem is that the cost random variable only depends on the random variables f, h, and/or d. In some scenarios—none of which are considered in this chapter—this is not the case, and it is appropriate to modify the EBF by adding some other variables to the space (e.g., g, or a hyperparameter). In other scenarios, although the standard EBF might suffice, a slight modification is more appropriate. (For example, if one is investigating the use of cross validation over a fixed set of generalizers it makes sense to have one of the random variables in the EBF be the generalizer one chooses. See Wolpert.[56]) The underlying feature that unites all such variations of the EBF is that the space being analyzed includes hypotheses as well as targets.

- All of PAC, all of SP, all of the VC framework, and all of Bayesian supervised learning can be cast in terms of the EBF. In particular, the "vanilla" versions of all four of the frameworks investigated in this chapter can be defined in terms of the abridged version of the EBF shown in Figure 1.

---

1. n and r are the number of elements in the input and output spaces, X and Y, respectively.
2. m is the number of elements in the (ordered) training set d. $\{d_X(i), d_Y(i)\}$ is the corresponding set of m input and output values. m' is the number of distinct values in d_X.
3. Outputs h of the learning algorithm are always assumed to be of the form of a function from X to Y, indicated by $h(x \in X)$. Any restrictions on h are imposed by $P(f, h, d, c)$.
4. The learning algorithm is given by $P(h|d)$. It is "deterministic" if the same d always gives the same h.
5. "Targets" f are always assumed to be of the form of a function from X to Y, indicated by $f(x \in X)$. Any restrictions on f are imposed by $P(f, h, d, c)$.
6. The "likelihood" is $P(d|f) = \delta(d \subset f)\Pi_i \pi(d_X(i))$, where "$\delta(d \subset f)$" equals 1 if d lies completely on f, 0 otherwise, and $\pi(x)$ is the "sampling distribution."
7. The "posterior" is $P(f|d)$. In this chapter, probability is *not* interpreted to mean "degree of personal belief," as some conventional Bayesians define it. Accordingly, it is not true that the researcher automatically knows $P(f|d)$. See Wolpert.[61]
8. $P(h|f, d) = P(h|d)$, $P(f|h, d) = P(f|d)$, and therefore $P(h, f|d) = P(h|d)P(f|d)$.
9. The cost c associated with a particular h and f is either given by $Er(f, h, d) = \sum_x \pi(x)[1 - \delta(f(x), h(x))]$ ("iid error function"), or by the "off-training-set error" function, $Er(h, f, d) = \sum_{x \notin d_X} \pi(x)[1 - \delta(f(x), h(x))]/\sum_{x \notin d_X} \pi(x)$.
10. The empirical misclassification rate $s \equiv \sum_{i=1}^{m}\{1 - \delta[h(d_X(i)), d_Y(i)]\}/m$.

FIGURE 1 Abridged version of the EBF.

• In what follows, it is implicitly understood that assumptions like the exact absence of noise are replaced by the presence of infinitesimal noise. More precisely, it is implicit that no distributions ever equal zero *exactly*, although they might be arbitrarily close to zero. This is to ensure that we will never divide by zero in evaluating conditional probabilities.

3. THE NO-FREE-LUNCH THEOREMS

3.1 PRESENTATION OF THE THEOREMS

The theorems presented in this section bound how well a learning algorithm can be assured of performing in the absence of assumptions concerning the real world. For the sake of space, no proofs that appear in other papers are presented. The interested reader is referred to Wolpert.[56,66]

THEOREM 1. $E(C|d)$ can be written as a (non-Euclidean) inner product between the distributions $P(h|d)$ and $P(f|d)$: $E(C|d) = \sum_{h,f} Er(h, f, d)P(h|d)P(f|d)$.

(Similar results hold for $E(C|m)$, etc.)

Theorem 1 says that how well you do is determined by how "aligned" your learning algorithm $P(h|d)$ is with the actual posterior, $P(f|d)$. This allows one to ask questions like "for what set of posteriors is algorithm G_1 better than algorithm G_2?" It also means that, unless one can somehow prove (!), from first principles, that $P(f|d)$ has a certain form, one cannot prove that a particular $P(h|d)$ will be aligned with $P(f|d)$ and, therefore, one cannot prove anything concerning how well that learning algorithm generalizes.

This impossibility of first-principles proofs can be formalized in a number of ways. One of them is as follows:

THEOREM 2. Consider the off-training-set error function. Let "$E_i(.)$" indicate an expectation value evaluated using learning algorithm i. Then for any two learning algorithms $P_1(h|d)$ and $P_2(h|d)$, independent of the sampling distribution,

i. Uniformly averaged over all f, $E_1(C|f, m) - E_2(C|f, m) = 0$;

ii. Uniformly averaged over all f, for any training set d, $E_1(C|f, d) - E_2(C|f, d) = 0$;

iii. Uniformly averaged over all $P(f)$, $E_1(C|m) - E_2(C|m) = 0$;

iv. Uniformly averaged over all $P(f)$, for any training set d, $E_1(C|d) - E_2(C|d) = 0$.

In other words, by any of the measures $E(C|d)$, $E(C|m)$, $E(C|f, d)$, or $E(C|f, m)$ (all generically known as "risks"), all algorithms are equivalent, on average.[3] Or to put it another way, for any two learning algorithms, there are just

---

[3]Note that one could argue that 2.i, for example, is misleading; different f's will have different probabilities in the real world, so a flat average over all f's is in some sense inappropriate. To correct this "misleading" nature of 2.i we are lead to consider averaging over all f according to a $P(f)$ which need not be uniform. Such an average equals $E(C|m)$. Now one can always construct a $P(f)$ to argue in favor of any particular learning algorithm. However, in almost all real-world supervised learning, we do not know $P(f)$. So we are led to ask if there is a $P(f)$ such that $E(C|m)$

as many situations (appropriately weighted) in which algorithm one is superior to algorithm two as vice versa, according to any of the measures of "superiority" in Theorem 2.

3.2 EXAMPLES

As an example, an algorithm that uses cross validation to choose amongst a pre-fixed set of learning algorithms does no better on average than one that does not. (However, since cross validation can only be viewed as a $P(h|d)$ if it is used to choose amongst a pre-fixed set of learning algorithms, Theorem 2 says nothing about cross validation "in general," when the set of generalizers is not pre-fixed. See also Schaffer.[39]) As another example of the no-free-lunch theorems, assume you are a Bayesian, and calculate the Bayes-optimal guess assuming a particular $P(f)$. (For example, you use the $P(h|d)$ that minimizes the data-conditioned risk $E(C|d)$, given your assumed $P(f)$.) You now compare your guess to that made by someone who uses a non-Bayesian method. Then 2.iv means (loosely speaking) that there are as many actual priors in which the other person has a lower data-conditioned risk as there are for which your risk is lower. Another set of examples is provided by all the heuristics that people have come up with for supervised learning: avoid "over-fitting," prefer "simpler" to more "complex" models, etc. Theorem 2 says that all such heuristics fail as often as they succeed.

Another example of Theorem 2 is given by the case where our learning algorithm is deterministic, and we have a particular training set d so the risk of interest is $E(C|d)$. The empirical misclassification rate s is fixed by d, since our algorithm takes d and gives h, which together with d gives s. Accordingly, $E(C|s,d) = E(C|d)$. Now assume that for our d, s happens to be very small (i.e., h and f agree almost always across the elements d_X). Assume further that our learning algorithm has a very low VC dimension. Since s is low, we might hope that that low VC dimension confers some assurance that our generalization error will be low. (This is one common way people try to interpret the VC theorems.) However, according to 2.iv, low s and low VC dimension provide no such assurances concerning off-training-set error. Either given d or (equivalently) given both d and s, no advantage is conferred as far as

---

is lower for algorithm one, and/or if there is also a $P(f)$ such that according to $E(C|m)$ algorithm two is superior. But this question has already been answered—in the affirmative—by 2.iii. In fact, given that we do not know $P(f)$, the obvious thing to do, if one wishes to compare two algorithms with the measure $E(C|m)$, is compare their averages over all $P(f)$—which 2.iii tells us makes all learning algorithms just as good as one another. Now one could try to "jump a level" yet again, and argue that some $P(f)$ are "more likely" than others, so one should not perform a flat average over all $P(f)$, etc. But the math responds the same way as it did to the objection to 2.i—in response to this new objection, one constructs new questions concerning probability distributions across the set of $P(f)$'s, questions whose answers again state that all algorithms perform the same, in the absence of information about the problem suggesting otherwise.

off-training-set behavior is concerned if s is low and one's algorithm happens to have low VC dimension.[4]

So all learning algorithms are the same in that: (1) by several definitions of "average," all algorithms have the same average off-training-set misclassification risk and, therefore, (2) no learning algorithm can have lower risk than another one for all f, for all $P(f)$, for all f and d, or for all $P(f)$ and d. However, learning algorithms can differ in that: (1) for particular (nonuniform) $P(f)$, different algorithms have different data-conditioned risk (and similarly for other kinds of risk), and (2) for some algorithms there is a distribution-conditioning quantity (e.g., f) for which that algorithm is optimal (i.e., for which that algorithm beats all other algorithms), but some algorithms are not optimal for any value of such a quantity; and, more generally, (3) for some pairs of algorithms the no-free-lunch theorems are met by having comparitively many cases in which algorithm A is just slightly worse than algorithm B, and a few cases in which algorithm A beats algorithm B by a lot.

It is interesting to speculate about the possible implications of point (3) for cross validation. Consider two algorithms α and β. α is identical to algorithm A, and β works by using cross validation to choose between A and B. α and β must have the same expected error, on average. However, the following might be the case for many choices of A, B, $\pi(x)$, etc.: For most situations (i.e., most f or $P(f)$, depending on which of Theorem 2's averages is being examined) A and B have approximately the same expected off-training-set error, but β usually chooses the worse of the two, so in such situations the expected cost of β is (slighly) worse than that of α. In those comparitively few situations where A and B have significantly different expected off-training-set error, β might correctly choose between them, so the expected cost of β is significantly better than that of α for such situations. In other words, it might be a common case that when asked to choose between two generalizers in a situation where they have comparable expected cost, cross validation usually fails, but in those situations where the generalizers have significantly different costs, cross validation successfully chooses the better of the two. In such a case, cross validation still has the same average off-training-set behavior as any other algorithm. And there are actually more situations in which it fails than in which it succeeds. However, in such a case, cross validation has desirable minimax behavior. (It's important to note though that one can explicitly construct cases where cross validation does not have this desirable minimax behavior. See Section 8.)

---

[4] This result is reconciled with the usual VC theorems in the VC section below. As an aside, it should be mentioned that the *only* reason this result might appear to be at odds with the VC theorems is because the usual statements of those theorems are guilty of sin number one from the introduction—the conditioning event is not specified. (In particular, Vapnik's seminal book on the subject[51] is, unfortunately, guilty of this sin.) Accordingly, it is not immediately clear to the first-time reader of the VC theorems that they do not concern any of the conditioning events discussed in Theorem 2. (It should also be noted that the VC theorems concern iid error, not off-training-set error.)

3.3 VARIANTS OF THEOREM 2

All of this applies to more than just off-training-set error. In general iid error can be expressed as a linear combination of off-training-set error plus on-training set error, where the combination coefficients depend only on d_X and $\pi(x \in d_X)$. So generically, if two algorithms have the same on-training-set behavior (e.g., they reproduce d exactly), the no-free-lunch theorems apply to their iid errors as well as their off-training-set errors.

In addition, there are a number of variants of Theorem 2, for example, dealing with noisy training sets, other conditional distributions (like $E(C|s, \{$other quantities$\}))$, etc. For the sake of space they are not detailed here; the purpose of this section is only to present a sample of the no-free-lunch theorems, sufficient to provide a context for scrutinizing the results of the four frameworks.

3.4 IMPLICATIONS OF THEOREM 2 FOR THE USE OF "TEST SETS"

Theorem 2 also has implications for the (very common) use of a single test set to estimate c. Consider splitting d into two parts, d_1 and d_2. Training is done on d_1, and d_2 is a "test set" used to measure the resultant performance. (Note that since there can be duplicates in d, d_1 and d_2 might share input-output pairs.) The simplest situation to set up is where our error function runs over both $(d_1)_X$ and $(d_2)_X$—in this simple-minded version of things, no attention is being paid to off-training-set considerations, and the no-free-lunch theorems do not apply.

As an alternative, consider the off-training-set $Er(f, h, d)$, where "off-training-set" means off all of d. Now the no-free-lunch theorems apply; they tell us that behavior on d, which includes behavior on d_2, can tell us nothing about c, on average. (If this were not the case, behavior on d could be used to successfuly choose between competing algorithms.) The implication is that as far as off-d behavior is concerned, the most common procedure used for evaluating algorithms—examining their behavior on test sets—fails as often as it succeeds, on average. (Although, in general the minimax properties of this procedure might not be so poor—see the discussion above on cross validation's minimax behavior.)

On the other hand, assume that once d_1 is fixed d_2 is set to be the remaining pairs in d that are not also found in d_1. One might view such a d_2 as "off-training-set" in the sense of having no overlap with d_1. (Which definition of off-training-set is appropriate depends on which of the reasons listed in Section 1 for being interested in off-training-set behavior applies.) If we adopt this definition, then it makes sense to redefine the error function to be the off-training-set error function for this training set d_1 (rather than for all of d). With this redefinition the error function runs over $(d_2)_X$ as well as $X - d_X$, and the no-free-lunch theorems do not apply; behavior on d_2 now can tell us something about the likely c value.

Indeed, for this scenario we might have d_2 be an iid sample of a process which, if infinitely repeated, gives our error function. We could then apply the usual variants of the central limit theorem to derive a confidence interval bounding the likely difference between empirical error on d_2 and the value of c. As with all confidence intervals though, this one comes with the major caveat that it does not directly give us what we want, which in this case is $P(\text{error on } d_2)$. See Sections 8 and 10. (These points concerning test sets grew out of a conversation with Manny Knill.)

3.5 INTUITION BEHIND THEOREM 2

The results presented above do not mean that the technique of cross validation does not work, or that the technique of using test sets to estimate error off of test sets does not work, or the like. Rather they mean that one can not formally justify these techniques (as far as expected off-training set error is concerned) without making assumptions. More practically, the results mean that if you are interested in off-training-set behavior, then using such a technique amounts to an assumption that $P(f)$ is not "typical," as measured by a uniform distribution over $P(f)$ (or that f is not typical as measured by a uniform distribution over f, or what have you.) As with all other assumptions, the validity of this one will vary from case to case.

Intuitively, it is not hard to see why an assumption must be implicit in techniques like cross validation. Consider the case where $P(f)$ is fixed and uniform over all f, and we are concerned with $E(C|d)$ for some particular d. Since $P(f)$ is uniform over all f, all f agreeing with d are equally probable. Accordingly, all possible patterns of f values outside of the training set are equally probable; the off-training-set world is essentially random. This means that building into a learning algorithm a preference for some particular outside-the-training-set pattern will not gain you anything; all algorithms are equal as far as off-training-set behavior is concerned when $P(f)$ is uniform.

To complete the intuitive justification for Theorem 2, note that since X and Y are finite, so is the set of all f's, and therefore $P(f)$ is a finite-dimensional real-valued vector living on the unit simplex. Accordingly, uniformly averaging all $P(f)$ results in a vector on that simplex all of whose components are equal—the uniform $P(f)$. So uniformly averaging all $P(f)$ (or uniformly averaging all f) is essentially equivalent to having a uniform $P(f)$. (The actual proof of Theorem 2 is a bit more complicated than this because we're not interested directly in the average of $P(f)$ but rather in the average of a distribution conditioned on $P(f)$ together with some quantity statistically coupled to $P(f)$. But the basic idea is the same.)

When considering things from the perspective of this particular argument, one should bear in mind that for $Y = \{0, 1\}$ and n large, uniform $P(f)$ (or equivalently a uniform prior over $P(f)$) means you are unlikely to find an f for which the proportion of all x such that $f(x) = 1$ differs much from .5. One might wish to instead have something like a uniform probability over the proportion of 1's in f (rather than over f's directly). For such a case, the f of all 1's is more probable than any particular f having both 1's and 0's in its outputs. A direct result is that if the training set has all its output values equal to 1, then the posterior favors the f having all 1's off the training set over the one having all 0's off the training set. (More generally, this is a situation that favors having more 1's off the training set than 0's.) In other words, in such a case we would have an automatic coupling between on- and off-training-set behavior. (Note though that Theorem 2 means that there also cases in which we have an automatic "anti-coupling" between on- and off-training-set behavior.) Similar arguments follow from the fact that for any fixed h, you are unlikely to find an f for which the number of x such that $f(x) \neq h(x)$ differs much from .5.

Another intuitive justification for the no-free-lunch theorems is based on viewing supervised learning "in reverse." Conventionally one views supervised learning as a process whereby $P(f)$ is sampled to set f, which is then sampled to get d, which is then used to get h. Viewed this way, it might seem odd that an h that agrees with d has no a priori correlation with f off of d. However, instead view the process as starting with a d, fitting h to it, and then considering all f going through those d. From this point of view, there is no reason at all to believe that h and f agree off of d.

As a final example of intuitive arguments supporting Theorem 2, simply note that it's very difficult to see how you could infer anything substantive about the likely c that accompanies use of a particular learning algorithm, unless you make an assumption for $P(c, f, d)$. And if you do make an assumption for $P(c, f, d)$, but it only concerns on-training set behavior (e.g., a noise-model), it's very difficult to see how you could infer anything substantive about the likely off-training-set c that accompanies use of a particular learning algorithm.

As these intuitive arguments suggest, there are many other aspects of off-training-set error which, although not actually no-free-lunch theorems, can nonetheless be surprising to those used to iid error. An example is the proof in Appendix 1 that in certain situations the expected off-training-set error grows as the size of the training set increases, even if one uses the best possible learning algorithm, the Bayes-optimal learning algorithm (i.e., the learning algorithm that minimizes $E(C|d)$—see the section below on the Bayesian framework). In other words, sometimes the more data you have, the less you know about the off-training-set behavior of f, on average.

3.6 ERROR FUNCTIONS OTHER THAN THE MISCLASSIFICATION RATE

Finally, it should be pointed out that things are a bit messier when error functions other than the misclassification rate are considered. (In that the vanilla versions of the four frameworks do not consider such error functions, such error functions are not considered in this chapter.) In particular, if the error function induces a geometrical structure over Y, then we can have a priori distinctions between learning algorithms.

An example is the case of quadratic error functions. For such functions, everything else being equal, an algorithm whose guessed Y values lie away from the boundaries of Y is to be preferred over an algorithm that guesses near the boundaries.[5] In addition, for such an error function, guessing an h that equals the Y-average of a stochastic algorithm's guess can never increase expected error beyond that of the original algorithm. Phrased differently, for a quadratic error function, given a series of experiments and a set of deterministic generalizers G_i, it is always preferable to use the average generalizer $G' \equiv \sum_i G_i / \sum_i 1$ for all such experiments rather than randomly to choose a new member of the G_i to use for each experiment. Intuitively, this is because such an average reduces variance without changing bias (see Wolpert[65] and Perrone[36]). (Note though that this in no way implies that using G' for all the experiments is better than using any particular single $G \in \{G_i\}$ for all the experiments.)

Though important, such geometry-based distinctions do not say much about generalization once their strictures are met. In essence they serve as a zero-point or a baseline to generalization.

3.7 SUMMARY

- Theorem 1 tell us that $E(C|d)$ is given by an inner product between the generalizer and the posterior.
- Theorem 2 tells us that if one is interested in off-training-set error, then any pair of generalizers perform the same on average, where performance is measured by one of the distributions $E(C|d), E(C|f,d), E(C|m), E(C|f,m)$, or $E(C|s,d)$, and the averaging is over all f or all $P(f)$ as is appropriate.
- Some of the implications of Theorem 2 are that as far as off-training-set behavior is concerned, even techniques like cross validation and the use of test sets to estimate generalization error are unjustifiable unless one makes assumptions. They fail in as many scenarios as they succeed, loosely speaking.
- Appendix 1 demonstrates that even for the Bayes-optimal algorithm (see Section 5), $E(C|m)$ can rise with m for off-training-set error. So can $E(C|f,m)$.
- These results must be modified when there is a natural metric structure in the error function, since that structure allows for the *a priori* superiority of one algorithm over another.

[5] As an aside, note that in certain circumstances, this kind of effect will mean that, everything else being equal, one should prefer an h that stays away from the borders of Y, and therefore one should prefer an h that is relatively smooth. This is an example of how the choice of error function can affect how one regularizes (i.e., can affect the "bias" one imposes that competes with fitting the training set in determining h). This issue is returned to in Section 5 below.

4. QUICK OVERVIEW OF THE FOUR FRAMEWORKS

4.1 HOW THE EBF VIEWS THE FOUR FRAMEWORKS

Each of the four frameworks is simply the evaluation of a certain conditional probability distribution over the space $\{h, f, d, c\}$, under the assumption that certain other distributions in that space have a particular form. For example, the distribution one wishes to evaluate might be of the form $E(C|\ldots)$, and the distributions one assumes might be things like $P(f)$, or $P(h|d)$. At the risk of a bit of hyperbole, one might say that all of the "thought" in the frameworks goes into this choice of the distribution that they wish to evaluate and of what other distributions they will assume. It is this stage of the analysis which distinguishes (and relates) the four frameworks on a foundational level. Again using a bit of hyperbole, one might say that subsequent steps of the analysis—which is all one finds discussed in most research papers—amount to turning a calculational crank.

For some of the results in the non-Bayesian frameworks, it takes more work to "turn the calculational crank" using the EBF than using the original framework. This is a drawback of the EBF. However, in part this extra work simply reflects the higher level of rigor demanded by the EBF. If one wants to be sure about what assumptions are being made—and, in particular, if one wants to be sure about what the conditioning events in the distributions are—one has no choice but to put in extra work like that entailed by using the EBF. (In fact, it was only by using the EBF that I convinced myself what the conditioning events are for the SP, PAC, and VC frameworks. The original papers describing those frameworks never say.) In addition, the proofs found in the original research papers usually serve as an outline around which the EBF-based proof is built; in this sense the extra work simply amounts to cleaning things up. Finally, one can justify the extra work of the EBF simply by pointing out that there is no alternative; if one wishes to reconcile the frameworks, one must express them in terms of the EBF.

4.2 THE "VANILLA" NATURE OF THE PRESENTATION

As was also mentioned in Section 1, only vanilla versions of the frameworks are being considered here. (Extensions beyond such a flavor can be found in the other papers in this proceedings and in the references at the end of this chapter.) This means no noise, countable X and Y, and a misclassification rate error function.

In the case of the VC framework this vanilla-ness also means that the learning algorithm is only capable of producing one hypothesis function h. The reason for this restriction is to highlight the foundational aspects of the VC framework; the "uniform convergence" insight of Vapnik and Chervonenkis constitutes a means of performing the calculations when the algorithm can produce more than one h, but does not have any implications for the underlying decision of what one wishes (!)

to calculate. In this sense it does not constitute a foundational aspect of the VC framework.

In the case of the SP framework the vanilla-ness means that we are considering the "zero-temperature" case (a.k.a. exhaustive learning—see Schwartz et al.,[41] Hertz et al.,[27] van der Broeck and Kawai,[50] and Wolpert[63]). Allowing the temperature to be nonzero constitutes a slight modification of the underlying axioms. However, this modification introduces extra complexity to the analysis, and does not really change how the SP framework relates to the other frameworks. Accordingly, such complexity is here ignored.

Other "vanilla" aspects of the four frameworks are described in those frameworks' sections below.

4.3 SECONDARY ISSUES NOT ADDRESSED IN THE PRESENTATION

In a number of cases, calculational intractability means that the frameworks do not actually evaluate the conditional distribution that I ascribe to them. Sometimes approximations are made instead (e.g., in the Bayesian framework). Other times the frameworks settle for a bound concerning the distribution of interest. For example, rather than evaluate $P(c|f, m)$, PAC evaluates $P(c > \varepsilon|f, m)$. The use of such calculational surrogates is of secondary importance for current purposes. Of primary importance is the underlying quantity that the framework in question is trying to get a handle on.

Another issue of secondary importance is whether the framework tries to characterize some aspect of the distribution-of-interest in terms of polynomial vs. exponential convergence properties, like PAC does. As is demonstrated below, such concerns can be superimposed on all of the frameworks. In this sense they do not constitute underlying axioms of the frameworks in question. In addition, they can sometimes lead to results sufficiently absurd as to call into question the meaningfulness of such convergence properties. (See the section on PAC below.)

4.4 WORST/AVERAGE/THIS-CASE ANALYSIS

Often one will encounter discussion of the "worst/average/this-case" features of a particular framework. In terms of the EBF, such a phrase refers to what is and is not fixed on the right-hand side of the framework's distribution of interest. If a particular random variable's value is specified on the right-hand side, then the framework is a "this-case" analysis for that variable. An example is that the Baysian framework is a this-case analysis for the training set d. (That frameworks' distribution of interest is $E(C|d)$—see below.) If a particular random variable in the event space is not specified (though perhaps some characteristic of it is), then the framework is an "average-case" analysis for that variable. As an example, all of the non-Bayesian frameworks discussed in this chapter are average-case analyses for training sets; only the training set size m is specified, not the training set it-self, and therefore the distribution of interest is an average over all training sets ($P(a|b) = \sum_d P(a|b,d)P(d|b)$).

Finally, worst-case analysis means that the distribution of interest fixes the variable in question and then max's over it. For example, worst-case Bayesian analysis over training sets would involve the distribution $\max_d E(C|d)$. Unfortunately, often worst-case analysis is performed over a "hyperparameter" governing some aspect of $P(f,h,d,c)$ rather than over one of f,h,d,c directly. For example, PAC is worst-case analysis over sampling distributions $\pi(x)$, which are hyperparameters governing the distribution $P(d_X|f)$. To deal with such cases with complete rigor, one should expand the event space to include such hyperparameters, and then introduce the max operator to make the analysis worst-case. For example, one could let π index the sampling distribution, have the event space be f,h,d,c,π, and then indicate a worst-case analysis over sampling distribution by fixing π in the distribution of interest and then prefixing that distribution with the $\max_\pi$ operator. (E.g., PAC's distribution of interest would be expressed as $\max_\pi P(c|f,m,\pi)$ rather than $P(c|f,m)$.)

However such an approach entails redefining the event space for every new worst-case analysis. To avoid this, in this chapter worst-case analysis over such non-f,h,d,c variables will be indicated simply by having the result presented that concerns the distribution of interest not depend on the variable in question. For example, the results presented for PAC will not depend on the sampling distribution. Accordingly those results must hold for all such distributions, and therefore are in particular worst-case for such distributions.

4.5 HIGHLIGHTS OF THE FOUR FRAMEWORKS

In Table 1 I use the EBF to present synopsized versions of (the vanilla versions of) the four frameworks. More details (including conditions for the foundational results to hold) are presented in the sections below. Note that all four frameworks assume the noise-free $P(d|f)$ of Section 2 and the iid error form of $P(c|f,h,d)$. (In other words, such assumptions are implicit in column three for all four frameworks.) The precise definitions of Gibbs generalizer and PAC generalizer are provided in the corresponding sections below. Both are learning algorithms which exactly reproduce (i.e., go through) d. Finally, $\rho(c,f,T)$ is a function defined in the discussion below of the SP formalism.

Loosely speaking, since it is the probability of what we want to know, conditioned on what we do know, $P(c|d)$ is the best choice for distribution of interest when one wishes to consider how to generalize from the actual training set at hand.

TABLE 1 Synopsized versions of the four frameworks. The variable ε is an arbitrary real number.

Frame-work	Distribution of Interest	Distribution Assumed	Foundational Result
Bayes	$P(c\|d)$	$P(f)$—can be arbitrary; $P(h\|d)$ optimal for $P(f\|d)$	Minimal $E(C\|d)$ set by $P(f\|d)$
SP	$P(c\|f,m)$	$P(h\|d)$—"Gibbs generalizer" with starting distribution $T(h)$.	$P(c = \varepsilon\|f,m) \propto (1-\varepsilon)^m \rho(\varepsilon, f, T)$
PAC	$P(c\|f,m)$	$P(h\|d)$—"PAC generalizer" with classes H' and H''.	$P(c > \varepsilon\|f \in H'', m) \leq (1-\varepsilon)^m \|H'\|$
VC	$P(\|c-s\| > \varepsilon\|f,m)$	$P(h\|d)$—here always giving the guess h'.	$P((\|c-s\| > \varepsilon\|f,m) < 2\exp(-2m\varepsilon^2)$

Unfortunately, since that distribution does not fix f, to evaluate it one must average over f. Therefore, one needs to make a strong assumption about the real world—one has to make an assumption for $P(f)$. This is the famous need of a Bayesian to make an assumption for the prior. (The Bayesian framework is the only one of the four whose distribution of interest is $P(c|d)$.) In contrast, the other three frameworks use poorer choices of the distribution of interest (again, if one's goal is to generalize from the training set at hand). However their distributions of interest all have f fixed. Accordingly, they do not need to average over f and, therefore, do not need to make an assumption about $P(f)$. (As an aside, note that cases where the distribution of interest is conditioned on f and m can be analyzed using what Berger[4] calls "(frequentist) risk functions" and the notion of "(in)admissible decision rules.")

Another quick point worth noting is that in the Bayesian framework, SP, and PAC, c is free to vary, and we are evaluating probabilities of various values of c. In SP and PAC, d_X is also free to vary. In contrast, in the VC framework, since h is fixed (to h') and so is f, c is also fixed. In the VC framework, the only variable on the left-hand side of the distribution of interest that is free to vary is d_X. Loosely speaking, in the VC framework we are looking for "the probability of a d_X such that...," as opposed to "the probability that c has a particular value given that...."

These and many other comparisons between the frameworks are described below. As an aside, note that conventional sampling theory statistics—involving bias, variance, and the like—is concerned with $P(c|f,m)$ and, in its most general form, makes no assumption concerning other distributions like $P(h|d)$. In addition, it assumes an iid error function, just like the four frameworks considered in this chapter do. However, in its bias-variance form, it is concerned with an iid quadratic error function, unlike (the vanilla versions of) those four frameworks. As such it is beyond the purview of this chapter. See Geman et al.,[19] Eubank,[18] Berger,[4] Moody,[31] Perrone,[36] and Wolpert.[65]

5. THE BAYESIAN FRAMEWORK

5.1 OVERVIEW OF THE BAYESIAN FRAMEWORK

The Bayesian supervised learning framework is in many ways the simplest of the four frameworks, from a foundational viewpoint. It is defined as follows: Assume one knows $P(f)$ and $P(d|f)$. (In this chapter $P(d|f)$ is the noise-free likelihood described in Section 2.) Then by Bayes' theorem we know $P(f|d)$; it equals $P(d|f)P(f)$, up to an overall proportionality constant that is independent of f. Given this, and any particular form for $Er(f, h, d)$, we have uniquely specified the value of $E(C|d)$ that accompanies any particular learning algorithm $P(h|d)$ (cf. Theorem 1) and, in particular, we can find the $P(h|d)$ that minimizes $E(C|d)$.

This is all there is to the Bayesian framework, as far as foundational issues are concerned. (See Berger,[4] Loredo,[26] MAXENT,[30] Neal,[33] Mackay,[29] Buntine and Weigend,[10] Wolpert and Strauss,[62] Wolpert,[57,58,61] Haussler et al.,[22] Opper and Haussler.[35]) Everything else one reads concerning the framework involves either philosophical or calculational issues.

The philosophical issues usually revolve around what $P(f)$ "means." In particular, some hard-core Bayesians like to interpret $P(event)$ as meaning one's personal degree of belief in *event*, no matter what *event* is. According to this rather extreme view, statistics is simply a calculus for forcing consistency in one's use of probability to manipulate one's subjective beliefs. Such a view makes some sense for example when one is trying to construct a Bayesian "belief net" (see Charniak,[11] Pearl, and references therein). In such a scenario the direct object of interest is in fact a human expert's beliefs, so it is quite reasonable to interpret all probabilities as degrees of belief. However, in the context of supervised learning we are primarily concerned with modeling some external Nature rather than a human expert. In this context, the degree of belief interpretation implies that the Bayesian practitioner fixes the distribution concerning Nature, $P(f)$. But it is not nearly as clear that a human's beliefs determine the probabilities concerning Nature as that they determine the probabilities concerning a human expert's behavior. In particular, such a view of Nature has the rather dubious implication that practitioners of non-Bayesian approaches to supervised learning are, by definition, always going to perform worse than a Bayesian (since the Bayesian "fixes" $P(f)$ and therefore $P(f|d)$ and can use that posterior to guess optimally—cf. Theorem 1). In addition, there are also some logical problems with the degree-of-belief interpretation (see Wolpert [61]). Due to all of this, that interpretation is not adopted in this chapter.

The calculational issues in the Bayesian framework arise in two ways. The first is through the modification of the framework to involve parameters of f (e.g., a neural net weight vector) and/or hyperparameters governing the probability distributions (e.g., a noise level). Such (hyper)parameters must be "marginalized out" or in some other way dealt with for one to get the posterior $P(f|d)$, and this can be difficult.

The second kind of calculational issue can arise even when one knows $P(f|d)$ directly. This occurs for example when one tries to "follow the advice" of the conventional Bayesian framework. The Bayesian framework (or more precisely, Bayesian decision theory) stipulates that one use the $P(h|d)$ that minimizes $E(C|d)$ for the $P(f|d)$ at hand. This $P(h|d)$ is known as the "Bayes-optimal" learning algorithm.[6] It is often extremely difficult to calculate, which is why people often settle for approximations like MAP (maximum a posteriori) quantities (see Section 9).

5.2 BAYES-OPTIMALITY

The Bayes-optimal learning algorithm is specified by the following well-known theorem (rederived in Appendix 2), which is actually more general than what is needed for this chapter:

THEOREM 3. Let $Er(f, h, d) = \sum_{x \in X} \pi'(x) G[h(x), f(x)]$ for some real-valued function $G(\ldots)$. The variable $\pi'(.)$ is a nowhere-negative function that may or may not equal th distribution $\pi(.)$ arising in $P(d|f)$. Then

  i.   The Bayes-optimal $P(h|d)$ always guesses the same function h^* for the same d.

  ii. h^* is the h that minimizes $E(C|d, h)$.

  iii. $h^* = \{x \in X \longrightarrow \arg\min_{y \in Y} \Omega(x, y)\}$,
       where $\Omega(x, y) \equiv \sum_f G(f(x), y) P(f|d)$.

  iv. The resultant value of $E(C|d)$ is $\sum_{x \in X} \pi'(x) \Omega(h^*(x), x)$.

There are no restrictions on whether the function $\pi'(.)$ may vary with d. In this sense, the results of the Bayesian framework are "robust" under substitution of off-training-set error for iid error.

Intuitively, Theorem 3 says that for any x, one should choose the $y \in Y$ that minimizes the average distance from y to $f(x)$, where the average is over all $f(.)$, according to the distribution $P(f|d)$, and "distance" is measured by $G(.,.)$. Note

[6] A bit confusingly, the term "Bayes-optimal" (and/or terms very similar to it) is sometimes used even when there is no data, so one is just trying to find the h that minimizes $E(C|h)$. Historically, this use has arisen from problems in which Bayes' theorem must be used simply to find $P(f)$, and in particular in such problems which have nonsingle-valued f. See Duda and Hart.[17]

that this result holds regardless of the form of $P(f)$, and regardless of what (if any) noise process is present; all such considerations are taken care of automatically, in the $P(f|d)$ term. Note also that h^* might be an f with zero-valued posterior; in the Bayesian framework, h does not really constitute a "guess for the f which generated the data,"

Example. 1. Let $G(.,.)$ be the "zero-one" $G(.,.)$, $G(a,b) = 1 - \delta(a,b)$, where $\delta(.,.)$ is the Kronecker delta function. With appropriately chosen $\pi'(.)$'s, this gives both of the $Er(f,h,d)$'s assumed in this chapter. For this $G(.,.)$, $h^* = \{x \in X \longrightarrow \arg\max_{y \in Y} \sum_f \delta(f(x), y)P(f|d)\}$. Note that unlike the h^* in the next example, the h^* here might not be unique (i.e., there might exist more than one y minimizing $\Omega(x,y)$ for some particular x). In fact, if $P(f)$ is flat and (as in this chapter) there is no noise, then for the zero-one $G(.,.)$ there is no minimum to $\Omega(x,y)$ for $x \notin d_X$; all guesses result in the same (expected) error. See Wolpert.[56] ∎

For pedagogical purposes, it is worth briefly considering the quadratic $G(.,.)$ case, even though that corresponds to a different $Er(f,h,d)$ from the one assumed in this chapter:

Example. 2. Let $G(a,b) = (a-b)^2$, and let Y be the set of real numbers. For this $G(.,.)$, by differentiating $\Omega(x,y)$ with respect to y we find that $h^* = \{x \in X \longrightarrow \sum_f P(f|d)f(x)\}$, i.e., $h^*(x) = \sum_f P(f|d)f(x)$. In other words, h^* is given by the average (according to the posterior) value of y. This is the source of the common claim that a Bayesian evaluates the posterior average. (As an aside, it is interesting to note that if instead $G(a,b) = |a-b|$, h^* is given by the median rather than the mean.) In the current context though (where Y is finite), this average y usually will not lie in Y. In such a case, we cannot find h^* by differentiating $\Omega(x,y)$ with respect to y, but must instead go through all the possible y values by hand. ∎

In general, one might wish to do something like find the $P(h|d)$ maximizing $P(c = 0|d)$ rather than the $P(h|d)$ minimizing $E(C|d)$. In general such a $P(h|d)$ will not be the same as the one given in Theorem 3. Such goals are usually not pursued in the Bayesian framework however, and therefore are pursued no further here.

5.3 THE PREDICTIVE DISTRIBUTION

The distribution $T(x,y) \equiv \sum_f \delta(f(x),y)P(f|d)$ arising in Example 1 tells us all we need to know about the posterior $P(f|d)$, as far as finding h^* is concerned, regardless of what $G(.,.)$ is. This is because $\Omega(x,h(x)) = \sum_y T(x,y)G(y,h(x))$. If two distinct posteriors have the same $T(x,y)$, we are assured that for the same $G(.,.)$ they will result in the same h^*. In this regard, note that for fixed d, $P(f|d)$ is specified by $|f|$ real numbers (minus 1 for overall normalization) $= r^n - 1$ real numbers. In contrast, $T(x,y)$ consists of $(n \times r)$ real numbers.

To "access" the variations in $P(f|d)$ not distinguished in $T(x,y)$, we would need to use an error function which is not simply the x-summed value of a function of $f(x)$ and $h(x)$. For example, if $Er(f,h) = 1$ if $f = h$ (for all x), 0 otherwise, then our optimal guess for h is the MAP (maximum a posteriori) f, $\arg\max_f P(f|d)$. For this case, so long as they have different MAPs, two different posteriors will result in a different h^* even if they have the same $T(x,y)$. (See Wolpert[58] for a more general discussion of the use of MAP estimators in the context of neural net backpropagation.)

Viewing f as an n-dimensional vector, $T(x,y)$ is $P(f|d)$ marginalized out over all components of f except for the one corresponding to input x, that component being fixed to the value y. It can also be viewed as $P(y|x,d)$, where x is a query input point and y is the corresponding output on the (unknown) target function. ($P(y|x,d) = \sum_f P(y|x,d,f)P(f|x,d) = \sum_f P(y|x,f)P(f|d)$.) As such, it constitutes the "predictive distribution" for y given x and d.

5.4 IMPLICATIONS OF THEOREM 3 FOR OFF-TRAINING-SET GENERALIZATION

Theorem 3 can help clarify certain counterintuitive aspects of generalization. For example, choose to use off-training-set error by setting $\pi'(x) = 0$ for $x \in d_X$. Now in our no-noise scenario, h^* will reproduce d. Therefore, it would seem that there is an *a priori* correlation between minimal expected off-training-set generalization error (associated with h^*) and minimal observed on-training-set error. (At least, this would seem to be the case if one were a Bayesian and presumed to know $P(f)$.) Such a correlation, if it existed, would seem to justify the common heuristic that, in the absence of noise, one should search for an h that agrees often with d.

Note that this conclusion does not violate the no-free-lunch theorems. This is because those theorems say that for any *fixed* $P(h|d)$, when averaged over all $P(f)$, $E(C|d)$ does not favor an algorithm that reproduces d over one that does not. However, in the scenario considered here, as one changes $P(f)$ one also changes $P(h|d)$, so Theorem 2 does not apply. After all, despite the no-free-lunch theorems, $E(C|d)$ for a Bayesian who knows $P(f)$ is $\leq E(C|d)$ for all other generalizers, even for off-training-set error. (Moreover, in general one only needs nonuniform $P(f)$—or more precisely, nonuniform $P(f|d)$—for $E(C|d)$ for a Bayesian who knows $P(f)$ to be strictly smaller than $E(C|d)$ for a uniformly random generalizer.)

Even so, the conclusion is flawed. An h which disagrees with d will have the same expected off-training-set generalization error as h^*, provided $h = h^*$ for those $x \in \{X - d_X\}$. Conversely, there exist hypothesis functions that agree with d exactly, and yet have maximal (off-training-set) error. Intuitively, the (off-training-set) error function does not see the behavior of $h(x)$ for those $x \in d_X$, and therefore it cannot have *a priori* dependence on that behavior.

Theorem 3 can clarify this apparent contradiction. The proof of Theorem 3 makes clear that if $\pi'(x) = \dot{0}$ for any particular x value x', then changing $h(x')$, even if it changes the value of $\Omega(x', h(x'))$, does not affect $E(C|h, d)$ (so long as $h(x \neq x')$ is left unchanged). In other words, Theorem 3 only sets the behavior of the optimal hypothesis function over those x such that $\pi'(x) \neq 0$; behavior over other x—and in particular over the elements of d_X—is completely free. This is true even if $P(f|d)$ strongly favors those f that reproduce d. Indeed, there might be an f_1 and an f_2 so that $P(f_1|d) \gg P(f_2|d)$ due to the likelihood of f_2 being extremely low, but, nonetheless, f_2 has much better off-training-set behavior than f_1 (for example, f_2 might agree with the Bayes-optimal f for $x \notin L_X$, whereas f_1 does not).

So: h^* agrees with d; optimal off-training-set behavior is exhibited by the hypothesis function h^* *amongst others*; therefore, one way to get optimal behavior involves finding an h that agrees with d; nonetheless, there are other h's that have the same optimal behavior, but that do not agree with d, and there are h's that agree with d but have poor off-training-set behavior. In general there is no *a priori* correlation between on-training-set accuracy and off-training-set error. (See also Wolpert.[56])

In a similar vein, in general when noise is allowed in generating d h^* does not exactly reproduce d; the prior pulls $h^*(d_X)$ away from d_Y. As above, this will be the case even if $\pi'(x) = 0$ for all $x \in d_X$, so that we are dealing with off-training-set error. This suggests that over-training (i.e., fitting h to d even when there is noise) is correlated with suboptimal off-training-set behavior *a priori*. But again, there cannot be any such *a priori* correlation, since we can take any h and modify its on-training-set behavior in any way we wish, while leaving its off-training-set behavior alone. Once more the apparent contradiction is resolved by the fact that Theorem 3 does not fix $h^*(x \in d_X)$ when $\pi'(x \in d_X) = 0$. When there is nonzero noise, one h with optimal off-training-set behavior, h^*, does not reproduce d. But one cannot use the fact that a particular h reproduces d to infer that its off-training-set behavior is likely to be suboptimal.

5.5 MISCELLANEOUS COMMENTS

Theorem 3 can be used to investigate the effects of incorrect assumptions for $P(f)$. For example, let's say we make a guess for $P(f|d)$, call it $Q(f|d)$, which does not exactly equal $P(f|d)$. For pedagogical simplicity, assume we have the quadratic $G(a, b)$ of Example 2. Then, if one used Theorem 3 to derive a "Bayes-optimal" hypothesis function based on the function Q, one gets $h^*(x) = \sum_f Q(f|d)f(x)$. However, the real $h^*(x)$ equals $\sum_f P(f|d)f(x)$. Therefore, for any $x \in X$, one's guess is off by the amount $\sum_f \{P(f|d) - Q(f|d)\}f(x)$. One can plug this into our assumed error function to get the error between the true Bayes-optimal h and our h: error $= \sum_{x \in X}[\sum_f \{P(f|d) - Q(f|d)\}f(x)]^2 \pi(x)$. In a similar manner one can calculate the gain in $E(C|d)$ accruing from using $Q(f|d)$ rather than $P(f|d)$. (Analysis of mis-specified priors is called "sensitivity analysis" or "robust Bayesian analysis" in the conventional statistics community. See Buntine,[8] and Berger.[4] Also see Haussler et al.[24,23] for work on misspecified priors.)

As a final demonstration of the Bayesian framework, note that many authors take Occam's razor to mean that $P(f)$ is larger for "simpler" f, where "simplicity" is usually defined in some ad hoc manner (e.g., as a description length). To see how this is related to Occam's razor, note that under this assumption, with everything else (e.g., the likelihood) being equal, if one had to choose between two functions f_1 and f_2, it would make sense to choose the "simpler" of the two, since that function has a higher posterior probability. This is exactly what Occam's razor tells us to do. However, Theorem 3 tells us that in many circumstances (i.e., for many $G(.,.)$) the Bayes-optimal h is neither f_1 nor f_2, but a combination of both. The implication is that despite its vaunted position at the core of the scientific method, Occam's razor is suboptimal. (See section 7.8 for some other comments concerning Occam's razor.)

It is important to realize that in many respects the version of the Bayesian framework presented here differs from Bayesian analysis as actually practiced. One reason for this is that the calculational difficulties alluded to earlier means that approximations must be used in practice. Another is that practicing Bayesians are often more "pragmatic" than to deal with priors in the strict manner dictated by pure Bayesian analysis. An example of such a pragmatic approach can be found in Spiegelhalter et al.[44] See also the discussion of empirical Bayes and ML-II in Section 9 below.

5.6 SUMMARY

- The Bayesian framework assumes that $P(f|d)$ is known. As the framework is usually constituted, $P(f|d)$ is derived via Bayes' theorem (hence the framework's name) from assumptions for $P(d|f)$ and $P(f)$.

- The Bayesian framework is concerned with $E(C|d)$. Given $P(f|d)$, $E(C|d)$ is determined by $P(h|d)$. The (single-valued deterministic) generalizer $P(h|d)$ that minimizes $E(C|d)$ is known as the Bayes-optimal generalizer; the formula for it is given in Theorem 3.

- One does not need to know $P(f|d)$ in full to find the Bayes-optimal generalizer; only the predictive distribution (which equals $\sum_f \delta(f(x), y)P(f|d)$ in the noise-free context of this chapter) is needed.

- The Bayesian framework might appear to imply that in the absence of noise one should search for an h which agrees often with d. Careful examination of Theorem 3 shows this not to be the case.
- Theorem 3 tells us that even if $P(f)$ is larger for "simpler" f, despite the advice of Occam's razor in many circumstances the optimal guess is not given by the simplest f consistent with the data.

6. THE SP FRAMEWORK

6.1 OVERVIEW OF THE SP FRAMEWORK

The SP framework is also relatively straightforward, from a foundational viewpoint. Unlike the Bayesian framework, it makes no assumptions concerning $P(f)$. Rather it assumes that one's learning algorithm is the "Gibbs generalizer,"

$$P(h|d) = k(d)T(h)\delta(d \subset h),$$

where $k(d)$ is an overall normalization constant (sometimes called a "partition function"), $T(h)$ is a nowhere-negative distribution, and $\delta(d \subset h) = 1$ if d lies on h, 0 otherwise. (The nonzero temperature version of the formalism relaxes the $\delta(d \subset h)$ restriction, replacing it with an exponential distribution.) Intuitively, this "Gibbs," or "Boltzmann" algorithm consists of the following rule: Given d, find all h that reproduce d; then choose randomly amongst those h, where "randomly" means according to the distribution $T(h)$.

There are no *a priori* restrictions on $T(h)$ in exhaustive learning. In particular, one is free to choose $T(h)$ to be nowhere zero. Alternatively, one could choose it to equal zero for many h's, so that the generalizer has limited expressive power. Of course, one should never set $T(h) = 0$ for $h = f$, since then the generalizer would be undefined if d equalled the set $\{x \in X, f(x)\}$. (Because of this, there are some practical difficulties associated with using a $T(h)$ that is not nowhere-zero; see below.)

The SP framework investigates the implications of the Gibbs generalizer for the distribution $P(c|f, m)$, for the iid error function.[27,41,46,47,50,63] Historically, the SP framework was presented in the context of neural nets, volumes over weight space, and the like. In point of fact there is no need to restrict consideration to cases where h and f are parametrized, in terms of neural nets or otherwise; the vanilla version of the framework is completely specified by the $P(h|d)$ and distribution of interest outlined above.

6.2 THE CENTRAL RESULT OF EXHAUSTIVE LEARNING

This vanilla version of the SP framework is known as "exhaustive learning." Its central result is that when $k(d) = k(d_X, f(d_X))$ is independent of d_X (a phenomenon sometimes called "self-averaging," and which is closely related to the "annealed approximation"),

$$P(c|f, m) \propto (1 - c)^m \rho_0(c, f, T),$$

where $\rho_0(c, f, T) \equiv \sum_h T(h) \delta(c, Er(f, h))$, and the normalization constant depends on m in general.

In particular, let $T(h)$ be uniform over all h (which does *not* correspond to having $T(.)$ uniform over all perceptron weight vectors **w** parametrizing h). Then, for $n \gg m$, self-averaging holds, and $P(c|f, m) \propto (1-c)^m \rho_0(c, f, T)$. Moreover, for this $T(h)$, $\rho_0(c, f, T)$ is independent of f. (However, for other $T(h)$, self-averaging does not hold, and this central result of exhaustive learning is not valid—see Wolpert,[63] Schwartz et al.,[41] and Hertz et al.[27])

The striking feature of this central result of exhaustive learning is that all of the m-dependence of $P(c|f, m)$ appears in the $(1-c)^m$ term. Since as m grows that term is concentrated more and more heavily about $c = 0$, we see that (under the assumptions of the framework) $P(c|f, m)$ itself becomes more and more strongly concentrated toward $c = 0$ as m grows. In other words, we appear to have an *a priori* argument that larger training sets lead to better generalization, for a Gibbs generalizer.

Workers in the SP community give the following intuition for this: With each new i-o pair in the training set, one restricts the "volume of function space consistent with the training set." Accordingly, each new pair allows us to "zero in" more on f. In particular, if the new pair excludes a large volume, we have just improved things a lot, and one might expect we have just gained a lot in generalization accuracy. Note that this restricting-of-volumes intuition is a likelihood-driven phenomenon (at least in exhaustive learning, where there is no noise), in that it is the likelihood that is excluding all those f, and not a prior. Note also that this intuition does not directly address off-training-set behavior. (This point is returned to below.)

6.3 THE SP FRAMEWORK AND BAYESIAN ANALYSIS

For some $T(h)$ (e.g., uniform $T(h)$), whether or not one has self-averaging does not depend on f. For such cases, exhaustive learning's *a priori* argument that larger training sets lead to better generalization holds for all f. One only needs to make assumptions about the generalizer; no assumptions whatsoever about f or $P(f)$ are needed. This is one of the strengths of the SP framework in comparison to other frameworks, and in particular in comparison to the Bayesian framework.

Unfortunately, this strength has been obscured by the fact that the SP framework was originally viewed as an instantiation of the Bayesian framework. This view is false.[7] This is immediately apparent from the fact that the Bayesian framework has a deterministic generalizer, unlike the stochastic Gibbs generalizer. Note also that the SP framework averages over data and keeps f fixed (in exact opposition to the Bayesian framework).

This mistaken view may have arisen from the fact that the Gibbs generalizer can be made to *look like* an instance of Bayes' theorem. To see this, confuse h and f; i.e., replace h with f in the rule defining the Gibbs generalizer. Then since $P(d|f) \propto \delta(d \subset f)$, if we assume $P_F(h) = T(h)$ our modified version of the rule defining the Gibbs generalizer is identical to the usual Bayesian expression giving $P(f|d)$ in terms of $P(d|f)$ and $P(f)$. Of course, conventional Bayesian analysis concerns itself with $P(f|d)$, not $P(h|d)$ (that is why we had to confuse h with f), $P_F(h) \neq T(h)$ in general, etc. So this formal parallel does not establish any causal connection between the Gibbs generalizer and conventional Bayesian analysis. It simply establishes that in certain circumstances the functional dependence of the Gibbs generalizers on h and d is the same as the posterior's dependence on f and d.

Perhaps reflecting this confusion between f and h and between $P(.)$ and $T(.)$, in some of the literature the extra assumption is made that $T(h) = P_F(h)$. However, this assumption is not needed for many of the more important results to go through, and it is not needed for any of the results of this chapter. Moreover, as mentioned above, making an assumption for $P(f)$ removes one of the major strengths of exhaustive learning. (As an aside, it should be noted that some important results do depend on this kind of an assumption, for example, a number of the results in Haussler et al.[22])

Another way to view the Gibbs generalizer so that it "looks Bayesian" is to simply write $P(h|d) \propto P(h)P(d|h) = P(h)P(d_Y|d_X,h)P(d_X|h)$. This holds for any generalizer. If we now restrict ourselves to generalizers producing h's that go through d, this expression can be rewritten as $P(h)\delta(d \subset h)P(d_X|h)$. Next recall (Section 2) our assumption that g is independent of f. This means that $P(d_X|f)$ is independent of f, and in particular it is indepenent of $f(x \notin d_X)$. If we could similarly say that $P(d_X|h)$ is independent of $h(x \notin d_X)$, then we *would* have Bayes' theorem proving that $P(h|d)$ is a Gibbs generalizer, under $T(h) \equiv P(h)$. Unfortunately, although taking $P(d_X|f)$ to be independent of $f(x \notin d_X)$ is not entirely

---

[7] For me to contest the views of some of the originators of the SP framework on how to interpret that framework might seem odd. One is tempted to argue that the SP framework is whatever its originators defined it to be. (Indeed, one might argue that about all four frameworks.) However the definitions in the original papers on the SP framework were not precise. In particular, they did not distinguish f from h. (See sin number two in the introduction.) Accordingly, one must consider a list of possibilities of what the framework might amount to formally, i.e., a list of possible interpretations of the framework. For each such interpretation, one must see if the derivations found in the original papers are valid. The interpretation of the SP framework that I am presenting here is the only one I know of that is consistent with those derivations. In addition, current practitioners that I have talked with agree with this interpretation.

unreasonable, $P(d_X|h)$ is almost never independent of $h(x \notin d_X)$ and d_X, even for a Gibbs generalizer.[8]

A final possible reason for why the SP framework is often confused with Bayesian analysis is that the Gibbs generalizer can be viewed as a Monte Carlo sampler of $P(f|d)$, given that $P_F(h) = T(h)$ and that we have a noise-free likelihood. In such a situation, the Gibbs generalizer can be used to calculate an approximation for the Bayes-optimal h. (For example, if we have quadratic $G(.,.)$, the average of the h's found by infinitely rerunning the Gibbs algorithm on d will equal the Bayes-optimal h.) However, as mentioned earlier, many of the SP framework's results hold even if one does not assume that $P_F(h) = T(h)$. Moreover the SP framework is not concerned with $E(C|d)$, the object defining Bayes-optimality; even in those cases where the generalizer of interest to the SP framework is a useful calculational tool to Bayesians, the way in which that generalizer is analyzed differs for the SP and Bayesian frameworks. (Thanks to Wray Buntine for illuminating conversation on this point.)

6.4 COMMONLY MISUNDERSTOOD SUBTLETIES OF THE SP FRAMEWORK

There are a number of other subtleties concerning the exhaustive learning scenario. For example, it turns out that few generalizers can be cast as Gibbs generalizers. In particular, no generalizer that produces the same h when trained on the same d—e.g., a nearest-neighbor generalizer—can be a Gibbs generalizer. (This is proven by Wolpert.[59])

Another subtlety involves $T(h)$. It is tempting to view $T(h)$ as "the prior beliefs of the learner about which target (function) it will be learning." However, in general, $P(h)$ need not be related to $P(f)$ (see above). In addition, $P(h)$ usually does not equal $T(h)$, even up to an overall proportionality constant. To see this, write $P(h) = \sum_d P(h|d)P(d)$, which for an exhaustive learning generalizer equals $T(h) \sum_d k(d)P(d)\delta(d \subset h)$. If the sum is independent of h (for example, if $k(d)P(d)$ is constant), it follows that $T(h)$ is proportional to $P(h)$. This means that one can always choose a $P(d)$ such that a particular $T(h) = P(h)$. However $P(d) = \sum_f P(d|f)P(f)$, and is set by the prior over f. And in general $P(f)$ will not give a $P(d)$ for which our sum is independent of h. In other words, if $P(h|d)$ is

---

[8] For example, assume that $P(f)$ allows only one f, f', so that if there is no noise in $P(d|f)$, only those d agreeing with f' are allowed. Now pick a training set d' (lying on f') and a guess h' that agrees with d', but does not agree with f' for any X values outside of d'_X. Pick another training set d'', which also agrees with f', but which has no overlap with d'. Choose an h'' which agrees with d'' but does not agree with f' for any X values outside of d''_X. Let both $T(h')$ and $T(h'')$ be nonnegligible. Then if we observe that the guess was h', we can infer that d_X was (contained in) d'_X. (If d_X extended beyond d'_X, the algorithm could not produce h', since h' would not agree with d for those X values outside of d'_X.) So $P(d'_X|h')$ is nonnegligible. On the other hand, if we observe that the guess was h'', we can infer that d_X has zero overlap with d'_X. So $P(d'_X|h'') = 0$; $P(d_X|h)$ is not independent of h.

held constant, changing $P(d)$ changes $P(h)$, and therefore this changes whether or not a particular function $T(h) \propto P(h)$.[9]

The fact that $P(h)$ need not equal $T(h)$ might seem counterintuitive. After all, $T(h)$ is the distribution determining h before any data is given, and we often think of such a distribution as the prior probability of h. The problem is that in the context of this chapter, where hypothesis functions are formed from training sets (and therefore *after* the training sets are given), this is actually a misconception of what a prior probability is. The variable $P(h)$ is *not* the probability of h, before having seen the training set. Rather it is the probability, evaluated before having seen the training set, that *after* seeing the training set the hypothesis will be h: $P(h) = \sum_d P(h|d)P(d)$, and varies with $P(d)$. See Wolpert[59] for a more detailed discussion of this issue.

6.5 OFF-TRAINING-SET ERROR AND THE SP FRAMEWORK

Other subtle points concerning exhaustive learning at first appear to constitute paradoxes. For example, according to Theorem 2.i, as far as off-training-set error is concerned, the Gibbs generalizer has the same value of $E(C|f, m)$ as any other generalizer, on average. In particular, it has the same value as the generalizer {always guess the same h no matter what the data}. For that generalizer the f average of $E(C|f, m) = 1/r$, independent of m. Therefore, averaged over all f, $E(C|f, m)$ for the Gibbs generalizer cannot vary with m, for the off-training-set error function.

Similarly, consider the case where f is fixed, and $T(h)$ is uniform over all h. Since such a uniform-distribution generalizer will guess any one of the h consistent with the training set with equal probability, its behavior is equivalent to guessing randomly off of the training set. In other words, it is the following rule: for a question equivalent to one of the inputs in the training set, answer with the corresponding output. For any other question, flip a uniformly weighted coin. Clearly the off-training-set (i.e., coin-flipping) behavior of such a generalizer cannot improve with m. (See also Wolpert.[63])

To see how these results are compatible with the central result of exhaustive learning, note that that central result concerns iid error, thereby allowing test set questions to range over the training set as well as off of it. In addition, $P(c|f, m)$ has a fixed (though unknown) f, and averages over h's. Because of all this, for a binary output space, and for a uniform-$T(h)$ generalizer, the average iid error c—the average number of errors for randomly chosen test questions—is trivially given by $(n - m')/2n$, where n is the size of the input space, and m' is the number of distinct pairs in the training set (this result assumes that the sampling distribution over

---

[9] It turns out that if $T(h) = P_F(h)$, the (unknown to the researcher) prior probability of target function $f = h$, then the sum is independent of h and $T(h)$ is proportional to $P(h)$ (which in turn implies that $P(h) \propto P_F(h)$). Of course, in the real world where we do not know $P(f)$, there is no way of knowing that $T(h) = P_F(h)$. In general however, when $T(h) \neq P(f)|_{f=h}$, the sum varies with h. See Wolpert[59,63] for a discussion of this issue.

the input space is uniform). Now exhaustive learning also uses an iid likelihood function, thereby allowing duplicates in the training set. This means that m' is usually less than m. Therefore, when expressed in terms of m, the straight line giving the average error develops a "tail"; it gets stretched and pulled towards $m = \infty$. It look like $1/m$, or e^{-m} for large m, in fact. And as Schwartz et al. demonstrated,[41] such average errors are precisely what one would expect if the probability of iid error c goes like $(1 - c)^m$.

Presumably this uniform-$T(h)$ generalizer is not the one that the originators of the exhaustive learning formalism had in mind. Indeed, subsequent analyses within the SP framework have considered cases where $T(h)$ is explicitly nonuniform, and related in some particular way to f and/or $P(f)$. Nonetheless, the general mathematical results that have been reported on exhaustive learning hold for the uniform-$T(h)$ generalizer. Accordingly, one can view the uniform-$T(h)$ case as a baseline; in a certain sense, only if one gets better behavior than that of the uniform-$T(h)$ case has one achieved something useful. In particular, only if $P(c|f, m)$ has better-than-exponential behavior as a function of m is one doing better than random off-training-set guessing. In addition, the mathematics associated with the uniform distribution serves as the "backbone" of the mathematics for other distributions. Changing the distribution changes the details of the analysis (e.g., constants changes, precise forms of monotonic functions change, etc.), and generally makes the analysis more complicated. But it does not affect the underlying behavior. (For example, the central result of exhaustive learning still holds, assuming self-averaging, and it is still true that $E(C|f, m)$ is independent of m, on average, for off-training-set c, so one still has the "paradox" mentioned above.)

Moreover, unless one wants to make extra assumptions (for example, concerning $P(f)$), then one should exercise care in setting $T(h)$ in the zero-temperature version of the SP framework. For example, consider applying the exhaustive learning generalizer in the real world. If, rather than being infinitesimal, $T(h')$ equalled zero exactly for some particular hypothesis h', use of the exhaustive learning $P(h|d)$ would implicitly set strong restrictions on $P(f)$. This is because if $T(h') = 0$, then the exhaustive learning $P(h|d)$ is undefined for $d = h'$. (Note the same problem can arise if we use a Bayes-optimal generalizer based on an (incorrect) assumed $P(f)$ that has limited support.) Since we have no noise in the likelihood, the only way we could be assured of avoiding such a d in the real world is if the corresponding f cannot occur (assuming $\pi(x)$ is nonzero for all x), i.e., if $P_F(h')$ equals 0 exactly. To avoid this dilemma, one should ensure that $T(h)$ never equals zero exactly. Unfortunately, this can be quite difficult if, as is conventional, functions h are viewed as neural nets, so that rather than writing down $T(h)$ directly, one writes down a sort of $T(\mathbf{w})$, where $\mathbf{w}$ is a weight vector parametrizing h.

6.6 THE CARDINALITY OF X IN THE SP FRAMEWORK

Another subtlety of the SP framework involves the cardinality of the input space X. In standard accounts of the SP framework X is often taken to be uncountable (a Euclidean vector space in fact). This would seem to distinguish it from the version of the SP framework presented here. In addition, the argument has been made that when X *is* uncountable, the measure of d_X is 0 and, therefore, there cannot be any difference between iid error and off-training-set error. Given the claims made above concerning the off-training-set behavior of exhaustive learning, this would seem to imply that the cardinality of X is in fact a crucial consideration.

However, if iid error and off-training-set error were indeed equal, it would be bizarre for the central result of exhaustive learning to hold. That result is not predicated on any assumptions concerning f or $P(f)$, but without such an assumption all the intuition behind the no-free-lunch theorems maintains that $E(C|f,m)$ must be independent of m for off-training-set c, regardless of the cardinality of X. So we seem to have a paradox.

In the event, the paradox does not apply because X is always *effectively* countable in exhaustive learning, even when it is written as $\mathbf{R}^k$ for some k. The formal proof of this can be found in Wolpert.[63] The general idea is as follows: In the standard accounts of exhaustive learning, the set of all allowed h is countable (e.g., we have sums over h's rather than integrals). For any countable set of h's taking $\mathbf{R}^k$ to Y, one can cover $\mathbf{R}^k$ with at most a countable number of regions over each of which none of the h's change their output values. (In other words, there are regions in $\mathbf{R}^k$ such that all the h's have constant output value over the region, and $\mathbf{R}^k$ is completely covered by a countable number of such regions.)

Now we must be assuming that f is such an h, since the generalizer must be able to produce an h agreeing with any d sampled from f. So f, too, is constant within any of our regions. Accordingly, being given the value of $f(x)$ at one point $x = x'$ fixes the value of $f(x)$ throughout the region containing x'. Therefore, *given our assumption about f*, the more points x at which one knows $f(x)$, the more regions of nonzero measure over which one knows $f(x)$. So if "off-training-set" is taken to mean the set of points in $\{\mathbf{R}^k - d_X\}$, then indeed as m grows, we know the value of $f(x)$ over portions of $\mathbf{R}^k$ with larger measure, and therefore off-training-set error necessarily goes down as m rises.

On the other hand, consider the case where we redefine X to be the countable set of regions covering $\mathbf{R}^k$, the logic being that since neither the generalizer nor f can access $\mathbf{R}^k$ on a finer scale, there is no reason to consider such finer scales. If we now define "off-training-set" in the same manner, as off the training set's *regions*, then we have no-free-lunch results, and off-training-set error need not improve as m grows.

As an aside, it is interesting to note that the definition of "regions" used here is a sort of inverse of the definition of VC dimension.[6,51,52] Rather than being concerned with the VC notion of "shattering" a set of input points, which means that you can get all possible sets of output values for the points in question, we are considering the opposite end of the spectrum, where you are not able to get any differing output values.

6.7 SUMMARY

- The exhaustive learning (zero-temperature) version of the SP framework considers the behavior of the "Gibbs generalizer" $P(h|d) = k(d)T(h)\delta(d \subset h)$ when there is no noise.
- Despite common intuition to the contrary, in general the prior over h does not equal $T(h)$.
- The central result of exhaustive learning is that when the "self-averaging" property holds, for the exhaustive learning scenario and iid error, $P(c|f,m) \propto (1-c)^m$. Note this distribution gets more peaked about lower c's as m increases.
- Although the SP framework is often said to be a variant of the Bayesian framework, this isn't the case. In particular, the central result of exhaustive learning holds even if $T(h) \neq P_F(f)$. In addition, the quantity of interest in exhaustive learning, $P(c|f,m)$, is this-case over targets and average-case over data, unlike the Bayesian framework. Note that since f is fixed, the SP framework doesn't have to make an assumption f or $P(f)$.
- As Theorem 2 suggests, the implication of the central result of exhaustive learning that error shrinks as m rises is critically dependent on the use of the iid error function; for off-training-set error, the result disappears.
- Although often couched as though its input space were uncountably infinite, due to the countability of the number of hypothesis functions the input space is effectively countable in exhaustive learning. This is important in understanding the relationship between the central result and off-training-set error.

---

7. THE PAC FRAMEWORK

7.1 OVERVIEW OF THIS SECTION

The PAC framework is very similar to the SP framework. Both are interested in the distribution $P(c|f,m)$, and both are predicated on iid error as well as the iid likelihood assumed throughout this chapter. In fact, their only important foundational difference is that PAC does not fix the generalizer, but rather makes an assumption about how the generalizer is related to f. This close relationship is often obscured by the fact that PAC superimposes on its foundation concern for convergence issues, concern that is absent in all non-PAC frameworks including the SP framework.

For current purposes, perhaps the most natural way to make these statements precise is to

i.   First present a scheme that arises frequently in PAC for characterizing generalizers.

ii. Then present a bound, central to much of PAC, involving PAC's distribution of interest, $P(c|f, m)$. The bound applies whenever the generalizer is related to f in a particular manner that involves the scheme in item i.

iii. Then show how PAC superimposes concerns with polynomial vs. exponential convergence on top of issue ii. It is argued here that such concerns are often only peripherally related to the issue of supervised learning in the real world.

The first step is done in subsection (2). The second is in subsection (3), with a discussion in subsection (4). The third step is carried out in subsection (5), with a discussion presented in subsections (6) through (10).

7.2 CHARACTERIZING GENERALIZERS

Let H'' be a "target class" of a set of $X \rightarrow Y$ functions, and H' a potentially distinct "hypothesis class" of such functions. (This is slightly different terminology from that usually found in the literature.) The vanilla version of PAC characterizes generalizers in terms of classes H' and H'' that are associated with one another in the following manner:

1. For any d consistent with an element of H'', the generalizer reproduces d (i.e., for such a d, $P(h|d) = 0$ for any h not going through d);
2. For any such d the generalizer's guess is assured of being in H' (i.e., for such a d, $P(h|d) = 0$ for all $h \notin H'$).

PAC concerns itself with $P(c|f, m)$ when $f \in H''$. So intuitively, one should think of the process underlying PAC as: f comes from the target class; it is noise-free sampled to get d; the h produced by the generalizer for this d agrees with d, and is assured of falling in the hypothesis class. There are a number of reasons why one might be interested in restriction h to H'. For example, one might view it as a way to try to avoid overtraining.

Much of PAC implicitly exploits the full generality of items 1 and 2 and allows H' and H'' to differ. For example, the application of Lemma 2.1 in the proof of Theorem 2.3 in Blumer et al.'s paper[5] has H'' be one set (the set of h with complexity $\leq$ (what in that paper is called) n), and H' another (the set of h with complexity $\leq$ (what in that paper is called) $n^c m^\alpha$). Nonetheless, many definitions of PAC, particularly early ones, do not explicitly concern themselves with H'. In particular, Lemma 2.1 in Blumer et al.'s paper[5] does not distinguish between H' and H'', despite the need to make that distinction in the use of the lemma to prove Theorem 2.3.

Condition 1 is usually relaxed in those versions of PAC that do not assume no noise. In such cases H'' only arises in that we are examining $P(c|f \in H'', m)$.

Accordingly, in such cases H'' does not "characterize" the generalizer in any way. (Such versions of PAC are not considered here.)

The number of elements in the hypothesis class is written as $|H'|$, and similarly for the target class. For *any* generalizer there will always be an H' satisfying Condition 2, namely the set of all possible h. Moreover, if m is allowed to equal n and if for no x does $\pi(x) = 0$, then d could completely specify f, and H' must be a superset of H''. By and large, PAC is concerned with the minimal H' for a particular H''. Even when H' is a limited set of h's though, Condition 2 imposes no restrictions on the guess of the generalizer for a training set not consistent with any element of H''.

7.3 A BOUND CONCERNING $P(c|f, m)$

Unfortunately, it is very difficult to say anything in general about $P(c|f, m)$, given only that one has a PAC generalizer with classes H' and H'' and that $f \in H''$. (Note that PAC does not specify the generalizer precisely, but only some characterizations it must meet.) Accordingly, PAC instead concerns itself with $P(c > \varepsilon|f, m)$ for arbitrary ε.[1,5,6,14,15,23,24,38,49]

More precisely, in Appendix 3 the following foundational PAC theorem is derived using the EBF:

THEOREM 4. If $f \in H''$, then $P(c > \varepsilon|f, m) \leq \{1-\varepsilon\}^m |H'|$ for all PAC generalizers with the given H' and H''.

Note that $P(c > \varepsilon|f, m)$ is an average over all training sets of size m. Define unique(d) to be the (unordered) set consisting of the m' distinct input-output pairs in d. Due to PAC's likelihood, duplicate input-output pairs are allowed in d (so that m' ranges over all values $\leq m$). As a result, for any two training set sizes m_1 and $m_2 > m_1$, the set of all the unique(d) which are averaged over to form $P(c > \varepsilon|f, m_1)$ is contained within the set of all the unique(d) which are averaged over to form $P(c > \varepsilon|f, m_2)$. This overlap in unique(d) space couples $P(c > \varepsilon|f, m_1)$ and $P(c > \varepsilon|f, m_2)$, and is directly reflected in the m-dependence in Theorem 4. (See Blumer et al.[5] for an intuitive argument supporting the precise form of the result in Theorem 4.)

Note that if we were calculating a quantity conditioned on m' rather than m, we would not have this coupling. On the other hand, for $n \gg m$, generically the set of all d's for which $m' < m$ collectively contribute very little to our sum. (See Wolpert.[63]) In general, for such a situation distributions conditioned on m are well approximated by distributions conditioned on m', which implies that, for such a situation, Theorem 4 is not driven by the coupling of unique(d)'s for different m's.

Since $f \in H''$ in Theorem 4, we are not completely ignorant about f. However, $P(f)$ does not arise in the calculation of $P(c|f, m)$, since f is fixed. In this sense it is more perhaps more accurate to interpret the condition on f as "these results will hold if $f \in H''$," rather than "we assume $P(f) = 0$ for $f \notin H''$." With f fixed, either f is or is not $\notin H''$; probabilistic quantities (e.g., $P(f)$) do not enter the issue.

On the other hand, the bound in Theorem 4 does not depend on f, so long as $f \in H''$. In other words, $\{1 - \varepsilon\}^m |H'|$ serves as an upper bound on $\max_{P(h|d)} P(c > \varepsilon|f, m)$ for all $f \in H''$ and, therefore, it also serves as an upper bound on $\sum_{f \in H''} \max_{P(h|d)} P(c > \varepsilon|f, m) P(f|m)$. However, this last expression is bounded below by $\max_{P(h|d)} \{\sum_{f \in H''} P(c > \varepsilon|f, m) P(f|m)\}$ (see Appendix 1). Assuming $P(f|m) = 0$ if $f \notin H''$, this equals $\max_{P(h|d)} P(c > \varepsilon|m)$. In other words, for such a $P(f|m)$, $\{1 - \varepsilon\}^m |H'|$ serves as a bound even if f is not only unknown, but also unfixed. (This argument is similar to the one presented by Buntine.[8])

Often PAC is characterized as being a "worst case" analysis. This characterization arises from what is not specified in Theorem 4: neither $\pi(x)$, the precise $P(h|d)$ (and therefore the precise H' and H''), nor the precise f is specified, even though providing such specifications usually results in tighter bounds. Indeed, PAC is sometimes called "distribution free," which explicitly refers to the fact that its result holds for all $\pi(x)$. In an important sense though, none of these "worst-case" aspects to PAC are really foundational; they simply reflect the fact that rather than hyperparameter-dependent results PAC's practitioners choose to derive hyperparameter-independent results like those in Theorem 4, results that concern what *is* a foundational issue, namely PAC's choice of distribution of interest, $P(c|f, m)$. (See Section 9.)

As an aside, note that since PAC is distribution-free its results hold if $\pi(x)$ has limited support. For such a $\pi(.)$ one could have $c = 0$ even though h and f differ for many x. In this sense, "PAC" is not really concerned with Probably Approximately *Correct* guesses, but rather probably approximately zero-error guesses.

7.4 PAC, SP, AND OFF-TRAINING-SET ERROR

Note the very close similarity between the bound involving $P(c|f, m)$ given in Theorem 4 and the value of $P(c|f, m)$ given by the central result of exhaustive learning; both involve a quantity of the form $(1 - \text{generalization error})^m$. This similarity directly reflects the fact that the frameworks are so closely related and, in particular, it reflects the fact that both use iid error and an iid noise-free likelihood.

As this close similarity suggests, PAC shares some of the subtleties of the SP framework. In particular, the results of PAC too are strongly modified by using the off-training-set error function. One cannot see this by invoking the no-free-lunch theorems, since PAC imposes restrictions on f. However, it is straightforward to construct particular examples that illustrate the breakdown of Theorem 4 when one goes to off-training-set error. In particular, the examples presented in Appendix 1 show how for a Bayes-optimal algorithm, when f is known to lie in a class H'' (called F in that appendix), expected off-training-set error can be flat with increasing m, or even rise as m rises. This does not necessarily indicate that Theorem 4 fails, since the bound in Theorem 4 is so weak for the case considered in Appendix 1. However it does indicate that the gain in generalization as m increases *suggested* by Theorem 4 does not hold.

An example where the bounds of Theorem 4 do fail for off-training-set error is provided by taking H' to be the entire set of input-output functions. If an off-training-set error function is used for such an H', then $\delta \equiv \max_P(h|d)P(c > \varepsilon|f, m)$ always equals 1, regardless of m or ε (so long as $\varepsilon < 1$). This is because for fixed f, one can always find a $P(h|d)$ such that for any $d \subset f$, $P(h|d) = 0$ for any h that intersects f on the points $X - d_X$. So, for example, with $Y = \{0,1\}$, $|H'| = 2^n$, and if $m = n/2$ while $\varepsilon = 15/16$, PAC's bounds on δ equals 2^{-n}, whereas δ for off-training-set error cannot be assured of being less than 1. (Note that even if $m > n$, since m' ranges over all values from 1 to m, m' can be $< n$, and therefore a distribution over off-training-set error values can be associated with such an m.)

As another example of how Theorem 4 can fail for off-training-set error, consider the following $m = 1$ case: Fix some $x' \in X$. Assume that whenever $d_X = x'$ the learning algorithm produces an h which has the value 0 for all $x \in \{X - x'\}$, whereas f has the value 1 for all $x \in \{X - x'\}$. Then if $d_X = x'$, off-training-set error $= 1$. Let π' be the value of $\pi(x')$. Then the probability that the off-training-set error $= 1$ for $m = 1$ is bounded below by π'. So, in particular, the probability that for $m = 1$ the error exceeds any $\varepsilon < 1$ is bounded below by π'. So for all ε such that $\pi' > (1 - \varepsilon)|H'|$, we have a case where Theorem 4 fails for off-training-set error; Theorem 4 necessarily fails for all $\varepsilon > 1 - \pi'/|H'|$.

Finally, note that in those cases where $H' \supseteq H''$, and H' is a strongly limited subset of the set of all input-output functions, then in a manner similar to that in exhaustive learning the input space is broken up into regions over which all hypotheses and (possible) target functions are fixed to the same value. As with exhaustive learning, in PAC one might modify the definition of off-training-set error to mean error over those regions containing no elements of the training set.

7.5 SUPERIMPOSING CONCERN WITH CONVERGENCE ISSUES

Unlike the other frameworks, PAC usually does not content itself with results like those in Theorem 4, nor is it concerned with the tighter bounds one can get for various choices of H' and H'', by and large. (Although some broadly applicable, tighter bounds have garnered interest.) Rather PAC "superimposes" a concern with issues of polynomial vs. exponential convergence of $P(c > \varepsilon|f, m)$.

To see how PAC does this, first rephrase Theorem 4: If δ is a value that is bounded below by $\{1 - \varepsilon\}^m|H'|$, then for any PAC generalizer $P(c \geq \varepsilon|f \in H'', m)$ is bounded above by δ. Accordingly, if $m > [\ln(\delta) - \ln(|H'|)]/\ln(1 - \varepsilon)$, then we are assured that for any associated PAC generalizer $P(c \geq \varepsilon|f \in H'', m)$ is bounded above by δ. In particular, we have such assurance if $m > (1/\varepsilon)\{\ln(|H'|) + \ln(1/\delta)\}$.

This means that for any PAC algorithm, regardless of H' and H'', such an m (needed to assure that $P(c \geq \varepsilon | f \in H'', m) \leq \delta$) is polynomial in $(1/\varepsilon)$ and $(1/\delta)$.

PAC considers the following issue: Given a "concept class" H consisting of a set of $X \to Y$ functions, can one find a PAC generalizer whose target class equals H such that whenever $f \in H$, (i) the generalizer produces its guess in time polynomial in m, and (ii) the value of m needed to assure that $P(c \geq \varepsilon | f \in H, m) \leq \delta$ is polynomial in $(1/\varepsilon)$, $(1/\delta)$, and Q, for some function Q whose value is determined by f and/or H? (That value of m is sometimes referred to as the "sample complexity.") If there is such a generalizer, then H is said to be "polynomially learnable." Note that, in determining that a class is polynomially learnable, one will often have a particular $P(h|d)$ in mind, in which case the worst-case nature of PAC (see above) only concerns $P(f)$ and $\pi(x)$.

Different versions of PAC make different choice of Q. For example, one version chooses Q to be the dimension of X (where X is represented as a multidimensional space or an approximation thereof). In such a case, to allow Q to vary, X must vary and, therefore, so must our learning algorithm (since both h and d are defined in terms of X). In such a situation, usually one implicitly presumes that the algorithm is parametrized by X, and that H is as well.

7.6 THE SIGNIFICANCE OF EXPONENTIAL LEARNABILITY

The presence of polynomial learnability has significance. However, the worst-case nature of PAC means that lack of polynomially learnability usually carries little significance. In this sense, a proof that a particular class is not polynomially learnable has little import.

To see this, let $P \equiv \{P_1, P_2, \ldots\}$ be a set of learning problems, all sharing the same support of $P(f)$ (namely that $f \in H$), but differing in the detailed nature of $P(f)$ and/or in $\pi(.)$. Now consider the (usual) situation where one knows more about $P(f)$ than just its support, and/or one knows something about $\pi(.)$; one knows that the actual problem at hand lies in some proper subset of P, P^*. Loosely speaking, lack of polynomial learnability means (for sufficiently large Q) that it would take a huge m to be assured that your algorithm provides decent generalization for *every* problem in P, including problems outside of P^*. This has no implications for whether you need a huge m to be assured of decent generalization for the problem at hand (i.e., for a problem in P^*). So lack of polynomial learnability has little significance for this situation.

On the other hand, since polynomial learnability implies (for sufficiently small Q) that a small m suffices for any of the problems in P, in particular it means a small m suffices for the problem at hand; presence of polynomial learnability does carry meaning. Phrased differently, a tight (i.e., polynomial) bound conveys information—we know we are somewhere in a small region. But a loose bound conveys no information.

7.7 THE SIGNIFICANCE OF POLYNOMIAL LEARNABILITY

The preceding discussion doesn't tell us exactly *what* information polynomial learnability conveys. In particular, it doesn't tell us what (if any) implications for how one conducts learning in the real world follow from considerations of such an issue. PAC's distribution of interest and assumptions seem to have real-world significance (see Section 9), but does its concern for polynomial learnability? (This is a separate question from whether polynomial learnability is mathematically interesting, or provides some theoretical insight. Those questions are not in dispute.) The rest of this section presents some reasons for doubting that real-world significance. (See Buntine[9] for some other reasons.) These reasons should be viewed as cautionary difficulties in ascribing such real-world significance, rather than as a definitive proof that there is none.

First, recall from the SP framework that in the limit of $n \gg m$, for random off-training-set guessing $P(c|f, m) \propto (1-c)^m \rho_0(c, f, T)$—rather encouraging functional convergence with m, despite the uselessness of the algorithm providing it. There is no obvious reason to believe that the convergence issues PAC concentrates on provide results that have more real-world significance than this. In fact there's a strong reason to believe the opposite: aside from a few special cases (see Section 10), no one has ever used PAC results in the real world.

Second, note that in almost all real-world supervised learning scenarios one has a particular d of size m. One's orders are to generalize, using d. To perform that generalization based on concerns with how things vary with d's and even m's different from the one in front of you is rather peculiar. To one degree or another, this objection applies to the application of all non-Bayesian frameworks to the real world—such frameworks concern learning curves, so to make use of them in the real world would mean somehow treating learning curves as prescriptive rather than descriptive. However, things are worse in PAC, since there Q must vary as well as d; PAC is not concerned with the single learning curve in front of you, but rather in an infinite series of curves, an infinite series of experiments (e.g., an infinite set of X's). How can behavior across that infinite set of experiments have implications for the one particular experiment in front of us?

As an example, say we have two separate learning experiments, A, and B, which share the same value of Q (e.g., the same X). Say that for that value of Q, the generalizer in A performs better than the one in B. However, A is part of a series of experiments in which there is no polynomial learnability (e.g., a series of experiments across different X's in which as one varies X there is no polyomial convergence in $|X|$). On the other hand, B is part of a series of experiments in which there is polynomial learnability. This polynomial learnability tells us that even in the limit of infinitely large Q, we will still get "good" generalization performance in B's series of experiments. But how is this relevant to a real-world statistician confronted with the value of Q in problems A and B?

Even if for some reason we are concerned with a series of experiments, usually we won't be concerned with whether that series exhibits polynomial or exponential convergence. The reason is that the type of convergence is determined solely by behavior in the limit of infinite $1/\delta, 1/\varepsilon$, and Q; the sample complexity for any finite range of these variables has no effect on whether we have polynomial learnability. So if we're interested only in behavior across such a finite range, we won't be interested in whether we have polynomial learnability.[10] (This criticism is essentially identical to the argument found in the conventional statistics community that asymptotics and "consistency" arguments are of little real-world significance, by themselves. See Breiman's first article in these proceedings.)

Another reason to doubt the real-world significance of PAC's notion of polynomial learnability is that PAC is not concerned with ε and $P(c > \varepsilon | f \in H'', m)$ directly, but rather with nonlinear functions of them. The quantities $1/\varepsilon$ and $1/\delta$ are a bit unnatural, from a statistical point of view, compared to $1 - \varepsilon$ and $1 - \delta$. As far as learning in the real-world is concerned, why if I want to consider increasing my "confidence" from $\delta \to \delta + k$ should I not concern myself with error-dependencies involving k, but rather with dependencies involving $k/\delta(\delta + k)$ (i.e., $1/\delta - 1/(\delta + k)$)? (Which dependence is less trivial mathematically is a different issue!) Or better yet, why concern myself with δ and ε at all, rather than more conventional measures of $P(c|f, m)$ that are mostly overlooked by the PAC community, like $E(c|f, m)$, the standard deviation of c, etc.?

Finally, note that if it weren't for the requirement of polynomial dependence on Q, all (finite) classes would be polynomialy learnable, since m is always polynomial in $(1/\varepsilon)$ and $(1/\delta)$. (This is true even if $\pi(x)$ is very highly peaked about one x, in which case on average a new training set element will not help one nail down f.) So if it were not for Q there would be no convergence issues to disentangle for such classes. In this sense, the reason to introduce Q—and thereby go beyond subsections (ii) and (iii) above—is to make the convergence analysis nontrivial, rather than to address issues of concern in real-world generalization.

7.8 BLUMER ALGORITHMS AND THE SIGNIFICANCE OF POLYNOMIAL LEARNABILITY

To illustrate this problematic nature of PAC's notion of learnability, consider another common choice for Q, in which it is defined as a function of f. In doing so it will be useful to consider "Blumer algorithms" for a concept class H, which are defined as follows:

---

[10] It's important to keep in mind that the issue is only whether the single bit of whether we have polynomial learnability is significant. Obviously if we have polynomial learnability with small constants in the polynomial, that will have significant implications for behavior over any finite range of $1/\delta, 1/\varepsilon$, and C. However, the same is also true if the sample complexity has non-polynomial bounds that involve small constants.

DEFINITION 1. Fix an arbitrary single-valued mapping from f to $\mathbf{R}$, $r(f)$. A *Blumer algorithm* with constants $\beta > 1$, $K > 1$, and $0 \leq \alpha < 1$ for the class H is one that

i.   is a PAC generalizer with target class H;

ii. has running time polynomial in m;

iii. when given any m-element d formed by sampling an $f \in H$ obeying $r(f) = Q$, produces an h lying in a set of h's, $A(m, Q)$; and

iv. $A(m, Q)$ contains at most $K^{\lceil Q^{\beta} m^{\alpha} \rceil}$ elements.

Note that for any fixed Q, if we take the set of f's obeying $r(f) = Q$ to be a target class, $A(m, Q)$ is an associated hypothesis class, although not necessarily a minimal one.

As an example, $r(f)$ might be the number of bits needed to encode f with some fixed binary encoding scheme. In such cases one might view $r(.)$ as a "complexity measure" of f. Similarly, log-base-2 of the number of h's in $A(m, Q)$ might be an upper bound on the number of bits needed to encode any h in the set $H' \cap A(m, Q)$. In this case, with $K = 2$, one might interpret $\ln_2[\|A(m, Q)\|] \leq Q^{\beta} m^{\alpha}$ as a bound on the "complexity" of any $h \in H' \cap A(m, Q)$. A Blumer algorithm for such a case is one that when trained on a sample of an f having complexity Q, produces an h of complexity at most $Q^{\beta} m^{\alpha}$. Such complexity-based Blumer algorithms are called "Occam algorithms" by Blumer et al.,[5,6] since such algorithms try to produce guesses with low complexity.

Using Theorem 4 (with H' and H'' being subsets of H), one can show that a Blumer algorithm for H learns H polynomially in $(1/\varepsilon)$, $(1/\delta)$, and the value Q of $r(f)$.[5] (Intuitively, this result holds because the $|H'|$ in Theorem 4 is growing slowly enough with m and Q.) Conventionally this result is interpreted as an *a priori* justification for Occam's razor by taking $r(.)$ and $\ln_2[\|A(.,.)\|]$ to be complexity measures. This is because the result says: if f lies in H; if my guesses always reproduce any d sampled from such an f; and if I can force the complexity of my guesses to be low enough for any d sampled from such an f; then I will have polynomial learnability of H.[11]

One could argue that by itself, this result does not fully justify Occam's razor. After all, it implies nothing about how well an Occam algorithm performs for $f \notin$

[11] As an aside, it should be noted that other disciplines interpret "Occam's razor" differently. For example, one variant of Bayesian analysis interprets the "complexity" dealt with by Occam's razor to refer to the model an algorithm uses (e.g., an H) rather than to a function expressed with that model (see MacKay,[28] and references therein; Jefferys and Berger[25]; and Wolpert[60]). Another version of Bayesian analysis agrees with PAC that "complexity" refers to a function, but also like PAC leaves the mapping from a function to its complexity value essentially arbitrary. (See Section 5, and also Sorkin.[43]) Finally, some approaches define "complexity" to refer to a function, but force the mapping from the function to its complexity value to take a form determined uniquely by the model, up to certain overall invariances. Using this last approach one can prove that a particular model has its off-training-set error diminish with m if and only if Occam's razor works for that model. See Wolpert.[60]

H''. Nor does it imply anything about how well algorithms that are only close to being Occam algorithms perform. (For example, it implies nothing about how well algorithms perform that try to guess h's with low complexity but do not quite meet part (iv) of Definition 1.) Things are worse than this though, because the result casts doubt on its own premises; it casts doubt on the premise that polynomial learnability (or lack thereof) is the correct way to characterize learning problems.

7.9 EXAMPLES OF BLUMER ALGORITHMS CASTING DOUBT ON THE SIGNIFICANCE OF CONVERGENCE ISSUES

The result does this because there are examples where the algorithm is Blumer for reasonable classes H—so we have polynomial learnability—but the algorithm is nonetheless of dubious utility for real-world learning. Intuitively, these examples exploit the fact that the polynomial learnability result holds for any $r(f)$. In particular, $r(f)$ could be the *opposite* of some standard notion of complexity—so that the algorithm prefers what would usually be considered more complex functions over simpler ones—and so as long as the conditions in Definition 1 are met the polynomial learnability implications would still hold. (In this, the result no more justifies Occam's razor than it justifies the opposite of Occam's razor.)

Example 1. Choosing between those theories that agree with the data according to the alphabetical listing of the author of the theory is a Blumer algorithm. The proof of this is in Appendix 4. ∎

So choosing between theories according to their alphabetical listing implies polynomial learnability. Few would claim that such a scheme is anything other than absurd in the real world. Yet when used in its convergence-issues form PAC not only does not disallow this scheme, it actually counsels it.

In other cases, polynomial learnability obtains if the learning algorithm ignores all patterns in the data (aside from assuring that its guess reproduces the data). For example, polynomial learnability obtains for algorithms that set their guesses outside of d_X in a way completely independent of the values in d_Y. The following two examples illustrate this.

Example 2. The space $Y = \{0, 1\}$. The space X can be either infinite or finite (although always countable). $H = H''$ is the set of all functions f for which Y takes the value 1 a finite number of times. Then if the learning algorithm reproduces d but guesses 0 for all $x \notin d_X$, the algorithm learns H'' polynomially. The proof of this is in Appendix 4. ∎

Example 3. X contains a finite number of elements. Y is the set of counting numbers, $\mathbf{N}$. H'' contains all functions from X to Y, and is therefore countably infinite. The algorithm that reproduces d but guesses 0 for all $x \notin d_X$ learns H polynomially. The proof that this case results in polynomial learnability is in Appendix 4. ∎

The reader should not be given that impression that it is always necessary to use an absurd algorithm to get polynomial learnability in terms of Q. In fact, the majority of the cases discussed in the literature where one has polynomial learnability in terms of Q involve quite reasonable learning algorithms. An example pertinent to this issue is given by a modification of Example 3, in which for appropriately chosen $r(f)$ essentially all algorithms result in polynomial learnability:

Example 4. X contains a finite number of elements. Y is the set of counting numbers, $\mathbf{N}$. H'' contains all functions from X to Y, and is therefore countably infinite. Let $r(f)$ be arbitrary, except that it results in real numbers > 1, and the number of f such that $r(f) = Q$ is bounded above by κ^Q for some constant $\kappa > 1$. Then a sufficient condition that an algorithm that reproduces d learns H polynomially is that the algorithm ignores all duplicate pairs in d. The proof of this is in Appendix 4. ∎

The point of these examples is not to prove that the notion of polynomial learnability is somehow "wrong." Rather it is to illustrate that one should be careful in ascribing real-world significance to whether or not one has polynomial learnability.

7.10 COUNTERARGUMENTS TO THE EXAMPLES

In fact, there are counterarguments to all of these examples, and the meaning of these examples could be debated at length. For instance, some might interpret Example 1 to mean simply that PAC should only be used with "reasonable" measures $r(f)$. By this reasoning, if one uses "unreasonable" measures, it's not surprising that one gets unreasonable conclusions.

As a response to this, one might point out that PAC never formally defines what a "reasonable" measure is; the point of Example 1 is to illustrate that convergence issues *considered in isolation*, without any formal specification of what constitutes an allowed measure, can be misleading. (Polynomial learnability holds for Blumer algorithms because the $|H'|$ term in Theorem 4 is growing slowly enough with m; the exact listing of members H' takes on as m grows—i.e., the exact measure used—doesn't matter to the proof.)

In any case, the measures $r(f)$ in the remaining examples are, by most lights, reasonable. What's more, if measures are formally restricted to some "reasonable" set, one immediately runs into the discomforting possibility that the measures we

164   D. H. Wolpert

humans find "reasonable" are precisely those that empirically (!) obey Occam's razor. If this is so, it means that the cost of modifying Theorem 4 so that it doesn't imply an anti-Occam's razor (by requiring that $r(f)$ be "reasonable") is to make Theorem 4's "proof" of Occam's razor circular.

As another example of a counterargument to the examples presented above, one might say in Example 2 that for the case where X is infinite, guessing all 0's outside of d_X is quite sensible, since there are an infinite number of 0's in f but only a finite number of 1's. One response here is to point out that this counterargument does not apply to the finite X example. And in any case, other than in the trivial sense of memorizing what it is seen so far, it is hard to see in what sense the learning algorithm of Example 2 "learns" H as it is provided more data, even though PAC refers to it as an algorithm that learns H optimally.

7.11 CONVERGENCE ISSUES AND THE OTHER FRAMEWORKS

Concern for the sorts of convergence issues investigated in PAC could be super-imposed on most of the other frameworks if one desired. One way to do this is to modify the distribution of interest to be of the form $P(c > \varepsilon | stuff, m)$, and investigate how this quantity is related to m, $1/\varepsilon$, and Q, where Q is some function of $stuff$. As an example, the SP framework has $stuff$ equal to $\{f\}$, just like PAC. Accordingly, one way of applying PAC's convergence concerns to the SP framework would be to analyze what (if any) concept classes are polynomially learnable by Gibbs algorithms of a particular form. (See Section 9 below.) One could compare Gibbs algorithms—or more generally any learning algorithms—by comparing the concept classes each can learn polynomially. (Note that if Q is $|X|$, for example, then to carry out such a comparison one would need an algorithmic way of defining $T(h)$ in terms of $|X|$.) Alternatively, for the case where the distribution of interest is $P(c > \varepsilon | m)$ rather than $P(c > \varepsilon | f, m)$ and Q is the dimension of X, one could analyze what $P(f)$'s are polynomially learnable by Gibbs algorithms of a particular form. (To do this, $P(f)$ must be parametrized by Q, just as H is in PAC.)

It is precisely because such concern for convergence issues can be superimposed on the other frameworks that such concern is not listed in Table 1 as a foundational distinction amongst the frameworks.

7.12 SUMMARY

• PAC is concerned with generalizers obeying the properties given in subsection (ii) above: Loosely speaking, for all samples generated from any target in a specified "target class," the guess of the generalizer is assured of being in a specified "hypothesis class." The generalizer is also required to reproduce such a training set.
• For such generalizers theorem (4) bounds $P(c > \varepsilon | f, m)$ (for varying ε) in the absence of noise. The result does not depend on the sampling distribution, and in this sense is "distribution-free." It is average-case over data, this-case over targets,

and worst-case over generalizers lying in a particular class (in addition to being worst-case over sampling distributions). It can be made average-case over targets if one knows the support of $P(f)$. Note that such knowledge is less restrictive than the specification of $P(f)$ required by the Bayesian framework.

- The close similarity between PAC and the SP framework is reflected in the formal similarity between PAC's bound on $P(c > \varepsilon | f, m)$ and the central result of exhaustive learning.

- As that formal symmetry suggests, conventional PAC results do not hold for off-training-set error.

- PAC superimposes a concern with polynomial versus exponential convergence of bounds concerning $P(c > \varepsilon | f, m)$. Such concerns could be superimposed on most of the other frameworks if one so desired.

- It can be argued that due to the worst-case nature of PAC, exponential convergence results have little real-world meaning.

- There are also reasons for doubting the real-world significance of polynomial learnability results. Among these are the fact that polynomial learnability is only an asymptotic property, and the fact that it concerns varying the supervised learning scenario (e.g., the size of the spaces involved) from the one at hand.

- Another reason to doubt the significance of polynomial learnability arises from analysis of Blumer algorithms. A result concerning Blumer algorithms and polynomial convergence has been presented as a first principles argument for Occam's razor. However, the result holds for any complexity measure. Accordingly, it "justifies" the use of "Occam's razor" in a number of rather absurd scenarios. This casts doubt on the real-world significance of polynomial learnability.

8. THE VC FRAMEWORK

8.1 OVERVIEW OF THE VC FRAMEWORK

The VC framework is in many ways similar to both the SP framework and PAC. All three are conventionally defined in terms of iid likelihood and iid error. They also all have m rather than d fixed in their distribution of interest (unlike the Bayesian framework), and all can be cast with f fixed in that distribution.

Cast in this f-fixed form, the (vanilla) VC framework investigates the dependence of the confidence interval $P(|c - s| > \varepsilon | f, m)$ on m, s, the learning algorithm, and potentially f (recall that s is the empirical misclassification rate). Sometimes the value of $P(|c - s| \leq \varepsilon | f, m)$ for one particular set of values for c, s, f, and m is called one's "confidence" that $|c - s| \leq \varepsilon$. Note that if one has a bound on $P(|c - s| > \varepsilon | f, m)$, then one also has a bound on $P(c > s + \varepsilon | f, m)$. Some researchers interpret this—incorrectly, as is demonstrated below—to be a bound on how poor c can be for a particular observed s. (Such misinterpretations are a direct result of the widespread presence in the VC literature of the first sin discussed in the introduction.)

The famous "VC dimension" is a characterization of $H\tilde{\,}$, the h-space support of a learning algorithm $P(h|d)$. For current purposes (i.e., $Y = \{0,1\}$ and our error function), it is given by the smallest m such that for any d_X of size m, all of whose elements are distinct, there is a d_Y for which no h in $H\tilde{\,}$ goes through d. (The VC dimension is this smallest number minus one.) It turns out that in a number of scenarios, as far as $P(|c - s| > \varepsilon|f, m)$ is concerned, what is important concerning one's algorithm is its VC dimension. In particular, one major result of the VC framework is an upper bound on $P(|c - s| > \varepsilon|f, m)$ in terms of m, s, and the VC dimension of the learning algorithm. (Accordingly, the challenge in the VC framework is usually to ascertain the VC dimension of one's generalizer.)

8.2 THE RELATIONSHIP BETWEEN THE VC FRAMEWORK AND THE OTHER FRAMEWORKS

This and other VC bounds on $P(|c - s| > \varepsilon|f, m)$ turn out to be independent of f. (See Baum and Haussler,[3] Cohn and Tesauro,[13] Haussler et al.,[22] Haussler,[24] Hertz et al.,[27] Vapnik,[51] Vapnik and Bottou,[52] COLT,[14] and Rivest.[38]) In contrast, PAC needs to make an assumption concerning f, and the SP framework's result depends on a function of f, ρ. Although this does not constitute a foundational distinction between the VC framework and those other frameworks, it can be turned into one: it means that unlike those other two frameworks, the VC framework can be recast as concerning a distribution that does not fix f (namely $P(c - s > \varepsilon|m)$), and this can be done without making any assumptions for $P(f)$.

However, like PAC, the VC framework only gives us bounds. Moreover, one can turn the results of both SP and PAC into bounds that are applicable independent of f; in SP replace ρ with a worst-case (over f) ρ and, in PAC, make the target class the entire space of possible f's. (These are not particularly *interesting* bounds, but that is another issue.) If one does this the question of whether f need be fixed no longer distinguishes the three frameworks.

Another difference between the frameworks is their choice of generalizer. The VC framework only assumes that the generalizer's guess lies in some "concept class" $H\tilde{\,}$, regardless of the data. This obviously differs from the investigations of the SP framework (and especially the investigations of that framework where the support of $T(h)$ is the entire set of allowed functions). And although similar to the PAC generalizer, this VC generalizer differs a bit from that generalizer as well. This is because VC results do not directly concern themselves with whether d is a sample of a target function from a particular target class; $H\tilde{\,}$ is the set of all possible guesses, no matter what the data. This differs from vanilla PAC, in which the generalizer is allowed to make guesses outside the hypothesis class if the data is not compatible with an element of the target class. Of course, one can modify the VC framework to remove this difference—just assume that f falls in some target class, and redefine $H\tilde{}$ to be the associated hypothesis class. However, this modification to the VC framework entails a restriction on f, which is unnecessary and is also often undesirable, given the power of f-independent results.

Another difference between the VC framework and the others is that the VC framework does not concern itself with the random variable c, but rather with $c - s$. It is in this sense that the VC results are confidence intervals. The other frameworks do *not* directly concern confidence intervals. (Except in the trivial sense that $c = c - s$ when $s = 0$, so distributions over c are "confidence intervals" for the $s = 0$ case.)

8.3 THE SINGLE-h CASE

To show all this in more detail, consider the case where $H\tilde{}$ consists of a single h, h'. In doing so we will be ignoring the major insight of Vapnik and his co-workers, which is how to use the concept of uniform convergence to derive confidence intervals when $H\tilde{}$ consists of multiple h. However, this case of a single-h $H\tilde{}$ makes the foundational issues clearer. Moreover, although it is crucially important in applying the VC results, in many respects the uniform convergence technique is not very important in relating the foundations of the VC framework to those of the other frameworks.

For such a single-h $H\tilde{}$, the VC framework consists of bounds on the confidence interval $P(|c - s| > \varepsilon | f, h', m)$. In particular, the following theorem is derived in Appendix 5 using the EBF:

THEOREM 5. Assume that there is an h' such that $P(h|d) = \delta(h - h')$ for all d. Then

$$P(c > s + \varepsilon | f, m) < 2e^{-m\varepsilon^2}.$$

For a more general class of generalizers than the single-h $H\tilde{}$, the bound of Theorem 5 applies if one multiplies the right-hand-side by $|H\tilde{}|$, the number of functions in $H\tilde{}$. Similar modifications obtain if one instead characterizes $P(h|d)$ by its VC dimension. Common to all such results is a rough equivalence (as far as c is concerned) between (i) lowering s; (ii) lowering the expressive power of $P(h|d)$ (i.e., shrinking its VC dimension, or shrinking $|H\tilde{}|$); and (iii) raising m.

A word of caution concerning Theorem 5: For our learning algorithm, $P(c > s + \varepsilon | f, m) = P(c > s + \varepsilon | f, h', m)$. Moreover, the quantity $P(c > s + \varepsilon | f, h', m)$ is perfectly well defined even for algorithms which can guess more than one h; it simply means the probability that $c > s + \varepsilon$ when m elements are sampled from f and we know that the resultant h is h'. However, for such an arbitrary learning algorithm, Theorem 5 does not bound $P(c > s + \varepsilon | f, h', m)$. The reason is that for an arbitrary learning algorithm, knowing that the algorithm guessed h' tells us something about d, and the derivation of Theorem 5 is based on having no restrictions on d other than those imposed by f. (Indeed, it is to get a bound valid for an arbitrary algorithm that one must introduce uniform convergence.)[12]

8.4 RELATING VC RESULTS TO THE REAL WORLD

Since (for our algorithm) h and f are fixed, c is also fixed, for iid error. This differs from all three of the other frameworks, which have c vary. In fact, in Theorem 5 what is varying is d_X (or more generally, when there is noise, d). So Theorem 5 is *not* giving us the probability that c lies in a certain region, but rather the probability of a d_X such that the difference between the fixed c and (the function of d_X) s lie in a certain region.

This issue of what is and is not varying is crucially important when one tries to apply the results of any of the frameworks to the real world. A detailed discussion of this issue is presented in Section 10. For now though, it is worth noting that since you can measure s and want to know c (rather than the other way around), one would like to have a bound on something like $P(c > k|s, m)$, perhaps with $k \equiv s + \varepsilon$. With such a bound, we could say that since we observe m and s to be such-and-such, with high probability c is lower than function(such-and-such).

It might seem that Theorem 5 can tightly restrict $P(c > k|s, m)$, since other than the extra 's' on the right-hand side of the conditioning bar, because it is f-independent the quantity bounded in Theorem 5 is precisely $P(c > k|s, m)$. Unfortunately this is not the case. Indeed, Appendix 5 shows that Theorem 5 can be written as a bound on the "inverse" of $P(c > k|s, m)$, $P(s < \kappa|c, m)$, where $\kappa \equiv c - \varepsilon$.

How does $P(s|c, m)$ relate to what we wish to know, $P(c|s, m)$? The answer is given by Bayes' theorem: $P(c|s, m) = P(s|c, m)P(c|m)/P(s|m)$. Unfortunately, this result has the usual problem associated with Bayesian results; it is prior-dependent. To try to "solve" this problem we might try fixing f in the conditioning event, just like PAC and SP do (i.e., we might try considering $P(c \mid s, f, m)$ rather than $P(c \mid s, m)$). However for our generalizer, $P(c|s, f, m) = P(c|s, f, h', m)$ is just a delta function about whether f and h' have error c, and as such is not interesting. (For fixed f it is only the absence of s on the right-hand side of the conditioning bar that makes the VC framework's results nontrivial.) So fixing the value of f is not a viable way of getting around the need to use the prior. Given that we wish to know the probability of c given an observed s, it appears that we must consider the prior.

---

[12] Interestingly, if d is fixed as well as h' (rather than just m and h'), then it usually does not matter how h' was generated; generically $P(stuff|h', d)$ is independent of $P(h|d)$, unlike $P(stuff|h', m)$. $(P(stuff|h', m) = \sum_{f,d} P(stuff|f, h', d)P(f|d)P(d|h'); P(stuff|h', d) = \sum_f P(stuff|f, h', d) P(f|d).)$ In other words, if you are looking at distributions conditioned on the actual d at hand in addition to h', quantities like the VC dimension of the generalizer that produced h' are irrelevant. It is only by pretending that we only know the training set's size (and not its actual elements) that it can matter what generalizer generated h'.

Does it somehow turn out though that that prior has little effect? Alas, no; $P(c > s+\varepsilon|s, m)$ can differ markedly from the bound on $P(s < c-\varepsilon|m|a|c)$ given in Theorem 5. Even if *given a truth c*, the probability of an s that differs substantially from the truth is small, it does not follow that *given an s*, the probability of a truth that differs substantially from that s is small.

As an example, consider the case where $\pi(x)$ is flat over all x and $P(f)$ is flat over all f. Together with our likelihood, these distributions sets $P(c \mid m)$. In Appendix B of Wolpert[56] it is proven that for this case, for iid error, and for the generalizer of Theorem 5,

$$P(c|s, m) = C_{sm}^m c^{sm}(1 - c)^{m-sm} C_{nc}^n (r - 1)^{nc}, \tag{8.1}$$
$$\text{where } C_b^a \equiv a!/[b!(a - b)!].$$

So for this case the c-dependence of $P(c|s, m)$ is given by the product $c^{sm}(1 - c)^{m-sm} C_{nc}^n (r - 1)^{nc}$. If we instead calculated $P(c|s, m)$ under the assumption that $P(c|m)$ were uniform over all possible c, the last pair of these factors in Eq. (8.1) would disappear. Clearly the prior we assume can have a major effect.

Equation (8.1) can be viewed as a sort of compromise between the likelihood-driven "something for nothing" results of the VC framework, and the no-free-lunch theorems. The first term in the product has no c-dependence. The second and third terms together reach a peak when $c = s$; they "push" the true misclassification rate towards the empirical misclassification rate, and would disappear if we were using off-training-set error. These two terms are closely related to the likelihood-driven VC bounds. However, the last two terms, taken together, form a function of c whose mean is $(r - 1)/r$. They reflect the fact that all f's are being allowed with equal prior probability, and are closely related to the no-free-lunch theorems (despite the fact that iid error is being used). In this sense, Eq. (8.1) is nothing other than a product of a no-free-lunch term with a VC-type term.

8.5 INTUITIVE PERSPECTIVES ON THE VC FRAMEWORK

When expressed as a bound on $P(s < \kappa|c, m)$, it becomes clear that Theorem 5 can be viewed as an exercise in coin flipping. You have a coin with probability c of heads, and you wish to know the probability that in m iid flips of the coin, you see ms heads. In supervised learning this becomes "you have a probability c that at a random x the values $f(x)$ and $h'(x)$ differ, and you want to know the probability that in m randomly chosen x values $f(x)$ and $h(x)$ differ ms times." That probability is $P(s|c, m)$ and (modulo the replacement of "s" with "$s < \kappa$") is given by Theorem 5. (See Appendix 5, result (5.3).) However, if instead of the probability of getting sm heads in m tosses of a coin with bias c, we wished to know the probability that the coin's bias is c given that there were sm heads in m tosses, then we would have to use Bayes' theorem. And, therefore, we would have to consider $P(c)$. In general, VC results say it is unlikely for s and c to differ greatly. By themselves, without information about $P(c)$, they do not say that it is unlikely that c is large in those cases where s is small.

In response to such formal admonitions, one is tempted to make the following intuitive reply: "If there is a fixed h' and a fixed f and you randomly sample X, and if you see that h' and f agree, would *you* not expect that h' and f agree on future trials?" Alternatively, consider the sequence where: a sample point is drawn from f; h' correctly predicts it; another sample point is drawn; h' correctly predicts it, etc. After going a decent distance into such a sequence, you guess that h' will correctly predict the next sample point. And lo and behold, it does (s is small). Continue along this way, until you have the entire training set in hand, and we see that the generalizer {guess h} has excellent cross-validation error. So if you believe that cross validation works, you should believe that it is unlikely for c to be large in the single-h case if it is observed that s is small.

To respond to these heuristic arguments, consider the case where h' is some extremely complex function (it appears to be random in fact). h' was fixed before any fixing of f, d, or anything else. We choose f by sampling $P(f)$. Then we iid pick sample points of f to make up d, and lo and be hold, f and h' agree on them. Despite the intuitive argument presented above, in such a situation our first suspicion might be that cheating has taken place, that h' is based on insider information of $P(f)$. After all, how else could the "essentially random" h' agree with f—h' was fixed without information of d (and therefore without any "coupling" with f).

If, however, we are assured that no cheating is going on, then "intuition" might very well just shrug its shoulders and say that the agreements between f and h are simple coincidence. *They have to be* since, by hypothesis, there is nothing that could possibly connect h and f.[13] So intuition need not proclaim that the agreements on the data set mean that f and h' will agree on future samples.

8.6 A SUGGESTIVE ANALOGY

In addition to coin flipping, another helpful analogy is the problem where you have a one-dimensional real-valued random variable x, and the probability distribution over x, $P(x)$, is a Gaussian of a certain width which you happen to know. All you do not know is the mode of the Gaussian, μ. $P(x)$ is now sampled m times, giving m x-values with sample mean $\bar{x}$. Based on this information, estimate μ. A confidence interval approach to this problem (like the VC framework) would say that regardless of what μ is, for large m it is quite unlikely that $\bar{x}$ differs much from μ. However, if we were to try to directly evaluate $P(\mu|\{x_i\})$—the probability of what we want to know conditioned on what we do know—then we would have no choice but to consider the distribution $P(\mu)$. If it so happens that that distribution

---

[13]Note that if this hypothesis is incorrect, then to formulate the problem correctly, we must know about the possible *a priori* "connection" between f and h. In such a situation, in essence, cheating *is* allowed. In which case yes, the number of agreements between f and h is meaningful.

is strongly peaked about a point far from $\bar{x}$, then we would not be able to conclude that μ differs little from our $\bar{x}$. Instead we would simply conclude that our sample is an unusual one.

An example of this effect in a learning context is the following scenario: Let $Y = \{0, 1\}$, let h' be the all-zeroes function $(h'(x) = 0 \,\forall\, x)$, and let f be a function that is almost all 1's and that has a small number K of zero-valued outputs (i.e., $n - K$ is much larger than K). For this scenario, c is close to 1 (assuming $\pi(x)$ is not strongly biased towards those K x's at which $f(x) = 0$). Let m be $< K$, and $s = 0$. A VC-type confidence interval analysis says it is unlikely that c and s differ a lot. This is true, but in fact though, they differ maximally, *for this s*. This is correctly reflected in $\mathrm{P}(c|s, f, m)$, which is a delta function about the correct value of c. So the confidence interval is doing a very poor job of estimating c. Rather what it is telling us (given the large value of c) is that the s value we happen to have is unlikely. (Small solace, if our s-based prediction is badly in error.)

Note that this example can be modified so that rather than low s being accompanied by high c, high s is accompanied by low c. In fact, let h'' be the all 1's function. Then whenever s' (the s for h') $= 0$, $s'' = 1$. However c'' is close to 0. So $P(c', c''|s' = 0, s'' = 1, m)$ is close to a delta function about c values that are the reverse of one might expect given that s' is low and s'' is high. Even if one averages over training sets, the values of s are serving as very poor predictors of the values of c.

This not only serves as a caution against using s to choose among hypotheses; it also serves as a caution against using error rate on a validation set to choose between generalizers (in that scenario "s" means empirical error on the validation set). The most natural way to avoid these unpleasant conclusions is to take into account the comparitive rarity of the values of s_1 and s_2 giving this c-is-the-reverse-of-s behavior. However, to do that we must change our distribution of interest so that it is no longer conditioned on the s values at hand. This means that even if we know the values of s_1 and s_2 (for example if we're using cross validation to choose between two generalizers, so s_1 and s_2 are the observed errors on validation sets), we must analyze a distribution of interest in which we pretend to not know those values.

The preceding examples suggest an analysis of the following issue: Are there broadly applicable bounds on the probability of an s such that the VC bound is over-optimistic, i.e., such that $P(c > s + \varepsilon|f, m) < P(c > s + \varepsilon|f, s, m)$? At present the answer to this question is not known. However, it appears that for some scenarios at least, such bounds do exist; there exist bounds on the VC bounds, so to speak. (See also the end of the discussion in Section 10 on We-Learn-It.)

8.7 THE VC FRAMEWORK, DATA-AVERAGING, AND ON- AND OFF-TRAINING-SET ERROR

There are other difficulties in trying to use the VC framework to aid real-world supervised learning, in addition to problems with which side of the conditioning bar "s" is on. One such problem—a problem shared by PAC and SP—is that only m is fixed in the distribution of interest, whereas in the real world we almost always know d in full, not just its size. To rectify this for the VC framework we might modify the distribution of interest to be $P(c > s + \varepsilon | f, d) = P(c > s + \varepsilon | f, h', d)$. However, $P(c > s + \varepsilon | f, h', d)$ is simply a delta function, since f and h' determine c, and h' and d determine s. Moreover, the dependence of this delta function on c will depend crucially on f, when in the real world we do not know f. Now if we leave out the f in the conditioning event—i.e., examine $P(c > s + \varepsilon | h', d)$—we do not get a trivial delta-function, nor does our result depend on f. However, now our result depends on $P(f)$; in essence, we are doing a modified Bayesian calculation. In fact, there is no reason to calculate $P(c > s + \varepsilon | h', d)$ rather than just $P(c > \kappa | h', d)$, since h' and d fix s. (For similar reasons, there is no point to putting "s" in the conditioning event.) But aside from the "$>$", $P(c > \kappa | h', d)$ is exactly the kind of quantity calculated in conventional Bayesian analysis.

At present, the issue of how VC results are affected by going to an off-training-set error function is not well understood. In particular, little is currently known about $P(|c - s| > \varepsilon | f, m)$ for an off-training-set c. However some illuminating examples can be constructed. For example, consider the h', h'', and f considered at the end of the preceding subsection. Change $\pi(x)$ from being uniform to being highly biased towards the K X values where $f(x) = 0$. Then we are *likely* to have a training set such that $s' = 0$ and $s'' = 1$. Moreover, if we are concerned with off-training-set cost, then for $m \sim K$, almost all off-training-set points are in that part of X for which $f(x) = 1$, and off-training-set errors are the reverse of the s values, *for a set of likely s values*. In this case $P(|c - s| > \varepsilon | f, m)$ is substantially worse for off-training-set error than for iid error.

In any case, the no-free lunch theorems always apply. So if we are comparing two deterministic algorithms, unless assumptions are made about $P(f)$, the fact that algorithm one has lower VC dimension and lower s than algorithm two implies *nothing whatsoever* about whether $E(C|s, d)$ is lower for algorithm one than for algorithm two, as far as off-training-set error is concerned. In this sense, the VC theorems do not justify "trading off" the expressive power of one's generalizer (i.e., its VC dimension) against how well it fits the training set, in an effort to optimize generalization.

The no-free-lunch theorems also have some implications for the scenario of interest in the VC framework even if one uses the iid error function. To see this, note that the iid error does not always equal a linear combination of off-training-set error and s where the coefficients are determined by m, m' and n; more information is needed. (E.g., if $m = 5$, $m' = 3$, $s = 2/5$, the resultant contribution to the iid error could either be 1 misclassification out of 3 x's (this is the case where both disagreements are on the same x) or 2 misclassifications out of 3 x's (this is the case where all 3 agreements are on the same x). However when $n \gg m$, we can generically take $m' = m$, in which case the iid error is given by the ratio $(ms + (n - m)C_{\text{off}})/n$ (C_{off} being the off-training-set error). So by the no-free-lunch theorems (which apply to C_{off}), in this limit, averaged over all $P(f)$ (or all f if you prefer), for any fixed d any two algorithms have the same s-conditioned expected iid error, even if their VC dimensions differ.

On the other hand, this formula for iid error also makes clear that for a fixed generalizer, averaged over all $P(f)$, the s-conditioned expected iid error shrinks if s decreases while m stays constant (if $n \gg m$). However, for particular $P(f)$ this need not be the case; increasing s while keeping the generalizer and m fixed may *decrease* expected (s-conditioned) iid error for certain $P(f)$. (Recall that we have no noise, so "overtraining" in the traditional sense isn't an issue.) As an example, say $P(f)$ is a delta function about the f $\{Y = 1\}$. Say that for certain d_X, the generalizer reproduces d, but guesses 0 everywhere off of d_X. For all other d_X, the generalizer doesn't reproduce d, but guesses 1 everywhere off of d_X. This gives the claimed behavior.

An open question is to characterize the relationship between $P(h|d)$ and the spread (as one varies $P(f)$) of how s-conditioned expected iid error changes as s increases. In particular, it would be interesting to characterize the relationship between $P(h|d)$ and the percentage of all $P(f)$'s such that for that $P(h|d)$, over some particular range of s values, expected s-conditioned iid error decreases as s rises.

8.8 SUMMARY

- The VC framework is concerned with confidence intervals $P(|c - s| > \varepsilon | f, m)$ for variable ε.

- The VC dimension is a characterization of the support (over all h's) of $P(h|d)$. Many theorems bounding $P(|c - s| > \varepsilon | f, m)$ depend only on m, the VC dimension of the generalizer, and ε. In particular, such results hold for all f and/or $P(f)$ and for any generalizer of the given VC dimension.

- In the illustrative case where $P(h|d)$ always guesses the same h, c does not vary in $P(|c - s| > \varepsilon | f, m)$; rather s does. (s is determined by the random variable d_X, which is unspecified in the conditioning event.) In all the other frameworks c varies.

- This motivates the analogy of viewing the VC framework as coin-tossing; one has a coin of bias-towards-heads c, and flips it m times. The VC frameworks' distribution of interest is analogous to the coin-flipping distribution $P(|c - s| > \varepsilon | c, m)$. This coin-flipping distribution concerns the probability that the empirically observed frequency of heads will differ much from s. This probability does not directly address the process of observing the empirical frequency of heads and then seeing if the bias-towards-heads differs much from that observed frequency. That process instead conerns $P(|c - s| > \varepsilon|s, m)$.

- Accordingly VC results do not directly apply to distributions like $P(c|s, m)$; one can not use VC results to bound the probability of large error based on an observed value of s and the VC dimension of one's generalizer. Note no recourse to off-training-set error is needed for this result.

- The behavior of $P(|c - s| > \varepsilon|f, m)$ for off-training-set error is not currently well understood.

9. EXTENDING THE FRAMEWORKS

This section discusses various extensions of the frameworks. In particular, a "Bayesian modification" to the familiar bias-plus-variance results is presented in Eq. (9.5) below.

9.1 EXAMPLES OF PREVIOUS WORK EXTENDING THE BAYESIAN FRAMEWORK

The oldest of the four frameworks, the Bayesian framework, is in many respects set in its ways. Variations of the framework do not concern the distribution of interest. Rather, variations tend to concern the form of what is assumed known, and how to manipulate what is known. For example, one can parametrize the distribution $P(d|f)$, say, by a noise level β, thereby getting $P(d|f, \beta)$. One can then impose a distribution over $P(\beta)$, so that $P(d|f) = \int d\beta P(d|f, \beta)P(\beta|f) = \int d\beta P(d|f, \beta)P(\beta)$ (assuming β and f are independent). This kind of procedure is known as "hierarchical Bayesian analysis," with β being a "hyperparameter."

As an alternative, one can set $P(d|f)$ to $P(d|f, \beta')$, where $\beta' \equiv \arg\max_\beta P(d|\beta)$, and $P(d|\beta) = \sum_f P(d|f, \beta)P(f)$. This procedure is known as ML-II, or generalized maximum likelihood.[4,53] It was recently rediscovered as the "evidence procedure."[21,28] Interestingly, even if $P(d|\beta)$ is very peaked in β, ML-II need not be a valid approximation to hierarchical Bayes—in this sense, choosing the most probable model is not always warranted (see Strauss et al.,[45] Wolpert and Strauss,[62] and Wolpert[57]).

As another example of a variation of the Bayesian framework, in empirical Bayesian analysis, one dispenses entirely with the notion of full formal rigor, and uses experience (i.e., previous data) to set one's distributions.[4] Finally, note that no matter how one arrives at the distribution $P(f|d)$, there is always the subsequent difficulty of finding the Bayes-optimal h^* based on $P(f|d)$. Often potentially severe approximations must be used, like replacing the Bayes-optimal guess with the MAP f, $\arg\max_f P(f|d)$, or (of even more dubious validity—see Wolpert[58]) replacing it with the f parametrized by the MAP value of a parameter θ describing f, $f_{\arg\max_\theta P(\theta|d)}$.

9.2 EXAMPLES OF PREVIOUS WORK EXTENDING THE SP AND PAC FRAMEWORKS

In contrast to the case with the Bayesian framework, there is a good deal of current work concerned with modifying and extending the other three frameworks. For example, a lot of work has been done extending SP to the case of noise, nonzero temperature generalizers, and/or assumed correspondences between the generalizer and the prior.[42,46,47] Often this is all done with f and h parametrized as neural nets. Sometimes the distributions involving f and d are referred to as "the teacher," and $P(h|d)$ as "the student."

As example of an extension of PAC is the "Probably Approximately Bayes" framework (PAB—see Anoulova[1]). Simplified a bit, PAB is PAC where f is no longer single-valued (i.e., where one allows noise), and where one modifies the error function accordingly to be $Er(f,h) = \sum_x \pi(x) \sum_y f_{x,y}[1 - \delta(h(x),y)]$. (Note that if h, too, could be stochastic, then $Er(f,h) = \sum_x \pi(x) \sum_{y,y'} f_{x,y} h_{x,y'}[1 - \delta(y',y)]$.) Since f is not single-valued, there is no h for which $Er(f,h) = 0$. Accordingly, rather than concentrate on $P(c > \varepsilon | f, m)$, in PAB one concentrates on $P(c - BO > \varepsilon | f, m)$, where $BO \equiv \min_h Er(f,h)$. The term "Bayesian" in the name of this framework is a bit misleading, since f is fixed and d varies; one does not compare to the h that is Bayes-optimal given d, but rather to the h that is "Bayes-optimal" given f. (See Section 5.) A similar framework is analyzed using VC-type results by Haussler.[23,24]

9.3 EXAMPLES OF PREVIOUS WORK EXTENDING THE VC FRAMEWORK

An example of an extension of the VC framework (which actually came to prominence about the same time as the original work) is "structural risk minimization." Loosely speaking, in structural risk minimization, one has several generalizers[51] forming a nested hypothesis class structure, $H_1^- \subset H_2^- \subset \ldots$. One parses the set of generalizers to find the generalizer that *by itself* (i.e., according to the VC dimension of the associated H_i^-) has the best confidence interval. The nesting means that to an often good approximation we can assign that "best confidence interval" to our generalizer's guess, despite our having parsed a set of generalizers to arrive at that guess. Without nesting, generically one would have to treat the entire parsing algorithm along with the underlying generalizers as simply a new learning algorithm, with VC dimension determined by the union of the H_i^-. In such a situation, one would not be able to exploit differences in the VC dimensions of the constituent H_i^-.

A more recent example of an extension of the VC framework is a "localized" version of that framework.[52] Loosely speaking, in this version, one modifies the error function to be an average of a loss which is very sensitive to guessing errors in some regions of the input space and less sensitive in others. With an appropriately modified definition of empirical misclassification rate, VC techniques can be used with such an error function to get "localized" confidence intervals.

A priori, we have no assurances that any particular localization is to an off-training-set region, so the technique cannot be used to give an off-training-set version of the VC framework. In addition, it often makes most sense to define a learning algorithm to be "local" based on properties of $P(h|d)$. For example, one might say an algorithm is local if the h guessed by the algorithm has the property that its value at x is most sensitive to those elements of d lying closest to x. (An alternative is to define an algorithm as "local" based on how it is implemented.) For such a definition, it is not clear that the localized VC framework is in any sense a modification of the conventional VC framework to make it better address local learning algorithms, as some have suggested.

9.4 NOVEL EXTENSIONS OF THE FRAMEWORKS

Many other extensions of the frameworks can be found in the literature. However, only a small fractions of all extensions have been thought of, never mind analyzed. One of the advantages of phrasing the four frameworks in terms of the EBF is that doing so provides a programmatic way to find extensions, combinations, and modifications of the frameworks: we can "mix and match" amongst the middle two columns of Table 1, make up new entries for those columns, etc. The rest of this section is a presentation of some such variations:

1. One can investigate what happens if one mixes and matches the conditioned event in the distribution of interest across the four frameworks (i.e., the left-hand side of the distribution of interest). For example, rather than $E(C|d)$, one might modify the Bayesian framework to analyze $P(c > \varepsilon|d)$, or $P(c > \varepsilon + s|d)$, or even something like $P(c = 0|d)$ (see Wolpert[56]).

The following variations all concern the conditioning event in the distribution of interest (i.e., the right-hand side of the distribution of interest). See Buntine[8] for some other interesting variations.

2. One can investigate what happens to $E(C|d)$ if we use a PAC generalizer. What kinds of bounds can we set on $E(C|d)$ if we know that $f \in H''$ and that our generalizer's guess is in H'? In general, the upper bound is given by $\max_{f,h} Er(f, h, d)$, where the maximum is over all h in H' and all f in H'' with nonzero posterior, and similarly for the lower bound. A bit less trivially, we can consider the minimax h'', that is the h'' that minimizes $\max_f Er(f, h'', d)$. But what if we do not only know that $f \in H''$ but know the entire $P(f)$; given $P(f)$, what $h \in H'$ minimizes $E(C|d)$? How much does restricting h to H' add to $E(C|d)$, above and beyond the lowest possible value of $E(C|d)$ over all h? What relationships between $P(f)$ and H' govern this?

These kinds of questions are of more than academic interest. It is *very* common to perform Bayesian analysis with a parametrized model, and such models almost always are incapable of expressing arbitrary h's. (One example of such a parametrized model used in Bayesian analysis is neural nets—see Buntine and Weigend,[10] and Wolpert.[58]) In other words, with such models $h \in H'$ where H' is a proper subset of the set of all possible functions. Despite the wide-spread popularity of such an approach, relatively little is known about its formal properties in the (extremely common) scenario where f can lie outside of H'. Usually one simply assumes that H' is a "good enough approximation" to the support of $P(f|d)$.

3. In a similar way, one can see what happens to $E(C|d)$ if one uses a Gibbs generalizer. That generalizer can either have $T(h) = \text{Pr}_F(h)$ or it can have a degree of "misfit" between $T(h)$ and $P(f)$. More generally, one can see what happens to $E(C|d)$ if one uses a generalizer that is Bayes-optimal, but for a different prior over f than the actual $P(f)$. (Some preliminary results concerning these issues have been achieved,[22,34,35] albeit primarily for a distribution of interest other than $E(C|d)$. See below.)

4. As another example of analyzing one framework's distribution of interest for another framework's generalizer, consider $\delta \equiv P(c > \varepsilon|f, m)$ for a Gibbs generalizer. For illustrative purposes set $T(h)$ constant, $\pi(x)$ constant, and $n \gg m$. Then it is straightforward (see Appendix 6) to prove that

$$\delta \leq \{1 - \varepsilon\}^m r^m. \tag{9.1}$$

(Another bound that is tighter but more cumbersome is also given in Appendix 6.)

This result should be compared to the PAC result, Theorem 4. In particular, it is instructive to compare the tighter bound given in Appendix 6 to the PAC bound (derived just before the end of Appendix 3) $\delta \leq \{1 - \varepsilon\}^m \sum_{h \in H'} \delta(Er(h, f) > \varepsilon)$, for the case where H' is the set of all input-output functions (since $T(h)$ is nonzero for all h). Doing so, we see that the bounds where we know our generalizer to be the Gibbs generalizer with constant $T(h)$ are exactly r^{m-n} times the associated PAC bounds, independent of $\pi(x)$. This ratio can be viewed as a factor giving the gain in generalization accuracy which accrues from specifying the generalizer (as an exhaustive learning generalizer with $T(h)$ constant) rather than using a worst-case generalizer.

5. Other interesting issues concern $E(C|f, m)$. Asking for the learning algorithm that minimizes $E(C|f, m)$ is silly—that "optimal" algorithm is the algorithm that guesses f independent of the training set. And since we do not know f in practice, one can also argue that this distribution's a bit pointless as far as aiding real-world generalization is concerned. However, consider the case where one hides the f, i.e., where one considers $E(C|m)$. In that only m is specified, not d, this distribution is like the distributions of interest in PAC, SP, and the VC framework. However, in that f is not fixed, this distribution is

like the distribution in the conventional Bayesian framework. In particular this distribution will depend on $P(f)$. So $E(C|m)$ can be seen as a "compromise" between these two kinds of distributions.

It turns out that in a number of respects the analysis of $E(C|m)$ is only a stone's throw away from conventional Bayesian analysis:

> The $P(h|d)$ that minimizes $E(C|m)$ is the Bayes-optimal learning, algorithm, $P(h|d) = \delta(h, h^*(d))$. $\qquad$ (9.2)

PROOF. Write $E(C|m) = \sum_c c \sum_d P(c|d, m)P(d|m) = \sum_d P(d) \sum_c cP(c|d)$, where the "$m$" condition is implicit. It is implicitly assumed that $P(d|f)$ and $P(f)$ are fixed. Accordingly, $P(d) = \sum_f P(d|f)P(f)$ is fixed and, therefore, does not depend on $P(h|d)$. This means that if we subtract $E(C|d)$ for the case where $P(h|d) = \delta(h, h^*(d))$ from $E(C|d)$ for the case where $P(h|d) = g(h, d)$, we get

$$\sum_d P(d) \sum_c c[P(c|d)|_{P(h|d)=g(h,d)} - P(c|d)|_{P(h|d)=\delta(h,h^*(d))}].$$

Now $P(d)$ is nonnegative for all d. However, by the first paragraph of Appendix 2, it is also true that

$$\sum_c cP(c|d)|_{P(h|d)=\delta(h,h^*(d))} \leq \sum_c cP(c|d)|_{P(h|d)=g(h,d)}$$

for all d, for any function $g(.,.)$. This means that the difference in values of $\sum_d P(d) \sum_c cP(c|d)\}$ must be nonnegative. QED.

6. Let's say that we do not know $P(f)$ though, so we cannot evaluate $h^*(d)$. Perhaps we only know the support of $P(f)$ (see point 2 above). So, for example, rather than questions like "given $P(f|d)$ exactly, what $P(h|d)$ should I use to minimize $E(C|m)$?" we can instead ask the related question, "Given only that $P(f|d) = 0$ for $f \notin H''$, can I choose a $P(h|d)$ so that $E(C|m)$ is bounded by a monotonically decreasing function of m?" With this slight modification to the Bayesian scenario, we have actually moved close to the PAC scenario; we are considering distributions of the form $E(C|f \subset H'', m)$, as opposed to PAC's $P(c > \varepsilon | f \in H'', m)$.

Now in the real world, one thing we *do* know about $P(f)$ is that its support is only very rarely restricted in any way. This means that unless H'' equals the entire space of functions from X to Y, the answer to our new question can, at best, only rarely have direct relevance for real-world generalization.

Nonetheless, it is a question worth investigating for pedagogical reasons. For example, consider the issue of how to modify the Bayesian framework to reflect the fact that we're unsure of $P(f)$. A conventional Bayesian approach to this issue is to use hierarchical Bayesian analysis (see beginning of this section). In such an approach one would introduce a hyperparameter α indexing the possible priors, and then let a nondelta function distribution over α reflect your uncertainty in the prior. Unfortunately, once both $P(\alpha)$ and $P(f|\alpha)$ are specified, a new $P(f)$ is specified exactly, with no uncertainty $(P(f) = \int d(\alpha)P(f|\alpha)P(\alpha))$. An alternative approach is to say that you're only sure of some characteristic of $P(f)$ that doesn't fully specify $P(f)$, and examine the consequences. An example of such a characteristic is the support of $P(f)$, exactly the characteristic specified in PAC. (This criticism of the importance of analyzing cases where we know $P(f)$'s support and the response to it also applies to the PAC frame work as a whole.)

The answer to our question, provided by PAC, is yes:

If the support of $P(f)$ is contained in H'', than for a PAC learning algorithm with target class H'', $E(C|m)$ is bounded by a shrinking function of m: $E(C|m) \leq 1 - [(m+1)|H'|]^{-1/m} + |H'|[(m+1)|H'|]^{-(m+1)/m}$.

$$(9.3)$$

PROOF. By Theorem 4, under our assumptions for $P(f)$ and $P(h|d)$, $P(c > \varepsilon|m) \leq |H'|(1-\varepsilon)^m$. Call the right-hand side of this inequality δ. The smaller the value of δ, the less of the probability distribution one can "dump" in high c values, i.e., the upper bound on $E(C|m)$ shrinks with shrinking δ. More precisely, given only that $P(c > \varepsilon|m) \leq \delta$, we know that $E(C|m)$ is bounded above by $\varepsilon(1-\delta) + \delta = \varepsilon + |H'|(1-\varepsilon)^{m+1}$. (This bound occurs when $P(c|m)$ is nonzero only for $c = 1$ and $c = \varepsilon$ – an infinitesimal constant.) Since this bound holds for any ε, $E(C|m)$ is in fact bounded by $\min_\varepsilon[\varepsilon + |H'|(1-\varepsilon)^{m+1}]$. Taking derivatives with respect to ε, the minimal value of the bound occurs for the ε satisfying $1 = (m+1)|H'|(1-\varepsilon)^m$; $\varepsilon = 1 - [(m+1)|H'|]^{-1/m}$. Plugging this in, the upper bound on $E(C|m)$ is $1-[(m+1)|H'|]^{-1/m}+|H'|[(m+1)|H'|]^{-(m+1)/m}$. That this is a shrinking function of m follows from the facts that: (1) for any allowed ε (i.e., any ε such that $0 \leq \varepsilon < 1$) the bound $\varepsilon(1-\delta) + \delta$ on $E(C|m)$ shrinks with shrinking δ; and (2) for any ε δ shrinks with growing m. Together, these facts mean that the minimum over all ε of $E(C|m)$ must shrink with growing m. QED.

Note that Eq. (9.3) applies even if our learning algorithm is allowed to make guesses that lie outside of H'', despite the fact that we know such guesses can not equal f. Note also that there are many other kinds of "partial information" concerning $P(f)$ that one might investigate besides its support. For example, one might not know $P(f)$, but have an ordering so that for any two f's, f_1, and f_2, one knows if $P(f_1) > P(f_2)$ or not. (In particular, one might relate a complexity measure like those used in PAC to such an ordering. See Section 7.) One could investigate how much such extra information can improve generalization, using any of the frameworks' definitions of "good generalization."

7. As an aside, $E(C|m)$ exhibits some particularly interesting behavior when we use the quadratic error function. Recall[19] that for that error function $E(C|f,m)$ is given by the sum of the variance and the square of the bias:

Define $Y_h(x,d) \equiv \sum_h P(h|d)h(x)$, the average output guessed by the

generalizer for input x. Then

$$E(C|f,m) = \sum_x \pi(x) \left\{ \sum_d P(d)[Y_h(x,d)]^2 - \left[\sum_d P(d)[Y_h(x,d)] \right]^2 \right\} \quad (9.4)$$

(the x-average of the "variance")

$$+ \sum_x \pi(x)[\sum_d P(d)[Y_h(x,d) - f(x)]]^2$$

(the x-average of the square of the "bias").

Bias tends to be low when variance is high, and vice versa. This is known as the "bias-variance trade-off." As an example, if one has lots of parameters in one's model, usually one can fit the data well, and the bias tends to be low. However, a large numbers of parameter usually makes the variance large.

For the reasons given above though (see beginning of point 5), one might prefer examining $E(C|m)$ rather than $E(C|f,m)$. After all, the generalizer always-guess-f has both 0 bias and 0 variance; there is not always a bias-variance tradeoff. (One might argue that in practice one can never use such a 0-bias, 0-variance generalizer, since one does not know f. However the obvious rejoinder is that if you are concerned about the fact that you do not know f, then you should calculate a distribution where f is not fixed.)

Replacing $E(C|f, m)$ with $E(C|m)$ results in a "correction" to the famous bias + variance result:

For a quadratic cost function, $E(C|m) = $ constant

$$+ \sum_x \pi(x) \left\{ \sum_d P(d)[Y_h(x,d)]^2 - \left[\sum_d P(d)[Y_h(x,d)] \right]^2 \right\}$$

$$+ \sum_x \pi(x) \left[\sum_d P(d)[Y_h(x,d) - Y_f(x,d)] \right]^2$$

$$+ 2 \sum_x \pi(x) \left[\sum_d P(d)Y_f(x,d) \sum_d P(d)Y_h(x,d) \right.$$

$$\left. - \sum_d P(d)Y_f(x,d)Y_h(x,d) \right], \tag{9.5}$$

where $Y_f(x,d)$ is defined in direct analogy to $Y_h(x,d)$,
and the constant term is independent of $P(h|d)$.

This result follows from simple algebra. A general proof, applicable even for nondeterministic generalizers and non-single-valued f and h, is in a paper by Wolpert.[65] Similar results carry over for the misclassification error rate case when $Y = \{0, 1\}$ (so that $(y_1 - y_2)^2 = 1 - \delta(y_1, y_2)$).

The first nonconstant term in Eq. (9.5) is exactly the same as the variance term in Eq. (9.4). The second nonconstant term is the same as the square of the bias term from Eq. (9.4), except that with f no longer fixed, $f(x)$ is replaced by $Y_f(x,d)$. The last term is the correction term. It measures how well "aligned" $P(h|d)$ is with $P(f|d)$, and is a direct reflection of Theorem 1. In that it depends on the posterior $P(f|d)$, the correction term has no analogue in conventional sampling theory statistics and, in that it depends on $P(h|d)$, it has no analogue in conventional Bayesian analysis.

Note that that correction term can be written as $-2 \sum_x \pi(x)$ $\sum_d [P(d)(Y_h(x,d) - Y_h^*(x))(Y_f(x,d) - Y_f^*(x))]$, where $Y_h^*(x) \equiv \sum_d P(d)Y_h(x,d)$ and similarly for $Y_f^*(x)$. In other words, the correction term formally reflects the degree of correlation between $Y_h(x,d)$ and $Y_f(x,d)$.

8. One can add some information we *do* have (we do not know f) to the "m" in our conditioning event. For example, we can investigate $P(c|s, m)$, under various conditions. Some very preliminary results concerning this distribution were presented in the last section, where it was also pointed out that a potentially interesting area of research is the relationship between $P(c|s, m)$ and $P(c - s|m)$.

9. In the same spirit of conditioning-on-what-you-know, one can analyze the case where $P(c|f, d)$ is one's distribution of interest rather than $P(c|f, m)$. Since f

is fixed, this differs from Bayesian analysis but, since d is fixed, it also differs from the other three frameworks. Such an analysis can be interesting when one does not precisely specify $P(h|d)$ (as in PAC) or when $P(h|d)$ is stochastic (as in SP). See Wolpert[63] for a preliminary investigation of $P(c|f,d)$ for the exhaustive learning scenario.

10. Again in the spirit of conditioning-on-what-you-know, one might investigate what happens if one interchanges target and hypothesis functions everywhere, including in the distribution of interest. (Unlike f, we almost always know the hypothesis we make, h.) The only (potential) formal differences between h and f lie in $P(h|d)$ and $P(f|d)$, since the error function is symmetric between h and f, since to every rule like $P(f|h,d) = P(f|d)$ there is a corresponding $P(h|f,d) = P(h|d)$, etc. (See Section 2.) Accordingly, we can interchange h and f in the distribution of interest, if we also send $P_{H|D}(h|d) \rightarrow P_{F|D}(h|d)$ and $P_{F|D}(f|d) \rightarrow P_{H|D}(f|d)$ (i.e., interchange $P(h|d)$ and $P(f|d)$). This means that we can use any calculation giving (some characteristic of) $P(c|f,m)$ to give us $P(c|h,m)$ under the interchange of distributions. In particular, if $P_{H|D}(h|d) = P_{F|D}(h|d)$, we do not need to interchange the distributions (they are already identical), and $P_{C|H,M}(c|h,m) = P_{C|F,M}(c|h,m)$.

$P(c|h,m)$ concerns the following scenario: Perform many different supervised learning experiments (having target functions set according to $P(f)$ and training sets chosen according to $P(d|f)$), all with the same generalizer. Collect all those experiments in which the hypothesis function guessed by the generalizer is h. The resultant generalization error distribution, as a function of m, is $P(c|h,m)$. In contrast, to calculate $P(c|f,m)$ one has f fixed (though in practice we do not know what it is fixed to), many times samples f to get d, trains the generalizer on d, and then sees what the resultant error distribution is.

In some recent papers,[59,63] $P(c|h,m)$ is analyzed in some scenarios where one can interchange $P(f|d)$ and $P(h|d)$, so that one can exploit $P(c|f,m)$ results. An open question is how $P(c|h,m)$ behaves in more general scenarios.

11. The "worst case" nature of some of the frameworks can be manipulated in a number of interesting ways. One is to make more attributes worst-case. As an example of such analysis, consider PAC modified to be worst-case over d. The resulting framework is trivial if the sampling distribution is not fixed. So long as there exists at least one $h \in H'$ that does not equal f but that intersects f, then there exists a $\pi(.)$ such that "$\forall \varepsilon, \max_{d \subset f; m, P(h|d)} [\sum_{c=\varepsilon}^{1} P(c|f,d)] = 1$; i.e.,

there exists a $\pi(.)$ such that $P(c|f,d)$ is arbitrarily close to the delta function $\delta(c,1)$.[14] On the other hand, if (for example) $\pi(.)$ is flat, $P(c|f,d)$ is not arbitrarily close to $\delta(c,1)$.[15] In general, if the sampling distribution is fixed, the behavior of a worst-case analysis over d depends on the precise form for $\pi(.)$, i.e., the worst-case analysis over d is not distribution-free.

Moving in the opposite direction, one might try to derive results based on everything that is relevant, rather than leaving some quantities (e.g., $\pi(x)$) unspecified. (Some work on this kind of issue has already been done in the PAC community.) With this modification, one is no longer doing a worst-case analysis over any variables. Alternatively, one can make some quantities worst-case (e.g., d), but have other quantities specified and therefore no longer worst-case (e.g., $P(f)$).

Other interesting issues involve manipulating the average-case attributes of the frameworks (e.g., $P(c|f,m)$ is average-case over d). For example, as was pointed out by C. Van der Broeck (personal communication), one might wish to calculate the higher-order moments with respect to training sets in addition to the first moment given by (for example) $P(c \mid f,m)$. In this way one could bound the probability of a particular training set giving a result which is very different from the average-case result. (See point 5 above.)

Finally, one can manipulate both average-case and worst-case attributes of the frameworks. One interesting way to do this is to mix the order of averages and maximums, since they do not commute in general. For example, bounds on PAC's distribution $P(c|f,m)$ which do not specify $P(h|d)$ are equivalent to $\max_{P(h|d)} \sum_d P(c|f,d)P(d|f)$, where the max is restricted to those $P(h \mid d)$ that reproduce any d sampled from f and that have support H'. This bound is the worst that one's average over d can be. As an alternative, since in any particular supervised learning problem d is fixed, one might analyze $\sum_d \max_{P(h|d)}[P(c|f,d)P(d|f)] = \sum_d P(d|f) \max_{h \in H';d}P(c|f,h,d)$, where the max is over those $h \in H'$ that reproduce d, and $\pi(x)$ is implicitly held fixed. As opposed to what PAC calculates, this quantity is the worst you can do for a particular d, on average. It can be viewed as a "half-way point" between $\max_{P(h|d)} P(c|f,m)$ and $\max_{P(h|d)} P(c|f,d)$. As other possibilities, one

[14]**Proof:** Let h' be the hypothesis function $\in H'$ which $\neq f$ but which intersects f. Let x' be an X value such that $h'(x') = f(x')$ (i.e., let x' be one of the intersection points). Let x^* be an X value such that $h'(x^*) \neq f(x^*)$. Let d' be a training set that consists of m copies of the same input-output pair, $\{x', f(x')\}$. d' meets the restrictions on training sets (i.e., $d' \subset f$, and d' has m elements). Let $P(h|d)$ be arbitrarily close to $\delta(h,h')$, for $d = d'$. Such a $P(h|d)$ does not violate any of the restrictions on generalizers. Let $\pi(x)$ be arbitrarily close to $\delta(x,x'')$. For such a sampling distribution, $Er(f,h')$ is arbitrarily close to 1. QED.

[15]This follows from the fact that $P(h|d)$ must reproduce $d = \{d_X, f(d_X)\}$: $Er(f,h) = 1 - \sum_{x \in X} \pi(x)\delta(f(x),h(x)) < 1 - \sum_{x \in d_X} \pi(x)\delta(f(x),h(x))$, and the reproduction of d requirement means that this bound equals $1 - \sum_{x \in d_X} \pi(x) = 1 - m'/n \neq 1$ (m' being the number of distinct input-output pairs in d).

can consider modifications involving the max over $\pi(x)$ or the max over $P(f)$ with support restricted to H'', etc.

12. As an aside, it is interesting to compare the worst-case and average-case attributes of conventional PAC and SP. To start, we can investigate the apparent distinction that PAC's Theorem 4 is independent of f (so long as $f \in H''$) whereas the exhaustive learning formula for $P(c|f,m)$ depends on f (see Section 6). This distinction is illusory. This is because in the PAC scenario, $\delta \leq \{1-\varepsilon\}^m \times \sum_{h \in H'} \delta(Er(h,f) > \varepsilon)$ (see Appendix 3). The sum (and therefore the bound on δ) *is* dependent on f. However, PAC replaces the sum with an upper bound ($|H'|$), which does not depend on f (assuming $f \in H''$). Exhaustive learning makes no such replacement; if it did, its result, too, would be independent of f.

As another example, it is often said that PAC is "distribution free," in that the bounds given by Theorem 4 on δ does not specify the sampling distribution $\pi(.)$. This would appear to contrast with the case in exhaustive learning, since through the $\pi(.)$ appearing in $Er(.,.)$, $\rho_0(.)$ (and therefore $P(c|f,m)$) is explicitly dependent on the sampling distribution. However, just as with independence with respect to f, independence with respect to $\pi(.)$ is simply a reflection of whether or not it is conventional to replace the appropriate sum ($\sum_{h \in H'} \delta(Er(h,f) > \varepsilon)$ in the case of PAC, $\rho_0(c,f,T)$ in the case of exhaustive learning) with an upper bound.[16] In PAC, such a replacement is conventional, whereas in exhaustive learning it is not. Nothing more profound than simple convention is involved in this distinction.

Knowing $\pi(.)$ modifies the results of both frameworks, although with PAC this modification is in the form of tighter bounds (the modification in exhaustive learning accruing from knowledge of $\pi(.)$ does not involve any "tightening of bounds"). This is also true of knowing $P(f)$. To illustrate all this with the EBF, consider the PAC case where we know that $\pi(.)$ is flat and H' equals the entire space of input-output functions. This H' allows us to replace $\sum_{h \in H'} \delta(Er(h,f) > \varepsilon)$ with $\sum_{c > \varepsilon} \sum_h \delta(Er(h,f) = c)$. The flat $\pi(.)$ assumption allows us to evaluate this directly, getting $\sum_{z=0}^{n(1-\varepsilon)} C_z^n (r-1)^{n-z}$ as the upper bound on the ratio $\delta/(1-\varepsilon)^m$, rather than Theorem 4's upper bound of $|H'|$, which here equals r^n. (Note that $\sum_{z=0}^{n(1-\varepsilon)} C_z^n (r-1)^{n-z}$ is the sum of

---

[16]Note that I am implicitly assuming that self-averaging holds for a broad range of functions $\pi(.)$ when I say that the $\pi(.)$ dependence in exhaustive learning occurs only in the $\rho_0(.)$ term. This is the case in the scenario where self-averaging holds because $n \gg m$ and $T(h)$ is constant (see Wolpert[63]). It is not known how common such behavior is for other self-averaging scenarios however. When the property of self-averaging itself depends on $\pi(.)$, the formal similarity between PAC and exhaustive is a bit more problematic, although one can still get "distribution free" exhaustive learning results by strategically replacing sums with bounds, just as in PAC.

the first $n(1 - \varepsilon)$ terms of the binomial expansion of $r^n = [(r - 1) + 1]^n$, and therefore is $\leq r^n$.)

In fact, in many circumstances knowing $\pi(.)$ means there is no need to focus on $\max_{P(h|d)}[\sum_{c=\varepsilon}^1 P(c|f,m)]$ at all, rather than on the more fundamental quantity $P(c|f,m)$. To get bounds on $P(c|f,m)$ directly, one can use the derivation of Theorem 4 in Appendix 3. That derivation goes through completely unchanged up to the final line, where instead of the quantity $\sum_{h \in H'} \delta(Er(h,f) > \varepsilon)$ (which in Appendix 3 was bounded by $|H'|$) we have the quantity $\sum_{h \in H'} \delta(Er(h,f) = \varepsilon)$. So for example, if $\pi(.)$ were flat and if H' equalled the set of all input-output functions, then we could write $P(c|f,m) \leq \{1 - c\}^m C_z^n (r - 1)^{n-z}$, where $z \equiv n(1 - c)$.

13. Some have advocated investigating other distributions of interest not found in Table 1. For example, in a paper by Haussler et al.,[22] one examines quantities like $P(c|d_X, q)$, where q is the test set input value. Such a distribution can be viewed as a compromise between distributions conditioned solely on m (like those in PAC, SP, and the VC framework) and distributions conditioned on d (like those in conventional Bayesian analysis). Note in this regard that $P(c|m)$ is an average (over d_X and q) of $P(c|d_X, q)$.

The distribution $P(c|d_X, q)$ is also interesting because it applies to situations where one can choose the values d_X before producing the sample, and might want to do so in an optimal way. (In conventional statistics, issues related to how best to choose d_X are known as "experimental design." See also Mackay[29] and references therein.)

All of the foregoing—extensions of the frameworks, novel distributions of interest, etc.—can be analyzed using the off-training-set error function rather than the iid error function. Indeed, many of the issues that have already been addressed by the four frameworks for the case of the iid error function have yet to be considered in the context of an off-training-set error function. Clearly there is a lot of "filling in" of Table 1 to do.

---

10. COMPARING THE REAL-WORLD UTILITY OF THE FRAMEWORKS

10.1 WHAT DISTRIBUTION SHOULD WE INVESTIGATE?

One reason to investigate a particular framework is pedagogical: one wants to learn about some particular aspect of supervised learning, whether or not it has anything to do with the real world. However, a potentially far more important reason to investigate a framework is that it might be able to assist you in real-world supervised learning. This second reason raises a crucial question: what distribution of interest should one be concerned with to facilitate one's real-world learning?

The answer to this question depends in part on how probability theory is related to reality. This has long been a controversial issue (see Cheeseman,[12] Good,[20] MAXENT,[30] Berger,[4] and Wolpert[61]). Fortunately, here we can bypass most of the controversial aspects of this issue.

All four frameworks agree that the conditioned variable in the distribution of interest should be c. In fact, we could almost define C as that function of the variables F, H, and D whose distribution we want to know. (For current purposes, it is not important if the precise form of the distribution is $P(c > \varepsilon | \ldots)$, or $P(c > \varepsilon + s | \ldots)$, or $E(C | \ldots)$, or even some other distribution like $P(c = 0 | \ldots)$; for the most part an aficionado of any of the four frameworks would agree that all such distributions are interesting, even if (s)he happens to concentrate on only one of them.)

Therefore, the only issue of contention is what the conditioning event should be if we want results concerning our distribution to help us in the real world. The rest of this section considers this issue in some depth. It does so be positing some *a priori* principles for applying supervised learning to the real world, and analyzing the implications of those principles. The saga of a ficticious company trying to apply the VC framework in the real world is used to illustrate that analysis. After this the analysis is abstracted and applied to the remaining frameworks. For simplicity, when the precise form of the error function needs to be specified, iid error is assumed.

10.2 THE HONESTY PRINCIPLE

To address the issue of what conditioning event to use one can invoke a simple "honesty" principle.

PRINCIPLE. Consider a particular real-world statistical scenario, defined by $P(f, h, d, c) = g(f, h, d, c)$. For a formal analysis to apply to this scenario, that analysis must not make assumptions for $P(f, h, d, c)$ that disagree with $g(f, h, d, c)$.

The honesty principle does not concern itself with whether we who conduct the analysis happen to know the exact forms of the distributions we are manipulating. In our analysis, when we specify $P(.)$, we can set all aspects of it that we do not know to equal "whatever they are in our particular statistical scenario" and the honesty principle will be met. For example, write $P(f, h, d, c) = P(c|f, h, d)P(h|d)P(d|f)P(f)$. Say we know that $P(c|f, h, d)$, $P(h|d)$, and $P(d|f)$ equal $L_1(c, f, h, d)$, $L_2(h, d)$, and $L_3(d, f)$ respectively, for a particular real-world scenario, but we do not know $P(f)$. In such a case, the honesty principle will be met if in our analysis we set $P(c|f, h, d)$, $P(h|d)$, and $P(d|f)$ to $L_1(c, f, h, d)$, $L_2(h, d)$, and $L_3(d, f)$ respectively, but indicate $P(f)$ simply as $P(f)$, value unspecified. On the other hand, the honesty principle will be violated if our analysis makes a precise assumption for $P(f)$ (rather than simply saying that $P(f)$ is "whatever it is in our real-world scenario") and that assumption happens to disagree with $g(.)$. (Cf. Theorem 1.)

The honesty principle does not say that an analysis based on the $g(f, h, d, c)$ of scenario one cannot tell you anything about the related but different scenario two. However, it does say that such an analysis of scenario one can tell you something about scenario two only if the analysis' results can be transformed so that they apply to scenario two's $g(.)$. Indeed, consider modifying our behavior by using a mathematical result concerning a particular statistical scenario. Such a modification means that some aspect of the statistical scenario (i.e., of $P(f, h, d, c)$) has been changed. For example, $P(h|d)$ will have been changed. (Hopefully such a change will bias the scenario towards lower c.) According to the honesty principle, this change in $g(.)$ means we must change our analysis. Ultimately any justification for our changed behavior must arise in what that new analysis says about the new $P(f, h, d, c)$.

10.3 THE HONESTY PRINCIPLE AND THE VC FRAMEWORK; THE EXAMPLE OF WE-LEARN-IT INC.

As an example, it is worth considering in detail the implications of the honesty principle, say there is a company, We-Learn-It Inc., who own the rights to (what they think is) a really nifty learning algorithm. They go around searching for training sets d for which people would like to know the "best" h. (For example, training sets involving OCR, speech recognition, protein folding, etc.) For each such d they create an h by training their algorithm on d. Then they use the resultant s, together with the VC dimension of their algorithm, to determine how tight the VC confidence intervals are. If s is lower than some threshold λ (which in general depends on m and the VC dimension of their generalizer), then the bounds are considered tight, and accordingly they go to market with their h. Otherwise, they do not. This whole algorithm will be called We-Learn-It's "operating procedure." Cost is some small number ζ if they do not go to market (reflecting man-hours expended getting d, running the algorithm on it, etc.). If they do go to market, cost is given by the misclassification rate of h and f. (More realistic—and therefore more complicated—go-to-market costs don't change the main results of this section.)

We-Learn-It's operating procedure is one of the most innocuous uses of the VC framework that one can imagine. However, to use the VC framework to try to justify that procedure violates the honesty principle.

To understand this intuitively, consider the frequentist perspective, in which a "real-world statistical scenario" is equivalent to a series of experiments consisting of iid samples of $P(f, h, d, c)$. We-Learn-It Inc. is concerned with a series of experiments in which they (i) many times generate f according to $P(f)$; (ii) many times sample f to create a d; (iii) many times produce an h by training on d using the nifty learning algorithm; (iv) based on the resultant s decide whether to go to market with h; (v) if you do go to market, reap financial consequences given by h's error; and (vi) otherwise the consequences are given by ζ. The VC result is not concerned with such a series of experiments. Rather it concerns the following series:

(i) many times generate f according to $P(f)$; (ii) many times sample f to create a d; (iii) many times produce an h by training on d using the nifty learning algorithm; (iv) always go to market with h (there is no "hook" in the VC framework for withholding h); and (v) reap financial consequences given by h's error. (The VC result then tells us how likely it is in any given one of these experiments that the financial consequences will differ from s by more than ε.) So a VC-based analysis concerns a different series of experiments than the series defining our statistical scenario. Accordingly the analysis is not directly applicable to the real-world scenario. The hope is that the analysis can be modified so that it is applicable.

10.4 FORMAL ANALYSIS OF HOW WE-LEARN-IT'S USE OF THE VC FRAMEWORK VIOLATES THE HONESTY PRINCIPLE

More formally, to see how We-Learn-It violates the honesty principle, first note that We-Learn-It's $P(c|f, h, d)$ is different from the VC framework's $P(c|f, h, d)$, due to the possibility of We-Learn-It not going to market. This is how in their attempt to use the VC framework, We-Learn-It created a statistical scenario different from the one addressed by the VC framework. (This is a common phenomenon with confidence-interval results like those of the VC framework; often as soon as you true to use such results, you change the statistical scenario so that the results no longer directly apply.)

This difference in $P(c|f, h, d)$ constitutes a violation of the honesty principle. To circumvent it, We-Learn-It must derive results concerning the new (i.e., actual) $P(c|f, h, d)$ that justify their claim concerning that $P(c|f, h, d)$, i.e., that justify their operating procedure. Since We-Learn-It's "justification" for their operating procedure is based on the old (i.e., VC-based) $P(c|f, h, d)$, the natural way to do this is to try to transform results for the old $P(c|f, h, d)$ into results for the new $P(c|f, h, d)$. In other words, transform VC results so that they apply to the actual statistical scenario.

Unfortunately, it turns out that if we carry out such a transformation, the VC results that We-Learn-It is trying to exploit do not justify We-Learn-It's operating procedure. To see this, begin by having $P(f, h, d, c)$ for We-Learn-It be $U(f, h, d, c)$, and for the VC framework be $V(f, h, d, c)$. $U(f, h, d, c)$ and $V(f, h, d, c)$ can be decomposed as $U(c|f, h, d)P(h|d)P(d|f)P(f)$ and $V(c|f, h, d)P(h|d)P(d|f)P(f)$ respectively (i.e., they share the same $P(h|d)$, $P(d|f)$, and $P(f)$).

We can relate $U(.)$ and $V(.)$:

$$U(c|f, h, d) = \theta(\lambda - s(d, h))V(c|f, h, d) + \theta(s(d, h) - \lambda)\delta(c, \zeta),$$

where $\theta(.)$ is the Heaviside step function, and the d-h dependence of s is made explicit. (This relation holds regardless of the form of $V(c|f, h, d)$.) Now $U(c|f, h, d) = \sum_{s,m} U(c|f, h, d, s, m)P(s, m|f, h, d)$. Since d and h determine both s and m, $P(s, m|f, h, d)$ is a delta function; $U(c|f, h, d) = U_{C|F,H,D,S,M}(c|f, h, d, s(d, h), m(d))$. Note that now "$S$" indicates a random variable, which in general might not be a function of D and H (although for us it is); to simplify the analysis we have implicitly gone to a new event space.

A similar result holds for $V(.)$. In addition, since $P(f, h, d)$ is the same for both $U(.)$ and $V(.)$, so is $P(f, h, d, s, m) = P(s, m|f, h, d)P(f, h, d)$. Accordingly,

$$U(c, f, h, d, s, m) = \theta(\lambda - s)V(c, f, h, d, s, m) + \theta(s - \lambda)\delta(c, \zeta)P(f, h, d, s, m). \quad (10.1)$$

Since all three probability distributions in this equality equal zero unless $s = s(d, h)$ and $m = m(d)$, the equality holds for all values of s and m, and not just for $s = s(d, h)$ and $m = m(d)$.

This equality is the fundamental transformation law relating our two statistical scenarios. Our task is to use it to transform the VC theorems from their usual statistical scenario to the scenario of We-Learn-It. In particular, this means performing such a transformation to the conditional distribution of interest arising in those VC theorems. Before discussing this transformation of that distribution, I will show how to transform a similar distribution.

Because $P(f, h, d, s, m)$ is the same for both $U(.)$ and $V(.)$, so is $P(s, m)$. Therefore, Eq. (10.1) holds if $P(f, h, d, s, m)$ is replaced by $P(f, h, d|s = \gamma, m)$, $U(c, f, h, d, s, m)$ is replaced by $U(c, f, h, d|s = \gamma, m)$, $\theta(s - \lambda)$ is replaced by $\theta(\gamma - \lambda)$, and similarly for $V(.)$ and $\theta(\lambda - s)$.

Now for either $U(.)$ or $V(.)$, we can write

$$P(c = \chi|s, m) = \sum_{f, h, d, c} P(f, h, d, c|s, m)\delta(c, \chi).$$

Accordingly,

$$U(c|s, m) = \theta(\lambda - s)V(c|s, m) + \theta(s - \lambda)\delta(c, \zeta). \quad (10.2)$$

(To see this, operate on both sides of our equation for $U(c, f, h, d|s = \gamma, m)$ with $\sum_{f, h, d, c} \delta(c, \chi)$, and then replace χ with c and γ with s.) So a result that concerns $P(c|s, m)$ for the $P(c|f, h, d)$ of the VC framework can be transformed to tell us something about the same distribution for We-Learn-It's $P(c|f, h, d)$. In particular, if Eq. (10.2) and a formula for $V(c|s, m)$ jointly mean that $E_U(c|s, m)$ is low for all s, then $E_U(c|m)$ is low, and We-Learn-It Inc. has a formal justification for their operating procedure. (However note that for pernicious $P(s)$ We-Learn-It never goes to market; with each new project they would incur cost ζ without any compensation.)

Unfortunately the result of the VC framework that We-Learn-It is trying to invoke concerns $V(c - s = \kappa|m)$, not $V(c|s, m)$. Accordingly, Eq. (10.2) does not justify their operating procedure. Are there equations similar to Eq. (10.2) that do rely on the result of the VC framework that We-Learn-It is trying to exploit and that thereby justify We-Learn-It's operating procedure?

To answer this question, note that it is crucial that Eq. (10.2) concerns distributions conditioned on the variable used by We-Learn-It to modify $P(c|f, h, d)$ (and therefore occuring in θ functions in Eq. (10.1)), s. In particular, it is because the distributions in Eq. (10.2) are conditioned on s that when we took a sum to arrive at Eq. (10.2) the sum was not over s.

Since it does not fix s, we cannot do this for the sum giving $P(c - s = \kappa|m)$. In particular, write $P(c - s = \kappa|m) = \sum_{f,h,d,c,s} P(f,h,d,c,s|m)\delta(c - s, \kappa)$. If we now try to use the variant of Eq. (10.1) giving $U(c, f, h, d|m)$, we get a mess:

$$U(c - s = \kappa|m) = \sum_{f,h,d,c,s} V(f, h, d, c, s|m)\delta(c - s, \kappa)\theta(\lambda - s)$$
$$+ \sum_{f,h,d,c,s} P(f, h, d, s|m)\theta(s - \lambda)\delta(c - s, \kappa)\delta(c, \zeta). \tag{10.3}$$

(Note the second sum reduces to $\theta(\zeta - \kappa - \lambda)P_S(\zeta - \kappa)$.) So we cannot simply relate the confidence intervals over $U(.)$ and $V(.)$. In particular, the confidence interval over $V(.)$, by itself, does not fix the confidence interval over $U(.)$.

These kinds of problems makes it difficult to try to use only results based on $V(c - s = \kappa|m)$—the results given by the VC framework—and thereby learn something of interest concerning $U(f, h, d, c)$. Intuitively, since s is not specified in $V(c - s = \kappa|m)$, that distribution does not fix whether s exceeds λ and, therefore, it does not help us know which "regime" of $U(c|f, h, d)$ we are in. Accordingly, it provides little useful information about the distribution over c in We-Learn-It's statistical scenario. We would have to known something like $P(s)$ (which in general means we need to know $P(f)$, something not considered in the VC framework) to get such regime information and thereby transform our result into something useful concerning U.

Another intuitive way to view things is to note that $P(c - s = \varepsilon|m) = \sum_s P(c - s = \varepsilon|s, m)P(s|m)$. Knowing the c-dependence of this sum over all s tells us little concerning the c-dependence of a partial sum extending only over a subset of all s. Yet it is precisely such a partial sum that determines the efficacy of We-Learn-It's procedure.

10.5 THE GHOST OF WE-LEARN-IT; TRYING AGAIN TO USE THE VC FRAMEWORK IN THE REAL WORLD

Given these problems with We-Learn-It's logic, it should be no surprise that they soon go bankrupt. Somewhat disgruntled, their CEO, Dr. Smith, decides to return to academia. Soon thereafter a government agency decides to hold a supervised learning competition. All the entrants in the competition are given the same data set, and are told to create an h. Each entrant must publicly report their h, the VC dimension of the algorithm they used, and their s. One week after the deadline for submitting an h, the agency will say how each h fared. Each entrant will be rewarded according to the misclassification rate of their h with respect to the underlying f.

The day after the close of submissions, a businessperson comes to Dr. Smith. This person asks Dr. Smith to use the publicly reported VC dimensions, s's and

h's to analyze each h and decide which one is best. The businessperson would then rush to market with a product based on that h.

"Aha!" says Dr. Smith, "I'm no fool. VC results concern statistical scenarios where d is allowed to vary. (That is, $P(f, h, d, c)$ is nonzero for more than one d.) Therefore, the simplest way to exploit those results and yet meet the honesty principle is to consider statistical scenarios in which d is allowed to vary. (The series of experiments corresponding to such a scenario would have the government agency run many competitions, each with a different d.) Now, in general, if the businessperson uses me to choose an h for each such experiment, they will be using a $P(h|d)$ that differs from the $P(h|d)$ of any one of the entrants in the competition, since they will be choosing different entrants' $P(h|d)$'s depending on d. Therefore, the businessperson would have a different statistical scenario—a different $P(f, h, d, c)$—from that of any of the entrants. In other words, $P(f, h, d, c)$ for the businessperson would differ from that of the cases the businessperson wants me to analyze with the VC framework. So I'd be violating that nasty honesty principle again if I used such analyses to justify the businessperson's action. Loosely speaking, as soon the businessperson tried to use the results of the VC framework this way, they'd be violating the assumptions of that framework." The VC results are inherently descriptive and you're trying to use them in a prescriptive manner.

"But wait!," says the businessperson, "Couldn't you come up with an algorithm for how to choose among the entrants' h's, and then use the VC framework to find the implications the entrants' VC dimensions, s's and h's have for the cost associated with your overall $P(f, h, d, c)$? In this regard, you should know that I have contacts who inform me that the hypothesis classes of the entrants' algorithms are all nested, exactly as they must be for structural risk minimization to apply. You should also know that I plan to go to market no matter what; holding back my product is not an option."

10.6 A VARIANT OF THE HONESTY PRINCIPLE. (THE SAGA OF WE-LEARN-IT CONTINUED)

The new form of the businesspersons' question is a very subtle one. The next several subsections present a preliminary investiation of some aspects of it. (By no means do they constitute an exhaustive analysis.) To start, here is Dr. Smith's reply:

"First off, since you plan to go to market no matter what, you aren't allowing yourself an option that is not available in the the scenario implicitly addressed by the VC theorems. Therefore it may be that you do not need to transform results concerning one probability distribution into results concerning another in order to obey the honesty principle. In other words, you won't automatically run into the same difficulties with the honesty principle that sunk my old company, We-Learn-It.

There's still the problem though that the question you pose is a bit vague. So to answer it I must first formalize it. There are a number of different such formalizations; here I will use one based on the following variant of the honesty principle:

Principle: Consider a particular real-world statistical scenario defined by $P_{A,B,...}(a, b, ...) = g(a, b, ...)$. Information about the implications that $B = b$ has for the value of A is information about the a-dependence of $P_{A|B}(a|b)$.

This can be taken as a definition of the phrase 'implications.' Information about implications can take many forms: a bound on $E(A|b)$, the value of $P(a > \varepsilon|b)$, etc. Note that if your information tells you that $B = b$ means that A is identical to some variable Z, but tells you nothing about Z, then your information tells you nothing about the a-dependence of $P_{A|B}(a|b)$; you have no information about the implications that $B = b$ has for the value of A.

Just as with the original honesty principle, in this variant close isn't good enough: Information about a distribution that's related to the one we want is nice. But until that information is transformed into information about the exact distribution we want (a process that might entail extra assumptions in general), it isn't relevant to the issue at hand.

Now the statistical scenario in front of me is the distribution $P(f, d, h_1, h_2, c_1, c_2, s_1, s_2, \gamma_1, \gamma_2, \text{choice})$, where f is the target function, d is the training set, h_i is entrant i's guess, c_i is the associated cost, s_i is the associated empirical misclassification rate, γ_i is a hyperparameter specifying the details of entrant i's learning algorithm (i.e., $P(h_i|d, \gamma_i) = \gamma_i(d, h_i)$), and 'choice' is whether I decide to use entrant one's h or entrant two's. (The full distribution $P(f, d, h_1, h_2, c_1, c_2, s_1, s_2, \gamma_1, \gamma_2, \text{choice})$ implicitly specifies how I make that choice.)

In the problem posed to me, the value b that I am directed to use is the value of the composite variable $\{s_1, s_2, h_1, h_2, VC(\gamma_1), VC(\gamma_2), \gamma_1\text{'s hypothesis class is contained inside } \gamma_2\text{'s}, m\}$. I was also directed to take A to be my cost, c_{choice}, which is determined by the composite variable $\{\text{choice}, c_1, c_2\}$. Accordingly, the variant of the honesty principle instructs me to try to glean something about the c_{choice}-dependence of $P(c_{\text{choice}}|s_1, s_2, h_1, h_2, VC(\gamma_1), VC(\gamma_2), \gamma_1\text{'s hypothesis class is contained inside } \gamma_2\text{'s}, m)$. (More generally, I might be interested in simply comparing the two generalizers rather than exploiting knowledge of their properties, which I could do with $A = \{c_1, c_2\}$. Conclusions similar to those presented below follow for such an A.)

The VC framework doesn't tell me how to evaluate this distribution. Even using structural risk minimization, the VC framework would instead provide me with information concerning a distribution like $P(c_{\text{choice}} - s_{\text{choice}}|VC(\gamma_1), VC(\gamma_2), \gamma_1\text{'s}$ hypothesis class is contained inside $\gamma_2\text{'s}, m)$. The variables s_1 and s_2 are not specified in the conditioning event of this distribution. Accordingly, in general I can not readily transform information concerning this distribution into information concerning the distribution I want to know unless I exploit extra information concerning s, i.e., concerning the individual learning algorithms and $P(f)$.[17]

In addition, consider the case where B is redefined to be the variable $\{VC(\gamma_1), VC(\gamma_2), \gamma_1$'s hypothesis class is contained inside γ_2's, $m\}$. In such a case, since s_{choice} is not specified, I could not readily infer anything concerning the variable I'm interested in—c_{choice}—from information concerning $c_{choice} - s_{choice}$, like the information provided by the VC theorems. (Again, unless I knew something concerning the likely values of s, or was willing to settle for bounds that are worst-case over all $P(s)$.) Accordingly, again, I would not be able to use the VC framework by itself to infer anything about my likely cost.

Of course, none of this means that it is not *reasonable* to use the algorithm having lowest VC dimension and s. What's at issue here is not what's reasonable, but what can be formally justified. Also I should be careful to note that I haven't proven that such formal justification is impossible. I've simply pointed out some of the difficulties that must be surmounted by any such putative justification."

After making this reply, Dr. Smith goes on to note that, though difficult, it might nonetheless be possible to glean some information of use to the business-person by using the VC framework. This is especially true if one uses a B that does not specify s, if one is willing to consider analysis that is worst-case over $P(s)$, and if one doesn't try to derive anything directly concerning the likely values of c_{choice}, but rather tries to derive something concerning likely differences in that value between two schemes for setting the variable "choice."

It is easiest to illustrate this for the case where one has a single hypothesis class, and must choose an hypothesis from it. Now in general it is not true that for any particular s_1 and s_2 (associated with h_1 and h_2 respectively), the h with the preferable VC bounds (i.e., the one with the lower s value) will perform better than the other one. This is true even if one averages over all training sets having the specified values of s_1 and s_2. (See the example in Section 8 of how for one hypothesis, having low s might imply very high c, and might also imply that for another hypothesis s is high while c is low.) However, it might still be that using the VC bounds to choose between the candidate hypotheses is better than randomly choosing amongst them if one averages over s values (in addition to averaging over

---

[17]This is not always the case. For example, if $P(c_{choice} - s_{choice} = \lambda|...) = 0$ for all values of λ above some threshold κ, then I know that $P(c_{choice}|s_{choice}...)$ must equal zero if $c_{choice} - s_{choice} \geq \kappa$. (This example is due to Tal Grossman.) This is an unusual situation though: $P(c_{choice} - s_{choice} = \lambda|...) = \sum_{c_{choice}, s_{choice}} \delta(c_{choice} - s_{choice}, \lambda)P(c_{choice}|s_{choice}...)P(s_{choice}|...)$, and it's only when the left-hand side equals 0 for some λ that we can conclude something substantial about the form of the distribution $P(c_{choice}|s_{choice}...)$ without knowing anything about $P(s_{choice}|...)$. (In particular, even if the left-hand side is very small (but non-zero), there can be a c_{choice} for which the value $P(c_{choice}|s_{choice} = \lambda - c_{choice}, ...)$ is large; just have $P(s_{choice} = \lambda - c_{choice}|...)$ be sufficiently small.) In the current discussion it is implicitly assumed that we are not in such an unusual situation.

training sets). This may in fact hold regardless of the priors setting setting those s values.

One way to establish such a result would be to bound the probability of an s such that the VC bounds strategy (of choosing the h with the smallest s) would be in error if one were to average only over all training sets with that s. (See the comment on "bounding the VC bounds" in Section 8.) It might even be possible to use the VC bounds themselves to do this. Given such a result, to derive our desired bound we would average over s, according to the worst possible $P(s)$. In other words, our program would be to try to take VC bounds concerning $P(|c_i - s_i| > \varepsilon | m)$ and from them alone (!) infer something concerning the expectation value which describes the effects of using the s-based strategy rather than randomly choosing between the hypotheses: $E((c_1 - c_2)\,\mathrm{sgn}(s_2 - s_1)|m)$. (A similar program would apply to the businessperson's question involving structural risk minimization.)

Although some preliminary work along these lines has been done, the analysis is far from complete.

10.7 ABSTRACTING THE LESSON OF WE-LEARN-IT INC

Properly speaking, to justify a strategy that uses the value of a variable to generate a hypothesis (e.g., a strategy that uses the values of some VC dimensions and/or empirical misclassificaton rates to choose between generalizers), one should specify the full statistical scenario involving that strategy, and prove that that scenario has certain desirable properties when compared to scenarios involving alternative strategies. Nonetheless some general (and, it must be admitted, frustratingly non-rigorous) comments can be made without detailed specification and analysis of the statistical scenario. These comments follow from the variant of the honesty principle.

If one has a result that concerns a distribution involving the (perhaps composite) variable T and the variable {my cost} (and no other variables), then it is possible that that result, by itself, could justify using the observed value of a variable B to determine one's behavior. For example, let B be the s value associated with a particular hypothesis. If the hypotheses under consideration are all contained in some pre-fixed hypothesis class, and the result in question is the appropriate VC result, then it might be that by itself, that result justifies choosing amongst the hypotheses according to their values of B.

However, say we define a "justification" for using the value of B to determine one's behavior to mean information about the implication that the observation $B = b$ has for the value of my cost, c. Furthermore, require that those implications vary with the values of b. In regard to this second stipulation, note that any variables fixed in the statistical scenario (e.g., hyperparameters like the γ_i) do not have to be explicitly listed in B. Since they are implicit in the statistical scenario and that scenario defines $P(c, b)$, they are implicit in the expression "$P(c \mid b)$." Accordingly, our second requirement does not mean that those kinds of variables must be varied.

If we accept this definition of "justification," then for a particular result to justify our b-based behavior that result would have to (be able to be extended) to concern the c-dependence and b-dependence of $P_{C|B}(c|b)$, by the variant of the honesty principle. In the absence of any extra statistical information, this usually means that the result must concern the distribution $P_{C|T}(c|t)$, with $T \supseteq B$ (i.e., with T determining B).

For example, if $T = \{B, Z\}$, then $P(c|b) = \sum_z P(c|t)P(z|b)$ and, therefore, we know that $\min_z P(c|t) \leq P(c|b) \leq \max_z P(c|t)$, so (some kinds of) information concerning $P_{C|T}(c|t)$ can be extended to concern the c-dependence of $P_{C|B}(c|b)$ for a particular observed b. On the other hand, if $T = \{B, Z\}$, and our result concerns $P(c, z|b)$, then we can get $P(c|b)$ directly from our result. This second case is a counterexample to our proposed rule that the result in question must concern $P_{C|T \supseteq B}(c|t)$. Note though that none of the frameworks have a distribution of interest of the form $P(c, z|b)$, so this counterexample is not germane.

Now consider the case where $B = \{T, Z\}$. In this situation $P(c|b) = P(c|t, z)$ and, in general, even knowing $P_{C|T}(c|t)$ exactly tells you little concerning the c-dependence of $P_{C|T,Z}(c|t, z)$ for one particular t and z. (And what it does tell you does not vary with the observed value of z.) This case illustrates the importance of the $T \supseteq B$ condition. As a final example, note that regardless of the relationship between T and B, any distribution of the form $P(t|c)$ has to be multiplied by (the assumed unknown) $P(c)/P(t)$ (perhaps along with other things) if it is to tell us something about $P(c|b)$. So this case too agrees with the general rule that the result in question must concern $P_{C|T \supseteq B}(c|t)$ if it alone is to justify our use of b.

Now examine the two possible B's Dr. Smith discussed in his reply to the businessperson. Dr. Smith's first B is $\{s_1, s_2, h_1, h_2, VC(\gamma_1), VC(\gamma_2), \gamma_1$'s hypothesis class is contained inside γ_2's, $m\}$. Since we need not have B contain those variables that are fixed by the statistical scenario, we can instead have $B = \{s_1, s_2, h_1, h_2, m\}$, if we wish. Accordingly, to justify our proposed use of b, our result would have to concern $P(c|t)$ where t fixes the values s_1, s_2, h_1, h_2, and m. VC results do not concern such a distribution however. In the case of the second B, $\{m\}$, things look a bit better. But now we're faced with the difficulty that VC results do not concern a distribution over C and some composite variable T (rather it's over $C - S$ and T).

So consider the general idea of having one's behavior governed by the value of a variable. The discussion above illustrates that it is difficult to justify such behavior with a result concerning the distribution of interest of a framework, unless the result concerns $P_{C|T \supseteq B}(c|t)$. Now note that usually you can not have a result concerning $P(c|t)$ that involves a variable u not specified by c and/or t. [18] Therefore the (non-cost) variables in our result should all (!) be contained in the distribution of

---

[18] For example "$P(c|t) < u$" is not meaningful if neither c nor t determines the value of u, since in that case $P(c|t) = \sum_u P_{C,U|T}(c, u|t)$ so there is no single u value associated with the left-hand side of the proposed inequality. (On the other hand, "$P(c|t) < u$" is meaningful if, for example, t determines the value of u.)

interest's conditioning event. (Intuitively, to use a result to tell us something about the likely values of c, we must be able to plug in for all the non-c variables in that result.) This is not the case in the VC framework; the variables in the result include s, so to use that result we would use the value of s, i.e., have $S \subseteq B \subseteq T$. However, s is not in the conditioning event in the distribution of interest in the VC framework.

10.8 APPLYING THE LESSON TO FRAMEWORKS OTHER THAN THE VC FRAMEWORK

In both PAC and SP we have similar problems. The conditioning event in our distributions of interest in those frameworks is $\{f, m\}$, or just $\{m\}$, if one assumes the prior is of a particular form. Since we never know f, we cannot use it to determine our behavior (i.e., we can not have it be part of B). Yet it is hard to see how using the variable m to govern our behavior can do us much good in the real world. So just as it is difficult to massage VC results to be applicable in the real world, so is it difficult to do so with PAC and/or SP results.

Difficult does not mean impossible however. In particular, it appears that PAC results can be "massaged" in this way. To see this, create a statistical scenario much like the one described by Dr. Smith, governed by $P(f, d, \text{choice}, c_1, c_2, h_1, h_2, \gamma_1, \gamma_2)$. Also let $B = T = \{h_1, h_2, \gamma_1, \gamma_2, m\}$. (In other words, create a new scenario with a new distribution of interest that replaces the $P(c \mid f, m)$ of standard PAC and SP.) Use the value of B to govern our behavior, by guessing either h_1 or h_2 depending on the values for γ_1 and γ_2. In particular, use h_i if γ_i is a "boosted" version of $\gamma_{j \neq i}$.[16,40] It turns out that in this situation PAC can provide us some information about $P(c_{\text{choice}} | T)$, if we declare our statistical scenario to have priors falling in a certain class (associated with "weak learnability" of the unboosted algorithm). Unfortunately, such results with real-world implications are hard to come by in PAC and SP. Nonetheless, because they do not involve confidence intervals, with these frameworks there is not the same kind of difficulty in having the results concern $P_{C|T \supseteq B}(c|t)$ that there is in the VC framework.

Things are more straightforward in the Bayesian framework. There the distribution of interest can be expressed as $P(c|h, d)$. The result of the framework concerning that distribution is the value of $E(C|h, d)$ as a function of h and d. Therefore, the conditioning event is the same as the set of variables in our result, so in theory we can use those variables to determine our behavior. (We can do so in practice since—unlike the variable f—we can actually observe the variables h and d.)

10.9 THE BAYESIAN FRAMEWORK VERSUS ALL THE REST

This would seem to be a major advantage of the Bayesian framework compared to the other three frameworks; we can directly use the values of the variables in its results to govern our behavior. However, in any such situation where those results can be used this way one must know (!) $P(f)$. (Otherwise we are making an assumption for $P(f)$, an assumption that may violate the original honesty principle.) Unfortunately, this is almost never the case. Accordingly, although it is *in theory* possible for one to use the results of the Bayesian framework and still obey both honesty principles, in practice this is rarely possible. (The same problem affects the boosting results mentioned above.) The implication is sobering: even if one sticks to iid (rather than off-training-set) error, one is rarely justified in using any of the frameworks to determine one's behavior as a statistician.

Despite this, the Bayesian framework does have a leg up on the other frameworks, as far as real-world applicability is concerned. For example, it always obeys the variant of the honesty principle. On the other hand, it is hard to have the VC framework obey that principle, and although there are cases in which the other two frameworks obey it (like boosting), they are hard to come by. It should also be pointed out that in a number of scenarios all interested parties would agree that a particular assumption for $P(f)$ is at worst a close approximation to the true $P(f)$ (e.g., when that prior is set via the technique of empirical Bayes, or when one assumes the priors over phase space implicit in statistical mechanics). One might say that by using the Bayesian framework in such a situation one "approximately" satisfies the honesty principle. (This also applies to PAC as used with boosting.) Alternatively one can consider a minimax variation of the Bayesian framework, in which one comes to conclusions only knowing that $P(f)$ falls in some class. (See the last section.) If you could be assured that $P(f)$ does indeed fall in that class in the real world, then this variation of the Bayesian framework would obey the honesty principle.

The Bayesian framework also has the nice property that its conditioning event is all that you (usually) know, and nothing else: d. In the other frameworks, the conditioning event in the distribution of interest averages away some of what we know (e.g., it replaces d by m). So to use such a framework, we would adopt the odd strategy of considering scenarios that we know differ from the actual statistical scenario at hand, in an effort to predict what happens for the actual scenario. It is hard to reconcile this with the (spirit of the) honesty principles. There are a number of practical advantages to the Bayesian framework as well. For example, it makes all assumptions explicit, and naturally provides error bars (see Wolpert[61]).

On the other hand, there are a number of shortcomings to the conventional Bayesian framework. Due to the possibility of error in the assumption for the prior, there is no guarantee that using the Bayesian framework minimizes expected cost in practice. In addition, there are many assumptions one could make to fix the full predictive distribution in addition to those concerning $P(f)$; i.e., one does not have to make a direct assumption for a prior—as conventional Bayesians do—even if one's goal is to evaluate the Bayes-optimal guess. (See Wolpert.[61]) In this sense conventional Bayesian practice is not logically necessary. More generally, in the non-Bayesian frameworks f is fixed in the conditioning event (or can be taken to be). This is why in those frameworks, no matter how one evaluates things, there is no need to make assumptions for $P(f)$. However in the VC framework one can formulate things so that f does not occur in the conditioning event and the result still does not rely on any assumptions concerning $P(f)$. For PAC to remove f from the conditioning event some assumptions about $P(f)$ must be made, but these are in some ways more benign than those used in the Bayesian framework. And in SP, for certain scenarios (e.g., when self-averaging holds due to $T(h)$ and the relative sizes of n and m), the central result does not depend on f. Accordingly, for such scenarios, $P(c|m) = \sum_f P(c|f,m)P(f) \propto \sum_f (1-c)^m \rho_0(f,T,m)P(f) = (1-c)^m \sum_f \rho_0(f,T,m)P(f)$, and the $(1-c)^m$ dependence is independent of f.

This ability to sidetrack the central difficulty with the Bayesian framework—the possibility of an incorrect assumption for $P(f)$—is the primary reason the other frameworks are interesting even in the context of real-world generalizing. However, there are a number of secondary reasons as well. For example, neither the VC framework nor PAC are concerned with the actual generalizer, but rather with some characteristics of it. This can be both a weakness (potentially useful information is being ignored) and a strength (results have general applicability). As another example, many of the empirical results concerning learning algorithms one finds reported in the literature implicitly have f fixed in the conditioning event. So the appropriate way to analyze such experiments is with a distribution of interest that has f fixed.

10.10 FINAL COMMENTS

Some other advantages and disadvantages of the particular frameworks are discussed in the sections presenting those frameworks. In particular, the implications of considering those frameworks in light of off-training-set error is considered in those sections.

There are also some aspects of all the frameworks that are not so much advantages or disadvantages as they are oddities. For example, due to our noise-free likelihood, there can be no conflicts in d: anytime an input value is repeated, the corresponding output values must agree. This means that any such repeat in d conveys no extra information concerning f to the generalizer. (Of course, it conveys information about other objects, like $\pi(x)$.) This points up an odd aspect of the non-Bayesian frameworks: They are concerned with probabilities conditioned on m. However, it is not m but rather m', the number of distinct pairs in d, which is most directly related to "the amount of information" in d directly concerning the distribution f.

APPENDIX 1: PEDAGOGICAL EXAMPLES OF OFF-TRAINING-SET ERROR BEHAVIOR FOR THE BAYES-OPTIMAL ALGORITHM

This appendix proves the claims, made at the end of Section 3, concerning how expected off-training-set error varies with m for the Bayes-optimal algorithm. Some of the examples presented here grew out of conversations with Tal Grossman and Manny Knill.

Let X consist of three values, 0, 1, and 2, and let Y consist of two values. Let $P(f) = 0$ for all f aside from those in some "target class" F. Assume that $P(f)$ is uniform over all f in F. Let F be the set of all (!) functions f such that $f(x) = 1$ for any $x \in X' \equiv \{0, 1\}$. Let $\pi(x) = 1/2 - \gamma/2$ for both $x \in X'$. Then for the Bayes-optimal algorithm

$$E(C|m = 1) - E(C|m = 0) = \gamma[(1 - \gamma)/(1 + \gamma) - 1/2],$$

which is positive for $\gamma < 1/3$.

PROOF. $E(C|m) = \sum_{f,d} E(C|f, d) P(d|f) P(f)$. For the Bayes-optimal algorithm, $E(C|f, d) = \sum_h E(C|f, h, d) P(h|d) = E(C|f, h^*, d) = Er(f, h^*, d)$ (see Section 5). For the off-training-set error function, this means $E(C|m) = \sum_{f,d} P(d|f) P(f) \{\sum_{x \notin d_X} \pi(x)[1 - \delta(f(x), h^*(x))] / \sum_{x \notin d_X} \pi(x)\}$. When $m = 0$, this equals $\sum_{f \subset F} P(f) \sum_x \pi(x)[1 - \delta(f(x), h^*(x))]$. Label the two f's in F f_0 and f_1 (corresponding to the values for $x = 2$). For both of those f's, for both x in X', $h^*(x) = f(x)$, and therefore $1 - \delta(f(x), h^*(x)) = 0$. For $x = 2$, $h^*(x)$ can equal either 0 or 1 (it makes no difference). Whichever value it equals, only one of the f's contributes, and we get $E(C|m = 0) = \gamma/2$. When $m = 1$, we instead get $(1/2) \sum_{d_X(1)} \sum_{f=f_1, f_2} \pi(d_X(1)) \times \{\sum_{x \notin d_X} \pi(x)[1 - \delta(f(x), h^*(x))] / \sum_{x \notin d_X} \pi(x)\}$. If $d_X(1) \in X'$, then again for only one of the two f's will the summand be nonzero; it does not matter which one, and the associated value of the sum is $\gamma(1 - \gamma)/2$, all divided by $(1 + \gamma)/2$ (the normalization constant in $Er(f, h, d)$). Multiplying by the 1/2, and summing over both possible $d_X(1)$'s lying in X', gives $\gamma(1 - \gamma)/(1 + \gamma)$. For the remaining $d_X(1)$, the sum equals 0. QED.

Intuitively, so long as m is smaller than the number of elements in X', due to the nonuniformity of $\pi(x)$ it is highly likely that d_X is restricted to X'. Now the off-training-set elements are the set $\{X - X'\} \cup \{X' - d_X\}$. The Bayes-optimal algorithm knows that $f(x)$ must equal 1 for any $x \in \{X' - d_X\}$, so the off-training-set error rate for those elements is 0. However, $f(x)$ for $x \in \{X - X'\}$ is completely unspecified, just given a training set lying within X'. Therefore, the Bayes-optimal generalizer will have an expected error rate of 1/2 for all $x \in \{X - X'\}$. The full off-training-set error rate will be a weighted sum of these two numbers, 0 and $1/2$. The weighting is according to the probability that an x randomly sampled from $X - d_X$ will like in the region $\{X' - d_X\}$ or $\{X - X'\}$, respectively. Now as the size of the training set increase, the size of $\{X' - d_X\}$ shrinks, while that of $\{X - X'\}$ stays constant. According, as m increases the weighting gets more strongly biased towards the elements in $\{X - X'\}$. Accordingly, as m increases, so does expected off-training-set error.

Now consider this same scenario, just with $\pi(x)$ uniform, $n \gg m$, and F consisting of all functions containing a single 1 in their set of n output values. Write $E(C|m) = \sum_f \sum_d E(c|f,d)P(d|f)P(f)$, where the inner sum is restricted to those d containing m elements. Up to an unimportant overall proportionality constant $(|F|^{-1})$, the sum can be rewritten as $n^{-m} \times \sum_{f \in F} \sum_{d \subset f} E(c|f,d)$, where the n^{-m} factor reflects the $m\,\pi(x)$ values. Now consider each element of the sum over f. It turns out that for any such f, the sum over d exactly equals $n^{(m-1)}$ and, therefore, for this scenario $E(C|m)$ is independent of m; the Bayes-optimal algorithm neither improves nor degrades as more data is provided.

Rather than prove this in full generality, I will simply demonstrate it for the first several m. First, note that for any d and any $q \in X$ that lies outside of d_X, so long as n is sufficiently larger than m the number of $f \subset F$ that are compatible with d and which have the value 0 at q is greater than the number that are compatible with d and that have the value 1 at q. Accordingly the Bayes-optimal function—the function h which minimizes $E(C|h,d)$—will agree with d, and equal 0 everywhere off of d. (See the section on the Bayesian framework.)

Since the Bayes-optimal algorithm always guesses this h, the task is to calculate $E(c|f, h_{\text{Bayes-optimal}}(d), d)$. However, this just equals $N(f,d)/(n-m')$, where $N(f,d)$ is the number of times $f(x) = 1$ for $x \notin d_X$. In other words, $\sum_{d \subset f} E(c|f,d) = \sum_{d \subset f} N(f,d)/(n-m') = \sum_{m'=1}^m \sum_{d \subset f; m', m} N(f,d)/(n-m')$.

Now since f contains one "1," the summand equals zero unless d_X avoids the x giving that one "1." When it is nonzero, the summand equals $1/(n-m')$. Therefore, the entire sum equals $\sum_{m'=1}^m \zeta(n-1,m,m')/(n-m')$, where $\zeta(a,b,c)$ is the number of ordered ways of choosing b elements out of a, where only c of those b elements are distinct. So:

$m = 1$: The sum $= \zeta(n-1,1,1)/(n-1) = 1$. (There are exactly $n-1$ ways of choosing a single element out of $n-1$ elements.)

$m = 2$: The sum $= \zeta(n-1,2,1)/(n-1) + \zeta(n-1,2,2)/(n-2)$. The second sum equals $n-1$; there are $(n-2)(n-1)$ ordered ways of choosing two distinct elements out of $n-1$. The first sum equals 1; there are $n-1$ ordered ways of choosing two elements out of $n-1$ when those elements are identical. Therefore the sum $= (n-1) + 1 = n$.

$m = 3$: The sum $= 1 + 3(n-1) + (n-1)(n-2) = n^2$.

$m = 4$: The sum $= 1 + 7(n-1) + 6(n-1)(n-2) + (n-3)(n-2)(n-1)$
$= n^3$.

This completes the demonstration.

There is a subtlety concerning this result which bears mentioning. By the no-free-lunch theorems, averaged over all f, $E(C|f,m)$ is independent of the learning algorithm, and equals $1/r$ for all m. But we have just shown that $\sum_{d \subset f} E(c|f,d) = E(C|f,m) \times n^m = n^{(m-1)}$, for all allowed f, for the case considered here (where $r = 2$). Therefore $E(C|f,m) = 1/n$, which is less than $1/r$. How is this possible? The resolution is the fact that the no-free-lunch theorem involves averaging over *all* f, whereas the set of "possible f" for which we here calculated $E(C|f,m)$ only consists of those f with a single "1" in their outputs. Indeed, unless one ascribes nonzero prior to the other f, the Bayes-optimal algorithm will not even be defined for arbitrary d.

In fact, in general $E(C|f,m)$ can increase with m, even when we have a Bayes-optimal generalizer, $n \gg m$, $\pi(x)$ is flat, and $f \in F$, the set of "allowed f" over which $P(f)$ is uniform and nonzero. As an example, let F be the union of the following three sets of f's:

1. all f's with a single 1 in their outputs;
2. all f's with a single 0 in their outputs; and
3. the "alternating" f, f_a, obeying $f(x_i) = i \bmod 2$.

For $m = 0$, the Bayes-optimal guess is f_a, and $E(C|f = f_a, m) = 0$. As soon as we get a training set, however, the Bayes-optimal guess can change. For example, if we have a two-element training set, and both the training set outputs are 1's, then the Bayes-optimal algorithm knows that f could not have been any of the functions consisting of a single "1." This will mean the Bayes-optimal guess will be all 1's, and therefore $E(C|f = f_a$, this training set) is just greater than a half. Since (given $f = f_a$) the probability of such a training set is nonzero (it equals $1/4$), we see that $E(C|f = f_a, m = 2) > 0$ and, therefore, expected error has risen.

APPENDIX 2: PROOF OF THEOREM 3

$E(C|d) = \sum_h E(C|d, h)P(h|d)$. Let h^* be the h value minimizing $E(C|d, h)$. Then our sum over h will be minimized if $P(h|d)$ is a delta function around h^*. This proves (i) and (ii). Now we must solve for h^*. To this end write

$$E(C|d, h) = \sum_c cP(c|h, d)$$

$$= \sum_c c \left\{ \sum_f P(f|d)\delta(Er(f, h, d), c) \right\} \quad \text{(since } P(f|h, d) = P(f|d))$$

$$= \sum_{c,f} cP(f|d)\delta(Er(f, h, d), c)$$

$$= \sum_f Er(f, h, d)P(f|d)$$

$$= \sum_f P(f|d) \sum_{x \in X} G(f(x), h(x))\pi'(x)$$

$$= \sum_{x \in X} \pi'(x) \sum_f G(f(x), h(x))P(f|d)$$

$$= \sum_{x \in X} \pi'(x)\Omega(x, h(x)).$$

We want to choose a hypothesis function h^* so as to minimize this sum over x. (Note that for this hypothesis function h^*, $\Omega(., .)$ is a mapping from $x \in X \to \mathbf{R}$.) Our hypothesis is that $h^* = \{x \in X \to \arg\min_{y \in Y} \Omega(x, y)\}$. First, since this function is a mapping from x to Y, it is a legal hypothesis function. Second, note that $\Omega(x, h^*(x)) = \Omega[x, \arg\min_{y \in Y}\{\Omega(x, y)\}] = \min_{y \in Y} \Omega(x, y) = \min_{h \in H} \Omega(x, h(x))$. Since $\pi'(x) \geq 0 \forall x \in X$,

$$\min_{h \in H} \sum_{x \in X} \pi'(x)\Omega(x, h(x)) \geq \sum_{x \in X} \pi'(x) \min_{h \in H} \Omega(x, h(x)),$$

which by the previous sentence just equals $\sum_{x \in X} \pi'(x)\Omega(x, h^*(x))$. QED.

APPENDIX 3: PROOF OF THEOREM 4

This appendix proves Theorem 4 of the text, noting a tighter bound along the way. The derivations is more complicated than the corresponding derivations in the PAC literature, but has the advantages that (i) it is fully rigorous; (ii) it illuminates exactly what approximations go into the bound, what aspects of the axioms of PAC are needed for the bound, etc.

We start with the definition $\Delta \equiv \max_{P(h|d)} \sum_{c>\varepsilon} P(c|f \subset H'', m)$, where the maximum is over all generalizers with the given H'' and a particular H'. Any upper bound on Δ obviously applies individually to each generalizer with the given H' and H''.

Our definition of Δ can be rewritten as

$$\Delta = \max_{P(h|d)} \sum_{c>\varepsilon} \sum_h P(h|f, m)\delta(c, Er(h, f))$$

$$= \max_{P(h|d)} \sum_h P(h|f, m) \sum_{c>\varepsilon} \delta(c, Er(h, f))$$

$$= \max_{P(h|d)} \sum_h P(h|f, m)\delta(Er(h, f) > \varepsilon)$$

(this $\delta(.)$ function is defined to equal 1 if its argument is true, 0 otherwise)

$$= \max_{P(h|d)} \sum_{h, d \subset f; m} P(h|d)\delta(Er(h, f) > \varepsilon)\Pi_{i=1}^{m}\pi(d_X(i))$$

(using PAC's likelihood, where "$d \subset f; m$" means the d that lie on f and that have m elements),

$$= \max_{P(h|d)} \sum_{h \in H', d \subset f; m} P(h|d)\delta(Er(h, f) > \varepsilon)\Pi_{i=1}^{m}\pi(d_X(i))$$

(since we are assuming that $f \in H''$ and therefore the likelihood means that d is consistent with an element of H'' and therefore by the second requirement of PAC generalizers the support over H of $P(h|d)$ is contained within H'.)

By the first requirement of PAC generalizers, for any training set d sampled from f one can write $P(h|d) = P(h|d)\delta(d \subset h)$. Since we sum only over those $d \subset f$, this allows us to write

$$\Delta = \max_{P(h|d)} \sum_{h \in H', d \subset f; m} P(h|d)\delta(Er(h, f) > \varepsilon)\Pi_{i=1}^{m}\pi(d_X(i))\delta[f(d_X(i)), h(d_X(i))].$$

(3.1)

Rewrite this symbolically as

$$\max_P \sum_{ij} P_{ij}K_{ij} = \max_P \sum_i \left[\sum_j P_{ij}K_{ij} \right]$$

$$\equiv \max_P \sum_i R_{P,i} \leq \sum_i [\max_P R_{P,i}]$$

$$= \sum_i \max_P \left[\sum_j P_{ij}K_{ij} \right].$$

Accordingly, we can rewrite Eq. (3.1) as

$$\Delta \equiv \sum_{d \subset f; m} \max_{P(h|d)} \left\{ \sum_{h \in H'} P(h|d)\delta(Er(h,f) > \varepsilon)\Pi_{i=1}^m \pi(d_X(i))\delta[f(d_X(i)), h(d_X(i))] \right\},$$

where the max means that for any value of d, we maximize the value of the expression inside the $\{\}$ over all possible $P(h|d)$. (Note that the maximizing $P(h|d)$ might vary with d.) Now since $P(h|d)$ must sum to 1 and is nonnegative for all h and d, the innersum is maximized when $P(h|d)$ is a delta function (over h's) centered on the maximum (over H') of the quantity $\delta(Er(h,f) > \varepsilon)\Pi_{i=1}^m \pi(d_X(i))$ $\delta[f(d_X(i)), h(d_X(i))]$. This means that

$$\Delta \leq \sum_{d \subset f; m} \max_{h \in H'}\{\delta(Er(h,f) > \varepsilon)\Pi_{i=1}^m \pi(d_X(i)) \times \delta[f(d_X(i)), h(d_X(i))]\}.$$

PAC uses a cruder approximation than this however. It starts with Eq. (3.1) and then pulls the $\max_{P(h|d)}$ through both the sum over d and the sum over h, getting

$$\Delta \leq \sum_{h \in H', d \subset f; m} \max_{P(h|d)} \{P(h|d)\delta(Er(h,f) > \varepsilon)\Pi_{i=1}^m \pi(d_X(i))\delta[f(d_X(i)), h(d_X(i))]\},$$

where the max now means that for any value of h and/or d, we maximize the value of the expression inside the $\{\}$ over all possible $P(h|d)$.

We can rewrite this expression as

$$\Delta \leq \sum_{h \in H', d \subset f; m} \max_{P(h|d)} \{P(h|d)\} \, \delta(Er(h,f) > \varepsilon)\Pi_{i=1}^m \pi(d_X(i))\delta[f(d_X(i)), h(d_X(i))],$$

$$= \sum_{h \in H', d \subset f; m} \delta(Er(h,f) > \varepsilon)\Pi_{i=1}^m \pi(d_X(i))\delta[f(d_X(i)), h(d_X(i))]$$

$$= \sum_{h \in H'} \delta(Er(h,f) > \varepsilon) \sum_{d \subset f; m} \Pi_{i=1}^m \pi(d_X(i))\delta[f(d_X(i)), h(d_X(i))]$$

$$= \sum_{h \in H'} \delta(Er(h,f) > \varepsilon) \sum_{d_X; m} \Pi_{i=1}^m \pi(d_X(i))\delta[f(d_X(i)), h(d_X(i))]$$

$$= \sum_{h \in H'} \delta(Er(h,f) > \varepsilon) \left\{ \sum_{x \in X} \pi(x)\delta(f(x), h(x)) \right\}^m$$

(by a result from Wolpert[63])

$$= \sum_{h \in H'} \delta(Er(h,f) > \varepsilon)\{1 - Er(h,f)\}^m,$$

(due to the error function used by PAC)

$$\leq \sum_{h \in H'} \delta(Er(h,f) > \varepsilon)\{1 - \varepsilon\}^m$$

$$=\{1 - \varepsilon\}^m \sum_{h \in H'} \delta(Er(h, f) > \varepsilon);$$

$$\Delta \leq \{1 - \varepsilon\}^m |H'|, \text{ where } |H'| \text{ is the number of functions in } H'. \tag{3.2}$$

QED

Note that these arguments can also be used to bound $P(c \leq \varepsilon | f, m)$. However, PAC usually concerns itself with $P(c > \varepsilon | f, m)$.

APPENDIX 4: PROOFS THAT THE EXAMPLES IN SECTION 6 RESULT IN POLYNOMIAL LEARNABILITY

For brevity, the examples presented here do not directly show that the learning algorithm runs in time polynomial in m. Such an issue, which gets into the details of how the algorithms are implemented, can usually be addressed without too much difficulty. But doing so merely obscures the central points of the discussion. Also, in the examples that follow, polynomial learnability would obtain even if "tighter" measures $r(.)$ were used (e.g., if one used the log of the $r(.)$ presented here). Again, these example are primarily illustrative.

EXAMPLE 1. More precisely, say that $r(f)$ is an ordered listing of all functions in $H = H''$ (i.e., an injective mapping taking $f \to \mathbf{N}$, the counting numbers). Assume that our learning algorithm knows H, and that it returns the $f \in H$ that agrees with the data while having the lowest value of $r(f)$. No matter what d is sampled from f, the h returned by the algorithm must obey $r(h) \leq r(f)$. Accordingly, we can take $A(m, Q)$ to be the set of all h such that $r(h) \leq Q$. Due to the injectivity of $r(.)$, this means that $|A(m, Q)| = Q$. Since $Q < 2^Q$ for all counting numbers Q, this in turn means that we have a Blumer algorithm for H and, therefore, polynomial learnability of H. QED.

EXAMPLE 2. Label elements of X by the integers. Define $r(f)$ by viewing f as an extended-binary number: $r(f) = \sum_{x \in X} f(x)g(x)$, where $g(0) = 1$, $g(1) = 2$, $g(-1) = 4$, $g(2) = 8$, $g(-2) = 16$, etc. By definition of H'', $r(f)$ is defined for all allowed f, even if X is infinite. Furthermore, $r(.)$ is injective. Now for our learning algorithm $r(h) \leq r(f)$ always. Therefore, just as in Example 1, we can take $A(m, Q)$ to be the set of all h such that $r(h) \leq Q$ which means that $|A(m, Q)| < 2^Q$ for all Q, which in turn means that we have a Blumer algorithm for H and, therefore, polynomial learnability of H. QED.

EXAMPLE 3. View $r(f)$ as a "label" determined according to the following algorithm: The label is 0 for the function with all outputs 0's. All functions with outputs consisting of either 0's or 1's are viewed as a binary number and given that binary number as their label. (This, of course, is consistent with the rule for the all 0's function.) Now consider functions with all outputs being 0, 1, or 2. View such a function as a ternary number. Starting with the lowest ternary number (namely, all 0's) and going up to all 2's, label the function if it has not already been labelled. (The functions that have been labelled being those consisting of all 0's and 1's.) The label the function gets is the lowest integer that has not yet been assigned to a function. So for example all 1's has label $2n - 1$; all labels from 0 to $2^n - 1$ are taken by functions with no twos. The lowest (in ternary decoding) function with a 2 is the one with all 0's except for the low bit ($x = n$) being a 2: $000\ldots002$. This function gets label 2^n. The next highest ternary number is $000\ldots0010$. This already has a label however (namely its binary value, two). The next highest ternary number is $000\ldots0011$. This also already has a label (namely its binary value, three). The next highest number is $000\ldots00012$, which does not already have a label and, therefore, is given the label $1 + 2^n$, etc. After going through all 2's, one goes on to 3's, and so on. Note that the mapping $f \to r(f)$ is one-to-one.

First, note that $r(h) \leq r(f)$. To see this, let k be the largest Y value in f. Then $r(f)$ is determined by looking at f in base k. Similarly, $r(h)$ is determined by looking at h in base $t \leq k$. If $t < k$, then it is immediate that $r(h) < r(f)$. If $t = k$, then both h and f are determined by ordering all k-ary numbers which are not of the form of a $(k-1)$-ary number. Since for all $x\, h(x) \leq f(x)$, that ordering cannot encounter f before it encounters h.

Since $r(h) \leq r(f)$, we can take $A(m, Q)$ to be the set of all h with $r(h) \leq Q$. Due to the injectivity of $r(.)$, this means that $|A(m, Q)| = Q$. Since $Q < 2^Q$ for all counting numbers Q, this in turn means that we have a Blumer algorithm for H and, therefore, polynomial learnability of H. QED.

EXAMPLE 4. Given any f, since X is finite, an upper bound on the number of h's that reproduce some d with m' distinct elements but take on fixed values $\Gamma(x)$ outside of d_X is given by $\binom{n}{m'}$. So for all f and all m, the number of h's that might be produced from an m-element sample of f is bounded above by $\sum_{m'=1}^{n} \binom{n}{m'}$. Therefore $|A(m, Q)|$, the number of h's that might be produced from a sample of a function of complexity Q, is bounded above by $\kappa^Q \sum_{m'=1}^{n} \binom{n}{m'}$. Take $\alpha = 0$, and $K = \kappa \sum_{m'=1}^{n} \binom{n}{m'}$. Since $Q = r(f) > 1$ for all f, $Q^\beta > 1$. Therefore $K^{Q^\beta m^\alpha} \geq\!> \{\kappa \sum_{m'=1}^{n} \binom{n}{m'}\}^Q > \kappa^Q \sum_{m'=1}^{n} \binom{n}{m'} \geq |A(m, Q)|$ always. QED.

APPENDIX 5: PROOFS RELATED TO THEOREM 5

Start by proving the following:

Given a learning algorithm for which $P(h|d) = \delta(h - h')$ for all d,

$$P(|c - s| > \varepsilon | f, m) = P(|c - s| > \varepsilon | f, h', m). \tag{5.1}$$

PROOF. $P(|c - s| > \varepsilon | f, m) = \sum_h P(|c - s| > \varepsilon | f, h) P(h|f)$, where the "$m$" condition is implicit for now. We can rewrite this sum as

$$\sum_{h,d} P(|c - s| > \varepsilon | f, h) P(h|d, f) P(d|f) =$$

$$\sum_{h,d} P(|c - s| > \varepsilon | f, h) P(h|d) P(d|f).$$

For our generalizer, this just equals $\sum_d P(|c - s| > \varepsilon | f, h') P(d|f) = P(|c - s| > \varepsilon | f, h')$. QED.

Since f and h' fix c (for the case of iid error), specifying c in addition to f and h' provides no new information. So by Eq. (5.1), $P(|c - s| > \varepsilon | f, m) = P(|c - s| > \varepsilon | f, h', c, m)$. (More formally, if the random variable C is fixed by the variable B, $Pr(A = a|B = b, C = c) = Pr(A = a, B = b, C = c)/Pr(B = b, C = c) = Pr(A = a, B = b)/Pr(B = b) = Pr(A = a|B = b)$.) Similar results hold if $|c - s| > \varepsilon$ is replaced by $c > s + \varepsilon$.

Using this, we can prove the following: $P(|c - s| > \varepsilon | f, m)$ is the same for all f, having the same error with h' and equals

$$1 - \sum_{t=-\varepsilon}^{\varepsilon} \sum_{\{x_i\}} \Pi_{i=1}^m \pi(x_i) \delta[\sum_{i=1}^m g(x_i), -m(t - 1 + c)],$$

where $g(x) = 1$ if $f(x) = h'(x)$, 0 otherwise, $\{x_i\}$ is a set of m X values, t is a set of real numbers separated from one another by integer multiples of $1/m$, the sum over t is implicitly understood to be from t equals the smallest possible value of $s - c$ which $\geq -\varepsilon$ to t equals the largest possible value which $\leq \varepsilon$, and the $\delta(.,.)$ is a Kronecker delta function.

PROOF. $P(|c - s| > \varepsilon|f, h', c, m) = 1 - P(|c - s| \le \varepsilon|f, h', c, m)$. Since c is fixed, $P(|c - s| \le \varepsilon|f, h', c, m) = \sum_{t=-\varepsilon}^{\varepsilon} P(s - c = t|f, h', c, m)$, with t as specified in the body of Eq. (5.2). Recall that $S(\{x_i\}, f, h)$ is defined as the number of times $f(x)$ and $h(x)$ agree over the set $\{x_i\}$. Using this to re-express our summand gives $P(|c - s| > \varepsilon|f, h', c, m) = 1 - \sum_{t=-\varepsilon}^{\varepsilon} P(S(d_X, f, h') = -m(t - 1 + c)|f, h', c, m)$.

Rewrite our summand as $\sum_{d_X} \delta(S(d_X, f, h') = -m(t-1+c))P(d_X|f, h', c, m)$, and recall that $P(d_X = \{x_i\}|f, h', c, m) = \Pi_{i=1}^m \pi(x_i)$. Rewrite $S(d_X, f, h')$ as $\sum_{i=1}^m g(x_i)$.

Pulling it together,

$$P(|c-s| > \varepsilon|f, h', c, m) = 1 - \sum_{t=-\varepsilon}^{\varepsilon} \sum_{\{x_i\}} \Pi_{i=1}^m \pi(x_i)\delta\left(\sum_{i=1}^m g(x_i), -m(t - 1 + c)\right),$$

where the sum over $\{x_i\}$ is understood to be the product of m sums, one for each i, each over all x. QED.

The result in Eq. (5.2) gives the following:

$P(|c - s| > \varepsilon|f, m)$ equals the probability that, for a coin with bias c flipped m times, the magnitude of the difference between cm and the observed number of heads is bounded below by $m\varepsilon$.   (5.3)

PROOF. Define $-m(t-1+c)$ to be z, the number of agreements between f and h' on d_X. Examine the term inside the sum over t in Eq. (5.2), $\sum_{\{x_i\}} \Pi_{i=1}^m \pi(x_i)$ $\delta[\sum_{i=1}^m g(x_i), z]$. Re-express the delta function as $\sum_{\{a_j\}} \Pi_{i=1}^m \delta(g(x_i), a_i)$, where $\{a_j\}$ is the set of all possible strings of m 0's and 1's which contain exactly z 1's. (That is, the delta function equals 1 if there is such a string that exactly equals the sequence $g(x_i)$, 0 otherwise.) There are C_z^m such strings. We can now write our sum over $\{x_i\}$ as

$$\sum_{\{a_j\}} \sum_{\{x_i\}} \Pi_{i=1}^m \delta(g(x_i), a_i)\pi(x_i) = \sum_{\{a_j\}} \Pi_{i=1}^m [\sum_x \delta(g(x), a_i)\pi(x)].$$

When $a_i = 1$,
$$\sum_x \delta(g(x), a_i)\pi(x) = \sum_x g(x)\pi(x).$$

When $a_i = 0$, $\sum_x \delta(g(x), a_i)\pi(x) = 1 - \sum_x g(x)\pi(x)$. Therefore $\Pi_{i=1}^m [\sum_x \delta(g(x), a_i)\pi(x)] = [\sum_x g(x)\pi(x)]^z \times [1 - \sum_x g(x)\pi(x]^{m-z}$, for all strings $\{a_j\}$. Therefore, if we define $p \equiv \sum_x g(x)\pi(x)$, the quantity being calculated is

$\sum_{t=-\varepsilon}^{\varepsilon} C_z^m p^z [1-p]^{m-z}$. Since z equals $-m(t-1+c)$, and since $p = 1 - c$, this quantity is simply the probability that, for a coin with bias c for heads flipped m times, the (magnitude of the) difference between $(1-c)m$ and the observed number of tails z is bounded above by $m\varepsilon$. This equals the probability that the diference between cm and the observed number of heads $< m\varepsilon$. QED.

If one now applies the Hoeffding inequality, one gets $P(|c - s| > \varepsilon | f, m) < 2e^{-m\varepsilon^2}$. This is the usual way the results are presented in the VC framework. (See Eq. (6.15) in Vapnik.[51])

We have already established that $P(|c - s| > \varepsilon | f, m) = P(|c - s| > \varepsilon | f, h', c, m)$, where the conditioning c is the error for f and h'. Using the same arguments that proved Eq. (5.1), we can next equate $P(|c - s| > \varepsilon | f, h', c, m)$ with $P(|c - s| > \varepsilon | f, c, m)$. Since by Eq. (5.2) for all f having the same error with h' $P(|c - s| > \varepsilon | f, m)$ is independent of f, this establishes that $P|c - s| > \varepsilon | f, c, m)$ must also be independent of f for all f having the same error with h'. Now for any variables a, b, c if $P(a|b, c)$ is independent of c, we can write it as $\gamma(a, b)$. For this case $P(a|b) = \sum_c P(a, c|b) = \sum_{c \text{ consistent with } b} P(a, c|b) = \sum_{c \text{ consistent with } b} P(a|b, c)P(c|b) = \sum_c \gamma(a, b)P(c|b) = \gamma(a, b) = P(a|b, c)$. (The '$c$' in this equality should not be confused with the error variable c.) This establishes the following:

Given a learning algorithm for which $P(h|d) = \delta(h - h')$ for all d,

$P(c > s + \varepsilon | f, m)$ is equivalent to $P(c > s + \varepsilon | c, m) = P(s < c - \varepsilon | c, m)$, (5.4)

where c is the error for f and h'.

If we define $\kappa \equiv c - \varepsilon$, Eq. (5.4) means that Eq. (5.3) is equivalent to $P(s < \kappa | c, m) < 2e^{-m(c-\kappa)^2}$ when $\kappa < c$. Note that these bounds are clearly suboptimal for $\kappa \le 0$, since we know s cannot be negative.

APPENDIX 6: PROOF OF EQUATION (9.1)

This appendix proves Eq. (9.1) of the text. First, see Appendix 3. Note in particular that Eq. (3.1) of that appendix still holds, if one simply drops the "$\in H'$" restriction on the sum and removes the $\max_{P(h|d)}$. Now since $n \gg m$, we can take m', the number of distinct pairs in d, to equal m (see Wolpert[59,63]). (Strictly speaking, this only holds if the summand in Eq. (3.1) does not grow vastly larger for smaller m'. However, for almost any $\pi(.)$ and $T(.)$, this restriction will be a concern only if ε is so large that there exist no h agreeing with d for which $c > \varepsilon$, unless m' is sufficiently small.) By a result from Wolpert,[63] together with our other two hypotheses concerning $T(.)$ and $\pi(.)$, $m' = m$ implies self-averaging; i.e., the three conditions together imply that $k(d, T(.))$, the normalization constant in the exhaustive learning $P(h|d)$, is independent of d. Therefore, our analogue of Eq. (3.1) of Appendix 3 becomes

$$\Delta = \sum_h \frac{T(h)}{k(.,T(.))} \delta(Er(h,f) > \varepsilon) \sum_{d_X;m} \Pi_{i=1}^m \pi(d_X(i)) \delta[f(d_X(i)), h(d_X(i))]$$

$$= \sum_h \frac{T(h)}{k(.,T(.))} \delta(Er(h,f) > \varepsilon)[1 - Er(f,h)]^m$$

$$\leq \sum_h \frac{T(h)}{k(.,T(.))} \delta(Er(h,f) > \varepsilon)[1 - \varepsilon]^m$$

$$= [1 - \varepsilon]^m \sum_h \frac{T(h)}{k(.,T(.))} \delta(Er(h,f) > \varepsilon)$$

$$= [1 - \varepsilon]^m \left\{ \frac{\sum_h T(h)\delta(Er(h,f) > \varepsilon)}{\sum_h T(h)\delta(d \subset h)} \right\},$$

where d is any training set sampled from f.

Using the constancy of $T(h)$, this results in $\Delta \leq [1 - \varepsilon]^m[\sum_h \delta(Er(h,f) > \varepsilon)]/[\sum_h \delta(d \subset h)]$. Given that we can take $m' = m$, the sum in the denominator equals r^{n-m}. The sum in the numerator equals $\sum_{c>\varepsilon} \sum_h \delta(Er(h,f) = c)$. Using the assumption of a uniform $\pi(.)$, this numerator sum equals $\sum_{z=0}^{n(1-\varepsilon)} C_z^n (r-1)^{n-z}$ (intuitively, z is the number of agreements between h and f), which is the sum of the first $n(1-\varepsilon)$ terms of the binomial expansion of $r^n = [(r-1)+1]^n$. This allows us to immediately write $\Delta \leq \{1 - \varepsilon\}^m r^m$. QED.

ACKNOWLEDGMENTS

This work was supported by the SFI, by NLM grant F37 LM00011, and by TXN Inc. I would like to thank the following people for interesting discussions on the topics discussed in this chapter: Manny Knill, Tal Grossman, Geoff Hinton, Wray Buntine, Christian Van der Broeck, Cullen Schaffer, Tali Tishby, Eric Baum, David Stork, Charlie Strauss, Brian Ripley, Alan Lapedes, David Haussler, John Denker, and David Rosen. I would especially like to thank Wray Buntine for help in the presentation of the chapter.

REFERENCES

1. Anoulova et al. "PAB-Decisions for Boolean and Real-Valued Features." In *COLT*, 353–362. ACM, 1992.
2. Anthony M., and N. Biggs. *Computational Learning Theory*. Cambridge University Press, 1992.
3. Baum, E., and D. Haussler. "What Size Net Gives Valid Generalization?" *Neural Comp.* **1** (1989): 151–160.
4. Berger, J. *Statistical Decision Theory and Bayesian Analysis*. New York: Springer-Verlag, 1985.
5. Blumer, A., A. Ehrenfeueht, D. Haussler, and M. Warmuth. "Occam's Razor." *Info. Proc. Lett.* **24** (1987): 377–380.
6. Blumer, A., A. Ehrenfeueht, D. Haussler, and M. Warmuth. "Learnability and Vapnik-Chervonenkis Dimension." *J. ACM* **36** (1989): 929–965.
7. Bridle, J. "Probabilistic Interpretation of Feedforward Classification Network Outputs, with Relationships to Statistical Pattern Recognition." In *Neuro-Computing: Algorithms, Architectures, and Applications*, edited by F. Fougelman-Soulie and J. Herault. New York: Springer-Verlag, 1989.
8. Buntine, W. "A Theory of Learning Classification Rules." Ph.D. Thesis, University of Technology, Sydney, Australia, 1990.
9. Buntine, W. "Myths and Legends in Learning Classification Rules." NASA Ames Research Center Report IA-90-05-08-1, 1990.
10. Buntine, W., and A. Weigend. "Bayesian Back-Propagation." *Complex Systems* **5** (1991): 603–643.
11. Charniak, E. "Bayesian Networks Without Tears." *AI Mag.* **Winter** (1991): 50–63.
12. Cheeseman, P. "In Defense of Probability." In *Proceedings of the Ninth International Joint Conference on Artificial Intelligence*. 1985.
13. Cohn, D., and G. Tesauro. "How Tight Are the Vapnik-Chervonenkis Bounds?" Technical Report TR 91-03-04, Department of Computer Science, University of Washington, 1991.
14. *COLT: The Workshop on Computational Learning Theory*, Cambridge, MA: MIT Press. A series of books over many years, with many editors.
15. Dietterich, T. "Machine Learning." *Ann. Rev. Comp. Sci.* **4** (1990): 255–306.
16. Drucker, H., R. Schapire, and P. Simard. "Improving Performance in Neural Networks Using a Boosting Algorithm." In *Neural Information Processing Systems 5*, edited by S. Hanson et al. San Mateo, CA: Morgan-Kauffman, 1993.
17. Duda, R., and P. Hart. *Pattern Classification and Scene Analysis*. New York: Wiley & Sons, 1973.
18. Eubank, R. *Spline Smoothing and Nonparametric Regression*. New York: Marcel Dekker, 1988.

19. Geman, S., E. Bienenstack, and R. Doversat "Neural Networks and the Bias/ Variance Dilemma." *Neural Comp.* **4** (1992): 1–58.
20. Good, I. "Kinds of Probability." *Science* **129** (1959): 443–447.
21. Gull, S. "Developments in Maximum Entropy Data Analysis." In *Maximum Entropy and Bayesian Methods*, edited by J. Skilling, 53–71. Boston: Kluwer, 1989.
22. Haussler, D., M. Kearns, and R. Schapire. "Bounds on the Sample Complexity of Bayesian Learning Using Information Theory and the VC Dimension." *Machine Learning* **14** (1994): 83–114.
23. Haussler, D. "The Probably Approximately Correct (PAC) and Other Learning Models." In *Foundations of Knowledge Acquisition: Machine Learning*, edited by edited by A. Meyrowitz and S. Chipman, Ch. 9. Boston: Kluwer, 1994.
24. Haussler, D. "Decision Theoretic Generalizations of the PAC Model for Neural Net and Other Learning Applications." These proceedings.
25. Jeffreys and Berger. *New Scientist* **80** (1992): 64–72.
26. Loredo, T. "From Laplace to Supernova 1987a: Bayesian Inference in Astrophysics." In *Maximum Entropy and Bayesian Methods*, edited by P. Fougere, 81–142. Boston: Kluwer, 1990.
27. Hertz, J., A. Krogh, and R. G. Palmer. *Introduction to the Theory of Neural Computation.* Santa Fe Institute Studies in the Sciences of Complexity, Lect. Notes Vol. I. Redwood City, CA: Addison-Wesley, 1991.
28. MacKay, D. "Bayesian Interpolation," and "A Practical Bayesian Framework For Back-Prop Networks." Papers presented at Neural Networks for Computing conference, Snowbird, Utah, 1991.
29. MacKay, D. "Information-Based Objective Functions for Active Data Selection." *Neural Comp.* **4** (1992): 590–604.
30. *MAXENT: Maximum Entropy and Bayesian Methods* Boston: Kluwer. A series of books over many years, with many editors.
31. Moody, J. "The Effective Number of Parameters: An Analysis of Generalization and Regularization in Nonlinear Learning Systems." In *Neural Information Processing Systems*, vol. 4, edited by J. Moody et al. San Mateo, CA: Morgan Kaufmann, 1992.
32. Natarajan, B. *Machine Learning: A Theoretical Approach.* San Mateo, CA: Morgan Kauffman, 1991.
33. Neal, R. "Priors for Infinite Networks." Technical Report CRG-TR-94-1, Department of Computer Science, University of Toronto, 1994.
34. Opper, M., and D. Haussler. "Calculation of the Learning Curve of Bayes Optimal Classification Algorithm for Learning a Perceptron with Noise." Technical Report, University of California, Santa Cruz, 1991.
35. Opper, M., and D. Haussler. "Generalization Performance of Bayes Optimal Classification Algorithm for Learning a Perceptron." *Phys. Rev. Lett.* **66** (1991): 2677–2680.

36. Perrone, M. "Improving Regression Estimation: Averaging Methods for Variance Reduction with Extensions to General Convex Measure Optimization." Ph.D. Thesis, Department of Physics, Brown University, 1993.
37. Ripley, B. "Neural Networks and Related Methods for Classification." *J. Roy. Stat. Soc. B.* (to appear).
38. Rivest, R. Course notes and reprints for Machine Learning course 18.428, Massachusetts Institute of Technology, 1989.
39. Schaffer, C. "Overfitting Avoidance as Bias." *Mach. Learning* **10** (1993): 153–178.
40. Schapire, R. "The Strength of Weak Learnability." *Mach. Learning* **5** (1990): 197–227.
41. Schwartz, D., V. Samalam, S. Solla, and J. Denker. "Exhaustive Learning." *Neural Comp.* **2** (1990): 374–385.
42. Seung, H., H. Sompolinsky, and N. Tishby. "Statistical Mechanics of Learning from Examples I, II." *Phys. Rev. A* **45** (1991): 6056.
43. Sorkin, R. "A Quantitative Occam's Razor." *Intl. J. Theor. Phys.* **22** (1983): 1091–1104.
44. Spiegelhalter, D., A. Dawid, S. Lauritzen, and R. Cowell. "Bayesian Analysis in Expert Systems." *Stat. Sci.* **8** (1993): 219–283.
45. Strauss, C., D. Wolpert, and D. Wolf "Alpha, Evidence, and the Entropic Prior." In *Maximum Entropy and Bayesian Methods*, edited by A. Mohammed-Djafari. Boston: Kluwer, 1993.
46. Tishby, N., E. Levin, and S. Solla "Consistent Inference of Probabilities in Layered Networks: Predictions and Generalization." In *International Joint Conference on Neural Networks*, Vol. II, 403–409. New York: IEEE, 1989.
47. Tishby, N. "Statistical Physics Models of Supervised Learning." These proceedings.
48. Titterington, A. F. Smith, and V. E. Makov. *Statistical Analysis of Finite Mixture istributions*. New York: Wiley & Sons, 1985.
49. Valiant, L. "A Theory of the Learnable." *Comm. ACM* **27** (1984): 1134–1142.
50. Van der Broeck, C., and R. Kawai. "Generalization in Feedforward Neural and Boolean Networks." Paper presented at AMSE International conference on neural networks, San Diego, CA, 1991.
51. Vapnik, V. *Estimation of Dependences Based on Empirical Data*. Berlin: Springer-Verlag, 1982.
52. Vapnik, V., and L. Bottou. "Local Algorithms for Pattern Recognition and Dependencies Estimation." *Neural Comp.* **5** (1993): 893–909.
53. Wahba, G. "A Comparison of GCV and GML for Choosing the Smoothing Parameter in the Generalized Spline Smoothing Problem." *Ann. Stat.* **4** (1985): 1378–1402.
54. Weiss, S. M., and C. A. Kulikowski. *Computer Systems that Learn*. San Mateo, CA: Morgan Kauffman, 1991.

55. Weigend, A., D. Rumelhart, and B. Huberman. "Generalization by Weight-Elimination with Application to Forecasting." In *Advances in Neural Information Processing 3 (NIPS '90)*, edited by R. P. Lippman et al. San Mateo, CA: Morgan Kauffman, 1991.

56. Wolpert, D. "On the Connection Between In-Sample Testing and Generalization Error." *Complex Systems* **6** (1992): 47–94.

57. Wolpert, D. "On the Use of Evidence in Neural Networks." In *Neural Information Processing Systems 5*, edited by S. Hanson et al. San Mateo, CA: Morgan-Kauffman, 1993.

58. Wolpert, D. "Bayesian Backpropagation over I-O Functions Rather than Weights." To appear in *Neural Information Processing Systems 6*, edited by S. Hanson et al. San Mateo, CA: Morgan-Kauffman, 1994.

59. Wolpert, D. "Filter Likelihoods and Exhaustive Learning." To appear in *Computational Learning Theory and Natural Learning Systems. Volume II: Natural Learning Systems*, edited by S. Hanson et al. Cambridge, MA: MIT Press, 1994.

60. Wolpert, D. "On the Bayesian 'Occam Factors' Argument for Occam's Razor." To appear in *Computational Learning Theory and Natural Learning Systems. Volume III: Natural Learning Systems*, edited by S. Hanson et al. Cambridge, MA: MIT Press, 1994.

61. Wolpert, D. "Reconciling Bayesian and Non-Bayesian Analysis." To appear in *Maximum Entropy and Bayesian Methods*, edited by G. Heidbreder. Boston: Kluwer, 1994.

62. Wolpert, D., and C. Strauss. "What Bayes Has to Say About the Evidence Procedure." To appear in *Maximum Entropy and Bayesian Methods*, edited by G. Heidbreder. Boston: Kluwer, 1994.

63. Wolpert, D. "On Exhaustive Learning." These proceedings.

64. Wolpert, D. "The Status of Supervised Learning Science Circa 1994: The Search for a Consensus." These proceedings.

65. Wolpert, D. "How Good Generalizers Need to Be: Beyond Bias plus Variance." In preparation.

66. Wolpert, D. "On Overfitting Avoidance as Bias." In preparation. (A preliminary version of this paper appeared as Santa Fe Institute Working Paper #93-03-016.

Naftali Tishby[1]
Institute of Computer Science and Center for Neural Computation, Hebrew University,
Jerusalem 91904, Israel; tishby@galaxy.huji.ac.il

Statistical Physics Models of Supervised Learning

The emergence of very large parametric models in machine learning suggests a natural analogy between the model parameter space and the config-
furation space of complex physical systems. This physical analogy provides
powerful methods and techniques, originally developed within the statistical mechanics of disordered systems, for the analysis of large learning
models. Whereas the standard statistical analysis is essentially asymptotic
in the sample size, statistical mechanics enables us to analytically study the
typical case, small sample behavior of a large system. It provides exact calculations of the training and generalization as functions of the sample size,
for some specific models, as well as improved rigorous upper bounds on the
average case generalization performance. This analysis reveals highly nontrivial learning curves in generic models, which exhibit anomalous power
laws and discontinuous phase transitions.

[1] Notes for the Los Alamos National Laboratory and Santa Fe Institute "The Future of Supervised
Machine Learning Workshop," August 6–7, 1992. Based partly on joint papers with Sebastian
Seung and Haim Sompolinsky, and with Géza Györgyi.

Here we review some general features of this theory, to which there are many contributors, and illustrate it with two simple perceptron learning models: learning a binary perceptron from noisy examples and learning with an undetermined input architecture. These models are analyzed within the *annealed approximation* which, though simpler, gives the correct qualitative behavior in many cases.

1. INTRODUCTION

The mathematics of supervised learning is related to various disciplines, as is clearly evident from this volume. To a physicist, there is a obvious similarity between the training cost function and the "energy" function of a physical system. In this analogy the parameters of the learning machine correspond to the physical "configuration variables" and the performance optimization done during the training is similar to a search for the "ground state" of the system. One may take this analagy further and relate the "ground state entropy" of the system with the number of hypotheses consistent with the training sample. Then one can describe the learning phenomena as a decrease in this hypotheses entropy rather than the training energy, as proposed by Carnevali and Patarnello,[5] and by Denker et al.[7] This physical picture can easily describe also inconsistent hypotheses and be related directly to the Baysian formulation of learning, by the inclusion of a "temperature", using the exponential "Gibbs distribution" over the parameter space[30,23]. In this extension it is a linear combination of the training entropy and energy, the "free energy", which turns out to be the function that governs the learning behavior. The natiral question that is raised at that point is if these rather simplistic analogies with statistical thermodynamics useful? Can they be further developed into a complete theory of some important aspects of the learning phenomena?

More specifically, one should ask the following important questions about these "naive" theories.

1. What are the assumptions about the learning system for those thermodynamic functions to be well defined?
2. Is this an average case description and in what sense precisely?
3. What is the role of the randomness in the training examples?
4. Is there a proper measure of generalization in this framework? The energy function seems to describe training and not generalization.
5. Can this formalism deal with approximations ("unrealizable rules")?
6. How is this formulation related to other theories such as PAC, information theory, and Bayesian statistics? Does it provide anything which is not already captured by these other frameworks?
7. Can the statistical mechanics formulation be made rigorous?

Answers to many of these questions are the result of an ongoing research which turns it into a fruitful interdisciplinary field.

The first step beyond the naive statistical-mechanical (SM) approach is the observation that if the training procedure is assumed to be stochastic it leads to a Gibbs distribution on the model parameters (e.g., network weights). The performances of the system on the training set as well as on novel inputs are calculated as appropriate *thermal* averages on the Gibbs distribution in the parameter space and *quenched* (i.e., fixed) averages on the random sampling of the examples. These averages provide an accurate account, with high probability, of the *typical* behavior of large networks due to the phenomena of *self-averaging*.

The currently dominant approach in computational learning theory is based on Valiant's learning model and on the notion of *Probably Almost Correct* (PAC) learning.[4,31] The main achievements of this approach, in our context, are general bounds on the probability of error on a novel example for a given size of the training set,[2,15] as well as classification of learning problems according to their time complexity.[20,31] Most of these (sample complexity) combinatorial bounds depend on the specific structure of the model and the complexity of the task only through a single number, known as the Vapnik-Chervonenkis (VC) dimension.[32] Generally, they are independent of the specific-learning algorithm or distribution of examples. The generality of the PAC approach is also its main deficiency, since it is dominated by the worst case, atypical behavior. The statistical mechanical approach thus differs considerably from the PAC learning theory in that it can provide precise quantitative predictions for the typical behavior of specific-learning models. The price is that there is no single papramter, such as the VC-dimension which determines the learning curve. It is rather determined by the competition between the energy and entropy, as functions of the training set size.

The SM formalism can also be applied to several learning models for which few PAC results are yet known. Despite recent works which extend the original PAC framework,[1,15] most PAC theorems apply to *realizable* tasks, namely tasks that can be performed perfectly by the network, given enough examples. In many real-life problems the target task can only be approximated by the assumed architecture of the network, so the task is *unrealizable*. In addition, many of the PAC learning results are limited to networks with threshold decision elements, although, in many applications, analog neurons are used. The SM approach is close in its spirit, though not in its scope and results, to the Bayesian information-theoretic approach, recently applied also to continuous networks.[1,6]

The technical key to the SM approach to learning is the seminal phase-space method of Gardner.[8] Using Gardner's approach, studies of learning a classification task in a perceptron were performed in papers by Hansel and Sompolinsky[19] and del Giudice et al.[10] Gardner and Derrida,[9] and Györgyi and Tishby[11] have used these methods for studying learning of a perceptron rule. An extensive study of learning curves and learning in perceptron models has been published by Seung, Sompolinsky, and Tishby,[29] and in more comprehensive reviews by Watkin, Rau, and Biehl,[34] and by Opper and Kinzel.[24] The statistical mechanics of learning is gradually incorporated into the rigorous computational learning theory. Important steps in this direction has been made by Haussler, Kearns, and Schapire,[17] and more recently by Haussler et. al.,[18] who obtained rigorous bounds from statistical mechanics which significantly improve the Vapnik Chervonenkis bounds for specific distributions.

In the present chapter we review the general statistical mechanics formulation of learning from examples in large layered networks. Then we focus on two new perceptron models, learning binary weights from noisy patterns, and learning with binary weights when the input architecture is undetermined. The latter model is solved using the fact that a mismatch in the architectures of the learner and the teacher introduces pattern noise, which in turn can be solved using the the results of the first model. A proper treatment of learning from examples requires the use of sophisticated techniques from the SM of disordered systems, in particular the replica method. In order to facilitate the understanding of the results, we solve these two models in an annealed framework, which provides the main features of the complete quenched calculation, but is much more transparent. The general validity of this approximation for learning and its relation to the high-temperature limit is also briefly discussed, but the reader is referred to the complete account by Sompolinsky.[29]

Somewhat surprisingly, even such simple models exhibit nontrivial generalization properties. Some of these properties have been shown to exist also in two-layer models[28] and for continuous weights.[27] Furthermore, the strong nonlinearities generated in multilayer systems can be mimicked to some extent by strong constraints on the range of values of the perceptron weights. Indeed, even the loading problem in a perceptron with binary weights is a hard problem both theoretically and computationally (see, e.g., Mézard and Krauth[22] and also Vapnik[33]).

In Section 2, the general statistical-mechanical theory of learning from examples is formulated. In Section 3 the annealed approximation and the high-temperature limit are reviewed. In Section 4, we describe the statistical mechanics of learning perceptron models and particularly learning perceptron with binary weights. In Section 5 we discuss learning from noisy patterns, and a simple model of learning with mismatched architectures is described in Section 6.

---

2. GENERAL THEORY

2.1 LEARNING FROM EXAMPLES

We consider a network with M input nodes S_i $(i = 1, \ldots, M)$, N synaptic weights W_i $(i = 1, \ldots, N)$, and a single output node $\sigma = \sigma(\mathbf{W}; \mathbf{S})$. The quantities $\mathbf{S}$ and $\mathbf{W}$ are M- and N-component vectors denoting the input states and the weight states, respectively. For every $\mathbf{W}$, the network defines a map from $\mathbf{S}$ to σ. Thus the weight space corresponds to a class of functions, a class that is constrained by the architecture of the network. Learning can be thought of as a search through weight space to find a network with desired properties.

In supervised learning, the weights of the network are tuned so that it approximates as closely as possible a target function $\sigma_0(\mathbf{S})$. One way of achieving this is to provide a set of *examples* consisting of P input-output pairs $(\mathbf{S}^l, \sigma_0(\mathbf{S}^l))$, with $l = 1, \ldots, P$. We assume that each input $\mathbf{S}^l$ is chosen at random from the entire input space according to some normalized a priori measure denoted $d\mu(\mathbf{S})$. The examples can be used to construct a *training energy*

$$E(\mathbf{W}) = \sum_{l=1}^{P} \epsilon(\mathbf{W}; \mathbf{S}^l) \, , \qquad (1)$$

where the *error function* $\epsilon(\mathbf{W}; \mathbf{S})$ is some measure of the deviation of the network's output $\sigma(\mathbf{W}; \mathbf{S})$ from the target output $\sigma_0(\mathbf{S})$. The error function should be zero whenever the two agree, and positive everywhere else. A popular choice is the quadratic error function

$$\epsilon(\mathbf{W}; \mathbf{S}) = \frac{1}{2}[\sigma(\mathbf{W}; \mathbf{S}) - \sigma_0(\mathbf{S})]^2 \, . \qquad (2)$$

Training is then accomplished by minimizing the energy, for example, via gradient descent

$$\frac{\partial \mathbf{W}}{\partial t} = -\nabla_{\mathbf{W}} E(\mathbf{W}) \, . \qquad (3)$$

Gradient descent schemes have been criticized because they tend to become trapped in local minima of the training energy. Putting such difficulties aside, there is another potential problem with the training energy: it measures the network's performance on a limited set of examples. The true goal of supervised learning is to find a network that performs well on all inputs, not just those in the training set. The performance of a given network $\mathbf{W}$ on the whole input space is measured by the *generalization function*. It is defined as the average error of the network over the whole input space, i.e.,

$$\epsilon(\mathbf{W}) = \int d\mu(\mathbf{S}) \, \epsilon(\mathbf{W}; \mathbf{S}) \, . \qquad (4)$$

Like the gradient descent procedure Eq.(3), any training algorithm determines some trajectory $\mathbf{W}(t)$ in weight space. The performance of a training algorithm can be monitored by following the generalization and training errors as functions of time. We expect that training procedures using the energy Eq. (1) should be able to achieve optimal generalization in the limit $P \to \infty$, i.e., the generalization curve will asymptotically approach the minimum possible value of $\epsilon(\mathbf{W})$ within the specified weight space.

We distinguish between learning of *realizable rules* and *unrealizable rules*. Realizable rules are those target functions $\sigma_0(\mathbf{S})$ that can be completely realized by at least one of the networks in the weight space. Thus, in a realizable rule there exists a weight vector $\mathbf{W}^*$ such that

$$\epsilon(\mathbf{W}^*, \mathbf{S}) = 0, \quad \text{for all } \mathbf{S} , \tag{5}$$

or equivalently, $\epsilon(\mathbf{W}^*) = 0$. An *unrealizable* rule is a target function for which

$$\epsilon_{\min} = \min_{\mathbf{W}} \epsilon(\mathbf{W}) > 0 . \tag{6}$$

Unrealizable rules occur in two basic situations. In the first, the data available for training are corrupted with noise, making it impossible for the network to reproduce the *data* exactly, even with a large training set. This case has been considered by several authors. Particularly relevant works are those by Györgyi,[11,12] which show that even with noisy data the *underlying target rule* itself can be reproduced exactly in the limit. A second situation, which we do not consider here, is when the network architecture is restricted in a manner that does not allow an exact reproduction of the target rule itself.

In this work we focus on a simple model where the network architecture can adapt to the correct structure, and see that this is closely related to learning with noise.

2.2 LEARNING AT FINITE TEMPERATURE

We consider a stochastic learning dynamics that is a generalization of Eq. (3). The weights evolve according to a relaxational Langevin equation

$$\frac{\partial \mathbf{W}}{\partial t} = -\nabla_{\mathbf{W}} E(\mathbf{W}) + \eta(t), \tag{7}$$

where η is a white noise with variance

$$\langle \eta_i(t)\eta_j(t') \rangle = 2T\delta_{ij}\delta(t - t') . \tag{8}$$

The above dynamics tends to decrease the energy, but occasionally the energy may increase due to the influence of the thermal noise. At $T = 0$, the noise term drops out, leaving the simple gradient descent Eq. (3). The above equations are appropriate for continuously varying weights. We will also consider weights that are constrained to discrete values. In such cases the analog of Eq. (7) is a discrete-time Monte Carlo algorithm, similar to that used in simulating Ising systems.[3]

In simulated annealing algorithms for optimization problems, thermal noise has been used to prevent trapping in local minima of the energy.[21] The temperature is decreased slowly so that eventually at $T \approx 0$ the system settles to a state with energy near the global energy minimum. Although thermal noise could play the same role in the present training dynamics, it may play a more essential role in achieving good learning. Since the ultimate goal is to achieve good generalization, reaching the global minimum of the training energy may not be necessary. In fact, in some cases training at fixed finite temperature may be advantageous, as it may prevent the system from *overtraining*, namely, finding an accurate fit to the training data at the expense of good generalization abilities. Finally, often there are many nearly degenerate minima of the training error, particularly when the available data set is limited in size. In these cases it is of interest to know the properties of the ensemble of solutions. The stochastic dynamics provides a way of generating a useful measure, namely a Gibbs distribution, over the space of the solutions.

In the present work, we consider only long-time properties. As is well known, Eq. (7) generate at long times a Gibbs probability distribution. In our case it is

$$\mathcal{P}(\mathbf{W}) = Z^{-1}e^{-\beta E(\mathbf{W})} \ . \tag{9}$$

The variance of the noise in the training procedure now becomes the temperature $T = 1/\beta$ of the Gibbs distribution. The normalization factor Z is the partition function

$$Z = \int d\mu(\mathbf{W}) \exp(-\beta E(\mathbf{W})). \tag{10}$$

Possible constraints on the values of the weights are incorporated into the *a priori* normalized measure in weight space, $d\mu(\mathbf{W})$. The powerful formalism of equilibrium statistical mechanics may now be applied to calculate thermal averages, i.e., averages with respect to $P(\mathbf{W})$. They will be denoted by $\langle \cdots \rangle_T$. In the thermodynamic (large N) limit, such average quantities yield information about the typical performance of a network, governed by the above measure, independent of the initial conditions of the learning dynamics.

Even after the thermal average is done, there is still a dependence on the P examples $\mathbf{S}_l$. Since the examples are chosen randomly and then fixed, they represent *quenched* disorder. Thus to explore the typical behavior we must perform a second, quenched average over the distribution of example sets, denoted by $\langle\!\langle \cdots \rangle\!\rangle \equiv \int \prod_l d\mu(\mathbf{S}^l)$.

The *average training* and *generalization errors* are given by

$$\epsilon_t(T, P) \equiv P^{-1}\langle\!\langle\, \langle E(\mathbf{W})\rangle_T \,\rangle\!\rangle \ , \tag{11}$$

$$\epsilon_g(T, P) \equiv \langle\!\langle\, \langle \epsilon(\mathbf{W})\rangle_T \,\rangle\!\rangle \ . \tag{12}$$

The free energy F and entropy S of the network are given by

$$F(T, P) = -T\langle\!\langle \ln Z \rangle\!\rangle, \tag{13}$$

$$S(T, P) = -\int d\mu(\mathbf{W})\langle\!\langle\, \mathcal{P}(\mathbf{W}) \ln \mathcal{P}(\mathbf{W}) \,\rangle\!\rangle. \tag{14}$$

They are related by the identity

$$F = P\epsilon_t - TS \ . \tag{15}$$

Knowing F the expected training error can be evaluated via

$$\epsilon_t = \frac{1}{P}\frac{\partial(\beta F)}{\partial \beta}, \tag{16}$$

and the entropy by

$$S = -\frac{\partial F}{\partial T}. \tag{17}$$

Formally our results will be exact in the thermodynamic limit, i.e., when the size of the network approaches infinity. The relevant scale is the total number of degrees of freedom, namely, the total number of (independently determined) synaptic weights N. For the limit $N \to \infty$ to be well defined, we envisage that the problem at hand as well as the network architecture allow for a uniform scale-up of N. However, our results should provide a good approximation to the behavior of networks with a fixed large size.

The correct thermodynamic limit requires that the energy function be extensive, i.e., proportional to N. The consequences of this requirement can be realized by averaging Eq. (1) over the example sets, yielding

$$\langle\!\langle E(\mathbf{W}) \rangle\!\rangle = P\epsilon(\mathbf{W}). \tag{18}$$

Hence, assuming that $\epsilon(\mathbf{W})$ is of order 1, the number of examples should scale as

$$P = \alpha N, \tag{19}$$

where the proportionality constant α remains finite as N grows. This scaling guarantees that both the entropy and the energy are proportional to N. The balance between the two is controlled by the noise parameter T, which remains finite in the thermodynamic limit.

Finally, using the definitions Eqs. (11) and (12) and the convexity of the free energy, one can show that

$$\epsilon_t(T, \alpha) < \epsilon_g(T, \alpha) \tag{20}$$

for all T and α (see Sompolinsky[29]). When the inequality holds, it reflects the phenomenon of overtraining of the examples mentioned above. However, as the number of examples P increases, the energy scales like P and becomes proportional to the generalization function, as in Eq. (18). This implies that, for any fixed temperature, increasing α yields the desired limits

$$\epsilon_g \to \epsilon_{\min}, \quad \epsilon_t \to \epsilon_{\min}, \quad \alpha \to \infty. \tag{21}$$

This result is derived by Sompolinsky[29] using the replica formalism.

2.3 HIGH-TEMPERATURE LIMIT

A simple and interesting limit of the learning theory is that of temperatures. This limit is defined so that both T and α approach infinity, but their ratio remains constant:

$$\beta\alpha = \text{finite} , \quad \alpha \to \infty , \quad T \to \infty . \tag{22}$$

In this limit, E can simply be replaced by its average Eq. (18), and the fluctuations δE, coming from the finite sample of randomly chosen examples, can be ignored. To see this we note that δE is of order $\sqrt{P}$. The leading contribution to βF from the term $\beta\delta E$ in Z is proportional to $\beta^2 \langle\!\langle (\delta E)^2 \rangle\!\rangle \approx N\alpha\beta^2$. This is down by a factor of β compared to the contribution of the average term, which is of the order $N\alpha\beta$. Thus in this limit, the equilibrium distribution of weights is given simply by

$$\mathcal{P}_0(\mathbf{W}) = Z^{-1} \exp(-N\beta\alpha\epsilon(\mathbf{W})) , \tag{23}$$

where

$$Z_0 = \int d\mu(\mathbf{W}) \exp(-N\beta\alpha\epsilon(\mathbf{W})) . \tag{24}$$

The subscript 0 signifies that the temperature limit is the zeroth-order term of a complete -temperature expansion.

In the -T limit, it is clear from Eq. (23) that all thermodynamic quantities, including the average training and generalization errors, are functions only of the *effective temperature* T/α. It should be emphasized that the present case is unlike most $T \to \infty$ limits in statistical mechanics, in which all states become equally likely, regardless of energy. Here the simultaneous $\alpha \to \infty$ limit guarantees non-trivial behavior, with contributions from both energy and entropy. In particular, as the effective temperature T/α decreases, the network approaches the optimal ("ground state") weights $\mathbf{W}^*$ that minimize $\epsilon(\mathbf{W})$. This behavior is similar to the $T = \text{finite}, \alpha \to \infty$ limit mentioned in Eq. (21).

It is sometimes useful to discuss the microcanonical version of the statistical mechanics of learning in the -T limit. Equation (24) can be written as

$$Z_0 = \int d\epsilon \exp(-N\beta\alpha f(\epsilon)) , \tag{25}$$

where the free energy per weight of all networks whose generalization error equals ϵ is

$$f(\epsilon) = \epsilon - \frac{T}{\alpha}s(\epsilon) . \tag{26}$$

The function $s(\epsilon)$ is the entropy per weight of all the networks with $\epsilon(\mathbf{W}) = \epsilon$, i.e.,

$$s(\epsilon) = N^{-1} \ln \int d\mu(\mathbf{W}) \delta(\epsilon(\mathbf{W}) - \epsilon) . \tag{27}$$

In the large N limit the expected generalization error is simply given by

$$\beta\alpha = \partial s/\partial\epsilon . \tag{28}$$

Thus the properties of the system in the -T limit are determined by the dependence of the entropy on generalization error.

From the theoretical point of view, the -T limit simply characterizes models in terms of an effective energy function $\epsilon(\mathbf{W})$ which is often a rather smooth function of $\mathbf{W}$. The smoothness of the effective energy function also implies that the learning process at temperature is relatively fast. One does not expect to encounter many local minima, although a few large local minima may still remain, as will be seen in some of the models below. Another feature of learning at temperature is the lack of *overtraining*. One measure of overtraining of examples is the difference between the expected training and generalization errors, i.e., $\epsilon_g - \epsilon_t$. From Eq. (23) and the definitions Eqs. (11)–(12), it follows that $\epsilon_t = \epsilon_g$ in the -T limit. Of course, the price that one pays for learning at temperature is the necessity of a large training set, as α must be at least of order T.

3. THE ANNEALED APPROXIMATION

Another useful approximate method for investigating learning in neural networks is the annealed approximation, or AA for short. It consists of replacing the average of the logarithm of Z, Eq. (13), by the logarithm of the average of Z itself. Thus the annealed approximation for the average free energy F_{AA} is

$$-\beta F_{\mathrm{AA}} = \ln\langle\!\langle Z \rangle\!\rangle . \tag{29}$$

Using the convexity of the logarithm function, it can be shown that the annealed free energy is a lower bound for the true quenched value,

$$F_{\mathrm{AA}} \leq F . \tag{30}$$

Whether this lower bound can actually serve as a good approximation will be examined critically in this work.

Using Eqs. (10) and (1) one obtains

$$\langle\!\langle Z \rangle\!\rangle = \int d\mu(\mathbf{W})e^{-PG_{\mathrm{AA}}(\mathbf{W})} , \tag{31}$$

$$G_{\mathrm{AA}}(\mathbf{W}) = -\ln\int d\mu(\mathbf{S})\,e^{-\beta\epsilon(\mathbf{W};\mathbf{S})} . \tag{32}$$

The generalization and training errors are approximated by

$$\epsilon_g = \frac{1}{\langle\!\langle Z \rangle\!\rangle} \int d\mu(\mathbf{W})\epsilon(\mathbf{W})e^{-PG_{\mathrm{AA}}(\mathbf{W})} , \tag{33}$$

$$\epsilon_t = \frac{1}{\langle\!\langle Z \rangle\!\rangle} \int d\mu(\mathbf{W})\frac{\partial G_{\mathrm{AA}}(\mathbf{W})}{\partial\beta}e^{-PG_{\mathrm{AA}}(\mathbf{W})} . \tag{34}$$

3.1 SINGLE BOOLEAN OUTPUT

A particularly simple case is that of an output layer consisting of a single Boolean output unit. In this case $\epsilon(\mathbf{W}; \mathbf{S}) = 1$ or 0 only, so that

$$G_{\mathrm{AA}}(\mathbf{W}) = -\ln[1 - (1 - e^{-\beta})\epsilon(\mathbf{W})] \,. \tag{35}$$

Since G_{AA} depends on $\mathbf{W}$ only through $\epsilon(\mathbf{W})$, which is of order 1, we can write a microcanonical form of the AA, analogous to what was done for the $-T$ limit in Eqs. (25)–(28). The annealed partition function $\langle\!\langle Z \rangle\!\rangle$ takes the form

$$\langle\!\langle Z \rangle\!\rangle = \int d\epsilon \, \exp N[G_0(\epsilon) - \alpha G_1(\epsilon)] \,, \tag{36}$$

where

$$G_1(\epsilon) = -\ln[1 - (1 - e^{-\beta})\epsilon], \tag{37}$$

$$G_0(\epsilon) = N^{-1} \ln \int d\mu(\mathbf{W})\delta(\epsilon - \epsilon(\mathbf{W})) \,. \tag{38}$$

The function G_1 is simply defined by $G_{\mathrm{AA}}(\mathbf{W}) \equiv G_1(\epsilon(\mathbf{W}))$. The function $NG_0(\epsilon)$ is the logarithm of the density of networks with generalization error ϵ. At finite temperature, it is different from the annealed entropy $S_{\mathrm{AA}} \equiv -\partial F_{\mathrm{AA}}/\partial T$, which is the logarithm of the density of networks with training error ϵ. However, since $\epsilon_t = \epsilon_g$ in the temperature limit, G_0 approaches s_{AA}.

In the thermodynamic limit ($N \to \infty$), the integral Eq. (36) is dominated by its saddle point (the saddle-point method is also known as the Laplace method of integration). Thus, at any given α and T the value of the average generalization error is given by minimizing the free energy $f(\epsilon)$, where $-\beta f \equiv G_0 - \alpha G_1$. This leads to the implicit equation

$$\left.\frac{\partial G_0}{\partial \epsilon}\right|_{\epsilon=\epsilon_g} = -\frac{\alpha(1 - e^{-\beta})}{1 - (1 - e^{-\beta})\epsilon_g} \,, \tag{39}$$

which is analogous to the $-T$ result Eq. (28). It is interesting to note that in this case the AA predicts a simple relation between the training and generalization errors. Using Eq. (34) above, one obtains

$$\epsilon_t = \frac{e^{-\beta}\epsilon_g}{1 - (1 - e^{-\beta})\epsilon_g} \tag{40}$$

where ϵ_g is the average generalization error given by Eq. (39), or equivalently

$$\epsilon_g = \frac{\epsilon_t}{e^{-\beta} + (1 - e^{-\beta})\epsilon_t}. \tag{41}$$

To the extent that the annealed approximation is valid, this relation could be used in actual applications to estimate the generalization error from the measured training error.

3.2 HOW GOOD IS THE ANNEALED APPROXIMATION?

First we note that $G_{AA} \to \beta\epsilon(\mathbf{W})$ as $\beta \to 0$. Thus the AA is valid at temperatures, since it reduces to the $-T$ limit described above. At lower temperatures the AA deviates significantly from the $-T$ theory. In particular, it does seem to incorporate the important effect of overtraining, in that ϵ_g, Eq. (33), is, in general, larger than ϵ_t, Eq. (34). On the other hand, except for the $-T$ limit, the results of the AA are, in general, not exact.

To obtain some insight into the quality of the AA at finite temperatures, we examine its behavior in the limit of large α. From Eq. (35) it follows that in the AA the asymptotic value of the generalization error is

$$\lim_{\alpha \to \infty} \epsilon_g(T, \alpha) = \epsilon(\mathbf{W}^\dagger) , \tag{42}$$

where $\mathbf{W}^\dagger$ minimizes G_{AA}. In general, this vector is not necessarily the same as the vector $\mathbf{W}^*$ which minimizes $\epsilon(\mathbf{W})$. Hence, there is no guarantee that the AA correctly predicts the value of the optimal generalization error or the values of the optimal weights, except for two special cases. One is the case of a realizable rule, for which $\epsilon(\mathbf{W}^*; \mathbf{S}) = 0$ for all inputs $\mathbf{S}$. Clearly the minimum of G_{AA} in Eq. (32) then occurs at $G_{AA}(\mathbf{W}^*) = 0$. The second is the case of a network whose output layer consists of a single Boolean output unit, as discussed above. From Eq. (35) it is evident that the minimum of G_{AA}, in this case, coincides with the minimum of ϵ_g and, hence, $\mathbf{W}^\dagger = \mathbf{W}^*$. To summarize, the correct limit $\epsilon_g \to \epsilon_{min}$, stated in Eq. (21), can be violated by the AA for unrealizable, non-Boolean rules.

With respect to the training error, the AA for unrealizable rules is also inadequate: the correct limit $\epsilon_t \to \epsilon_{min}$ is typically violated, even for the Boolean case, and the limit $\epsilon_t \to \epsilon_g$ does not hold either. For Boolean rules, this can be seen from the expression (40) for the training error. In the $T \to 0$ limit, the annealed training error approaches

$$\lim_{T \to 0} \epsilon_t(T, \alpha) = \min_{\mathbf{W}, \mathbf{S}} \epsilon(\mathbf{W}; \mathbf{S}) \tag{43}$$

since annealing both the weights and examples at zero temperature minimizes the training energy with respect to all variables. Often the right-hand side is zero, so that the AA predicts $\epsilon_t(T = 0, \alpha) = 0$ for all α. This is a gross violation of the correct limit $\epsilon_t \to \epsilon_{min}$. Hence the fact that the inequality $\epsilon_t < \epsilon_g$ holds in the AA may be a spurious effect of the distortion of the input distribution, rather than an honest accounting for the randomness of the examples.

In short, we expect the AA to be adequate at least qualitatively for realizable rules but inadequate, except at temperatures, for unrealizable ones.

4. LEARNING OF A PERCEPTRON RULE

In this section we apply the AA to study several models of a single-layer perceptron learning a realizable perceptron rule. We then demonstrate the breakdown of the AA in a simple model of an unrealizable perceptron rule. Discussion of the full quenched theory of these models can be found in papers by Sompolinsky.[27,28]

4.1 GENERAL FORMULATION

The perceptron is a network which sums a single layer of inputs S_j with synaptic weights W_j, and passes the result through a transfer function σ

$$\sigma = g\left(N^{-\frac{1}{2}} \sum_{j=1}^{N} W_j S_j\right) = g\left(N^{-\frac{1}{2}} \mathbf{W} \cdot \mathbf{S}\right) \tag{44}$$

where $g(x)$ is a sigmoidal function of x. The normalization $1/\sqrt{N}$ in Eq. (44) is included to make the argument of the transfer function be of order unity. Learning is a search through weight space for the perceptron that best approximates a target rule. We assume for simplicity that the network space is restricted to vectors that satisfy the normalization

$$\sum_{i=1}^{N} W_i^2 = N \ . \tag{45}$$

The *a priori* distribution on the input space is assumed to be Gaussian,

$$d\mu(\mathbf{S}) = \prod_{i=1}^{N} DS_i \ , \tag{46}$$

where Dx denotes the normalized Gaussian measure

$$Dx \equiv \frac{dx}{\sqrt{2\pi}} e^{-x^2/2} \ . \tag{47}$$

In most of this section we consider only the case of a perfectly realizable target rule. This means that it is another perceptron of the form

$$\sigma^0(\mathbf{S}) = g\left(\frac{1}{\sqrt{N}} \mathbf{W}^0 \cdot \mathbf{S}\right) \ , \tag{48}$$

where $\mathbf{W}^0$ is a fixed set of N weights W_i^0. We assume that the teacher weights $\mathbf{W}^0$ also satisfy the normalization condition Eq. (45).

Training is performed by a stochastic dynamics of the form Eq. (7) with the training energy function Eq. (1). For each example the error function is taken to be

$$\epsilon(\mathbf{W};\mathbf{S}) = \frac{1}{2}\left[g\left(N^{-\frac{1}{2}}\mathbf{W}\cdot\mathbf{S}\right) - g\left(N^{-\frac{1}{2}}\mathbf{W}^0\cdot\mathbf{S}\right)\right]^2 . \tag{49}$$

The generalization function is

$$\begin{aligned}\epsilon(\mathbf{W}) &= \int D\mathbf{S}\,\epsilon(\mathbf{W};\mathbf{S}) \\ &= \int Dx \int Dy\,\frac{1}{2}\left[g(x\sqrt{1-R^2}+yR) - g(y)\right]^2\end{aligned} \tag{50}$$

where R is the overlap of the student network with the teacher network, i.e.,

$$R = \frac{1}{N}\mathbf{W}\cdot\mathbf{W}^0 \tag{51}$$

(see Appendix). The relationship between Eqs. (49) and (50) is plain, since in both cases the arguments of g are Gaussian random variables with unit variance and cross-correlation R.

It is important to note that in perceptron learning, the generalization function of a network depends only on its overlap with the teacher, i.e., $\epsilon(\mathbf{W}) = \epsilon(R)$. Learning can be visualized very easily since $R = \cos\theta$, where θ is the angle between $\mathbf{W}$ and $\mathbf{W}^0$. The generalization function goes to zero as the angle between the student and the teacher weight vectors vanishes. Perfect learning corresponds to an overlap $R = 1$ or $\theta = 0$.

In the following, we discuss only perceptrons with binary weights (*Ising*), and *Boolean* output, i.e., the output function is $g(x) = \text{sign}(x)$, which corresponds to the original perceptron model studied by Rosenblatt.[25] The *Boolean perceptron* $\sigma = \text{sign}(\mathbf{W}\cdot\mathbf{S})$ separates the input space in half via a hyperplane perpendicular to the weight vector. The error function, Eq. (49), is (up to a factor of 2)

$$\epsilon(\mathbf{W};\mathbf{S}) = \Theta[-(\mathbf{W}\cdot\mathbf{S})(\mathbf{W}^0\cdot\mathbf{S})] , \tag{52}$$

which is 1 when the student and teacher agree, and 0 otherwise. The generalization error

$$\epsilon(\mathbf{W}) = \frac{1}{\pi}\cos^{-1}R \tag{53}$$

is simply proportional to the angle between the student and teacher weight vectors.

4.2 THE ANNEALED APPROXIMATION FOR PERCEPTRON LEARNING

The annealed free energy of perceptron learning is shown in the Appendix to be

$$-\beta f = G_0(R) - \alpha G_1(R) , \tag{54}$$

$$G_0(R) = N^{-1} \ln \int d\mu(\mathbf{W}) \, \delta(R - N^{-1}\mathbf{W} \cdot \mathbf{W}^0) , \tag{55}$$

$$G_1(R) = -\ln \int Dx \int Dy \exp\left(-\frac{\beta}{2} \left[g\left(x\sqrt{1-R^2} + yR\right) - g(y)\right]^2\right) . \tag{56}$$

Since $G_{AA}(\mathbf{W})$, like $\epsilon(\mathbf{W})$, depends on $\mathbf{W}$ only through the overlap R, we have made the definition $G_{AA}(\mathbf{W}) \equiv G_1(N^{-1}\mathbf{W} \cdot \mathbf{W}^0)$. The function $NG_0(R)$ is the logarithm of the density of networks with overlap R, so we will sometimes refer to it as the "entropy," even though it is not the same as the thermodynamic entropy $s = -\partial f/\partial T$. The properties of the system in the large N limit are obtained by minimizing the free energy f, which yields

$$\frac{\partial G_0(R)}{\partial R} = \alpha \frac{\partial G_1(R)}{\partial R} . \tag{57}$$

Solving for R, one then evaluates the average generalization error via Eq. (50). Likewise the average training error is evaluated by differentiating G_1 with respect to β, as in Eq. (34).

The *Ising* perceptron corresponds to a network with binary valued weights $W_i = \pm 1$, or

$$d\mu(\mathbf{W}) \equiv \prod_{i=1}^{N} dW_i[\delta(W_i - 1) + \delta(W_i + 1)] . \tag{58}$$

The entropy of Ising networks with an overlap R is given by

$$G_0(R) = -\frac{1-R}{2} \ln \frac{1-R}{2} - \frac{1+R}{2} \ln \frac{1+R}{2} , \tag{59}$$

a result derived in the Appendix. It approaches zero as $R \to 1$, meaning that there is exactly one state with $R = 1$. This nondivergent behavior is typical of discrete weight spaces.

To conclude, the picture emerging from the AA is extremely simple. The properties of the system can be expressed in terms of a single order parameter, namely, the overlap R. The stochastic fluctuations in the value of R can be neglected in the limit of large N. Hence, the system almost always converges to a unique value of R given by the minimum of the free energy $f(R)$. Depending on the details of the problem, $f(R)$ can have more than one local minimum. If this happens, the equilibrium properties of the system are determined by the unique global minimum of f.

The Boolean/Ising perceptron model, first studied by Gardner and Derrida,[9] exhibits a striking feature. It has a first-order transition from a state of poor generalization to a state of perfect generalization,[13,27] which persists at all temperatures. The occurrence of this remarkable transition can be understood using the temperature limit.

THE -TEMPERATURE LIMIT. In the $-T$ limit the energy of the system is given simply as $N\alpha\epsilon(R)$. Hence, using Eq. (53) for $\epsilon(R)$, the free energy is simply

$$-\beta f = -\frac{\alpha\beta}{\pi}\cos^{-1} R - G_0(R) .$$ (60)

In contrast to previous model, the state $R = 1$ is a local minimum of f for all values of T and α. For small values of α/T the state $R = 1$ is only a local minimum of f. The global minimum is given by the solution of the equation

$$R = \tanh\left(\frac{\alpha\beta}{\pi\sqrt{1 - R^2}}\right) .$$ (61)

This state of poor learning $R < 1$ is the equilibrium state for $T > 0.59\alpha$. In this regime the optimal state $R = 1$ is only metastable. If the initial network has R which is close to 1, the learning dynamics will converge fast to the state $R = 1$. However, starting from a random initial weight vector $R \approx 0$, the system will not converge to the optimal state.

For $T/\alpha < 0.59$ the equilibrium state is $R = 1$ although there is still a local minimum, i.e., a solution of Eq. (61) with $R < 1$. Finally for $T < 0.48\alpha$, there is no solution with $R < 1$ to Eq. (61). In this regime (beyond the spinodal), starting from any initial condition the system converges fast to the optimal state.

The collapse of the system to the energy ground state at finite temperature is an unusual phenomenon. The origin of this behavior is the square root singularity of $\epsilon(R)$ at $R = 1$. This singularity implies that a state characterized by $\delta R \equiv 1 - R \ll 1$ has an energy which is proportional to

$$E \propto N\sqrt{\delta R} .$$ (62)

This big increase in energy cannot be offset by the gain in entropy, which is proportional to

$$\delta N G_0(R) \propto N(\delta R)\ln(\delta R) .$$ (63)

This effect can be nicely seen using the microcanonical description. According to Eq. (28) above, a smooth low temperature limit exists provided that

$$\lim_{\epsilon \to \epsilon_{\min}} \frac{\partial s(\epsilon)}{\partial \epsilon} = \infty .$$ (64)

On the other hand, Eq. (59) implies that in the present case

$$\frac{\partial s(\epsilon)}{\partial \epsilon} \approx -\epsilon \ln \epsilon , \quad \epsilon \to 0 . \tag{65}$$

Thus the rate of increase in entropy is too small to give rise to thermal fluctuations below some critical temperature.

It is instructive to apply the above argument to the case of states that differ from the ground state by a flip of a single weight. According to Eq. (62) the energy of such states is

$$E \propto \sqrt{N} , \tag{66}$$

whereas the entropy associated with such an excitation is only

$$\delta N G_0(R) \propto \ln N . \tag{67}$$

It should be emphasized, however, that examining the spectrum of the first excitations is generally not sufficient for determining the large N behavior at any finite T, where the relevant states are those with energy of order NT.

ANNEALED APPROXIMATION. Inserting Eq. (93) for the Boolean output into Eq. (56) we obtain for the annealed free energy

$$-\beta f = \alpha \ln \left(1 - \frac{1 - e^{-\beta}}{\pi} \cos^{-1} R \right) - G_0(R) . \tag{68}$$

The behavior of this free energy is qualitatively similar to that of the -T limit. The state $R = 1$ is always a local minimum of f with $f = 0$. In addition there is another local minimum at temperatures and small α, given by the equation

$$\tanh^{-1} R = \frac{\alpha(1 - e^{-\beta})}{\pi \sqrt{1 - R^2}} \left(1 - \frac{1 - e^{-\beta}}{\pi} \cos^{-1} R \right)^{-1} . \tag{69}$$

At temperatures, the transition to $R = 1$ approaches the one given by the -T theory above. At $T = 0$ the AA predicts that for $0 < \alpha < 1.45$ there are states with small values of R that have zero training energy; i.e., there are degenerate ground states. For $\alpha > 1.45$ these poor learning states are no longer compatible with the examples and, hence, their training error is positive. They are, however, still metastable and, starting from random initial weights ($R \approx 0$), the system will converge to one of these states. At $\alpha > 1.73$ the poor learning states lose their stability and the only metastable state is $R = 1$. This low temperature behavior is in qualitative agreement with the results of the exact theory of this model. The annealed result $\alpha_c = 1.448$ was already derived by Gardner and Derrida.[9]

5. LEARNING WITH NOISY PATTERNS

The perceptron-learning model described in the previous section can be easily extended to learning with noisy patterns (see Györgyi[11,12]). Assuming that the real-valued patterns are assumed to be corrupted with Gaussian noise with variance σ_S, the joint distribution of the noisy pattern $\mathbf{S}'$ and the original pattern $\mathbf{S}$ is given by

$$P(S_j, S_j') = \frac{1}{2\pi\sigma_S} \exp\left[-\frac{(S_j - S_j')^2}{2}\sigma_S^2 - \frac{S_j^2}{2}\right]. \tag{70}$$

Only the noisy patterns, $\mathbf{S}'$, are provided to the student, whereas the labels are generated by the teacher $\mathbf{W}^0$ from the original patterns $\mathbf{S}$.[2] When averaging over the noise distribution as well, the only modification to the previous perceptron theory is that the value of the overlap R is replaced by γR, where the noise parameter γ is given by

$$\gamma = \frac{1}{\sqrt{1 + \sigma_S^2}}. \tag{71}$$

Thus the generalization error is modified to

$$\epsilon(\mathbf{W}) = \frac{1}{\pi}\cos^{-1}\gamma R. \tag{72}$$

Equation (72) also has a simple geometrical interpretation: denoting $\sigma_S = \tan^2\phi$, and $R = \cos\pi\epsilon$, the noisy generalization error is the angle $\pi\epsilon_n$ such that

$$\cos(\pi\epsilon_n) = \cos\phi\,\cos(\pi\epsilon). \tag{73}$$

The annealed free energy is now given by

$$-\beta f = G_0(R) - \alpha G_1(\gamma R), \tag{74}$$

where G_0 and G_1 are given by Eq. (55) and Eq. (56), respectively.

The properties of the system in the large N limit are obtained by minimizing the free energy f, which yields

$$\frac{\partial G_0(R)}{\partial R} = \alpha\frac{\partial G_1(\gamma R)}{\partial R}. \tag{75}$$

Solving for R, one then evaluates the average generalization error via Eq. (72).

---

[2] A similar noise model where there is a noisy version of the teacher rather than the patterns leads to the same results.[13,14]

5.1 LEARNING CURVES WITH NOISY PATTERNS

Learning from noisy patterns is different from learning in the absence of noise. The rule is now unrealizable, or the data nonlinearly separable, and the learning process can only fit the examples up to a finite asymptotic error. It is interesting to understand how the first-order transition to perfect generalization, exhibited in the noise-free case, is altered due to the noise. The parameter γ allows us to study the full range between the noise-free case, $\gamma = 1$, and the random label case (i.e., the capacity problem) in the infinite noise limit, $\gamma = 0$. We have studied the replica symmetric solution for various values of the parameter γ, and the results are depicted in Figure 1. The first-order transition from poor to perfect generalization disappears for almost all values of the noise, and the learning curve undergoes a crossover from a fast exponential drop to a power law near the original critical transition. For a -noise level the asymptotic learning curve is gradual, approaching the finite-asymptotic error value, similar to other nonrealizable cases.[29] In the replica symmetric solution there exist a *finite* critical noise, $\gamma_c = 0.997$, below which the first-order transition to good generalization remains.

Notice, however, that for any $\gamma < 1$ there is a finite number of examples beyond which the replica symmetric solution is incorrect. This phenomenon is signaled by the vanishing of the replica symmetric entropy, and is known to yield very different critical values both for the capacity problem at $\gamma = 0$,[22] and for the noiseless learning case at $\gamma = 1$.[12,27] We conjecture that this is the case for all intermediate values of γ, and that a replica symmetry breaking occurs along the vertical solid line in Figure 1. Nevertheless, the asymptotic qualitative behavior beyond this line remains a nontrivial power law, as in the generic nonrealizable learning case.

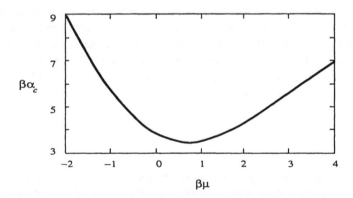

FIGURE 1   The overlap R as a function of the number of examples per weight, α, for various values of the noise parameter γ, for the binary perceptron learning problem at zero temperature. The replica symmetric critical line ($q \to 1$) is beyond the replica-symmetric zero entropy line (solid) on which the RSB transition occurs. Notice, however, that there is a first-order transition from a low to a R beyond some critical value of γ.

6. LEARNING WITH UNDETERMINED NUMBER OF INPUT NODES

Understanding the learning with noisy patterns allows us to study an even more interesting learning model, i.e., learning when the architecture of the student doesn't match the one of the teacher. This model addresses an important issue, very common in real-life applications, when the either the relevant inputs or the correct structure of the teacher are not known to the student. A very simple model of this phenomenon is given by the following perceptron learning problem, which we treat here using the annealed approximation.

Let N be the size of an input pattern $\mathbf{S}$, and λ_0 the probability that the input coordinate S_j is actually observed by the teacher perceptron. Thus only $N_0 = \lambda_0 N$ inputs are relevant to the output of the teacher perceptron. As before, we assume that both the teacher and student perceptrons have only binary weights.[3] Similarly, let the the probability that the S_j coordinate is observed by the student be λ, so λN is the expected number of input nodes of the students. Denoting by λ_1 the probability of a node being in the input of the student but not the teacher, the entropy of the possible input configurations per input node is given by

$$
\begin{aligned}
G_a = \; & \lambda_0 \log \lambda_0 + (1 - \lambda_0) \log(1 - \lambda_0) \\
& - \lambda_1 \log \lambda_1 - (\lambda_0 - \lambda_1) \log(\lambda_0 - \lambda_1) \\
& - (\lambda - \lambda_1) \log(\lambda - \lambda_1) \\
& - (1 - \lambda_0 - (\lambda - \lambda_1)) \log(1 - \lambda_0 - (\lambda - \lambda_1)),
\end{aligned}
\tag{76}
$$

which is simply $1/N$ the logarithm of the number of possible input configurations. Only the $(1 - \lambda_1)N$ inputs that are shared between the student and teacher carry information and, thus, the student and the teacher can be aligned only on that component of the input. Denoting by R the overlap between the teacher and student on that part of the input, and introducing the usual auxiliary variable $\hat{R}$, adds the following term to the free energy

$$
G_b = \lambda_1 \left(R\hat{R} - \log 2 \cosh \hat{R} \right).
\tag{77}
$$

---

[3] Clearly, one can always include the value zero for the weights. Here we would like to treat the model using a more general *Grandcanonical* formulation, where zero weights are excluded.

6.1 NODE PRICE AND A GRANDCANONICAL FORMULATION

The interesting aspect of this model is the ability to study the generalization problem when the number of input nodes is not fixed. The analogy with statistical mechanics can be taken even further if we assume that there is a given price per nonzero weight, analogous to the chemical potential in thermodynamics. Such a potential simply adds one parameter to the problem, denoted by μ, indicating the tendency of the system to add or eliminate input nodes. Positive μ indicate a tendency for growing small networks, while negative μ pushes toward larger networks. (This can also be viewed from an information theoretic view as the code length needed per connection.) Since μ is a Lagrange multiplier attached to the average number of nodes, optimizing the generalization with respect to μ is equivalent to optimizing over the size of the network.

Adding this term to the free energy yields the final form of G_0

$$G_0 = G_a + G_b - \beta\mu\lambda . \tag{78}$$

6.2 MISMATCH NOISE

Finally, we need to modify the energy term, G_1 due to the mismatch between the inputs. Since $\lambda_1 N$ student input nodes are not observed by the teacher, their contribution to the student is similar to pattern noise. It can be easily verified that the magnitude of this noise results in a noise parameter

$$\gamma_m = \frac{\lambda_1}{\sqrt{\lambda - \lambda_0}} , \tag{79}$$

resulting in the annealed G_1 of the form

$$G_1 = \alpha \ln\left[1 - \frac{\tau}{\pi}\cos^{-1}\gamma_m R\right] , \tag{80}$$

with $\tau = 1 - e^{-\beta}$.

The learning curve can now be calculated from the saddle point equations

$$\frac{\partial G}{\partial \lambda} = \frac{\partial G}{\partial \lambda_1} = \frac{\partial G}{\partial R} = \frac{\partial G}{\partial \hat{R}} = 0 , \tag{81}$$

where the noise parameter γ_m is taken to be a function of λ and λ_1 as well. The resulting equations are

$$\hat{R} = \frac{\alpha\beta\gamma_m}{\pi\lambda_1\sqrt{1 - (\gamma_m R)^2}}, \tag{82}$$

$$R = \tanh \hat{R}, \tag{83}$$

$$\lambda = \frac{AB\lambda_0}{1 + AB} + \frac{A(1 - \lambda_0)}{1 + A}, \tag{84}$$

$$\lambda_1 = \frac{\lambda}{1 + \frac{(1-\lambda_0)(1+AB)}{\lambda_0(1+A)B}} , \tag{85}$$

where

$$A = 2\exp\left[\frac{\alpha\beta R}{\pi\sqrt{1 - (\gamma_m R)^2}} \times \frac{\partial\gamma_m}{\partial\lambda} - \beta\mu\right] \tag{86}$$

$$B = \cosh\hat{R}\exp\left[\frac{\alpha\beta R}{\pi\sqrt{1 - (\gamma_m R)^2}} \times \frac{\partial\gamma_m}{\partial\lambda_1} - R\hat{R}\right].$$

These equations are solved self-consistently for R, $\hat{R}$, λ, and λ_1 with α, β, and μ as parameters.

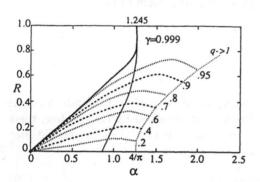

FIGURE 2   Generalization error as a function of the price per node for various values of α. Unlike the fixed noise case, here there is a first-order transition to perfect generalization at a critical number of examples for any μ and at any training temperature. The noise parameter γ jumps to the value 1 at the transition, when the two architectures match perfectly. Here $\lambda_0 = 0.5$ and the critical $\alpha = 3.48$ at an optimal $\mu = 0.82$.

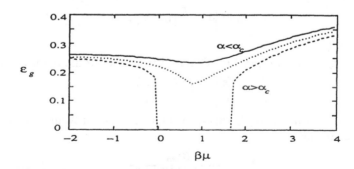

FIGURE 3   The critical α vs. μ for the $\lambda_0 = 0.5$. There is a unique optimal μ at which perfect generalization occurs. This corresponds to the optimal *penalty* per parameter for the best generalization performance.

6.3 DISCUSSION

The main features of this model are described in Figures 2 and 3. In Figure 2 the generalization error is given as function of the price per node. For every value of μ there is a critical at which the system undergoes a transition to perfect generalization. The fastest learning is obtained at an optimal value of $\beta\mu$, which is a function of λ_0 only. This value correspond to the optimal growth rate of the network. The number of examples for which the transition occurs, for different $\beta\mu$, is given in Figure 3 and is in agreement with some empirical results which show such an optimum in the generalization as a function of the network size. It is interesting to notice that not all learning experiments are consistent with this observation.

ACKNOWLEDGMENTS

I wish to thank the organizers of this workshop for this good opportunity to exopse these ideas beyond the statistical physics community. A special acknowledgment is due to Haim Sompolinsky for his enormous contributions to the development of this field, and for numerous discussions of the specific models; to Sebastian Seung and Géza Györgyi, whose works on the statistical mechanics of learning form the basis of these notes; and to David Haussler and Micheal Kearns who are taking this sufficiently seriously to turn it into decent mathematics.

APPENDIX

In the appendix we give a fuller account of the results that were outlined in Section 4. We begin with a derivation of Eq. (50) for the average generalization error, which illustrates many of the calculational techniques of this paper. Integrating the error function Eq. (49) over the *a priori* input measure Eq. (46), we obtain

$$\epsilon(\mathbf{W}) = \int D\mathbf{S}\, \epsilon(\mathbf{W}; \mathbf{S}) \tag{87}$$

$$= \int dx \int dy\, \frac{1}{2}[g(x) - g(y)]^2$$
$$\times \int D\mathbf{S}\, \delta\left(x - N^{-\frac{1}{2}}\mathbf{W}\cdot\mathbf{S}\right) \delta\left(y - N^{-\frac{1}{2}}\mathbf{W}^0\cdot\mathbf{S}\right) \tag{88}$$

$$= \int \frac{dx\,d\hat{x}}{2\pi} \int \frac{dy\,d\hat{y}}{2\pi} e^{ix\hat{x}+iy\hat{y}} \frac{1}{2}[g(x) - g(y)]^2$$
$$\times \int D\mathbf{S}\, \exp\left[-iN^{-\frac{1}{2}}(\mathbf{W}\hat{x} + \mathbf{W}^0\hat{y})\cdot\mathbf{S}\right] . \tag{89}$$

The two auxiliary variables x, y are introduced to remove $\mathbf{S}$ from the argument of the g functions, and $\hat{x}, \hat{y}$ are introduced to transform the delta functions into exponentials using the identity

$$\delta(x) = \int \frac{d\hat{x}}{2\pi} e^{ix\hat{x}} . \tag{90}$$

When the Gaussian average over $\mathbf{S}$ is now performed, a simple Gaussian integral in $\hat{x}$ and $\hat{y}$ is left. These variables can in turn be integrated out, leaving

$$\epsilon(\mathbf{W}) = \int \frac{dx\,dy}{2\pi\sqrt{1-R^2}} \exp\left(-\frac{x^2+y^2-2xyR}{2(1-R^2)}\right) \frac{1}{2}[g(x)-g(y)]^2. \tag{91}$$

This result has a simple interpretation. It is the average of $(1/2)[g(x)-g(y)]^2$, where x and y, like $\mathbf{W}\cdot\mathbf{S}/\sqrt{N}$ and $\mathbf{W}^0\cdot\mathbf{S}/\sqrt{N}$ in Eq. (49), are Gaussian variables with unit variance and cross-correlation R. A simple change of variables in Eq. (91), a shift followed by a rescaling, yields the form Eq. (50).

To derive the annealed approximation for perceptron learning, we begin by evaluating Eq. (32) for $G_{\mathrm{AA}}(\mathbf{W})$. The calculation is essentially the same as the previous one for $\epsilon(\mathbf{W})$, and results in a similar formula,

$$\begin{aligned} G_{\mathrm{AA}}(\mathbf{W}) = &-\ln \int \frac{dx\,dy}{2\pi\sqrt{1-R^2}} \\ &\times \exp\left(-\frac{x^2+y^2-2xyR}{2(1-R^2)} - \beta/2[g(x)-g(y)]^2\right) . \end{aligned} \tag{92}$$

Again, the answer depends on $\mathbf{W}$ only through the overlap R. We denote the function of R on the right-hand side by G_1, so that $G_{\mathrm{AA}}(\mathbf{W}) \equiv G_1(N^{-1}\mathbf{W}\cdot\mathbf{W}^0)$. A change of variables in the integral Eq. (92) yields the form Eq. (56). Evaluating the integral for the cases of $g(x) = x$ (linear perceptron) and $g(x) = \mathrm{sign}(x)/\sqrt{2}$ (Boolean perceptron), we obtain

$$G_1 = \begin{cases} \frac{1}{2}\ln\left[1+2\beta(1-R)\right], & \text{linear,} \\ -\ln\left[1-(1-e^{-\beta})\pi^{-1}\cos^{-1} R\right], & \text{Boolean.} \end{cases} \tag{93}$$

Although derived using the Gaussian a priori input distribution Eq. (46), the above results for $\epsilon(\mathbf{W})$ and $G_{\mathrm{AA}}(\mathbf{W})$ apply also to the case of the discrete inputs $S_i = \pm 1$ in the thermodynamic (large N) limit. This insensitivity to input distribution is explained by the central limit theorem, which guarantees that $\mathbf{W}\cdot\mathbf{S}/\sqrt{N}$ and $\mathbf{W}^0\cdot\mathbf{S}/\sqrt{N}$ are Gaussian variables (in the $N \to \infty$ limit) with very weak assumptions about the distribution of $\mathbf{S}$. This assertion may be verified by a straightforward calculation for discrete $\mathbf{S}$, which yields Eqs. (93) as the leading terms in a saddle point expansion in $1/N$.

Since $G_{AA}(\mathbf{W})$, Eq. (32), depends only on the overlap R, the annealed partition function (31) can now be rewritten as an integral over R,

$$\langle\!\langle Z \rangle\!\rangle = \int dR \exp N[G_0(R) - \alpha G_1(R)] \,, \qquad (94)$$

where

$$NG_0(R) = \ln \int d\mu(\mathbf{W})\, \delta(R - N^{-1}\mathbf{W}\cdot\mathbf{W}^0) \qquad (95)$$

is the logarithm of the density of networks with overlap R. In the thermodynamic limit ($N \to \infty$), the integral can be evaluated as

$$-\beta f(T,\alpha) \equiv \frac{1}{N}\ln\langle\!\langle Z \rangle\!\rangle = \mathrm{extr}_R\{G_0(R) - \alpha G_1(R)\} \,. \qquad (96)$$

Hence the thermodynamic free energy $f(T,\alpha)$ is determined by extremizing the free-energy function $-\beta f(R) \equiv G_0(R) - \alpha G_1(R)$. Differentiating $f(R)$ with respect to R, we obtain the stationarity condition Eq. (57).

We can write the stationarity equations in a more revealing form by proceeding further in the evaluation of G_0. The delta function in Eq. (95) can be expanded by introducing another order parameter $\hat{R}$,

$$G_0 = \frac{1}{N}\ln \int_{-i\infty}^{i\infty} \frac{d\hat{R}}{2\pi i} \exp\left(-NR\hat{R} + \ln \int d\mu(\mathbf{W})\, e^{\hat{R}\mathbf{W}\cdot\mathbf{W}^0}\right) \,. \qquad (97)$$

In the thermodynamic limit, this reduces to

$$G_0 = -R\hat{R} + \frac{1}{N}\ln \int d\mu(\mathbf{W})\, e^{\hat{R}\mathbf{W}\cdot\mathbf{W}^0} \,, \qquad (98)$$

where the right-hand side must be stationary with respect to the saddle point parameter $\hat{R}$. The free energy can now be written as a saddle point over two order parameters

$$-\beta f = \mathrm{extr}_{R,\hat{R}}\left\{-R\hat{R} + \frac{1}{N}\ln \int d\mu(\mathbf{W})\, e^{\hat{R}\mathbf{W}\cdot\mathbf{W}^0} - \alpha G_1(R)\right\} \,. \qquad (99)$$

The saddle point equations are

$$\hat{R} = -\alpha\frac{\partial G_1}{\partial R} \,, \qquad (100)$$

$$R = \frac{1}{N}\langle\mathbf{W}\rangle_{\hat{R}}\cdot\mathbf{W}^0 \,, \qquad (101)$$

where

$$\langle\mathbf{W}\rangle_{\hat{R}} \equiv \frac{\int d\mu(\mathbf{W})\,\mathbf{W}\exp(\hat{R}\mathbf{W}\cdot\mathbf{W}^0)}{\int d\mu(\mathbf{W})\exp(\hat{R}\mathbf{W}\cdot\mathbf{W}^0)} \,. \qquad (102)$$

The order parameter $\hat{R}$ has a natural interpretation: it is the strength of a local field pushing $\mathbf{W}$ in the direction of $\mathbf{W}^0$. Since it increases with α, it forces $\mathbf{W}$ toward $\mathbf{W}_0$ as $\alpha \to \infty$. Upon eliminating $\hat{R}$, these equations reduce to Eq. (57).

Equation (98) can be evaluated quite easily for the case of the Ising constraint $W_i = \pm 1$. Then we can make the replacement

$$\int d\mu(\mathbf{W}) \to \sum_{W_i = \pm 1} , \tag{103}$$

which leads finally to

$$G_0(R) = -R\hat{R} + \ln 2 \cosh \hat{R} , \tag{104}$$

assuming that the teacher weights also satisfy the Ising constraint. Extremizing with respect to $\hat{R}$, we obtain the equation

$$R = \tanh \hat{R}. \tag{105}$$

Eliminating $\hat{R}$, one can finally derive the result Eq. (59), which is purely a function of R, and is the familiar result for the entropy of the Ising model as a function of the magnetization R.

REFERENCES

1. Barron, A. R. "Statistical Properties of Artificial Neural Networks." In *Proceedings of the 28th Conference on Decision and Control*, Vol. 1, 280–285. New-York: IEEE Control Systems Society, 1989.
2. Baum, E., and D. Haussler. "What Size Net Gives Valid Generalization?" *Neural Comp.* 1 (1989): 151–160.
3. Binder, K., and D. W. Heerman. *Monte Carlo Simulation in Statistical Mechanics*. Berlin: Springer-Verlag, 1988.
4. Blumer, A., A. Ehrenfeucht, D. Haussler, and M. K. Warmuth. "Learnability and the Vapnik-Chervonenkis Dimension." *JACM* 36 (1989): 929–965.
5. Carnevali, P., and S. Patarnello. "Exhaustive Thermodynamical Analysis of Boolean Learning Networks." *Europhys. Lett.* 4 (1987): 1199–1204.
6. Devroye, L. "Automatic Pattern Recognition: A Study of the Probability of Error." *IEEE Trans. on Pattern Anal. and Mach. Intelligence* 10 (1988): 530–544.
7. Denker, J., D. Schwartz, B. Wittner, S. Solla, R. Howard, L. Jackel, and J. Hopfield. "Automatic Learning, Rule Extraction and Generalization." *Complex Systems* 1 (1987): 877–922.
8. Gardner, E., "The Space of Interactions in Neural Network Model." *J. Phys.* **A21** (1988): 257–270.

9. Gardner, E., and B. Derrida. "Three Unfinished Works on the Optimal Storage Capacity of Networks." *J. Phys.* **A22** (1989): 1983–1994.

10. Giudice, P. D., S. Franz, and M. A. Virasoro. "Perceptron Beyond the Limit of Capacity." *J. Phys.* (France) **50** (1989): 121–134.

11. Györgyi, G. "First-Order Transition to Perfect Generalization in a Neural Network with Binary Synapses." *Phys. Rev.* **A41** (1990): 7097–7100.

12. Györgyi, G. "Inference of a Rule by a Neural Network with Thermal Noise." *Phys. Rev. Lett.* **64** (1990): 2957–2960.

13. Györgyi, G., and N. Tishby. "Statistical Theory of Learning a Rule." In *Neural Networks and Spin Glasses*, edited by W. K. Theumann and R. Köberle, 3–36. Singapore: World Scientific, 1990.

14. Haussler, D. "Decision Theoretic Generalizations of the PAC Model for Neural Net and Other Learning Applications." *Information and Computation*, **100**(1) (1992): 78–150.

15. Hertz, J. A. "Statistical Dynamics of Learning." Preprint 90/34 S, Nordita, 1990.

16. Haussler, D., M. Kearns, and R. Schapire. "Bounds on the Sample Complexity of Bayesian Learning." In *Proceedings of the IVth Annual Workshop on Computational Learning Theory (COLT91)*, 61–74. Santa Cruz, CA, 1991.

17. Haussler, D., M. Kearns, H. S. Seung, and N. Tishby. "Rigorous Learning Curve Bounds from Statistical Mechanics." In *Proceedings of the VIIth Annual Workshop on Computational Learning Theory (COLT94)*, 1994.

18. Hansel, D., and H. Sompolinsky. "Learning from Examples in a Single-Layer Neural Network." *Europhys. Lett.* **11** (1990): 687–692.

19. Judd, S. "On the Complexity of Loading Shallow Neural Networks." *J. Complexity* **1** (1988): 177–192.

20. Kirkpatrick, Jr., S., C. D. Gelatt, and M. P. Vecchi. "Optimization by Simulated Annealing." *Science* **220** (1983): 671–680.

21. Krauth, W., and M. Mézard. "Storage Capacity of Memory Networks with Binary Couplings." *J. Phys.* (France) **50** (1989): 3057–3066.

22. Levin, E., N. Tishby, and S. A. Solla. "A Statistical Approach to Learning and Generalization in Layered Neural Networks." Special Issue on Neural Networks. *Proc. IEEE* **78** (1990): 1568–1578.

23. Opper, M., and W. Kinzel. *Physics of Neural Networks*, edited by J. S. van Hemmen, E. Domany, and K. Schulten. Berlin: Springer-Verlag, 1994.

24. Rosenblatt, F. *Principles of Neurodynamics*. New York: Spartan Books, 1962.

25. Seung, H. S., H. Sompolinsky, and N. Tishby. "Statistical Mechanics of Learning from Examples." *Phys. Rev. A* **45**(8) (1992): 6056–6091.

26. Sompolinsky, H., and N. Tishby. "Learning in a Two-Layer Network of Edge Detectors." *Europhys. Lett.* **13** (1990): 567–573.

27. Sompolinsky, H., N. Tishby, and H. S. Seung. "Learning from Examples in Large Neural Networks." *Phys. Rev. Lett.* **65**(13) (1990): 1683–1686.

28. Tishby, N., E. Levin, and S. Solla. "Consistent Inference of Probabilities in Layered Networks: Predictions and Generalization." In *IJCNN International Joint Conference on Neural Networks*, Vol. 2, 403–409.

29. Valiant, L. G. "A Theory of the Learnable." *Comm. ACM* **27** (1984): 1134–1142.

30. Vapnik, V. N. *Estimation of Dependences Based on Empirical Data.* New York: Springer-Verlag, 1982.

31. Venkatesh, S. S. "On Learning Binary Weights." In the *Proceedings of the IVth Workshop on Computational Learning Theory (COLT91)*. San Mateo, CA: Morgan Kaufmann, 1991.

32. Watkin, L. H., A. Rau, and M. Biehl. "The Statistical Mechanics of Learning a Rule." *Rev. Mod. Phys.* **65** (1993): 499.

David H. Wolpert† and Alan S. Lapedes‡
†The Santa Fe Institute, 1399 Hyde Park Road, Santa Fe, NM, 87505;
e-mail: dhw@santafe.edu
‡ Complex Systems Group, MS B213, Los Alamos, NM, 87545; e-mail: asl@t13.lanl.gov

On Exhaustive Learning

This chapter uses the extended Bayesian formalism to investigate the noise-free "exhaustive learning" scenario, first introduced by Schwartz et al. in the context of the statistical physics of learning. This scenario can be viewed as the zero-temperature limit of the work of Tishby et al. This chapter proves that the crucial "self-averaging" assumption invoked in the conventional analysis of exhaustive learning does hold in a particular trivial implementation of exhaustive learning. However, it does not hold in the simplest nontrivial implementation of exhaustive learning. Therefore, the central result of that analysis, concerning how generalization accuracy improves as training set size is increased, is not generic. More importantly, this chapter shows that if one uses an off-training-set error function rather than the usual error function, then this central result does not hold even when the self-averaging assumption is valid, and even in the limit of an infinite input space. This implies that the central result is a reflection of the following simple phenomenon: if you add an input/output pair to the training set, the number of distinct input values on which you know exactly how you should guess has either increased or stayed the same and, therefore, your generalization accuracy will either increase or stay the same.

INTRODUCTION

This chapter is an investigation of the "exhaustive learning" scenario introduced by Schwartz et al.,[4] Van der Broeck and Kawai,[8] and Hertz et al.[3] This scenario is perhaps the simplest possible supervised learning scenario. It is identical to the noise-free "Gibbs learning" scenario studied by Haussler et al.,[2] and can also be viewed as the zero-temperature limit of the "statistical mechanics" work of Tishby et al.[5,6,7] It consists of examining the noise-free performance of generalizers (also known as "learning algorithms") of the following type: exclude all hypothesis functions not consistent with the training set, and guess randomly amongst the rest. The central result of the conventional analysis of exhaustive learning (Schwartz et al.[4]) is that in this scenario, when "self-averaging" holds, $P(c|f, m) \sim (1 - c)^m$, where c is a value of the generalization error, f is a target function, and m is a training set size. The implication of this result is that as m increases, generalization improves.

This chapter uses the extended Bayesian formalism (EBF), introduced by Wolpert.[9] Readers unfamiliar with the EBF are directed to the self-contained overview of the EBF in Wolpert,[12] Section 2. As used in this chapter to investigate exhaustive learning, the EBF reduces to the following:

1. n and r are the number of elements in the input and output spaces, X and Y respectively.
2. m is the number of elements in the (ordered) training set L. $\{L_X(i), L_Y(i)\}$ is the corresponding set of m input and output values. m' is the number of distinct values in L_X.
3. Outputs h of the learning algorithm are always assumed to be of the form of a function from X to Y, indicated by $h(x \in X)$. Any restrictions on h are imposed by the probability distribution $P(f, h, L, c)$.
4. The learning algorithm sets the distribution $P(h|L)$. In exhaustive learning, $P(h|L) = k(L)T(h)\delta(L \subset h)$, where $k(L)$ is a normalization constant, $T(h)$ is a nowhere-negative distribution defining "random" guessing of h, and

$$\delta(L \subset h) = \begin{cases} 1 & \text{if } h \text{ goes through } L \\ 0 & \text{otherwise,} \end{cases}$$

   (The notation is motivated by viewed h as a set of input-output pairs, just as L is.) This is called the "Gibbs generalizer."
5. "Targets" f are always assumed to be of the form of a function from X to Y, indicated by $f(x \in X)$. Any restrictions on f are imposed by $P(f, h, L, c)$.
6. The "likelihood" is $P(L|f) = \delta(L \subset f) \times \Pi_i \pi(L_X(i))$, where

$$\delta(L \subset f) = \begin{cases} 1 & \text{if } L \text{ lies completely on } f \\ 0 & \text{otherwise.} \end{cases}$$

and $\pi(x)$ is the "sampling distribution." The $\delta(L \subset f)$ term imposes noise-free training-set generation.

7. The "posterior" is $P(f|L)$. In this chapter, probability is *not* interpreted to mean "degree of personal belief," as some conventional Bayesians define it. Accordingly, it is not true that the researcher automatically knows $P(f|L)$. (See Wolpert.[13])

8. $P(h|f,L) = P(h|L), P(f|h,L) = P(f|L)$, and therefore $P(h,f|L) = P(h|L) \times P(f|L)$.

9. The cost c associated with a particular h and f is either given by $Er(f,h,L) = \sum_x \pi(x)[1 - \delta(f(x), h(x))]$ ("iid error function"), or by the "off-training-set error" function, $Er(h,f,L) = \sum_{x \notin L_X} \pi(x)[1 - \delta(f(x), h(x))]/\sum_{x \notin L_X} \pi(x)$, where $\delta(.,.)$ is the Kronecker delta function.

10. The empirical misclassification rate $s \equiv \sum_{i=1}^m \{1 - \delta[h(L_X(i)), L_Y(i)]\}/m$.

11. The count function $S(g_1, g_2, \{x_i\}) \equiv \sum_i \{\delta(g_1(x_i), g_2(x_i))\}$ measures the number of times function g_1 agrees with function g_2 over the (non-duplicate) input values in $\{x_i\}$. For example, if L_X contains no duplicates $s = m - S(L, h, L_X)$.

Using the EBF this chapter proves that the crucial "self-averaging" assumption invoked in the conventional analysis of exhaustive learning holds in a particular trivial implementation of exhaustive learning, but does not hold in the simplest nontrivial implementation of exhaustive learning. Therefore, the central result of that analysis (hereafter called "the central result"), with its implication that generalization accuracy necessarily rises exponentially as training set size m is increased, is not generic. (This might be part of the explanation for the results reported by Cohn and Tesauro.[1])

More importantly, this chapter proves that a reasonable change in the definition of what one calls the "generalization error," to reflect only the error on inputs *outside* of the training set, negates the central result even when self-averaging *does* hold. Perhaps surprisingly, this is true even in the limit of infinite input spaces and small training sets. In other words, the central result is a reflection of the following phenomenon: If you add an input/output pair to the training set, the number of distinct input values on which you know exactly how you should guess has either increased or stayed the same and, therefore, your generalization accuracy will either increase or stay the same. This phenomenon would, taken naively, appear to only result in a linear rise in generalization accuracy with training set size in the exhaustive learning scenario, where generalization accuracy reflects on-training-set as well as off-training-set errors. However, as this chapter shows, it actually results in exhaustive learning's *exponential* rise ($P(c|fm) \sim (1 - c)^m$).[1]

Phrased differently, exhaustive learning's central result is exactly what you would expect for a simple memorization engine that guesses arbitrarily outside

---

[1]Generically, exponential $P(c|fm)$ results in either e^{-m} or $1/m$ behavior for the average error, $\sum_c c \times P(c|fm)$. (See Hertz et al.,[3] page 152.) To parallel the PAC formalism, rather than concentrate on the average error, this chapter concentrates on $P(c|fm)$ and the issue of whether or not $P(c|fm)$ is exponential.

of the training set.[2] This serves as a caution against reading too much meaning into exponential gains in generalization accuracy when that accuracy reflects on-training-set as well as off-training-set errors.

In addition to elaborating the foregoing, this chapter also does the following:

1. It rederives the mathematics of exhaustive learning in a context more general than that of neural nets. It is by doing this that one discerns that the conditional probability of interest in exhaustive learning is $P(c|f,m)$.

2. This chapter also clarifies the following: Exhaustive learning is sometimes couched as though its input space were uncountably infinite (e.g., a subset of $\mathbf{R}^K$). However, because in exhaustive learning the number of distinct hypothesis functions possibly output by the generalizer is countable, the input space is effectively discrete. The mathematics cannot "access" the full freedom inherent in an uncountably infinite input space.

3. In this chapter, we prove that if $T(h)$ is independent of h, then, as is claimed in the conventional analysis of exhaustive learning,[4] "self-averaging" obtains in the limit where the input space is infinitely bigger than the training set. Accordingly, for such a $T(h)$, the central result of exhaustive learning applies. This result holds despite the fact that for some schemes in which h is parametrized in terms of a vector $\mathbf{w}$ (e.g., perceptrons), self-averaging does not hold when $T(\mathbf{w})$ is uniform. (Uniformity over $\mathbf{w} \not\Rightarrow$ uniformity over h.)

4. This chapter shows how to calculate things exactly, without recourse to limits and any consequent "self-averaging," for the $T(h)$ constant case mentioned in item 3.

5. This chapter shows that self-averaging, and therefore the central result of exhaustive learning, does *not* hold for the simplest nonconstant $T(h)$, no matter what limit one operates in.

6. Exhaustive learning calculates P(error value $= c$ | target function f, training set size m). This chapter shows how to calculate P(error value $= c$ | target function f, training set L), where L has m elements. The result differs from P(error value $= c$ | target function f, training set size m), even if P(error value $= c$ | target function f, training set L) is independent of what m input-output pairs make up L. However, in the limit of $n \gg m$, the two probability distributions agree.

7. Often one would like to consider the conventional iid error function modified to only measure off-training-set performance. (For example, see the discussion by Wolpert,[12] Section 1.) This chapter proves that with an off-training-set error function, the improvement in generalization with increasing training set size characteristic of exhaustive learning goes away. Indeed, for the limit of large errors, when the output space is binary the generalization gets worse exponentially with m. Perhaps surprising, this is true regardless of the size of the input space or the size of the training set.

---

[2]See Wolpert,[12] Section 6, and Schwartz et al.[4] for a discussion of the intuitive reasons for this.

8. Regardless of the error function one uses, usually in formalisms like exhaustive learning, one is ultimately interested in examining "learning curves" which show how generalization will change, on average, as one adds new elements to one's *already extant* training set. In general, this quantity of interest need not be given by the quantity calculated in exhaustive learning, $P(c|f, m)$, which simply averages over all training sets without regard to how they could arise from smaller ones. This chapter finds a sufficient condition for when this quantity of interest is given by $P(c|f, m)$.

9. Exploiting the symmetry of the EBF under $f \leftrightarrow h$, this chapter derives formulas for $P(c|h, m)$ directly from the exhaustive learning formulas for $P(c|f, m)$.

In Section 1, we discuss some of the axioms defining the exhaustive learning scenario. In Section 2, we reproduce and extend the conventional analysis of exhaustive learning for the case which it addresses, where the testing set can run over the training set. Section 3 contains an analysis of exhaustive learning when an off-training-set error function is used. This section proves that the characteristic $(1 - c)^m$ behavior of exhaustive learning disappears with such an error function, even in the limit of an infinite input space. Finally, using the linearity of probability theory, In Section 4, we formally justify interpreting exhaustive learning in terms of "learning curves." In Section 4, we also exploit some symmetries in the EBF to extend exhaustive learning's results to other scenarios. In particular, we use those symmetries to extend the results of exhaustive learning from the distribution $P(c|f, m)$ to the distribution $P(c|h, m)$.

1. THE EXHAUSTIVE LEARNING SCENARIO

This section discusses the axioms (presented in the previous section) that define the exhaustive learning scenario.

As conventionally formulated (e.g., Hertz et al.[3]) it appears that exhaustive learning can have an uncountable input space (e.g., X is a subset of $\mathbf{R}^k$ for some k). However, in fact, its input space is *effectively* discrete and can be made explicitly discrete without any loss of generality. This is because there are only a countable number of hypothesis functions in the exhaustive learning scenario (see Hertz et al.[3]) and therefore the mathematics cannot "access" the full freedom of the $\mathbf{R}^k$ input space. See Appendix A.

In general, no restrictions are set on the distribution $T(h)$ defining the Gibbs generalizer used in exhaustive learning (see the review of the EBF in the previous section). In particular, $T(h)$ can equal zero for some h. However, it is only when $T(h)$ is nowhere-zero that the Gibbs generalizer is well defined for any L, and can therefore be used with impunity in the real world, without any regard for f or the L which are sampled from f.[3]

On the other hand, many real-world generalizers are not expressible as Gibbs generalizers, no matter what choice one makes for $T(h)$. In particular, no generalizer that makes the same guess whenever presented with the same data can be expressed as a Gibbs generalizer. (See Wolpert.[10])

Although $P(h|$ empty training set$) = T(h)$, in general this does not equal $P(h)$. (This is because $P(h) = \sum_L P(h|L)P(L) = T(h) \sum_{L,f} k(L)\delta(L \subset h)P(L|f)P(f)$, and depending on $P(f)$, the sum over L and f may not be independent of h.) Rather than $P(h|$ empty training set$)$, $P(h)$ is the distribution, evaluated *before* seeing any training examples, of what h the generalizer will guess *after* learning has taken place. See Wolpert[12] and Schwartz et al.[4]

In exhaustive learning, repeats are allowed in L_X; i.e., the same input value can occur more than once in L_X. (See the review of the EBF in the previous section.) This means that conditions like $L \subset f$ are being interpreted a bit loosely in this chapter; formally, they should be taken to mean that L is consistent with f, rather than that the set of input-output pairs $\{L_X(i), L_Y(i); 1 \le i \le m\} \subset$ the set of input-output pairs $\{x, f(x); x \in X\}$. Also, due to our noise-free likelihood, there can be no conflicts in L: anytime an input value is repeated, the corresponding output values must agree. This means that any such repeat in L conveys no extra information concerning f to the generalizer. (Of course, it conveys information about other objects, like $\pi(x)$.) This points up an odd aspect of exhaustive learning: as already mentioned, in exhaustive learning, probabilities conditioned on m will be calculated. However, it is not m but rather m', the number of distinct pairs in L, which is directly related to "the amount of information" in L directly concerning the distribution f.

The conventional analysis of exhaustive learning is concerned with evaluating the average value of the iid error function when the target function is held fixed, one has the noise-free iid likelihood given in the introduction, and the Gibbs learning algorithm is used. In other words, it is concerned with the quantity $\langle P(c|f, L) \rangle$ for iid error c, the expectation value being over all possible training sets L of size m

---

[3] If, rather than being infinitesimal, $T(h')$ equalled zero exactly for some particular hypothesis h', the use of the exhaustive learning $P(h|L)$ would implicitly set strong—and in the real world unverifiable—restrictions on f. This is because if $T(h') = 0$, then the exhaustive learning $P(h|L)$ is undefined for $L = h'$. Since we have no noise in the likelihood, the only way we could be assured of avoiding such an L in the real world is if the corresponding f cannot occur (assuming $\pi(x)$ is nonzero for all x); i.e., if $P(f)|_{f=h'}$ equals 0 exactly. In other words, if we set $T(h')$ to zero, either we must somehow know that it is physically impossible for the target function $f = h'$ to occur, or we must be willing to have our $P(h|L)$ undefined for certain L. To avoid this dilemma, one should ensure that $T(h)$ never equals zero exactly (although it can be arbitrarily small). Unfortunately this can be quite difficult if, as is conventional, functions h are viewed as neural nets, so that rather than writing down $T(h)$ directly one writes down a sort of $T(\mathbf{w})$, where $\mathbf{w}$ is a weight vector parametrizing h.

chosen from the (fixed but unknown) target function f, and the $P(h|L)$ and $P(L|f)$ of the introduction being assumed.

That the target function is assumed fixed in exhaustive learning's distribution of interest can be seen by noting that exhaustive learning is concerned with the "learning curves" of how average error changes as more samples of the target function are added to L. Since f is held fixed, the prior $P(f)$ never enters the analysis; exhaustive learning is not concerned with prior probabilities, as Bayesian analysis is. In fact, although exhaustive learning has sometimes been viewed as a variant of Bayesian analysis, the two approaches are completely distinct; see Wolpert[12] and Schwartz et al.[4] Note also that because X and Y are both finite, the set of all possible c values is also finite; the expectation value of interest in exhaustive learning must equal 0 for all but a finite number of c values.

It will be useful to rewrite:

$$\langle P(c|f, L)\rangle = P(c|f, m).\tag{1}$$

PROOF Writing it out, $\langle P(c|f, L)\rangle \equiv \sum_{L\subset f;m} P(c|f, L)P(L|f, m)$ where "; m" means of size m. (Note that the "$\subset f$" condition is actually superfluous; since there is no noise, it is automatically enforced by the probabilities in the summand. Nonetheless, it will be kept for clarity.). Now $P(L|f, m) = P(L, f, m)/P(f, m) = P(L, f)/P(f, m)$. Therefore

$$\langle P(c|f, L)\rangle = \sum_{L\subset f;m} \frac{P(c, f, L)}{P(f, m)} = \frac{P(c, f, m)}{P(f, m)} = P(c|f, m).$$

QED.

Although conventionally couched in terms of neural nets, volumes in weight space, etc., in point of fact the underlying concepts of exhaustive learning have nothing to do with neural nets. The exposition given above defines exhaustive learning exactly, and in a completely general manner. The precise scenario considered by the conventional analysis of exhaustive learning is simply a special case, where both h and f are explicitly parametrized neural nets. Of course, everything discussed in this chapter applies just as well to such a special case.

2. CALCULATION OF GENERALIZATION ERROR WHEN THE TESTING SET CAN INCLUDE THE TRAINING SET

This section shows how to replicate the conventional analysis of exhaustive learning[4] without assuming that one is using a neural net generalizer. This section also begins the exposition of how to go beyond the conventional analysis.[4]

The following discussion will make reference to the version of exhaustive learning presented by Hertz et al.[3] For those readers interested in translating between Hertz et al.[3] and the exposition presented here, note that in the Hertz book what are written as functions f are actually hypothesis functions, written here as h. Loosely speaking, what is written there as "$\mathbf{R}_0(f)$" is written here as $T(h)$, and in general what are treated as "volumes in weight space" by Hertz et al.[3] are here generalized to probabilities of hypothesis functions. Furthermore, Hertz et al.[3] give p as the size of the training set, which is here written as m. The fixed (though unknown to the experimenter) target function, referred to as $\overline{f}$ by Hertz et al.,[3] is here written as f. Finally, rather than the "generalization accuracy $g(.)$" used by Hertz et al.,[3] as was mentioned in the previous section we here use the "exhaustive learning error function" $Er(f, h)$, which is linearly related to $g(.)$.

Expanding in terms of the triples making up our event space,

$$P(c|f,m) = \sum_h \sum_{L \subset f;m} \frac{P(f,h,L)\delta(c,Er(f,h))}{P(f,m)}$$

(as above "$;m$" means "of size m"). This can be rewritten as

$$P(c|f,m) = \sum_h P(h|f,m)\delta(c,Er(f,h)). \tag{2}$$

The analogous formula given by Schwartz et al.'s analysis is

$$\rho_p(g) = \sum_f P_p(f)\delta(g - g(f))$$

from Eq. (6.67) of Hertz et al.[3] (N.b. this chapter uses probabilities, implicitly viewing errors as taking on only one of a finite set of possible values, whereas Schwartz et al. use probability densities; i.e., implicitly view errors as taking on continuous values, and therefore require Dirac delta functions.) So in particular, the term "$P_p(f)$" from Hertz et al.[3] means $P(h|f,m)$ in the notation of this chapter.

---

[4]After this chapter was written we received a technical report by Seung et al., which also extends the analysis by Schwartz et al.[4] In particular, they restricted themselves to the case where input-output functions are explicitly parametrized (as neural nets). Then they analyzed certain aspects of an approximation (the "annealing approximation") intimately related to the self-averaging assumption made by Schwartz et al.[4]

To proceed further we need to evaluate $P(h|f,m)$. Doing this will take us back to Eqs. (6.64) through (6.66) given by Hertz et al.[3] First, note the following lemma, which holds because training sets are ordered with repeats allowed:

$$\sum_{L_X;m} \Pi_{i=1}^m g_i(L_X(i)) = \Pi_{i=1}^m \sum_{x \in X} g_i(x), \text{ for any function } g(.). \tag{3}$$

PROOF We can prove the lemma using induction over m. To base the induction, note that the formula obviously holds when $m = 1$:

$$\sum_{L_X;1} \Pi_{i=1}^1 g_i(L_X(i)) = \sum_{L_X(1) \in X} g_1(L_X(1)) = \sum_{x \in X} g_1(x).$$

Now we assume that the formula holds for $m = c$, and check to see that it holds for $m = c + 1$. To do this, write

$$\sum_{L_X;c+1} \Pi_{i=1}^{c+1} g_i(L_X(i)) = \sum_{x \in X} \sum_{L_X;c} [g_{c+1}(x) \times \Pi_{i=1}^c g_i(L_X(i))]$$

$$= \sum_{x \in X} g_{c+1}(x) \sum_{L_X;c} \Pi_{i=1}^c g_i(L_X(i)),$$

which equals

$$\sum_{x \in X} \left\{ g_{c+1}(x) \times \Pi_{i=1}^c \sum_{x' \in X} g_i(x') \right\},$$

by the inductive hypothesis. This last expression can be rewritten as

$$\left[\sum_{x \in X} g_{c+1}(x) \right] \times \Pi_{i=1}^c \left[\sum_{x' \in X} g_i(x') \right] = \Pi_{i=1}^{c+1} \sum_{x \in X} g_i(x),$$

which completes the proof. QED.

Equation (3) allows us to evaluate $P(h|f,m)$:

$$P(h|f,m) = \sum_{L \subset f;m} P(h|L)\Pi_{i=1}^m \pi(L_X(i)). \tag{4}$$

PROOF

$$P(h|f,m) = \frac{\sum_{L \subset f;m} P(f,h,L)}{\sum_{h,L' \subset f;m} P(f,h,L')}$$

$$= \frac{\left\{ \sum_{L \subset f;m} P(h|L)P(f,L) \right\}}{\sum_{L' \subset f;m} P(f,L')}$$

$$(\text{since } P(h|f,L) = P(h|L))$$

$$= \frac{\left\{ \sum_{L \subset f;m} P(h|L)P(L|f) \right\}}{\sum_{L' \subset f;m} P(L'|f)}$$

$$= \frac{\left\{ \sum_{L \subset f;m} P(h|L)\Pi_{i=1}^{m}[\pi(L_X(i))] \right\}}{\sum_{L' \subset f;m} \Pi_{i=1}^{m}[\pi(L'_X(i))]}$$

(due to the exhaustive learning likelihood).

Now rewrite

$$\sum_{L \subset f;m} \Pi_{i=1}^{m} \pi[L_X(i)] = \sum_{L_X;m} \Pi_{i=1}^{m} \pi[L_X(i)]$$

$$= [\sum_{x \in X} \pi(x)]^m = [1]^m = 1.$$

(In writing down the second equality, use has been made of the fact that $\sum_{L_X;m} \Pi_{i=1}^{m} g(L_X(i)) = [\sum_{x \in X} g(x)]^m$ for any function $g(.)$, a fact which is a special case of the more general formula (3).) QED.

This equality shows how $P(h|f,m)$, which is called $P_p(f)$ by Hertz et al.,[3] can be an expectation value over training sets, as it is presented in Eq. (6.66) by Hertz et al.[3]: simply rewrite $\sum_{L \subset f;m} P(h|L)\Pi_{i=1}^{m} \pi(L_X(i))$ as $\langle P(h|L) \rangle$ (up to an overall proportionality constant), the expectation value being over all possible training sets of size m chosen from the target function.

To complete the evaluation of $P(h|f,m)$, we must plug the exhaustive learning $P(h|L)$ into Eq. (4). After doing this we will plug the resultant expression for $P(h|f,m)$ into Eq. (2).

$$P(h|f,m) = T(h) \sum_{L_X;m} k(L_X, f(L_X)) \; \Pi_{i=1}^{m}\delta(h(L_X(i)), f(L_X(i))) \; \pi(L_X(i)). \quad (5)$$

PROOF

$$P(h|f,m) = \sum_{L \subset f;m} P(h|L)\Pi_{i=1}^{m}[\pi(L_X(i))]$$

$$= \sum_{L_X;m} P(h|L_X, L_Y = f(L_X))\Pi_{i=1}^{m}[\pi(L_X(i))]$$

$$= \sum_{L_X;m} k(L_X, f(L_X))T(h)\delta(S(f,h,L_X),m')\Pi_{i=1}^{m}[\pi(L_X(i))]$$

(due to our assumption concerning $P(h|L)$; recall that m' is the number of distinct elements in L, and $[k(L_X, f(L_X))]$ is the normalization constant for $P(h|L)$.)

$$= \sum_{L_X;m} k(L_X, f(L_X))T(h)\Pi_{i=1}^{m}\delta[h(L_X(i)), f(L_X(i))]\Pi_{i=1}^{m}\pi(L_X(i))$$

$$= \sum_{L_X;m} k(L_X, f(L_X))T(h)\Pi_{i=1}^{m}\delta[h(L_X(i)), f(L_X(i))]\pi(L_X(i)).$$

QED.

Equation (5) exhibits "the contraction...(emphasizing) regions of configuration space with intrinsically high generalization ability" (Schwartz et al.,[4] page 379) of exhaustive learning. As m grows, the delta functions in Eq. (5) force h to agree with f for an increasing number of inputs if that h is to have nonzero probability.

It is after deriving Eq. (5) that the conventional analysis of exhaustive learning makes an approximation; it assumes that the $k(L_X, f(L_X))$ term in Eq. (5) can be absorbed into an overall m-dependent proportionality constant, with negligible effect on $P(c|f,m)$. In other words, it assumes that $k(L_X, f(L_X))$ is independent of L_X for all L_X that contribute substantially to $P(h|f,m)$, for all h that contribute substantially to $P(c|f,m)$ (see Eq. (2)). Schwartz et al. call this approximation "self-averaging" and claim that one would expect it to hold whenever $n \gg m$. Self-averaging is similar to the "annealed approximation"[5] which, loosely speaking, consists of replacing $k(.,.)$ with the L_X-average of $k(.,.)$.

Given self-averaging and Eq. (3),

$$P(h|f,m) \propto T(h)\left[\sum_{x \in X} \delta(h(x), f(x))\pi(x)\right]^m = T(h)[1 - Er(f,h)]^m.$$

Substituting this result into Eq. (2), collecting all constants into an overall normalization constant, and defining $\rho_0(c) \equiv \sum_h T(h)\delta(c, Er(f,h))$ (in loose agreement with the terminology used by Hertz et al.[3]), one derives the central result of the conventional analysis of exhaustive learning: $P(c|f,m) \propto [1 - c]^m \rho_0(c)$ (Eq. (6.68) given by Hertz et al.[3]). In other words, $P(c|f,m) = [1 - c]^m \rho_0(c)/$

$\{\sum_{E'}[1 - E']^m \rho_0(E')\}$. (Note that for every value of $E' \in [0,1]$, the summand of the expression $\sum_{E'}[1 - E']^m \rho_0(E')$ decreases as m is raised and, therefore, the entire sum is a decreasing function of m.)

Note that $\rho_0(c)$ will equal 0 for all but a finite number of c values. This is why the normalization constant giving $P(c|f,m)$ is a sum rather than an integral. Note also that in general the set of allowed c values is not uniformly distributed across $[0,1]$. For example, there will always be an upper bound on c which is less than 1 so long as no $\pi(x)$ value equals 0. (This point is returned to in Section 3.) As another example, when $r = 2$ and $\pi(x) = 1/n$ for all x, there are n h's which go into the sum for $\rho_0(c = (n-1)/n)$ (i.e., there are n h's having this error value). However, there are $C^n_{n/2}$ h's that go into the sum for $\rho_0(c = 1/2)$. Unless $T(h)$ is highly skewed, this phenomenon will induce a large hump in $\rho_0(c)$ favoring c values near $1/2$.

One can calculate a number of interesting quantities from this "central result." For example, one could calculate the average error with f and m fixed, $\sum_c cP(c|f,m)$. (Generically, this goes like $1/m$ or like $e^{-\alpha m}$; see Hertz et al.[3]) Or, if the prior $P(f|m)$ is specified, one could calculate the probability of error c when f is unspecified (which calculation allows one to find the average error when f is unspecified): $P(c|m) = \sum_{f \in F} P(c|f,m)P(f|m))$.

The central resul of the conventional analysis of exhaustive learning is taken to imply that as m increases, lower c's become more likely. (Though note that if m increases too much, then it's no longer true that $n \gg m$, and one should not expect the central result to apply.) In other words, it is taken as an argument that "inductive inference works," if one uses the exhaustive learning generalizer. Or as put by Schwartz et al.,[4] there is a "monotonic increase of the average generalization ability with increasing m."

One can proceed further than this conclusion by using the EBF. In particular, using the EBF one can directly test whether or not self-averaging holds (in the limit of $n \gg m$ or otherwise) for various $T(h)$ and $\pi(x)$. For example, one can prove the following:

Let ϵ and σ be fixed, where σ is an upper bound on the ratio

$$\frac{\max_x \pi(x)}{\min_x \pi(x)},$$

and $\epsilon < 1$ is an upper bound on the c we are considering. Then one special case where self-averaging does hold in the limit of $n \gg m$ is when $T(h)$ is independent of h.

$$(6)$$

The proof of Eq. (6) is in Appendix B. Note that as n increases, the space X changes and, therefore, $\pi(x)$ becomes a distribution over a different space. The σ

requirement in Eq. (6) limits how much $\pi(x)$ can change in this process, as n gets large. The requirement on ϵ serves a similar function, for c.

It is important to realize that in Eq. (6) (and everywhere else in this chapter) "constant $T(h)$" need not mean $T(h_1) = T(h_2)$ for all h_1, h_2. Rather it suffices to have $T(h_1) = T(h_2)$ for all h_1, h_2 such that both h_1 and h_2 are consistent with at least one L chosen from f. In other words, $T(h)$ can have nonuniformities and Eq. (6) still holds. Note also that the "constant $T(h)$" case referred to in Eq. (6) is *not* the same as the constant $T(\mathbf{w})$ case, where functions h are parametrized as perceptrons with weight vector $\mathbf{w}$, or some such.

This constant $T(h)$ case can be viewed as the "MDL," or "MaxEnt," most parsimonious exhaustive learning generalizer. However, it is not necessary to use the central result of the conventional analysis of exhaustive learning for this $T(h)$; by using the EBF, we can solve for $P(h|f, m)$ (and therefore for $P(c|f, m)$) for this case *exactly*, regardless of the relative sizes of n and m. Interestingly, the result explicitly depends on r, the size of the output space.

It is simplest to state the result when $\pi(x)$ is uniform. The proof is in Appendix C:

When $T(h)$ is independent of h and $\pi(x)$ is uniform, for those c values that can occur

$$P(c|f, m) \propto \left\{ \sum_{i=1}^{\min(m, n(1-c))} [r^{(i-n)} \times \varphi(m, i) \times C_i^{n(1-c)}] \right\} \tag{7}$$
$$\times C_{n(1-c)}^m \times (r-1)^{nc},$$

where $\varphi(.,.)$ is a combinatoric function defined in Appendix C.

In addition to allowing exact calculations when n is not $\gg m$, the EBF also allows the straightforward evaluation of $P(h|f, m)$ (and therefore of $P(c|f, m)$) in a number of situations where $k(L_X, f(L_X))$ is not independent of L_X, so that self-averaging does not apply. In Appendix D it is shown that what is perhaps the simplest such case explicitly results in a different result than the self-averaging-based "central result" of the conventional analysis of exhaustive learning. Therefore, that central result is not generic. This means that even if one accepts that the behavior of $P(c|f, m)$ is a direct reflection of "whether or not inductive inference works," the conventional analysis of exhaustive learning does not prove that inductive inference works with the exhaustive learning generalizer.

One peculiar aspect of exhaustive learning is that the probability it is interested in is $P(c|f, m)$ rather than $P(c|f, L)$. What is odd about this is that usually what goes on the right-hand side of a conditional probability should be the set of *everything* that is fixed for the problem. This set is a superset of the set of all that is known for the problem. So by evaluating $P(c|f, m)$ rather than $P(c|f, L)$, the conventional analysis of exhaustive learning is making the odd assumption that we know the size of the training set but not the actual elements making it up. Another drawback to concentrating on $P(c|f, m)$ rather than $P(c|f, L)$ is that it is always possible to calculate $P(c|f, m)$ from $P(c|f, L)$, in theory at least (just use Eq. (1)), but the reverse does not hold.

By using the EBF however, we can evaluate $P(c|f, L)$. For convenience, assume that $T(h)$ is uniform across h, and that $\pi(x)$ is also uniform and equals $1/n$, n being the cardinality of X. (As was mentioned previously, in such a case Eq. (6.68) given by Hertz et al.[3] holds exactly when $n \gg m$.) Under these conditions, assuming that the error c can occur,

$$P(c|f, L) = \frac{C_z^{(n-m')}(r-1)^{(n-m'-z)}}{r^{(n-m')}}, \tag{8}$$

where $z \equiv [n - m' - nc]$, and is the number of agreements between f and any hypothesis function having error c for questions outside of the training set.

PROOF

$$P(c|f, L) = \sum_h P(h|f, L)\delta(Er(f, h), c)$$

$$= \sum_h P(h|L)\delta(Er(f, h), c)$$

$$= \frac{\sum_h T(h)\delta(Er(f, h), c)\delta(S(L, h, L_X), m')}{\sum_h T(h)\delta(S(L, h, L_X), m')}$$

$$= \frac{\sum_{h \supset L} T(h)\delta(Er(f, h), c)}{\sum_{h \supset L} T(h)}.$$

Under the condition $h \supset L$, the uniformity of $\pi(x)$ allows us to rewrite the error function $Er(f, h)$ as $[n - m' - S(f, h, X - L_X)]/n$. We can then use the uniformity of $T(h)$ to cancel it out from the numerator and the denominator:

$$P(c|f, L) = \frac{\sum_{h \supset L} \delta(S(f, h, X - L_X), n(1 - c) - m')}{\sum_{h \supset L} 1}.$$

Note that the value of $P(c|f, L)$ depends only on m', the number of distinct elements in the training set, and not on the particular training set chosen. In other words, for this simple scenario $P(c|f, L)$ is independent of L, depending only on the number of distinct elements in L. So this expression we are calculating gives the probability of error c, no matter what X values are chosen to sample f. As such, it is extremely similar to $P(c|f, m)$. However, the expression derived here is simply the number of ways to pick a Y-valued function of a variable (that variable taking $(n - m')$ possible values) such that the function equals some pre-fixed values on some $n - m' - nc$ of the values of the variable and nowhere else, divided by the number of ways to pick a Y-valued function of a variable, that second variable taking on $(n - m')$ possible values. Carrying through the combinatorics gives the result in Eq. (8). QED.

Note that z must be a positive integer for Eq. (8) to be meaningful. This is a direct reflection of the fact that only certain c values can occur.

Equation 8 tells us that the form of $P(c|f, L)$ can be vastly different from the form of $P(c|f, m)$, even when (as here) $P(c|f, L)$, like $P(c|f, m)$, does not depend on the m actual pairs making up L. Part of this difference between the two conditional probabilities is a reflection of the fact that $P(c|f, m)$ is a function of m, whereas $P(c|f, L)$ is independent of m and depends only on m'.

This distinction between $P(c|f, L)$ and $P(c|f, m)$ notwithstanding, we would expect the two expressions to exhibit the same kind of m dependence in the limit of $n \gg m$ where we can take $m = m'$ (see the proof of Eq. (6)). This is indeed the case:

$$\text{When } n \gg m \text{ and both } T(h) \text{ and } \pi(x) \text{ are constant,} \qquad (9)$$
$$P(c|f, L) \propto (1 - c)^m \times \rho_0(c).$$

PROOF First, rewrite Eq. (8) as $P(c|f, L) = C_{n-m'-nc}^{n-m'} \times (r - 1)^{nc}/r^{n-m'}$. This means that the ratio $P(c|f, L')/P(c|f, L)$, where L is a training set of m' distinct elements and L' is a training set of $m' + 1$ distinct elements, equals $[r \times C_{z'}^{(n-m'-1)}]/C_z^{(n-m')}$, z' being the z value for the training set L', which has $m'+1$ distinct elements; $z' = z-1$. (Note that it is assumed that for both L and L' it is possible to have error c; i.e., it is being assumed that both $c < (n-m')/n$ and $c < (n-m'-1)/n$. If this assumption is not made, then z and/or z' might be negative.) The ratio can be rewritten as $r \times [1 - nc/(n - m')]$. If we now assume that $n \gg m \geq m'$, we can approximate this ratio as $r(1 - c)$. So in this approximation, $P(c|f, L) = r^{m'} \times [1-c]^{m'} \times \omega(c)$, for some function $\omega(.)$. Now for $n \gg m$, the probability that L has a repeat in it is vanishingly small. Therefore we can replace m' with m; $P(c|f, L) \propto [1 - c]^m \times \omega(c)$. Now using Eq. (1) and the fact that $\pi(x)$ is being assumed constant, for $m = m' = 1$ $P(c|f, m)$ must equal $P(c|f, L)$ for any L of size m. Therefore $\omega(c) = \rho_0(c)$. QED.

The interesting implication is that the increase in generalization accuracy with m embodied in Eq. (6.68) given by Hertz et al.[3] arises from the same causes behind the increase in generalization accuracy in Eq. (8): To get an error value c, one must have f and h disagree for a certain number of the off-training-set X values (they will always agree on all of L_X). The probability of that error value c, loosely speaking, is determined by how many distinct h there are that disagree with f for that "certain number of the off-training-set X values." Increasing the number of distinct elements in the training set decreases the number of X values off of the training set. This, in turn, changes the counting; this change in the combinatorics is the sole cause of the increase in generalization accuracy accompanying an increase in the size of the training set. Nothing more mysterious is going on. The increase in Eq. (8)—and by implication the increase in Eq. (6.68) as well—occurs simply from counting how many hypothesis functions exist that agree with a (fixed but unknown) target function a certain number of times for questions off of the training set. This point is emphasized by the analysis in the following section.

3. CALCULATION OF GENERALIZATION ERROR WHEN THE TESTING SET IS DISTINCT FROM THE TRAINING SET

In addition to its insistence on looking at probabilities conditioned on m rather than on L, there are a number of other peculiar aspects of exhaustive learning. In particular, the error function it uses, although it is conventional in much supervised learning research, has a number of major disadvantages. These arise from the fact that the exhaustive learning error function is based on looking at *all* X values for errors, including those in the training set. In other words, the error function gives a generalizer credit simply for reproducing a training set. Use of this error function forces one to calculate quantities which are completely oblivious of the natural distinction between error on the training set and error off of it. Especially when, as in exhaustive learning, a generalizer is explicitly constructed so that it will never have any errors on the training set, it is odd to fail to distinguish such (never occurring) on-training-set errors from (occurring quite often) errors off of the training set. After all, when comparing noise-free generalizers, it is *only* the off-training-set behavior that can distinguish the generalizers.

There is another, more concrete problem with the exhaustive learning error function, which turns out to be related to the on/off training set dichotomy. This problem is the fact that because the exhaustive learning generalizer always perfectly reproduces the training set, as m' is increased the upper bound on the error function shrinks. The immediate question is how much of Eq. (6.68) simply reflects this fact that exhaustive learning uses an error function whose upper bound shrinks with m.

To address this issue, we must replace the old error function $Er(f, h)$ with the normalized error function, $Er(h, f, L) \equiv \sum_{x \in X} \pi(x)[1 - \delta(f(x), h(x))] / \sum_{x \notin L_X} \pi(x)$. Given the noise-free nature of $P(h|L)$ and $P(L|f)$, this error function will always have the same lower and upper bound, regardless of m'. Moreover, it directly measures off-training-set error; since there is no noise in exhaustive learning, the normalized error function can be rewritten as the off-training-set error function, $\sum_{x \notin L_X} \pi(x)[1 - (\delta f(x), h(x))] / \sum_{x \notin L_X} \pi(x)$.

For the same reasons mentioned in the discussion preceding Eq. (8), we will calculate $P(c|f, L)$ rather than $P(c|f, m)$. For simplicity, again make the assumptions used to derive Eq. (9). We get the following:

When both $T(h)$ and $\pi(x)$ are constant and one uses the off-training-set error function:

i. For low c, $P(c|f, L) \propto [(r-1)^{-c}(1-c)]^{m'} \times w(c)$, for some function $w(.)$, and

ii. For high c, $P(c|f, L) \propto [(r-1)^{-c}c]^{m'} \times w'(c)$, for some function $w'(.)$. (10)

PROOF First continue along with the proof of Eq. (8). We get $P(c|f, L) = \sum_{h \supset L} \delta(Er(h, f, L), c) / \sum_{h \supset L} 1$, which holds independent of the error function used. Whereas in the proof of Eq. (8), we next replaced the error function with the expression $[n - m' - S(f, h, X - L_X)]/n$, our new error function must instead be replaced by $[n - m' - S(f, h, X - L_X)]/[n - m']$. Plugging this in, exactly as in the proof of Eq. (8), gives $P(c|f, L) = \sum_{h \supset L} \delta(S(f, h, X - L_X), (1-c)(n-m'))/\sum_{h \supset L} 1$. Carrying through the combinatorics gives the same result as in Eq. (8), $P(c|f, L) = C_z^{(n-m')}(r-1)^{(n-m'-z)}/r^{(n-m')}$, except that here $z \equiv (1-c)(n-m')$ rather than $(1-c)n - m'$. Now we must follow along with the proof of Eq. (9). First, rewrite our formula for $P(c|f, L)$ as $C_{(1-c)(n-m')}^{n-m'} \times (r-1)^{(n-m')c}/r^{n-m'}$. Now we want to calculate the ratio $P(c|f, L')/P(c|f, L)$, where L is a training set of m' distinct elements and L' is a training set of $m' + 1$ distinct elements. Our answer is

$$r(r-1)^{-c}(n-m')^{-1}\frac{(n-m'-cn+cm')!}{(n-m'-1-cn+cm'+c)!}\frac{(nc-m'c)!}{(nc-m'c-c)!}.$$

Now notice that since $c \in [0, 1]$, it is impossible for all of these factorials to be defined. This reflects the fact that since c is always normalized properly, as one changes m' and therefore the number of elements in $X - L_X$, one changes the possible values of c. To get around this problem, note that it is the shape of the curve connecting the allowed values of $P(c|f, L)$ (as a function of c) which interests us. To get this curve we must analytically extend the factorials. The most natural thing to extend them to is gamma functions, but the precise extension will not matter so long as we restrict our attention to the low c and the high c regimes. For low c, $(n-m'-1-cn+cm'+c) \cong (n-m'-cn+cm')-1$. This means our first ratio of factorials $\cong (n-m'-cn+cm') = (n-m')(1-c)$. Similarly, our second ratio of factorials $\cong 1$ for low c. Therefore, for low c, $P(c|f, L) = r^{m'}(r-1)^{-cm'}(1-c)^{m'}w(c)$ for some function $w(.)$. This holds regardless of the relative sizes of n and m. (As usual, for $n \gg m$, we could replace m' with m if we wanted to.) For large c, instead of setting c to 0 everywhere it appears by itself in a factorial, we can set it to 1. The result is $P(c|f, L) = r^{m'}(r-1)^{-cm'} c^{m'} \omega'(c)$ for some function $\omega'(.)$. QED.

In particular, for $r = 2$ and c large, $P(c|f, L) \propto c^m \omega'(c)$ (for $n \gg m$). This result should be contrasted with the $(1-c)^m$ behavior of Eq. (9) and Eq. (6.68) given by Hertz et al.[3]; if one goes to the off-training-set error function, then in certain regimes the $(1-c)^m$ generalization behavior of exhaustive learning not only disappears, it is actually reversed. Phrased differently, $(1-c)^m$ generalization behavior is not as significant as it appears. For a random (i.e., exhaustive learning) generalizer, this behavior is expected whenever the error measure counts as successful "generalization" the correct reproduction of the input-output pairs in the training set. Loosely speaking, the gain in "generalization accuracy" with increased m implied by the relation $P(c|f, m)(1-c)^m$ simply reflects the fact that as m increases, there are more points on which you're assured of guessing the output correctly (since you're assured of reproducing the training set correctly). In other words, the apparently exponential increase of Eq. (6.68) given by Hertz et al.[3] is a reflection of what naively might be thought to be a linear phenomenon.

This might be a somewhat surprising result. After all, in the limit of infinite n, one might expect that the probability that a randomly chosen X value $\in L_X$ is essentially 0. Accordingly, so this line of reasoning goes, it should not matter if one averages over all X values (as in the conventional exhaustive learning error function) or if one averages only over X those outside of L_X (as in the off-training-set error function). Equation 10 is a formal proof that this line of reasoning does not hold.

To help understand this result intuitively, first recall from Section 1 that it does not matter if X is originally defined to be (a subset of) $\mathbf{R}^K$, because X is always *effectively* discrete; in exhaustive learning X can be partitioned into a countable set of regions such that there is no h or f which changes value across any such region (see Appendix A). Accordingly, with any new training set pair there is always a gain in generalization accuracy simply due to the fact that we now know $f(x)$ for all x in the region containing that new pair. To exclude this artifact, in the definition of off-training-set error one implicitly redefines X so that each of its elements is a label of one of the regions in $\mathbf{R}^K$ rather than a point in $\mathbf{R}^K$. Once this is done, X is countable.

Now recall that for a countable X, if $P(f)$ were uniform over all f, then it would make no difference how one guesses as far as off-training-set error is concerned, since for such a $P(f)$ "the universe" is completely random off the training set (see Wolpert[9]). Similarly, it does not matter what f is if $T(h)$ is uniform over all h. Given all this, it would be quite surprising if somehow $(1-c)^m$ behavior *were* preserved when one shifts to an off-training-set error function.

Finally, it is worth noting that for the case considered in Eq. (10), the average off-training-set error $\sum_c cP(c|fm) = 1/r$, independent of m.[5] In certain respects, given that $T(h)$ is uniform, this is more "intuitively reasonable" than the generic e^{-m} or $1/m$ behavior associated with the error function of the conventional analysis of exhaustive learning (see Hertz et al.,[3] page 152). Other intuitive arguments related to this issue can be found in Wolpert.[12]

It is important to note that even though Eq. (10) was derived under very specific assumptions for $T(h)$ and $\pi(x)$, the general result—$(1-c)^m$ behavior goes away for an off-training-set error function—holds generically. Broadly applicable results are not in hand, but some salient arguments can be made which should illustrate the point. In particular, note from the proof of Eq. (10) that for uniform $\pi(x)$, $P(c|f,L) = \sum_{h \supset L} T(h)\delta(Er(f,h),c)/\sum_{h \supset L} T(h)$. Allowing $T(h)$ to be nonuniform means the results of Eq. (10) might not obtain. However, allowing nonuniform $T(h)$ also means that we can have essentially arbitrary $P(c|fL)$. In particular, if $T(h)$ is infinitesimal for all h except for one special h, h', then by appropriate choice of h' we can make $P(c|f,L)$ essentially a delta function about any of the allowed values of c. In such a case off-training-set error will not change with m.

Equation 10 points up two dangers inherent in many supervised learning formalisms. First, if one uses an error function which roams over the training set as well as off of it (i.e., if one confuses generalization with learning), one can have the illusion of impressive generalization when, in fact, nothing useful has been achieved. Second, especially when using those kinds of error functions, possessing "exponential" generalization behavior might have little (if any) significance.

4. DISCUSSION

The idea behind calculating $P(c|f,m)$ is to fix the target function, many times choose a random training set of size m, and measure the resultant distribution over c. Do this for several m. Eq. (6.68) given by Hertz et al.[3] says how this distribution varies with m. In practice though, often we are not going to many times take m

---

[5]From the proof of Eq. (10), we see that the average number of disagreements between f and h off of the training set equals $\sum_{z=0}^{n-m'} [C_z^{n-m'} z(r-1)^{n-m'-z}/r^{n-m'}]$. Now $(x+y)^{n-m'} = \sum_{z=0}^{n-m'} [C_z^{n-m'} x^z y^{n-m'-z}]$. Therefore

$$x \times \partial_x[(x+y)^{n-m'}] = \sum_{z=0}^{n-m'} [C_z^{n-m'} z x^z y^{n-m'-z}].$$

If we plug in $(r-1)$ for y and 1 for x, we see that the average number of disagreements equals $(n-m')/r$. Therefore the average error equals $1/r$.

samples of the target function, collect statistics, and then many times do the same thing for *completely novel* training sets of size $m+1$. Rather, we are going to take m samples of the target function, and then add another sample point (to the training set already in hand) to increase the size of the training set to $m+1$. In practice, we usually want to know something about how the probability distribution over c will change when we add this next sample point, "on average." As it turns out, the answer to this question is given precisely by $P(c|f,m)$. More formally,

> Assume that for all L, $P(L|f,m)$
> $= P($first m pairs in the training set
> $= L|f$, the training set has size $m+1)$.
> Then the average value of the change in the probability of   (11)
> error c which accompanies the addition of an input/output
> pair to a training set of size m is given by
> $P(c|f,m+1) - P(c|f,m)$.

PROOF Let λ indicate a generic training set, and let $\lambda = \{L\}^+$ mean that we have a training set, size $m+1$, whose first m pairs are given by L (nothing else is known about the training set). The quantity we want to calculate is $\sum_{L;m} P(L|f,m)[P(c|f,\lambda = \{L\}^+) - P(c|f,L)]$. Expanding $P(c|f,\lambda = \{L\}^+)$, we get

$$\frac{\sum_{L'=\{L\}^+} P(c,f,L')}{\sum_{L'=\{L\}^+} P(f,L')} = \frac{\sum_{L'=\{L\}^+} P(c|f,L')P(L',f)}{\sum_{L'=\{L\}^+} P(L',f)}.$$

We can rewrite this as

$$\frac{\sum_{L'=\{L\}^+} P(c|f,L')P(L',f, \text{ the size of } L' = m+1)}{\sum_{L'=\{L\}^+} P(L',f, \text{ the size of } L' = m+1)}.$$

Dividing top and bottom by $P(f, \text{ the size of } L' = m+1)$, we get

$$\frac{\sum_{L'=\{L\}^+} P(c|f,L')P(L'|f, \text{ the size of } L' = m+1)}{\sum_{L'=\{L\}^+} P(L'|f, \text{ the size of } L' = m+1)}.$$

The denominator can be rewritten as $P(\omega = \{L\}^+|f, \text{ the size of } \lambda = m+1)$. Under our assumption concerning probabilities of training sets, this equals $P(L|f,m)$. Therefore the quantity we want to calculate is

$$\sum_{L;m} \sum_{L=\{L\}^+} P(c|f,L')P(L'|f, \text{ the size of } L' = m+1) - \sum_{L;m} P(c|f,L)P(L|f,m)$$

Because training sets are ordered, we can rewrite the double sum as $\sum_{L';m+1}$, giving us $\sum_{L';m+1} P(c|f, L')P(L'|f, m+1) - \sum_{L;m} P(c|f, L)P(L|f, m)$. Using Eq. (1), we see that this just equals $P(c|f, m+1) - P(c|f, m)$. QED.

The assumption in Eq. (11) can be violated whenever training sets are generated "en masse" according to schemes that vary depending on the total number of elements in the training set. It seems that for many reasonable schemes for creating training sets, however, the assumption in Eq. (11) holds and, therefore, it suffices to calculate $P(c|f, m)$. For example, assume that $P(L_Y|f, L_X, m) = \Pi_{i=1}^m t[f(L_X(i)), L_Y(i)]$ for some function $t(., .)$ obeying $\sum_y t(z, y) = 1 \forall z \in Y$ (as when we have Gaussian noise, for example). Assume further that $P(L_X|f, m) = \Pi_{i=1}^m \pi(L_X(i))$ for some function $\pi(.)$ obeying $\sum_x \pi(x) = 1$ (as in iid sampling). Now it is always true that $P(L|f, m) = P(L_Y|f, L_X, m)P(L_X|f, m)$. Moreover, the probability P(first m pairs in the training set $= L|f$, the training set has size $m+1$) just equals $\sum_{x,y} P(\text{training set} = \{L, (x, y)\}|f, m+1)$. Using our assumptions, this, in turn, equals

$$\sum_{x,y} t(f(x), y)\pi(x)\Pi_{i=1}^m t[f(L_X(i)), L_Y(i)]\pi(L_X(i))$$

$$= \Pi_{i=1}^m t[f(L_X(i)), L_Y(i)]\pi(L_X(i)).$$

This last term equals $P(L|f, m)$ however, corroborating the assumption made in Eq. (11).

As was mentioned in connection with Eq. (8), in general all of one's knowledge about the supervised learning problem at hand should go on the right-hand side of the conditional probability of interest. Therefore, in a certain sense $P(c|h, L)$ is what we are really interested in; in any single supervised learning experiment, it is usually the case that our provided information is the training set we used along with the hypothesis function we fit to that training set. In this regard it is interesting to note that the expression on the right-hand side of Eq. (8) is identical to the expression worked out by Wolpert[9] for $P(c|h, L)$ under the assumption of a "maximum-entropy universe." [6]

This should not be too surprising because for many scenarios there is a formal symmetry under interchange of f and h (see Wolpert[9]). In particular, because the error function is symmetric under such an interchange, this is true in exhaustive learning if $T(h) = P(h) = P(f)|_{f=h}$. Accordingly, under these conditions, Eq. (6.68)

---

[6]There is one small difference. In the expression given by Wolpert[9] for $P(c|h, L)$, $z = (n - m) \times (1 - c)$ rather than $n(1 - c) - m$. This difference follows from the fact that the error function used by Wolpert[9] differs by a factor of $n/(n - m)$ from the one used in Eq. (8); Wolpert[9] uses the off-training-set error function introduced in this chapter in Section 3.

of Hertz et al.[3] holds with hypothesis and target functions reversed, i.e., with the appropriate self-averaging approximation,

$$P(c|h,m) = \frac{[1-c]^m \times \rho_0(c)}{\sum_{c'}[1-c']^m \times \rho_0(c')} \tag{12}$$

(where $\rho_0(c)$ is now defined as $\sum_f k\ P(f)\ \delta(c, Er(f,h))$ rather than as $\sum_h k\ T(h)\ \delta(c, Er(f,h))$).

PROOF It is always true that both $P(f|h,L) = P(f|L)$ as well as $P(h|f,L) = P(h|L)$. In addition, the error function in exhaustive learning is symmetric under $f \leftrightarrow h$. Therefore, since $P(h|L)$ and $P(L|f)$ are the only remaining quantities which are specified in exhaustive learning, we only have to prove that both the pair $\{P(f|L)$ and $P(h|L)\}$ and the pair $\{P(L|f)$ and $P(L|h)\}$ are interchangeable in exhaustive learning. Once we have done this, there will be no formal distinction in the math between f and h. Wolpert[10] proves that whenever $P(L_X|f)$ is independent of f and there is no noise in the sampling (both of which conditions hold in exhaustive learning), $P(f|L) \propto P(f)\delta(L \subset f)$, where the proportionality constant depends on L as usual. Given the definition of the Gibbs learning algorithm and our assumption for $T(h)$, this establishes the calculational symmetry under $f \leftrightarrow h$ of the pair $\{P(f|L)$ and $P(h|L)\}$. It is also proven by Wolpert[10] that if $P(h|L) \propto P(h)\delta(L \subset h)$ (as in exhaustive learning, if $T(h) = P(h)$), then $P(L_X|h) = P(L)k(L)$, where $k(.)$ is the usual normalizing constant $[\sum_h T(h)\delta(L \subset h)]^{-1} = [\sum_h P(h)\delta(L \subset h)]^{-1}$ and $L_y = h(L_x)$. Now $P(L) = \sum_f P(L|f)P(f)$, and $P(L|f) = P(L_Y|L_X, f)P(L_X|f)$. Given our likelihood, this means that $P(L) = \sum_f \Pi_{i=1}^m \pi(L_X(i))\delta(L \subset f)P(f)$, which in turn equals $\Pi_{i=1}^m \pi(L_X(i))\sum_f P(f)\delta(L \subset f)$. But by assumption $P(h) = P(f)|_{f=h}$, which means that $\sum_h P(h)\delta(L \subset h) = \sum_f P(f)\delta(L \subset f)$. Therefore $P(L_X|h) = \Pi_{i=1}^m \pi(L_X(i))$, and $P(L|h)$ is the same function of h as $P(L|f)$ is of f. QED.

In Eq. (12), instead of one target function and many possible hypothesis functions, we have a single hypothesis function and many possible target functions. The distinction from the conventional exhaustive learning scenario can be understood as follows.

In exhaustive learning, we conduct many learning experiments, with target functions chosen according to the probability distribution $P(f)$ and training sets L chosen from the target function according to $P(L|f)$. Look at all those cases where the training set has size m and the target function is some pre-fixed function f. Then Eqs. (2) and (5) give the probability that any particular one of those cases has error value c.

In the scenario considered in Eq. (12), we also conduct many learning experiments, with target functions chosen according to the probability distribution $P(f)$ and training sets L chosen from the target function according to $P(L|f)$. Again we look at all those cases where the training set has size m. However, rather than restricting our attention to those cases with a particular pre-fixed target function f, we restrict our attention to those cases where the hypothesis function chosen by the generalizer is some particular pre-fixed hypothesis function h. Then Eq. (12) gives the probability of error c. Intuitively, taken as a function of m, the result in Eq. (12) can be viewed as telling us something about the efficacy of sticking with the same hypothesis function h as the training set size changes.

APPENDIX A

WHY IN EXHAUSTIVE LEARNING X IS EFFECTIVELY COUNTABLE

Sometimes exhaustive learning is presented as though it could have an uncountably infinite input space, e.g., as though X is a subset of $\mathbf{R}^K$ for some K. In truth however, in exhaustive learning the input space is effectively discrete, and can be made explicitly discrete without any loss of generality. This follows from the fact that in exhaustive learning, even though the number of weight vectors $\mathbf{w}$ might be uncountable, the number of possible functions parametrized by those vectors—i.e., the number of possible h—is countable. In fact, if it were not countable, the usual mathematical exposition of exhaustive learning could not apply (c.f. Eq. (6.67) of Hertz et al.[3] and Eq. (2.10) of Schwartz et al.[4]).

To prove that X is effectively discrete, assume for simplicity that $K = 1$. Assume further, also for simplicity, that for any finite interval Σ in X, and any h for which $P(h|f) \neq 0$, the number of changes in the output value of $h(x)$ as x moves across Σ is finite. (Recall that Y is discrete. Note also that $P(h|f) = \sum_L P(h|L,f)P(L|f) = \sum_L P(h|L)P(L|f)$.) This property will certainly hold for a typical neural net h, and it will hold for most of the other functions usually used in supervised learning as well. Accordingly, although it is relatively straightforward to generalize to the case where the number of changes can be infinite, for the purposes of this chapter there is no need to add such a complication.

Consider the x interval $[0, 1)$. Define $ch(h)$ as the set of all $x \in [0, 1)$ for which either $\exists \delta > 0$ such that $\forall \varepsilon$ where $0 < \varepsilon < \delta, h(x - \varepsilon) \neq h(x)$ or $\exists \delta > 0$ such that $\forall \varepsilon$ where $0 < \varepsilon < \delta, h(x + \varepsilon \neq h(x)$; $ch(h)$ is the set of all x values at which $h(x)$ changes, in the interval $[0, 1)$. By hypothesis, $ch(h_i)$ is finite $\forall h_i$ with which we are concerned (i.e., for all h_i such that $P(h_i|f) \neq 0$). Therefore, $\{\cup_i ch(h_i)\}$, a countable union of countable sets, is itself countable. Call this union $Ch\{[0, 1)\}$. Since it is a countable union of countable sets, $\cup_j Ch\{[j, j + 1)\}$, where j runs over all integers, is a countable set.

Now by construction, the values of all of the h_i with which we are concerned are constant over each of the (countably many) regions demarcated by an adjacent pair of points from the set $\cup_j Ch\{[j, j+1)\}$. In other words, for any one of the countable number of regions in $\mathbf{R}$ lying in between adjacent demarcating points, $[a, b)$, it suffices to specify the values of the h_i at only a single point from within $[a, b)$. (See Hertz et al.,[3] Figure 6.14.)

Furthermore, by hypothesis the Gibbs generalizer always reproduces the training set. Since there are no *a priori* restrictions on $\pi(x)$, this means that there must be an h which reproduces any training set L containing input-output pairs i and $j \neq i$ such that both $L_X(i)$ and $L_X(j) \neq L_X(i)$ are contained in $[a, b)$. Since training sets are formed by noise-free sampling of f, the only way this can be assured is if f also does not change across $[a, b)$. Generalizing, we have shown that neither any of the h_i nor f can change over any of the (countably many) regions demarcated by an adjacent pair of points from the set $\cup_j Ch\{[j, j+1)\}$. In other words, for all function of interest, it suffices to specify the value of such a function at only a single point from within each demarcated region.

Accordingly, it suffices to treat X as discrete, by associating each separate value $x \in X$ with a separate one of the demarcated regions in $\mathbf{R}$. So, for example, the expression "$\pi(i \in X)$" implicitly means the integrated probability $\int dx\, \pi(x \in \mathbf{R})$, where the integral extends over the region associated with the index i. Similarly, to say that one X value does not "equal" another implicitly means that we are talking about two separate demarcated regions. So in particular, saying that an X value lies "outside of the training set" implicitly means that the region it is associated with has zero intersection with any of the regions containing the elements of the training set. In this chapter, it does *not* only mean that the exact, infinite precision X value differs from all of the exact, infinite precision X values in the training set.

More precisely, let X be our original input space, and let B be a set of n labels, delineating all the regions in $\cup_j Ch\{[j, j+1)\}$ that lie in between a pair of adjacent demarcating points. The idea is to show that we can replace the input space $X = \mathbf{R}$ with the input space $X = B$. To this end, define $B(x)$ as the region (i.e., B value) demarcated by those adjoining points in $\cup_j Ch\{[j, j+1)\}$ which contains the point $x \in \mathbf{R}$.

Let h be any hypothesis function such that $P(h|L)$ is not zero. By the argument given above, h is fully specified by giving its Y values for every point b in B. Accordingly, $T(h)$ is fully specified if its argument h is only specified to that accuracy. Similarly, $\delta(L \subset h)$ is fully specified if h is only specified to that accuracy and if every value $L_X(i)$ is replaced by $B(L_X(i))$. Accordingly, as far as $P(h|L)$ is concerned, one can replace $\mathbf{R}$ with B.

Similarly, in this chapter we will only deal with error functions like

$$\int dx\, \pi(x)[1 - \delta(f(x), h(x)] = \sum_{b \in B} \int_{(B^{-1}(b))} dx\, \pi(x)[1 - \delta(f(x), h(x)]$$

$$= \sum_{b \in B} [1 - \delta(f[B^{-1}(b)], h[B^{-1}(b)])] \int_{(B^{-1}(b))} dx\, \pi(x).$$

(The subscript on the integrals delineates a region of integration.) So here, too, we can replace $\mathbf{R}$ with B, assuming we also replace $\pi(x)$ with $\pi(B(x)) \equiv \int_{x \in B^{-1}(b)} dx\,\pi(x)$. (For the case of an off-training-set error function, the $\pi(x)$ in this expression is taken to be the same as the $\pi(x)$ occurring in the conventional error function, except that it is forced to zero for all x having the same b as an element of the training set, and is then renormalized.)

Finally, the other quantity specified in exhaustive learning is $P(L|f) = P(L_Y|L_X, f)P(L_X|f)$, where $P(L_Y|L_X, f) = \delta(L \subset f)$. Accordingly, since f is constant within the regions indexed by $b \in B$, $P(L_Y|L_X f)$ is fully specified if we only specify f by its value in each region indexed by $b \in B$ and if we also replace the $L_X(i)$ with $B(L_X(i))$. In other words, to give the value of $P(L_Y|L_X f)$, we only need to know f and L_X to the accuracy of B. (This is exactly the same justification as the one we used to replace $\mathbf{R}$ with B in $\delta(L \subset h)$.)

On the other hand, $P(L_X|f) = \Pi_{i=1}^{m} \pi(L_X(i))$, and this quantity does *not* allow replacing $X = \mathbf{R}$ with $X = B$. To know the value of $P(L_X|f)$, we need to know more about L_X than is given by $B(L_X)$.

This is not a problem however. To see this, first note that we can use Bayes' theorem to write

$$P(f|L) = \frac{P(L_Y|L_X, f)P(L_X|f)P(f)}{\sum_f P(L_Y|L_X, f)P(L_X|f)P(f)}.$$

Since $P(L_X|f)$ is independent of f (see the introduction), this equals

$$\frac{P(L_Y|L_X f)P(f)}{\sum_f P(L_Y|L_X, f)P(f)};$$

the $P(L_X|f)$ term has dropped out. So as far as $P(f|L)$ is concerned, we can replace $\mathbf{R}$ with B.

Note also that any quantity of the form

$$\int dL_X\, P(L_X|f)\ \mathrm{func}[B(L_X(1)), \ldots, B(L_X(m))]$$

can also be expressed as

$$\sum_{b_1, \ldots, b_m} \int_{B^{-1}(b_1)} dx_1 \ldots \int_{B^{-1}(b_m)} dx_m\ \Pi_{i=1}^{m} \pi(x_i)\ \mathrm{func}[B(x_1), \ldots, B(x_m)]$$

(the b_i in the sum each run over all possible B values). This, in turn, equals $\sum_{b_1, \ldots, b_m} \Pi_{i=1}^{m} \pi(b_i)\ \mathrm{func}[b_1, \ldots, b_m]$. This expression is of the form $\sum_{L_X} P(L_X|f)\ \mathrm{func}[L_X(1), \ldots, L_X(m)]$, if $X = \mathbf{R}$ is replaced with $X = B$ in the sum over training set inputs. So as far as quantities like $\int dL_X\, P(L_X|f)\ \mathrm{func}[B(L_X(1)), \ldots, B(L_X(m))]$ are concerned, we can replace $\mathbf{R}$ with B.

Now in this chapter, we will not be concerned with any distribution $P(c\,|\,\text{stuff})$ which depends on L_X in any way other than via $P(f|L)$ or via $\int dL_X P(L_X|f)$ func$[B(L_X(1)),\dots,B(L_X(m))]$. Accordingly, although $P(L_X|f)$ could in theory prevent us from replacing $X = \mathbf{R}$ with $X = B$, it does not do this for any of the calculations of interest. Since none of the other quantities specified by exhaustive learning (the error function and $P(h|L)$) can prevent the replacement of $\mathbf{R}$ with B under any circumstances, we are allowed to make the substitution.

In practice, "countable" $\cup_j Ch\{[j,j+1)\}$ will almost always be finite, and we can set the cardinality of X to $n < \infty$. The countably infinite case is recovered in the limit of $n \to \infty$.

APPENDIX B

PROOF OF SELF-AVERAGING WHEN BOTH $T(h)$ IS UNIFORM AND $\pi(x)$ IS BOUNDED, IN THE LIMIT OF $n \gg m$

The factor $k(L_X, f(L_X))$ is not independent of L_X, even when $T(h)$ is uniform. However, for such a $T(h)$, $k(L_X, f(L_X))$ only depends on m', the number of distinct elements in L_X. This implies that our task in proving self-averaging is to show, in effect, that we can ignore all terms in the sum giving $P(h|f,m)$ which have $m' \neq m$. We will be able to do this because the number of terms in the sum which have $m' = m$ vastly exceeds the number of terms which have $m' \neq m$.

Start by combining the equation in the middle of the proof of Eq. (5) with Eq. (2):

$$P(c|f,m)$$
$$= \sum_h \delta(c, Er(f,h))T(h) \sum_{L_X;m} k(L_X, f(L_X))\delta(S(f,h,L_X),m')\Pi_{i=1}^m[\pi(L_X(i))].$$

As an aside, note that if $c = 1$, the $\delta(c, Er(f,h))$ factor means that $Er(f,h)$ must equal 1, which in turn means that f and h must disagree everywhere (since $\pi(x)$ is nowhere zero—see Eq. (6)). This means that $\delta(S(f,h,L_X),m')$ equals 0 for all h in the sum. Consequently, $P(c = 1|f,m) = 0$, in agreement with the $(1-c)^m$ result of self-averaging.

First we must derive a lower bound on N, the number of X on which f and h agree. To do this we again use the fact that the $\delta(c, Er(f,h))$ term in the sum over h means we can set $Er(f,h)$ to c. Now $c < \varepsilon$ and ε is a number which is independent of n and < 1. Therefore, in the limit of $n \gg m$ (i.e., with m fixed, in the limit of $n \to \infty$), $nc \ll n \cong n - m$; $nc \ll n - m$. This means that

$$n \times \sum_x \pi(x)[1 - \delta(f(x), h(x))] \ll n - m;$$

$$\sum_x \pi(x)\delta(f(x), h(x)) \gg \frac{m}{n}.$$

Now $\sum_x \pi(x)\delta(f(x), h(x)) \leq \max_x \pi(x) \sum_x \delta(f(x), h(x))$. Therefore

$$\sum_x \delta(f(x), h(x)) \gg \frac{m}{n \max_x \pi(x)}.$$

However, by hypothesis, $\max_x \pi(x) / \min_x \pi(x) \leq \sigma$, which means that $\max_x \pi(x) \leq \sigma/(\sigma+n-1)$. Plugging this in, $\sum_x \delta(f(x), h(x)) \gg m/[n\sigma/(\sigma+n-1)]$. In our limit of $n \to \infty$, the right-hand side becomes m/σ; $N \gg m/\sigma$. Since σ is just a preset finite constant, in fact we have $N \gg m$.

Now the inner sum over L_X in the formula given above for $P(c|f, m)$ can be rewritten as $\sum_{L_X; m' < m} k(.,.)\Pi\delta(S(.,.,.), m') + \sum_{L_X; m'=m} k(.,.)\Pi\delta(S(.,.,.), m')$, with obvious notation. Using the bound on N we will prove that the first of these sums is insignificant compared to the second for all h of interest (i.e., for all h for which $Er(f, h) = c$), and can therefore be ignored. This will allow us to set $m' = m$, which we will use to prove self-averaging. (See Wolpert,[10] Appendix B, for an intuitive version of this argument.)

It is fairly straightforward to prove $m' = m$ using the uniformity of $T(h)$. To keep the argument as general as possible however, we will prove $m' = m$ without making any assumptions for $T(h)$. (Only after proving this will we exploit the uniformity of $T(h)$ to deduce self-averaging.)

The first step in the proof is to get rid of the Π factors; bound the ratio of our two sums by

$$\frac{\sum_{L_X; m' < m} k(.,.)\Pi\delta(S(.,.,.), m')}{\sum_{L_X; m'=m} k(.,.)\Pi\delta(S(.,.,.), m')} \leq \sigma^m \times \frac{\sum_{L_X; m' < m} k(.,.) \times \delta(S(.,.,.), m')}{\sum_{L_X; m'=m} k(.,.)\delta(S(.,.,.), m')}.$$

Since σ is some preset constant, if the ratio of sums on the right-hand side of our expression goes to 0 as $n \to \infty$, we can take $m' = m$, as desired. The fact that we are currently allowing nonuniform $T(h)$ affects things via the $k(.,.)$ factors occurring in the summands.

Let X' be the set of NX values on which f and h agree. Due to the δ functions in the summands, every nonzero term in both of our two sums over L_X equals the normalization constant k for $P(h|L = \{L_X, f(L_X)\})$, for an L_X all of whose elements come from X'. Due to the lower bound on N, we know that $N \gg$ the size of any of these L_X's.

To prove that the ratio of sums on the right-hand side of our expression goes to 0, we will construct a set of one-to-one mappings between (nonzero) terms in the denominator sum and (nonzero) terms in the numerator sum. (That is, between k of an ordered sets of m elements chosen from X', all of which are distinct, and k of an ordered sets of m elements from X', with $m' < m$ of the elements distinct.) Each such mapping will range over all of the terms in the numerator sum, but only a subset of the terms in the denominator sum. Indicate these mappings by Γ_i (a denominator term) = (a numerator term), where i indexes the mappings. Indicate the domain of Γ_i (which is a set of terms occurring in the denominator sum) by $D(i)$, and the range (the set of all terms in the numerator sum) by R. We will construct the mappings so that $D(i)$ has zero intersection with $D(j)$ for $i \neq j$.

It will turn out that for any i and any term $d \in D(i), d \geq \Gamma_i(d)$. (In other words, the $k(.,.)$ term associated with the $L_X d$ is larger than the $k(.,.)$ term associated with the $L_X \Gamma_i(d)$.) This means that $\sum_{d \in D(i)} d \geq \sum_{r \in R} r$. Since R is the set of all the terms in the numerator sum, we can rewrite this inequality as $\sum_{d \in D(i)} d \geq \sum_{L_X; m' < m} k(.,.) \delta(S(.,.,.), m')$. With κ the number of mappings (i.e., the number of Γ_i), $\sum_i \sum_{d \in D(i)} d \geq \kappa \sum_{L_X; m' < m} k(.,.) \delta(S(.,.,.), m')$. Since $D(i) \cap D(j \neq i) = \emptyset$, the double sum on the left is less than (or equal to) our denominator sum, $\sum_{L_X; m' = m} k(.,.) \delta(S(.,.,.), m')$. Therefore the denominator sum is more than κ times as big as the numerator sum. We will show that $\kappa \to \infty$ as $n \to \infty$, which will complete the proof that we can take $m' = m$.

There are five properties we need of the Γ_i:

1. they are one-to-one;
2. their range is the set of all terms in the numerator sum;
3. their domains have zero intersection;
4. the number of them goes to infinity as n does; and
5. for all $d \in D(i), d > \Gamma_i(d)$.

In what follows, we assume without loss of generality that X is the set of the first n integers, $\{1, 2, \ldots, n\}$, and X' the first N integers, $\{1, 2, \ldots, N\}$ (i.e., whenever referring to "an X value," we really mean the ordinal index of that X value, where the X values have been arranged so that the values on which f and h agree come first).

Rather than directly construct the Γ_i, we will construct their inverses, Γ_i^{-1}. We start with Γ_0^{-1}. Let Λ_X be a particular L_X chosen with $m' < m$. It consists of an ordered set of m X values (at least two of which are identical), each chosen from X'. In other words, it consists of an m-dimensional vector, each component of which can equal any one of the integers from 1 to N. There are a total of $m - m'$ of the m components of Λ_X each of which is identical to one of the remaining m' components. We map Λ_X to Λ'_X, an L_X with $m' = m$ (i.e., a vector all of whose components are distinct), via Γ_0^{-1}, as follows:

1. Let Ξ_0 be the set of m' distinct values found in the components of Λ_X. So components of the vector Λ_X are all contained in the set Ξ_0.

2. Let S be an m-dimensional vector, m' of whose components are identical to those from Λ_X. (The remaining $m - m'$ components of S equal zero—a value not found in X.) These m' components are chosen to be the first m' components of Λ_X all of which have distinct X values. In other words, the first component of S is identical to the first component of Λ_X; the second component of $S = 0$ if $\Lambda_X(2) = \Lambda_X(1)$; otherwise, it equals $\Lambda_X(2)$, etc.
   Note that every value in Ξ_0 is found as a nonzero component of S and vice versa, and that all these values are found in X'. However, since $N \gg m$, there are many values in X' not found in Ξ_0.

3. Let T be the vector $\Lambda_X - S$.
   Note that every nonzero component of T has a value already found in one of S's nonzero components. In other words, if $T_i \neq 0, T_i = S_j$ for some $j < i$. Moreover, if $T_i \neq 0, T_i = \Lambda_X(i)$, and $\Lambda_X(i) = \Lambda_X(j)$ for some $j < i$. Intuitively, T is a list of the duplicates in Λ_x.

4. Let t_1 be the first nonzero component of T, t_2 the next nonzero component of T, etc. To create Λ'_X, start with Λ_X, and replace the component t_1 with t'_1, the lowest X value contained in $\{X' - \Xi_0\}$. Then replace t_2 with t'_2, the lowest X value contained in $\{X' - \Xi_1\}$, where $\Xi_1 \equiv \{\Xi_0 \cup t'_1\}$. Repeat this procedure until all nonzero components of T but the very last one have been replaced in Λ_x.
   We have now fixed all but one of the components of Λ'_X and in so doing created $(m - m' - 1)\Xi_i$. Note that the values of those components of Λ'_X fixed so far are, by construction, distinct from one another. Let the unfixed component be component number J.

5. We will fix that last component of Λ'_X to a value chosen so that the sum of all the components of Λ'_X, $\sum_{i=1}^m \Lambda'_X(i)$, is a number which codes for the pattern of duplications of component values in Λ_X. Given this pattern and the values of the other components of Λ'_X, we can uniquely recover Λ_X from Λ'_X.
   More precisely, let $F(m, m')$ be the number of possible duplication patterns of the components of an m-dimensional vector, m' of whose components are distinct. (A "duplication pattern" is a set of rules like "the value in component 3 is identical with the values in components 7 and 8.") The number of such possible patterns for Λ_X is given by $F(m) \equiv \sum_{m'=1}^{m-1} F(m, i)$. Fix a one-to-one mapping Q from the integers $\{1, \ldots, F(m)\}$ to the set of possible duplication patterns. Define $q \equiv Q^{-1}$ (duplication pattern of Λ_X), the number coding for the duplication pattern of Λ_X.

6. Consider the smallest integer q' such that setting $\Lambda'_X(J) = q'$ means that $\sum_{i=1}^m \Lambda'_X(i)$, evaluated $\mod(F(m))$, equals q. (Note that this q' is bounded by $F(m)$.) If $q' \in \Xi_{m-m'-1}$, keep adding $F(m)$ to it until we get a number not contained in $\Xi_{m-m'-1}$. (Note that we will have to add $F(m)$ at most $m - 1$ times, since that is the number of elements in $\Xi_{m-m'-1}$.) Set $\Lambda'_X(J)$ to this value of q'. (Since $N \gg m$, $N \gg (m - 1) \times F(m)$ and, therefore, this value of $q' \in X'$.)

7. This completes the specification of Γ_0^{-1}. Γ_1^{-1} is identical, except that it starts trying to set $\Lambda_X'(J)$ at the sum [the smallest integer q' such that setting $\Lambda_X'(J) = q'$ means that $\sum_{i=1}^m \Lambda_X'(i)$, evaluated $\mathrm{mod}(F(m))$, equals q] + $[m \times F(m)]$. Similarly for all the other Γ_i^{-1}. We have as many Γ_i^{-1} as there are contiguous blocks of size $[m \times F(m)]$ which we can fit into N, the number of possible values of $\Lambda_X'(J)$.

i.   First we must prove that Γ_i as defined this way is one-to-one. To do this, note that Γ_i^{-1} as defined above is single-valued. To show that the inverse mapping, Γ_i, is single-valued, it suffices to show how, using only Λ_X', one can find the single Λ_X which, when mapped through Γ_i^{-1}, gives that Λ_X'. This is done as follows: find the sum of the components of Λ_X'. Evaluate this $\mathrm{mod}(F(m))$. This gives the duplication pattern in Λ_X; i.e., it tells us which of the components of Λ_X are duplicates of which other components of Λ_X. All the nonduplicated components of Λ_X are identical to the corresponding components of Λ_X'. To find the values of the duplicated components, if, as an example, $\{a, b, c, \ldots\}$ is a set of component indices which (are known to) have identical component values, with a the smallest, then set the values of Λ_X's components with indices $a, b, c, \ldots$ all to the value of component a of Λ_X'.

ii.   We must show that the range of Γ is all L_X with $m' < m$. This follows immediately from the fact that Γ^{-1} is defined for all such L_X, having as image an L_X with $m' = m$.

iii. We must show that the domains of the Γ_i (i.e., ranges of Γ_i^{-1}) are distinct for different i. To do this, consider the hypothesis that there is overlap: for some pair of values i and $j \neq i$, there exists a vector Λ_{X_i} and a vector Λ_{X_j} such that $\Gamma_i^{-1}(\Lambda_{X_i}) = \Gamma_j^{-1}(\Lambda_{X_j})$. By the procedure outlined in item (i) and the properties of the mod operator, $\Gamma_i^{-1}(\Lambda_{X_i}) = \Gamma_j^{-1}(\Lambda_{X_j})$ implies that $\Lambda_{X_i} = \Lambda_{X_j}$. However, by the definition of Γ^{-1}, $\Gamma_i^{-1}(\Lambda_X)$ cannot equal $\Gamma_j^{-1}(\Lambda_X)$ for $i \neq j$.

iv.   We must show that the number of Γ_i goes to infinity as N does. This follows directly from the statement at the end of item 7.

v.   We must show that "for all $d \in D(i), d > \Gamma_i(d)$"; i.e., any normalization constant $d \equiv k(\Lambda_x', f(\Lambda_x')) \in D(i)$ is greater than $k(\Gamma_i(\Lambda_x'), f(\Gamma_i(\Lambda_x')))$. This follows immediately from the definition of $k(.,.)$: Since the set of distinct X values in Λ_x' properly contains the set of distinct X values in $\Gamma_i(\Lambda_x')$, the set of h agreeing with f for the X values in Λ_x' is a proper subset of the set of h agreeing with f for the X values in $\Gamma_i(\Lambda_x')$. Substituting $k(L_X, f(L_X)) = [\sum_h T(h) \times \delta(S(f(L_X), h, L_X), m')]^{-1}$, we see that $k(\Lambda_x', f(\Lambda_x')) \geq k(\Gamma_i(\Lambda_x'), \Gamma_i(f(\Lambda_x')))$, as required.

In this way, using only $n \gg m$ and the bounds on $\pi(x)$ and c, we have proven that one can replace

$$\sum_{L_X; m' \leq m} k(.,.) \Pi \delta(S(.,.,.), m')$$

with

$$\sum_{L_X; m' = m} k(.,.) \Pi \delta(S(.,.,.), m'),$$

thereby incurring arbitrarily small error in $P(c|f, m)$.

Now note that with $T(h)$ constant, we can immediately write $k(L_X, f(L_X)) \propto r^{m-n}$ for all L_X for which $m' = m$. This value of $k(.,.)$ is independent of L_X and, therefore, it can be replaced with a constant in $\sum_{L_X; m'=m} k(.,.) \Pi \delta(S(.,.,.), m')$, giving $\sum_{L_X; m'=m} \Pi \times \delta(S(.,.,.), m')$ (up to an overall multiplicative constant).

Note, however, that self-averaging requires that we can replace $k(.,.)$ with an overall multiplicative constant in the sum $\sum_{L_X; m' \leq m} k(.,.) \Pi \delta(S(.,.,.), m')$. So we are not quite done; we must now "go backwards" and prove that $\sum_{L_X; m'=m} \Pi \delta(S(.,.,.), m')$ (note there is no $k(.,.)$ in the summand) can be approximated arbitrarily well by $\sum_{L_X; m' \leq m} \Pi \delta(S(.,.,.), m')$ (in the large n limit, for those h of interest). To do this, we proceed as before: examine

$$\frac{\sum_{L_X; m' < m} \Pi \delta(S(.,.,.), m')}{\sum_{L_X; m'=m} \Pi \delta(S(.,.,.), m')}.$$

This ratio is bounded by

$$\frac{\sigma^m \sum_{L_X; m' < m} \delta(S(.,.,.), m')}{\sum_{L_X; m'=m} \delta(S(.,.,.), m')}.$$

The sum in the denominator is the number of ways one can choose m distinct ordered X values so that f and h agree on those values. Since f and h agree with each other N times and L_X is an ordered set (see the introduction), this is just $N!/(N-m)$. The numerator sum can be written as $\sum_{m' < m} \varphi(m, m') C_{m'}^N$. ($C_{m'}^N$ is the number of ways of choosing m' distinct L_X values from out of N. $\varphi(m, m')$ is the number of ordered ways of picking m objects from m' distinct bins, making sure that each bin is picked at least once. This quantity is also discussed in Appendix C.) Therefore, the ratio of sums is

$$\sum_{m' < m} \frac{\varphi(m, m')}{m'!} \frac{(N-m)!}{(N-m')!}.$$

This is strictly less than $(N - m + 1)^{-1} \sum_{m' < m} \varphi(m, m')/m'!$. Since the sum is independent of n, as $n \to \infty$, $N \gg m$, and the sum gets swamped; our ratio of sums goes to 0. Therefore, we can replace the sum $\sum_{L_X; m'=m} \Pi \delta(S(.,.,.), m')$ with the sum $\sum_{L_X; m' \leq m} \Pi \delta(S(.,.,.), m')$ and, therefore, we have self-averaging. QED.

APPENDIX C
CALCULATING $P(c|f, m)$ EXACTLY, WITHOUT INVOKING SELF-AVERAGING

Note that $\pi(x)$ is (often) under our direct control; for simplicity, assume it is a constant (and therefore equals $1/n$). Plugging this, and the assumption that $T(h)$ is constant, into Eq. (4) gives

$$P(h|f, m) \propto n^{-m} \sum_{L_X;m} \left[\frac{\delta(S(f, h, L_X), m')}{\sum_h \delta(S(f, h, L_X), m')} \right]$$

$$= \sum_{L_X;m} \{ [\delta(h(.))\}$$

agrees with $f(.)$ for all X values in L_X] / [number of hypothesis functions which agree with $f(.)$ for all X values in L_X]}. Now the denominator inside in the sum depends only on m'; it equals $r^{(n-m')}$. Therefore it makes sense to break up the sum into a sum of sums, each of the inner sums having m' fixed:

$$P(h|f, m) \propto \sum_{m'=1}^{m} r^{(m'-n)} \sum_{L_X;m,m'} [\delta(h(.)$$

agrees with $f(.)$ for all X values in L_X], where by $\sum_{L_X;m,m'}$ (C.1)

is meant the sum over all m-element L_X where the m $L_X(i)$

take on only m' distinct values.

We must now evaluate $\sum_{L_X;m,m'} \delta(h(.)$ agrees with $f(.)$ for all X values in L_X). To do this, define $z \equiv n(1 - Er(f, h)) = S(f, h, X)$. Without loss of generality, reorder the X values so that the z agreements between $h(.)$ and $f(.)$ occur for the first z values of X. Our task is to count the number of ways to pick an ordered set of m X values, which take on $m' < m$ distinct values all together, so that all of those X values they take on are contained in the first z values of X.

First note that if $m' > z$, this "number of ways" equals 0. Therefore, the upper limit in the first sum in Eq. (C.1) can be replaced by $\min(m, z)$, and we can always assume that $m' \le z$ for purposes of counting the "number of ways."

The "number of ways" can be written as the product of two numbers. The first number gives the number of ways to choose m' distinct elements from amongst z possibilities; i.e., $C_{m'}^z$. The second number is the number of ways of assigning one of m' labels to each of m distinct elements, so that for each of the labels there is at least one element with that label. Write this second number as $\varphi(m, m')$. We have just proven that

$$P(h|f,m) \propto \sum_{m'=1}^{\min(m,n[1-Er(f,h)])} r^{(m'-n)} C_{m'}^z \varphi(m, m')].$$

If we plug this into Eq. (2), and collect all normalization constants, then by definition of z, for any c such that $n(1-c)$ is an integer between 0 and n,

$$P(c|f,m) \propto \left\{ \sum_{i=1}^{\min(m,n(1-c))} r^{(i-n)} \varphi(m,i) C_i^{n(1-c)} \right\} \left\{ \sum_h \delta(c, Er(f,h)) \right\}$$

$$= \left\{ \sum_{i=1}^{\min(m,n(1-c))} r^{(i-n)} \varphi(m,i) C_i^{n(1-c)} \right\} C_{n(1-c)}^n (r-1)^{nc}. \qquad (C.2)$$

($P(c|f,m)$ for all other c is zero, since those c cannot occur.)

APPENDIX D
AN EXAMPLE OF WHEN SELF-AVERAGING DOES NOT HOLD

Consider the situation where $T(h)$ factors: $T(h) = \Pi_{x \in X} V(h(x), x)$, for some everywhere nonnegative function $V(.,.)$. This is perhaps the simplest way one can have a nonconstant distribution over a set of functions. Intuitively, this form for $T(h)$ arises from viewing the functions h as a set of n pairs, $\{x_i, h(x_i)\}$; the value of $T(\{x_i, h(x_i)\})$ is simply the product of the n values $V(h(x_i), x_i)$ for some function V.

Write $d(x) \equiv \sum_{y \in Y} V(y,x)$. $d(x)$ is finite and nonzero for all $x(\{d(x) = 0\} \Rightarrow \{V(y,x) = 0 \forall y \in Y\} \Rightarrow \{T(h) = 0 \forall h\})$. Now define $V'(y,x) \equiv V(y,x)/d(x)$. $T(h) = \Pi_{x \in X} d(x) \Pi_{x \in X} V'(h(x), x) \equiv DT'(h)$.

For $n \gg m$ and $T(h) = \Pi_{x \in X} V(h(x), x)$, $P(h|f,m)$

$$= T'(h) \left[\frac{\sum_{x \in X} \delta(h(x), f(x)) \pi(x)}{V'(f(x), x)} \right]^m. \qquad (D.1)$$

PROOF Note that we can cancel a D in the numerator of Eq. (5) with a D in the denominator, k_{-1}; i.e., Eq. (5) still holds if we replace $T(h)$ everywhere with $T'(h)$. Making this substitution, instead of $[k(L_X, f(L_X))]^{-1}$ in the denominator in Eq. (5), we now have

$$\sum_{\{h(x)\};x\notin L_X} \Pi_{x\notin L_X} V'(h(x),x)\Pi_{x\in L_X} V'(f(x),x),$$

where the sum is understood to extend over all possible sets of $n - m'$ values $\{h(x)\}$, and where x is extending over the $n - m'X$ values not contained in L_X (there are $r^{n-m'}$ terms in the sum). By pulling the $\Pi_{x\in L_X}$ term out of the sum we can rewrite this as the product of two terms:

$$\left\{ \sum_{\{h(x)\};x\notin L_X} \Pi_{x\notin L_X} V'(h(x),x) \right\} \{\Pi_{x\in L_X} V'(f(x),x)\}.$$

Now note that $\sum_{\{h(x)\};x\notin L_X}$ is a sum over events all of which are themselves an ordered set of numbers (there are $n - m'$ such numbers), all those numbers coming from a definite range (which here is Y), with repeats allowed in that set of numbers. Therefore this sum is formally equivalent to the $\sum_{L_X;m}$ found in Eq. (3) (where the value m in Eq. (3) is $n - m'$ here.) Continuing with the formal parallel between $\sum_{\{h(x)\};x\notin L_X} \Pi_{x\notin L_X} V'(h(x),x)$ and what is written as $\sum_{L_X;m} \Pi_{i=1}^m g_i(L_X(i))$ in Eq. (3), we can use Eq. (3) to write

$$\sum_{\{h(x)\};x\notin L_X} \Pi_{x\notin L_X} V'(h(x),x) = \Pi_{x\notin L_X} \sum_{y\in Y} V'(y,x).$$

Since $\sum_{y\in Y} V'(y,x) = 1$, this sum-of-a-product term just equals 1. This means that we have reduced the $[k(L_X, f(L_X))]^{-1}$ term in the denominator of Eq. (5) to $\Pi_{x\in L_X} V'(f(x),x)$.

Now assume that $n \gg m$ and the minimal value of $\pi(x)$ is bounded, so that just as in the Appendix B we can here take $m' = m$ in the sum over L_X giving $P(h|f,m)$. Doing this allows us to replace $\Pi_{x\in L_X} V'(f(x),x)$ with $\Pi_{i=1}^m V'(f(L_X(i)),L_X(i))$. This, in turn, allows us to write the equality

$$P(h|f,m) = T'(h) \sum_{x\in X} \left[\frac{\delta(h(x),f(x))\pi(x)}{V'(f(x),x)}\right]^m.$$

QED.

The formula for $P(c|f,m)$ resulting from Eq. (D.1) will not, in general, be expressible as in Eq. (6.68) given by Hertz et al.[3] Accordingly, self-averaging can not hold. In particular, unless $V(.,.)$ is flat, you do *not* get the central result of the conventional analysis of exhaustive learning. The problem is the V' term in the denominator inside the sum. To try to follow along with the derivation of Eq. (6.68), instead of

$$\rho_0(c) \equiv \sum_h T(h)\delta(c, Er(f, h)),$$

you get

$$\rho_0(c, m) \equiv \sum_h T'(h)\delta(c, Er(f, h))[\eta(h)]^m,$$

where

$$\eta(h) \equiv \frac{\sum_{x \in X} \delta(h(x), f(x))\pi(x)/V'(f(x), x)}{\sum_{x \in X} \delta(h(x), f(x))\pi(x)}$$

and varies with h.

Finally, it is interesting to note that we *could* say that Eq. (6.68) holds, but only if we redefine the error function so that it uses a different sampling distribution from the one used by the likelihood ($\pi(x)/V'(f(x), x)$ vs. $\pi(x)$). One is extremely hard pressed, however, to justify such a change in the definition of the error function.

ACKNOWLEDGMENTS

This work was done under the auspices of the Department of Energy. DHW would also like to thank the Santa Fe Institute and TXN, Inc. Some of the results presented here (e.g., the counter-example to self-averaging) appeared earlier in Wolpert and Lapedes.[11]

REFERENCES

1. Cohn, D., and G. Tesauro. "How Tight are the Vapnik-Chervonenkis Bounds?" Technical Report 91-03-04, University of Washington, Department of Computer Science, 1991.
2. Haussler, D., M. Kearns, and R. Shapire. "Bounds on the Sample Complexity of Bayesian Learning Using Information Theory and the VC Dimension." *Machine Learning* **14** (1994): 83–113.
3. Hertz J., A. Krogh, and R. Palmer *Introduction to the Theory of Neural Computation*. SFI Studies in the Sciences of Complexity, Lect. Notes, Vol. I. Redwood City, CA: Addison-Wesley, 1991.

4. Schwartz, D., V. Samalam, S. Solla, and J. Denker. "Exhaustive Learning." *Neural Comp.* **2** (1990): 374-385.

5. Seung, H. "Statistical Mechanics of Learning from Examples I, II." *Phys. Rev. A* **45** (1991): 6056.

6. Tishby, N., E. Levin, and S. Solla. "Consistent Inference of Probabilities in Layered Networks: Predictions and Generalization." In *IJCNN International Joint Conference on Neural Networks*, Vol. II, 403-409. New York: IEEE, 1989.

7. Tishby, N. "Statistical Physics Models of Supervised Learning." This volume.

8. Van der Broeck, C., and Kawai, R. "Generalization in Feedforward Neural and Boolean Networks." To be presented at *AMSc International Conference on Neural Networks*, San Diego, CA, 1991.

9. Wolpert, D. "On the Connection Between In-Sample Testing and Generalization Error." *Complex Systems* **6** (1992): 47-94.

10. Wolpert, D. "Filter Likelihoods and Exhaustive Learning." To appear in *Compuational Learning Theory and Natural Learning Systems: Volume II Natural Learning Systems,* edited by S. Hanson, et al. MIT Press: Cambridge, 1992.

11. Wolpert, D, and A. Lapedes. "A Rigourous Investigation of Exhaustive Learning." Santa Fe Institute Working Paper #92-04-020. Santa Fe, New Mexico, 1992.

12. Wolpert, D. "The Relationship Between PAC, the Statistical Physics Framework, the Bayesian Framework, and the VC Framework." These proceedings.

13. Wolpert, D. "Reconciling Bayesian and Non-Bayesian Analysis." To appear in *Maximum Entropy and Bayesian Analysis*, edited by G. Heidbreder. Boston, MA: Kluwer, 1994.

Hussein Almuallim† and Thomas G. Dietterich‡
†Department of Information and Computer Science, King Fahd University of Petroleum and
Minerals, Dhahran 31261, Saudi Arabia; e-mail: facp076@saupm00.bitnet
‡Department of Computer Science, Oregon State University, Corvallis, OR 97331; e-mail:
tgd@cs.orst.edu

A Study of Maximal-Coverage
Learning Algorithms

The *coverage* of a learning algorithm is the number of concepts that can
be learned by that algorithm from samples of a given size. This paper asks
whether good learning algorithms can be designed by maximizing their
coverage. The paper extends a previous upper bound on the coverage of
any Boolean concept learning algorithm and describes two algorithms—
Multi-Balls and Large-Ball—whose coverage approaches this upper bound.
Experimental measurement of the coverage of the ID3 and FRINGE algorithms shows that their coverage is far below this bound. Further analysis
of Large-Ball shows that although it learns many concepts, these do not
seem to be very interesting concepts. Hence, coverage maximization alone
does not appear to yield practically useful learning algorithms. The paper
concludes with a definition of coverage *within a bias*, which suggests a way
that coverage maximization could be applied to strengthen weak preference
biases.

1. INTRODUCTION

Research in computational learning theory,[8,6,10,15,17,18] has provided many insights into the capabilities and limitations of inductive learning from examples. However, an important shortcoming of most work in this area is that it focuses on learning concepts drawn from *prespecified classes* of concepts (e.g., linearly separable functions, k-DNF formula). This style of research begins by choosing a restricted class of concepts and then finding a polynomial bound—called the *sample complexity*—such that if a sample of size larger than the sample complexity is available, any concept from the concept class that is consistent with the sample will be approximately correct with high probability.

Work of the above type usually leads to a learning algorithm that is *specialized* in learning the prescribed class of concepts, and an upper bound on the number of training examples required by the algorithm to guarantee successful learning. For real-world applications, such findings can be viewed as follows: A learning algorithm L designed to learn a class of concepts C is guaranteed to succeed[1] in application domains in which the target concept belongs to C (i.e., the restrictions used to define C are satisfied by the target concept), provided that a sufficient number of training examples is given to the algorithm. Of course, no such guarantees are given if the target concept is not in C.

This naturally means that one should seek algorithms that learn concept classes that are as large (i.e., less restricted) as possible. Rivest,[14] for instance, mentions this goal most explicitly by saying:

> "One goal of research in machine learning is to identify the largest possible
> class of concepts that are learnable from examples."

This goal is also declared (although less explicitly) in many papers in the related literature.[6,10,17,18]

Nevertheless, it is a well-known fact that learning larger classes of concepts necessarily requires a larger number of training examples.[4] Such tradeoff between the size of the class of concepts being learned and the required number of training examples dictates how far one can go in attempting to learn larger and larger classes of concepts.

Traditionally, this issue has been addressed by identifying new classes of concepts that are as large as possible but still require a training sample of size bounded by some polynomial. Such an approach, however, does not enjoy great practical merit. In fact, the idea of learning prescribed classes of concepts in general suffers from two important problems:

---

[1] The guarantees are on being approximately correct with high confidence.

- Training examples are usually hard to obtain. In a typical inductive learning task, one has only a limited number of training examples, much less than the polynomial bounds provided by learning theory.
- The concept class is usually unknown. In most application settings, there is often considerable flexibility (and concomitant lack of prior knowledge) concerning the choice of which concept class to explore. In fact, many of the concept classes studied in computational learning theory have never been supported by any practical justification.

Due to these difficulties, the learning algorithms and sample complexity bounds developed in computational learning theory have rarely been of practical value.

Recently, an alternative theoretical framework was introduced.[7] Instead of fixing a class of concepts and then deriving the sample complexity, this framework turns the problem around by asking: Given a *fixed number* of training examples, what is the largest collection of concepts that some algorithm can learn? The intuition behind this framework is that, in the absence of additional information, one should prefer the learning algorithm that has the highest chance of learning the unknown concept—that is, the algorithm that can learn the largest number of concepts. In short, this framework could provide an approach to discovering an "optimal" bias for inductive learning in the absence of prior knowledge.

The goal of this paper is to explore this approach. We define the *coverage* of a learning algorithm to be the number of concepts learnable by the algorithm from a given sample size (and other relevant parameters). There are three questions raised by this approach:

1. For given sample size m, accuracy parameter ϵ, and confidence parameter δ, what is the largest possible coverage that any algorithm can achieve?
2. Can we design a learning algorithm that attains this optimal coverage?
3. What is the coverage of existing learning algorithms?

This paper contributes to answering each of these questions. First, we generalize the upper bound on coverage given previously by Dietterich.[7] Next, we present two learning algorithms and determine their coverage analytically. The coverage of the first algorithm, Multi-Balls, is shown to be quite close to the upper bound. The coverage of the second algorithm, Large-Ball, turns out to be even better than Multi-Balls in many situations. Third, we considerably improve upon Dietterich's limited experiments for estimating the coverage of existing learning algorithms. We find that the coverage of Large-Ball exceeds the coverage of ID3[13] and FRINGE[11] by more than an order of magnitude in most cases.

These results are very thought provoking, because, upon careful analysis, it becomes clear that the Large-Ball algorithm is rather trivial and uninteresting. In the final part of the paper, we conclude that coverage analysis does not—by itself—provide a framework for deriving an optimal inductive bias. It does, however, provide a framework for designing optimal-coverage algorithms within a given bias.

2. DEFINITIONS AND NOTATION

We consider the space of Boolean concepts defined on n Boolean features. Let U_n be the set of all the 2^n truth assignments to the n features. A *concept* is an arbitrary set $c \subseteq U_n$. An *example* of a concept c is a pair $\langle X, c(X) \rangle$ where $c(X) = 1$ if $X \in c$ and 0 otherwise. The example is called *positive* in the first case, and *negative* in the second.

We assume the uniform distribution over U_n. However, all our results can be easily extended to the distributions where the probability is 0 on a subset of U_n and uniform on the rest. This is done by substituting the number of instances in U_n having nonzero probability in place of every occurrence of 2^n in the results.

A *training sample* of a concept c is a collection of examples drawn randomly from U_n and labeled according to c. The number of examples in this collection is called the *sample size*, denoted by m. Except in our experimental work, we assume that examples in a sample are drawn independently (i.e., with replacement), and thus, a sample of size m does not necessarily contain m distinct examples. It should be noted that, assuming $m \ll 2^n$, the difference between sampling with and without replacement is not significant.

The *disagreement* between a training sample and a concept is the number of examples in the sample that are incorrectly classified by the concept.

The *distance* between two concepts, c and h, is the number of assignments $X \in U_n$ such that $c(X) \neq h(X)$. The *error* between c and h is the distance divided by 2^n, which is equivalent to the probability that a randomly chosen X will be classified differently by the two concepts. For any $0 < \epsilon < 1$, we say that h is ϵ-close to c if the error between the two concepts is at most ϵ. We let $Ball(c, \epsilon)$ denote the set of concepts that are ϵ-close to c. Note that for any concept $c' \in Ball(c, \epsilon)$, the distance between c' and c is at most $\epsilon 2^n$. Therefore, the number of concepts in $Ball(c, \epsilon)$ is given by $|Ball(c, \epsilon)| = \sum_{i=0}^{\lfloor \epsilon 2^n \rfloor} \binom{2^n}{i}$. We call c and $\lfloor \epsilon 2^n \rfloor$ the *center* and *radius* of the ball, respectively.

A *learning algorithm* is a mapping from the space of samples to the space of concepts. The output of the algorithm is called a *hypothesis*. A hypothesis is *consistent* if it has no disagreement with the training sample.

We adopt PAC learning[4] as the criterion for successful learning, but we restrict this to learning under the uniform distribution only. We say that an algorithm L *learns* a concept c for given m, ϵ, and δ, if, with probability at least $1 - \delta$, L returns some hypothesis h that is ϵ-close to c when given a randomly drawn sample of c of size m. Formally, let S_c^m denote a random sample of c of size m. We say that L learns c with respect to m, ϵ, and δ if

$$Pr[error(h, c) \leq \epsilon] \geq 1 - \delta$$

where h is the hypothesis returned by L given S_c^m, and where the probability is computed over all the samples of c of size m. The variables ϵ and δ are called the *accuracy* and *confidence* parameters, respectively.

In general, ϵ and δ are in the range $0 < \epsilon, \delta < 1$. In practice, however, only values that are close to 0 are interesting. For this reason, we will sometimes explicitly assume for instance that $0 < \epsilon < 1/4$ and/or $0 < \delta < 1/2$, with the understanding that these are reasonable assumptions in practice. Further, to simplify our results, we will only consider the values of ϵ such that $\epsilon 2^n$ is a positive integer. Clearly, this is not a serious assumption when n is sufficiently large.

For given n, m, ϵ, and δ, the *coverage* of a learning algorithm is the number of concepts the algorithm learns with respect to these parameters.

3. UPPER BOUND ON COVERAGE

We begin by proving an upper bound on the best coverage that any algorithm can attain. An upper bound of this type has been proven for the case where the training sample is drawn randomly *without* replacement.[7] In the following, we generalize Dietterich's result and show that the same upper bound also holds for the case where sampling is done with replacement. In addition, we provide a closed-form expression for this bound.

THEOREM 1. Assuming that $m \le (1 - 2\epsilon)2^n$, the coverage of any learning algorithm under the uniform distribution cannot exceed

$$\frac{1}{1-\delta} \, 2^m \sum_{i=0}^{\epsilon 2^n} \binom{2^n - m}{i}$$

concepts, for sample size m, accuracy parameter ϵ, and confidence parameter δ.

PROOF. The proof of this theorem uses Lemmas 6 and 7 which are given in the appendix. Let L be any learning algorithm and let $C \subseteq 2^{U_n}$ be the set of concepts learned by L for the given learning parameters. Let us denote by p_c the probability that a sample of size m for a concept c be mapped by L to some hypothesis within ϵ of c. By definition, for any $c \in C$, p_c must be at least $1 - \delta$. Thus, it must be true that

$$\sum_{c \in C} p_c \ge |C|(1 - \delta)$$

and, therefore,

$$|C| \le \frac{1}{1-\delta} \sum_{c \in C} p_c \le \frac{1}{1-\delta} \sum_{c \in 2^{U_n}} p_c$$

since $C \subseteq 2^{U_n}$. In the following, we let

$$W = \sum_{c \in 2^{U_n}} p_c \ .$$

The theorem holds by proving an upper bound on W as follows.

Let S be the set of all possible outcomes of randomly drawing m objects from U_n. For any $s \in S$, let $\sharp s$ denote the number of distinct objects in s. Note that $\sharp s = m$ for every $s \in S$ in the case of sampling without replacement, and that $1 \le \sharp s \le m$ in the case of sampling with replacement. Also, let $Pr[s]$ denote the probability of the outcome s.

Now, for some $s \in S$, suppose t is a training sample obtained by arbitrarily labeling the objects in s as positive or negative, and let h be the hypothesis returned by L when given t. Mapping t to h contributes the amount of $Pr[s]$ to p_c for every concept c that is (i) consistent with t and (ii) within ϵ of h. Therefore, we can compute W by summing up this contribution for all possible outcomes $s \in S$ and all possible ways of labeling these. That is,

$$W = \sum_{s \in S} Pr[s] \sum_{t \in Labeling(s)} N(t)$$

where $Labeling(s)$ is the set of all training samples obtained by labeling s, and $N(t)$ is the number of concepts that are consistent with t and within ϵ from the hypothesis returned by L when given t. By Lemma 6, for any $t \in Labeling(s)$

$$N(t) \le \sum_{i=0}^{\epsilon 2^n} \binom{2^n - \sharp s}{i}$$

which means that

$$
\begin{aligned}
W &\le \sum_{s \in S} \left\{ Pr[s] \times |\ Labeling(s)\ | \times \sum_{i=0}^{\epsilon 2^n} \binom{2^n - \sharp s}{i} \right\} \\
&= \sum_{s \in S} \left\{ Pr[s] \times 2^{\sharp s} \times \sum_{i=0}^{\epsilon 2^n} \binom{2^n - \sharp s}{i} \right\} \\
&= \sum_{s \in S} \left\{ Pr[s] \times \sum_{i=0}^{\epsilon 2^n} 2^{\sharp s} \times \binom{2^n - \sharp s}{i} \right\} \ .
\end{aligned}
$$

For both cases of sampling with and without replacement, $\sharp s$ is at most m. Therefore, using Lemma 7 we can write

$$W \leq \sum_{s \in S} Pr[s] \sum_{i=0}^{\epsilon 2^n} 2^m \binom{2^n - m}{i}$$

$$= 2^m \sum_{i=0}^{\epsilon 2^n} \binom{2^n - m}{i} \sum_{s \in S} Pr[s]$$

$$= 2^m \sum_{i=0}^{\epsilon 2^n} \binom{2^n - m}{i}$$

which proves the theorem. $\square$

The following theorem puts the upper bound of Theorem 1 in closed form.

THEOREM 2. For $0 < \epsilon < 1/4$ and $m < 1/4(2^n)$, the quantity of Theorem 1 is further bounded above by

$$\frac{2^{(1 - \epsilon \log_2 e)m + 1} \binom{2^n}{\epsilon 2^n}}{1 - \delta} \leq \frac{2^{(1 - 1.44\epsilon)m + 1 + H(\epsilon)2^n}}{1 - \delta}$$

where e is the base of the natural logarithm, and

$$H(\epsilon) = \epsilon \log_2 \left(\frac{1}{\epsilon}\right) + (1 - \epsilon) \log_2 \left(\frac{1}{1 - \epsilon}\right).$$

PROOF. The first inequality follows directly from Lemma 9. The second inequality follows from Lemma 5. $\square$

This result shows that given a training sample of a reasonable size, any learning algorithm can learn only a small proportion of the concept space. As a numerical example, consider the case where $n = 20$, $m = 100,000$, and $\delta = \epsilon = 0.05$. Note that while having 20 features in a practical domain is not unusual, 100,000 examples is a rather large sample size. In this case, $H(\epsilon) \approx 0.286$. The above result states that no learning algorithm can learn more than

$$\frac{2^{92,788 + 0.286 \times 2^{20}}}{0.95}$$

concepts. This is less than $1/600,000$ of the $2^{2^{20}}$ possible concepts definable over 20 features—a strikingly small fraction.

For the extreme case where $\delta = 0$, we can derive a much tighter bound:

THEOREM 3. If $\delta = 0$ and $\epsilon < 1/4$, then the coverage of any learning algorithm is at most

$$\sum_{i=0}^{\epsilon 2^n} \binom{2^n}{i}$$

concepts, for accuracy parameter ϵ and confidence parameter δ.

PROOF. Let L be any learning algorithm. Any two concepts c_1, c_2, such that $c_1 \neq \neg c_2$, must share some samples in common. Since $\delta = 0$, for L to learn both concepts, L must map *all* of these samples to some hypothesis h that is within ϵ of both concepts. Such an h exists only if $distance(c_1, c_2) \leq 2\epsilon 2^n$. Now, suppose that C is the set of concepts learned by L for $\delta = 0$, $\epsilon < 1/4$ and some particular m. One of the following two cases must hold:

- There exists no concept c such that both c and $\neg c$ are in C. This implies that the distance between every pair of concepts in C is at most $2\epsilon 2^n$, and hence, $|C| \leq \sum_{i=0}^{\epsilon 2^n} \binom{2^n}{i}$ as desired.
- There exists some concept c such that $c, \neg c \in C$. Let c' be any concept such that $c' \neq c$ and $c' \neq \neg c$. c' must, therefore, share some samples with c and some samples with $\neg c$. Thus, for c' to be learned by L, it must be within $2\epsilon 2^n$ of both c and $\neg c$. However, this is impossible since $\epsilon < 1/4$. This implies that C contains no concepts other than c and $\neg c$ which means a coverage of only 2.

This proves the theorem. □

This suggests that Theorem 1 is not tight when δ is very small. It also implies that the degree of freedom provided by the confidence parameter, δ, in the PAC definition is very important. Any algorithm that does not exploit this freedom to output (with probability within δ) a totally incorrect hypothesis can have only very limited coverage.

4. THE MULTI-BALLS LEARNING ALGORITHM

Given these upper bounds on coverage, can we design algorithms that achieve these coverages?

Let c_1 and c_2 be two concepts with distance d, and suppose that we desire to construct a learning algorithm L that learns both c_1 and c_2. To do this, we must consider how L should treat every possible training sample consistent with c_1, c_2, or both.

ALGORITHM Two-Balls (*Sample*)
1. If *disagreement*(*Sample*, c_1) < *disagreement*(*Sample*, $\neg c_1$), then return c_1.
2. Else, return $\neg c_1$. Break ties arbitrarily.

FIGURE 1 The Two-Balls Algorithm. The variable c_1 is a built-in constant concept.

Obviously, any sample that is consistent with only one of c_1 or c_2 can be mapped to some hypothesis that is within ϵ of the consistent concept. The key question is how to map a sample that is consistent with both c_1 and c_2. Now, if $d/2^n \leq 2\epsilon$, then there exists some concept h that is within ϵ of both c_1 and c_2. In this case, all we need to do is to map the sample to h. However, if $d/2^n > 2\epsilon$, then there exists no concept that is within ϵ of both c_1 and c_2 and, therefore, we must map the sample either in favor of c_1 or in favor of c_2, but not both. In these cases, if the correct concept is c_2 and we map the sample in favor of c_1, we will commit a *mistake*, and we can only afford to do this with probability δ. The probability of getting a sample that is consistent with both c_1 and c_2 is $((2^n - d)/2^n)^m$, where m is the sample size. This quantity is decreasing as d increases, so if we choose d sufficiently large, we can keep the probability of a mistake below δ.

In short, if c_1 and c_2 are close together, then there is no problem, because we can choose an h ϵ-close to both. Conversely, if they are far apart, there is also no problem, because the probability of a mistake can be bounded by δ. This suggests that a good strategy for designing learning algorithms with high coverage is to choose a collection of concepts that is as large as possible, such that the distance between each pair of concepts in the collection is either

- sufficiently large to suppress the probability of getting a sample consistent with both concepts, or
- within $2\epsilon 2^n$, so that we can find concept(s) within ϵ of both concepts.

This means that the concepts to be learned must be clustered as one or more balls in the space of concepts. What we need to do in order to construct an appropriate algorithm is to keep the radius of each ball small enough, and at the same time, make the distance between the centers of the balls large enough.

As a trivial case, suppose that we want to learn the set of all concepts that are within ϵ of a fixed concept c—the set $Ball(c, \epsilon)$. This is achieved simply by returning c as the hypothesis regardless of the training sample. This leads to a coverage of $\sum_{i=0}^{\epsilon 2^n} \binom{2^n}{i}$.

A less trivial case is to learn 2 ϵ-balls of concepts. This is accomplished by the "Two-Balls" algorithm given in Figure 1. This algorithm returns, as the hypothesis, some fixed concept c_1 or its complement, whichever is *closer* to the training sample. For any concept c in $Ball(c_1, \epsilon)$, the probability of drawing an example of c that disagrees with c_1 (and thus, agrees with $\neg c_1$) is at most ϵ. The same argument applies to the concepts in $Ball(\neg c_1, \epsilon)$. Therefore, if $\epsilon \leq \delta$, then a sample of size 1 (that is, $m = 1$) is sufficient to learn these two balls. The following theorem gives the sample size sufficient to achieve this goal in the general case.

THEOREM 4. For accuracy and confidence parameters ϵ and δ, the coverage of the Two-Balls algorithm is

$$2 \sum_{i=0}^{\epsilon 2^n} \binom{2^n}{i}$$

when given a sample of size

$$m > \frac{12\epsilon}{(1 - 2\epsilon)^2} \ \ln \frac{1}{\delta}$$

assuming that sampling is with replacement under the uniform distribution and that $\epsilon < \frac{1}{2}$.

PROOF. Without loss of generality, assume that c_1 in the algorithm is just the *nil* concept. Let c be a concept with distance at most $\epsilon 2^n$ from the *nil* concept. It is enough to show that the given sample size is sufficient to make the algorithm learn c.

Let Z denote the number of positive examples in a sample of c of size m. Since c has at most $\epsilon 2^n$ positive examples, Z can be viewed as a binomial random variable of m trials and at most ϵ as the ratio of success. A sample of c is mapped to the *true* (instead of the *nil*) concept only if it has more positive examples than negative examples. The probability of this is bounded above by

$$Pr[Z \geq \frac{m}{2}] \leq \sum_{i=\lceil \frac{m}{2} \rceil}^{m} \binom{m}{i} \epsilon^i (1 - \epsilon)^{m-i} \ .$$

Using Chernoff's bound (Lemma 4), this is at most $e^{-(1-2\epsilon)^2 m/12\epsilon}$. Therefore, c is learned if

$$e^{-(1-2\epsilon)^2 m/12\epsilon} < \delta \ ,$$

which is satisfied if

$$m > \frac{12\epsilon}{(1 - 2\epsilon)^2} \ \ln \frac{1}{\delta}$$

as desired. $\square$

According to this thorem, even when δ is as small as 0.001 (i.e., 99.9% confidence), for any ϵ in the range $0 < \epsilon \leq 0.1$, the required sample size is only 13 examples, *independent* of n.

Since the sample size is usually much larger than this, a direct generalization of the above trivial cases is to attempt to learn as many balls of concepts as permitted by the sample size. The idea is to choose a collection of well-separated concepts (the centers of the balls) and attempt to learn all the concepts clustered around each of these centers. More specifically, we start by fixing two positive integers d and k, and then construct a set $\mathcal{H} = \{h_1, h_2, h_3, \cdots, h_k\}$ of k concepts such that the distance between each pair of concepts in $\mathcal{H}$ is at least d. Then given a training sample, the concept in $\mathcal{H}$ that has the minimum disagreement with the sample is returned as the hypothesis.

The goal of the algorithm is to learn all the concepts in $\bigcup_{h \in \mathcal{H}} Ball(h, \epsilon)$ which will give a coverage of $k \sum_{i=0}^{\epsilon 2^n} \binom{2^n}{i}$, provided that the intersection between the balls is empty (that is, $d > 2\epsilon 2^n$). The question is, of course, how to determine the appropriate values of d and k such that this goal is accomplished. Particularly, we need to worry about the following:

1. As explained earlier, d must be large enough so that the interaction between concepts in different balls is kept within what is allowed by the confidence parameter δ.
2. The number of concepts that we can construct such that the minimum distance between each pair is d drops sharply as d increases. Therefore, making d too large causes k (and hence, the coverage of the algorithm) to be too small.

The following two lemmas show how to choose appropriate values for d and k.

LEMMA 1. For $0 < \epsilon < 1/4$ and $2\epsilon < \alpha < 1/2$, let $\mathcal{H} = \{h_1, h_2, \cdots, h_k\}$ be a set of concepts such that the distance between each pair $h_i, h_j \in \mathcal{H}$ is at least $d = \lceil \alpha 2^n \rceil$, and let $c \in Ball(h, \epsilon)$ for some $h \in \mathcal{H}$. Assume that L is a learning algorithm that on any sample S outputs some hypothesis $h_i \in \mathcal{H}$ that has minimal disagreement with S. Then, under the uniform distribution, the probability that a sample of c of size m is mapped by L to an hypothesis other than h is at most

$$\min_{0 < \beta < 1} k \cdot \left\{ e^{-2(1-\beta)^2 \alpha^2 m} + e^{-\beta \frac{(\alpha - 2\epsilon)^2}{2\alpha} m} \right\} . \tag{1}$$

PROOF. Let $g \neq h$ be a *specific* concept in $\mathcal{H}$, and let us bound the probability that a sample of c is mapped to g (instead of h). We consider the case in which this probability is maximized. First, we assume that g is as close as possible to h—that is, $distance(h, g) = d$. Second, we place c to be as far as possible from h and as close as possible to g—that is, $distance(c, h) = \epsilon 2^n$ and $distance(c, g) = d - \epsilon 2^n$.

Let A be the set of objects where c agrees with g but not with h and let B be the set of objects where c agrees with h but not with g. Thus, $|A| = \epsilon 2^n$ and $|B| = d - \epsilon 2^n$. A sample S of c is mapped to g only if $disagreement(S, g) \leq disagreement(S, h)$. That is, only if S has more examples drawn from A than from B. Now, let us define the following three random variables:

- X: the number of examples in S drawn from A.
- Y: the number of examples in S drawn from B.
- $Z = X + Y$: the number of examples in S drawn from $A \bigcup B$.

Then, the probability that S is mapped to g is just $Pr[X \geq Y]$.

Note that Z is just a binomial random variable with m trials and $d/2^n$ as the ratio of success. Given that, out of the m examples in S, there are i examples drawn from $A \bigcup B$ (i.e., given that $Z = i$), the event $X \geq Y$ is equivalent to $Y \leq i/2$. In this case, Y can be viewed as a binomial random variable of i trials and $(d - \epsilon 2^n)/d$ as the ratio of success. Therefore, we can write

$$Pr[X \geq Y] = \sum_{i=0}^{m} Pr[Z = i] Pr[X \geq Y | Z = i]$$

$$= \sum_{i=0}^{m} Pr[Z = i] Pr[Y \leq \frac{i}{2} | Z = i]$$

$$= \sum_{i=0}^{m} \left[\binom{m}{i} \left(\frac{d}{2^n}\right)^i \left(\frac{2^n - d}{2^n}\right)^{m-i} \sum_{j=0}^{\frac{i}{2}} \binom{i}{j} \left(\frac{d - \epsilon 2^n}{d}\right)^j \left(\frac{\epsilon 2^n}{d}\right)^{i-j} \right].$$

Applying Hoeffding's bound (Lemma 3) on the inner summation, this is at most

$$\sum_{i=0}^{m} \left[\binom{m}{i} \left(\frac{d}{2^n}\right)^i \left(\frac{2^n - d}{2^n}\right)^{m-i} e^{-2i\left(\frac{d - 2\epsilon 2^n}{2d}\right)^2} \right]$$

$$\leq \sum_{i=0}^{\lfloor \beta \frac{d}{2^n} m \rfloor} \left[\binom{m}{i} \left(\frac{d}{2^n}\right)^i \left(\frac{2^n - d}{2^n}\right)^{m-i} e^{-2i\left(\frac{d - 2\epsilon 2^n}{2d}\right)^2} \right]$$

$$+ \sum_{i=\lceil \beta \frac{d}{2^n} m \rceil}^{m} \left[\binom{m}{i} \left(\frac{d}{2^n}\right)^i \left(\frac{2^n - d}{2^n}\right)^{m-i} e^{-2i\left(\frac{d - 2\epsilon 2^n}{2d}\right)^2} \right]$$

for any $0 < \beta < 1$. Letting $i = 0$ in the first exponential and $i = \beta(d/2^n)m$ in the second, the above is at most

$$\sum_{i=0}^{\lfloor \beta \frac{d}{2^n} m \rfloor} \left[\binom{m}{i} \left(\frac{d}{2^n} \right)^i \left(\frac{2^n - d}{2^n} \right)^{m-i} \right]$$

$$+ \; e^{-2\beta \frac{d}{2^n} \left(\frac{d - 2\epsilon 2^n}{2d} \right)^2 m} \sum_{i=\lceil \beta \frac{d}{2^n} m \rceil}^{m} \left[\binom{m}{i} \left(\frac{d}{2^n} \right)^i \left(\frac{2^n - d}{2^n} \right)^{m-i} \right]$$

$$\leq \sum_{i=0}^{\lfloor \beta \frac{d}{2^n} m \rfloor} \left[\binom{m}{i} \left(\frac{d}{2^n} \right)^i \left(\frac{2^n - d}{2^n} \right)^{m-i} \right] + e^{-2\beta \frac{d}{2^n} \left(\frac{d - 2\epsilon 2^n}{2d} \right)^2 m} \; .$$

Applying Hoeffding's bound (Lemma 3) again on the first term, the above is at most

$$e^{-2(1-\beta)^2 (\frac{d}{2^n})^2 m} + e^{-\beta \left((\frac{d}{2^n} - 2\epsilon)^2 / 2 \frac{d}{2^n} \right) m} \; .$$

Since the above bound holds for any $0 < \beta < 1$, we can, of course, minimize over all β in that range and get

$$\min_{0 < \beta < 1} \left\{ e^{-2(1-\beta)^2 (\frac{d}{2^n})^2 m} + e^{-\beta \left((\frac{d}{2^n} - 2\epsilon)^2 / 2 \frac{d}{2^n} \right) m} \right\} \; .$$

This quantity is decreasing in $d/2^n$ when $d/2^n > 2\epsilon$. Thus, we can replace $d/2^n$ by α since $2\epsilon < \alpha \leq d/2^n$. Since g can be any of the k concepts in $\mathcal{H}$, multiplying the above probability by k gives the desired result. $\square$

Note that d in this lemma is expressed as a fraction α of 2^n. This result says that if the conditions of the lemma are met, and if c is a concept in one of the k ϵ-balls, then the samples of c are usually mapped by L to the center of that ball (which is ϵ-close to c) except with a probability that is bounded by the quantity of Eq. (1). Thus, α and k must be chosen so that this probability is at most δ. It is important to note that this probability is diminishing in α and m and independent of n.

The problem of finding a collection of bit vectors that maintain a given minimum pairwise distance is well studied in the field of Error-Correcting Coding Theory. Specifically, the following theorem states how large k can (at least) be for a given separation distance d.

LEMMA 2. (The Gilbert-Varshamov Bound) There exist at least 2^{l-r} bit vectors of length l and minimum pairwise distance d, where r is any integer satisfying

$$\binom{l-1}{0} + \binom{l-1}{1} + \binom{l-1}{2} + \cdots + \binom{l-1}{d-2} < 2^r.$$

A proof of this lemma, in addition to a method of actually constructing the l-bit vectors as given above, can be found in many references in the literature of Error-Correcting Coding Theory.[12]

For our purposes here, it is convenient to draw the following corollary from the Gilbert-Varshamov bound given above.

COROLLARY. 1 For any α, $0 < \alpha < 1/2$, and any even positive integer l, there exist at least $2^{l-\lceil H(\alpha)l\rceil}$ bit vectors of length l and pairwise distance at least αl, where $H(\alpha) = \alpha \log_2 1/\alpha + (1-\alpha)\log_2 1/(1-\alpha)$.

ALGORITHM Multi-Balls $(Sample, \epsilon, \delta)$

1. Find α in the range $2\epsilon < \alpha < 1/2$ such that
$$1 - H(\alpha) = 2[1 - \beta(\alpha)]^2 \alpha^2 \frac{m}{2^n} \log_2 e$$
where
$$H(\alpha) = \alpha \log_2 \frac{1}{\alpha} + (1-\alpha)\log_2 \frac{1}{1-\alpha} \text{ and}$$
$$\beta(\alpha) = 1 + \frac{(\alpha - 2\epsilon)^2}{8\alpha^3} - \sqrt{\left[1 + \frac{(\alpha - 2\epsilon)^2}{8\alpha^3}\right]^2 - 1}.$$

2. Let $k = \lfloor 2^{2^n - H(\alpha)2^n} \times \frac{\delta}{2}\rfloor$.

3. Construct $\mathcal{H} = \{h_1, h_2, h_3, \ldots, h_k\}$ such that
$$\forall_{h_i, h_j \in \mathcal{H}} \ distance(h_i, h_j) \geq \lceil \alpha 2^n \rceil.$$

4. Return some hypothesis in $\mathcal{H}$ that has minimal disagreement with $Sample$ (break ties arbitrarily).

FIGURE 2   The Multi-Balls Algorithm.

PROOF. We just need to show that when $d = \lceil \alpha l \rceil$ and $r = \lceil H(\alpha)l \rceil$, the inequality of Lemma 2 is satisfied. The left-hand side of the inequality is just

$$\sum_{i=0}^{d-2} \binom{l-1}{i} \quad \begin{array}{c} < \sum_{i=0}^{\alpha l} \binom{l}{i} \\ \leq 2^{H(\alpha)l} \quad \text{(by Lemma 5)} \\ \leq 2^{\lceil H(\alpha)l \rceil} \end{array}$$

which equals the right-hand side. This proves the corollary. $\square$

Using Lemma 1 and Corollary 1, one can search for the appropriate value for d that leads to learning k different ϵ-balls, for k as large as possible. Let's now compute a lower bound on the coverage that can be achieved by this approach.

In Figure 2 we show the "Multi-Balls" algorithm in which we give a specific way of choosing the values of α (and, hence, d) and k. To be able to give a lower bound on the coverage of this algorithm, we need the following definition.

DEFINITION. For $0 \leq \theta \leq 1$ and $0 < \epsilon < 1/4$, define $\rho(\theta, \epsilon)$ as

$$\rho(\theta, \epsilon) = \frac{1 - H(\hat{\alpha})}{\theta}$$

for $\hat{\alpha}$ being the solution of the equation

$$1 - H(\alpha) = 2[1 - \beta(\alpha)]^2 \alpha^2 \theta \log e \qquad (2)$$

in the range $2\epsilon < \alpha < 1/2$, where

$$\beta(\alpha) = 1 + \frac{(\alpha - 2\epsilon)^2}{8\alpha^3} - \sqrt{\left[1 + \frac{(\alpha - 2\epsilon)^2}{8\alpha^3}\right]^2 - 1}$$

and

$$H(\alpha) = \alpha \log_2 \frac{1}{\alpha} + (1 - \alpha) \log_2 \frac{1}{1 - \alpha}.$$

Lemma 10 shows that the above $\rho(\theta, \epsilon)$ function is well defined. Although this function is not provided in closed form, it is easily computed for any given values of θ and ϵ using standard numerical methods. For illustration, in Figure 3 we plot this function over the range $0 \leq \theta \leq 1/2$ for $\epsilon = 0.01, 0.05$, and 0.10.

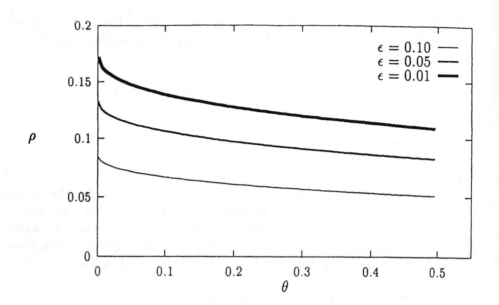

FIGURE 3 The function $\rho(\theta, \epsilon)$.

Using the above definition of ρ, a lower bound on the coverage of Multi-Balls can be stated as follows:

THEOREM 5. For $0 < \epsilon < 1/4$ and $0 < \delta < 1$, the coverage of the Multi-Balls algorithm under the uniform distribution is at least

$$\left\lfloor \frac{\delta}{2} \ 2^{\rho(\frac{m}{2^n}, \epsilon)} \ m \right\rfloor \sum_{i=0}^{\epsilon 2^n} \binom{2^n}{i}$$

for sample size m, accuracy parameter ϵ, and confidence parameter δ.

PROOF. First, the existance of α in the range $2\epsilon < \alpha < 1/2$ satisfying the equation of Step 1 is guaranteed by Lemma 10. It is sufficient to show the following three claims for the values of α and k chosen by the algorithm in Steps 1 and 2:

CLAIM. 1. Step 3 is executable. That is, for α and k as chosen in Steps 1 and 2, we can actually construct a set of k concepts such that the minimum pairwise distance is $\lceil \alpha 2^n \rceil$.

CLAIM. 2. For any $h \in \mathcal{H}$ and any $c \in Ball(h, \epsilon)$, the probability that a sample of c of size m is mapped to an hypothesis other than h is at most δ.

CLAIM. 3. The number of balls, k, is at least $\lfloor \frac{\delta}{2} 2^{\rho(\frac{m}{2^n}, \epsilon) m} \rfloor$.
Claim 1 follows from Corollary 1 since k is set to $\lfloor 2^{2^n - H(\alpha)2^n} \times \delta/2 \rfloor$ and thus

$$
\begin{aligned}
k &\leq 2^{2^n - H(\alpha)2^n} \times \frac{\delta}{2} \\
&= 2^{2^n - H(\alpha)2^n - \log_2 \frac{2}{\delta}} \\
&\leq 2^{2^n - \lceil H(\alpha)2^n \rceil + 1 - \log_2 \frac{2}{\delta}} \\
&< 2^{2^n - \lceil H(\alpha)2^n \rceil}
\end{aligned}
$$

since $\log_2 2/\delta$ is necessarily greater than 1.
For Claim 2, we use Lemma 1 which states that the probability of a mistake is bounded above by

$$
k \cdot \left\{ e^{-2(1-\beta)^2 \alpha^2 m} + e^{-\beta((\alpha - 2\epsilon)^2 / 2\alpha))m} \right\} \tag{3}
$$

for *any* β in the range $[0, 1]$. Let us choose β such that

$$
e^{-2(1-\beta)^2 \alpha^2 m} = e^{-\beta \frac{(\alpha - 2\epsilon)^2}{2\alpha} m}
$$

which means that

$$
2(1 - \beta)^2 \alpha^2 m = \beta \frac{(\alpha - 2\epsilon)^2}{2\alpha} m . \tag{4}
$$

Solving this quadratic equation for β gives the two solutions

$$
1 + \frac{(\alpha - 2\epsilon)^2}{8\alpha^3} \pm \sqrt{\left[1 + \frac{(\alpha - 2\epsilon)^2}{8\alpha^3} \right]^2 - 1} . \tag{5}
$$

It can be checked that Eq. (4) has two roots, one in the range $[0, 1]$ and the other larger than 1. Thus, the smaller root in Eq. (5), namely

$$
1 + \frac{(\alpha - 2\epsilon)^2}{8\alpha^3} - \sqrt{\left[1 + \frac{(\alpha - 2\epsilon)^2}{8\alpha^3} \right]^2 - 1} ,
$$

is between 0 and 1 and, thus, is an appropriate value for β. Note that this is exactly the quantity $\beta(\alpha)$ in Step 1 of the algorithm.

Now, substituting for k and β in the bound of Lemma 1, the probability that Multi-Balls commits a mistake is at most

$$\min_{0<\beta<1} k \times \left\{ e^{-2[1-\beta]^2\alpha^2 m} + e^{-\beta\frac{(\alpha-2\epsilon)^2}{2\alpha}m} \right\}$$

$$\leq \frac{\delta}{2} \times 2^{2^n[1-H(\alpha)]} \times \left\{ e^{-2[1-\beta(\alpha)]^2\alpha^2 m} + e^{-\beta(\alpha)\frac{(\alpha-2\epsilon)^2}{2\alpha}m} \right\}$$

$$= \frac{\delta}{2} \times 2^{2^n[1-H(\alpha)]} \times 2\, e^{-2[1-\beta(\alpha)]^2\alpha^2 m} .$$

The value of α is chosen in Step 1 such that

$$1 - H(\alpha) = 2[1 - \beta(\alpha)]^2\alpha^2 \frac{m}{2^n} \log_2 e .$$

Therefore, continuing from above, the desired probability is at most

$$\delta \times 2^{2^n 2[1-\beta(\alpha)]^2\alpha^2 \frac{m}{2^n} \log_2 e} e^{-2[1-\beta(\alpha)]^2\alpha^2 m}$$

$$= \delta \times e^{2[1-\beta(\alpha)]^2\alpha^2 m} e^{-2[1-\beta(\alpha)]^2\alpha^2 m}$$

$$= \delta ,$$

which shows Claim 2.

Finally, Claim 3 follows immediately from the definition of ρ. Since the value of α found in Step 1 is equivalent to $\hat{\alpha}$ in the definition of ρ, we can write

$$\rho\left(\frac{m}{2^n}, \epsilon\right) m = \frac{1 - H(\alpha)}{\frac{m}{2^n}} m$$

$$= 2^n[1 - H(\alpha)] .$$

Thus, k is equivalent to $\lfloor \delta/2\ 2^{\rho(m/2^n,\epsilon)m} \rfloor$, which completes the proof. $\square$

More specific bounds on the coverage of Multi-Balls can be obtained from Theorem 5 if upper bounds on $m/2^n$ and ϵ are assumed. For example, if we are interested only in the range $0 < \epsilon \leq 0.05$ and $0 \leq m/2^n \leq 0.25$ (which are reasonable assumptions in practice), then the coverage of Multi-Balls as given by Theorem 5 is at least

$$\left\lfloor \frac{\delta}{2}\ 2^{0.094\, m} \right\rfloor \sum_{i=0}^{\epsilon 2^n} \binom{2^n}{i}$$

where the constant 0.094 is just the value of $\rho(0.25, 0.05)$.

Note that the upper bound of Theorem 2 can be written as

$$\frac{1}{1-\delta}\ 2^{(1-1.44\epsilon)m+1} \binom{2^n}{\epsilon 2^n} .$$

The main difference between this upper bound and the above lower bound is the coefficient of m in the exponent of 2. Therefore, for any fixed δ, this lower bound indicates that to achieve a given coverage, Multi-Balls requires a sample size that is within a constant factor of that required by an optimal learning algorithm.

5. THE LARGE-BALL ALGORITHM

So far we have been trying to maximize the coverage by learning as many balls of concepts as possible, while fixing the radius of each ball at $\epsilon 2^n$. An alternative approach to increase the coverage is to learn ball(s) of concepts with larger radius. Because of the extremely high dimensionality of the space of concepts, any small increment in a ball's radius results in a huge increase in the number of concepts contained in the ball.

It turns out that learning a single ball of radius larger than $\epsilon 2^n$ is a surprisingly easy task that can be achieved by the simple algorithm "Large-Ball" given in Figure 4.

This algorithm works by modifying a *default* hypothesis c_1 so that the final hypothesis fully agrees with the training sample. For example, suppose c_1 is the *nil* concept. This algorithm classifies all examples as negative unless they appeared as positive examples in the training sample! The coverage of this algorithm is computed by the following theorem.

THEOREM 6. For sample size m, accuracy parameter ϵ, and confidence parameter δ, the coverage of the Large-Ball algorithm under the uniform distribution is

$$\sum_{i=0}^{\epsilon 2^n + \beta} \binom{2^n}{i}$$

ALGORITHM Large-Ball (*Sample*)
1. Let c_1 be a constant concept.
2. Define the concept s as:
$$s(X) = \begin{cases} 1 & \text{if } X \in Sample; \\ 0 & \text{otherwise.} \end{cases}$$
3. Define the concept p as:
$$p(X) = \begin{cases} 1 & \text{if } X \in Sample \text{ and } X \text{ is a positive example}; \\ 0 & \text{otherwise.} \end{cases}$$
4. Return $h = (\neg s \wedge c_1) \vee p$.

FIGURE 4 The Large-Ball Learning Algorithm.

where β is the largest integer such that

$$\sum_{k=0}^{\beta-1} \binom{\epsilon 2^n + \beta}{k} \left(\frac{2^n - \epsilon 2^n - \beta + k}{2^n} \right)^m$$

$$\left\{ 1 - \sum_{i=1}^{k} \binom{k}{i} \left(\frac{2^n - \epsilon 2^n - \beta + k - i}{2^n - \epsilon 2^n - \beta + k} \right)^m (-1)^{i+1} \right\} \leq \delta.$$

PROOF. Without loss of generality, assume that c_1 of the algorithm is the *nil* concept. Suppose c is a concept such that $distance(c, c_1) = \epsilon 2^n + \beta$ for some integer β. That is, c has exactly $\epsilon 2^n + \beta$ distinct positive examples. The algorithm classifies all examples as negative except those appearing in the training sample as positive examples. Therefore, any sample that has β or more positive examples is mapped by the algorithm to an hypothesis within ϵ of c. Thus, to prove the theorem, it is enough to show that the probability of getting a sample with less than β *distinct* positive examples is just the left-hand side of the inequality of the theorem.
Let

- F be the set of negative examples of the concept c. Thus, $|F| = 2^n - \epsilon 2^n - \beta$.
- T be the set of positive examples of the concept of c. Thus, $|T| = \epsilon 2^n + \beta$.
- D be the (random) number of distinct positive examples in a random sample of size m of the concept c.

What we need to compute is just

$$Pr[D < \beta] = \sum_{k=0}^{\beta-1} Pr[D = k] \ .$$

Now, let $S \subset T$ be some *specific* set of k distinct positive examples of the concept c, and let R be the event of getting a sample of the concept c of size m that contains exactly S and any number of negative examples, but no positive examples not in S. Then

$$Pr[D = k] = \binom{\epsilon 2^n + \beta}{k} Pr[R] \ .$$

Now, R occurs if and only if both

- event $R1$: No positive examples from $T - S$ are in the sample, and
- event $R2$: Each positive example in S is present in the sample

are true. Therefore, $Pr[R] = Pr[R1\cdot R2] = Pr[R1]Pr[R2|R1]$, and so finding $Pr[R1]$ and $Pr[R2|R1]$ is all what we need.

First, it is not difficult to see that

$$Pr[R1] = \left(\frac{2^n - \epsilon 2^n - \beta + k}{2^n}\right)^m .$$

Second, to find $Pr[R2|R1]$, let $A_i, i = 1, 2, \cdots k$, be the event that the positive example number i of S is *not* included in a sample of size m drawn from $S \cup F$. Then

$$Pr[R2|R1] = 1 - Pr\left[\bigcup_{i=1}^{k} A_i\right] .$$

By the principle of inclusion and exclusion,[16] we get

$$Pr\left[\bigcup_{i=1}^{k} A_i\right] = \sum_{i} Pr[A_i] - \sum_{i \neq j} Pr[A_iA_j] + \cdots + (-1)^{k+1} Pr[A_1A2\cdots A_k] .$$

Now,

$$Pr[A_i] = \left(\frac{2^n - \epsilon 2^n - \beta + k - 1}{2^n - \epsilon 2^n - \beta + k}\right)^m , 1 \leq i \leq k$$

$$Pr[A_iA_j] = \left(\frac{2^n - \epsilon 2^n - \beta + k - 2}{2^n - \epsilon 2^n - \beta + k}\right)^m , 1 \leq i,j \leq k \text{ and } i \neq j$$

$$\vdots$$

$$Pr[A_1A_2\cdots A_k] = \left(\frac{2^n - \epsilon 2^n - \beta}{2^n - \epsilon 2^n - \beta + k}\right)^m .$$

Therefore,

$$Pr\left[\bigcup_{i=1}^{k} A_i\right] = \sum_{i=1}^{k} \binom{k}{i} \left(\frac{2^n - \epsilon 2^n - \beta + k - i}{2^n - \epsilon 2^n - \beta + k}\right)^m (-1)^{i+1} .$$

Assembling, we get

$$Pr[D < \beta] = \sum_{k=0}^{\beta-1} P(D = k)$$

$$= \sum_{k=0}^{\beta-1} \binom{\epsilon 2^n + \beta}{k} Pr[R]$$

$$= \sum_{k=0}^{\beta-1} \binom{\epsilon 2^n + \beta}{k} \left(\frac{2^n - \epsilon 2^n - \beta + k}{2^n}\right)^m$$

$$\times \left\{1 - \sum_{i=1}^{k} \binom{k}{i} \left(\frac{2^n - \epsilon 2^n - \beta + k - i}{2^n - \epsilon 2^n - \beta + k}\right)^m (-1)^{i+1}\right\} ,$$

and the theorem follows. □

It can be easily shown that a sample of size $1/\epsilon \ln 1/\delta$ is sufficient to make β at least 1. The following theorem gives a lower bound on the value of β in general.

THEOREM 7. If $\epsilon < 1/2$, $\epsilon 2^n > 1$, and $m \le 1/2 \log_2 e 2^n \approx 0.347 \times 2^n$, then setting

$$\beta = \left| \frac{\epsilon m \log_2 e - \log \frac{1}{\delta} -}{n - \log \frac{1}{\epsilon}} \right|$$

satisfies the inequality of Theorem 6.

PROOF. The left-hand side of the inequality of Theorem 6 is at most

$$\sum_{k=0}^{\beta-1} \binom{\epsilon 2^n + \beta}{k} \left(\frac{2^n - \epsilon 2^n - \beta + k}{2^n} \right)^m \tag{6}$$

since $\left\{ 1 - \sum_{i=1}^{k} \binom{k}{i} ((2^n - \epsilon 2^n - \beta + k - i)/(2^n - \epsilon 2^n - \beta + k))^m (-1)^{i+1} \right\}$ is just the probability $Pr[R1|R2]$ as defined in the proof of Theorem 6 and, hence, can be at most 1. Also, since k is at most $\beta - 1$,

$$\left(\frac{2^n - \epsilon 2^n - \beta + k}{2^n} \right)^m \le \left(\frac{2^n - \epsilon 2^n - \beta + (\beta - 1)}{2^n} \right)^m$$

$$= \left(\frac{2^n - \epsilon 2^n - 1}{2^n} \right)^m$$

$$< (1 - \epsilon)^m$$

$$\le e^{-\epsilon m} .$$

This is independent of k and, thus, can be moved outside the summation of Eq. (6).

Next, we need to bound $\sum_{k=0}^{\beta-1} \binom{\epsilon 2^n + \beta}{k}$. Here, we can use Lemma 8 in addition to the fact that

$$\binom{\epsilon 2^n + \beta}{\beta - 1} = \frac{\beta(\epsilon 2^n + \beta)}{\epsilon 2^n (\epsilon 2^n + 1)} \binom{\epsilon 2^n + \beta - 1}{\beta} . \tag{7}$$

These give

$$\sum_{k=0}^{\beta-1} \binom{\epsilon 2^n + \beta}{k} = \sum_{k=0}^{\frac{\beta-1}{\epsilon 2^n + \beta}(\epsilon 2^n + \beta)} \binom{\epsilon 2^n + \beta}{k}$$

$$\leq \frac{1 - \frac{\beta-1}{\epsilon 2^n + \beta}}{1 - 2\frac{\beta-1}{\epsilon 2^n + \beta}} \binom{\epsilon 2^n + \beta}{\beta - 1}$$

(by Lemma 8 provided that $\beta < \epsilon 2^n + 2$) $\qquad$ (8)

$$= \frac{\epsilon 2^n + 1}{\epsilon 2^n - \beta + 2} \binom{\epsilon 2^n + \beta}{\beta - 1}$$

$$= \frac{\beta(\epsilon 2^n + \beta)}{\epsilon 2^n (\epsilon 2^n - \beta + 2)} \binom{\epsilon 2^n + \beta - 1}{\beta} \quad \text{(by Eq. (7))}$$

$$\leq \frac{\beta(\epsilon 2^n + \beta)}{\epsilon 2^n (\epsilon 2^n - \beta + 2)} (\epsilon 2^n)^\beta$$

$$\leq \frac{3}{2}(\epsilon 2^n)^\beta \quad \left(\text{provided that } \beta \leq \frac{1}{2}\epsilon 2^n\right). \qquad (9)$$

Therefore, the left-hand side of the inequality of Theorem 6 is at most

$$\frac{3}{2}(\epsilon 2^n)^\beta (1 - \epsilon)^m \leq \frac{3}{2}(\epsilon 2^n)^\beta e^{-\epsilon m}.$$

Setting this to be at most δ, and taking the logarithm of both sides of the inequality gives

$$\log_2 \frac{3}{2} + \beta(\log_2 \epsilon + n) - \epsilon m \log_2 e \leq \log_2 \delta.$$

This is certainly satisfied when

$$\beta = \left\lfloor \frac{\epsilon m \log_2 e - \log_2 \frac{3}{2\delta}}{n - \log_2 \frac{1}{\epsilon}} \right\rfloor.$$

Clearly, this is less than $\epsilon m \log_2 e$. Given the assumption that $m < 2^n/(2\log_2 e)$, the above value of β is at most $\epsilon 2^n/2$, which satisfies the conditions of Eqs. (8) and (9). This completes the proof. $\square$

For fixed δ, this theorem says that the radius of the ball of concepts learned by the Large-Ball algorithm grows linearly in ϵm. Note that a unit increment in the radius of the ball corresponds to a very large increment in coverage. The coverage of Large-Ball, therefore, grows quite rapidly as the sample size increases.

The coverage lower bound obtained for Large-Ball appears to overlap the bound for Multi-Balls, although Multi-Balls gives higher coverage for small values of ϵ (e.g., 0.01). In any case, the fact that a trivial algorithm like Large-Ball achieves such high coverage suggests that coverage analysis alone is not strong enough to derive good inductive biases. Before considering this point further, let us measure the coverage of some popular learning algorithms.

6. COVERAGE OF CURRENT LEARNING ALGORITHMS

The last problem investigated in this paper is the evaluation of the coverage of existing learning algorithms. Due to the difficulty of performing coverage analysis for empirical algorithms, one may consider measuring the coverage experimentally by running learning algorithms on every possible training sample.[7] However, this involves an immense amount of computation. Specifically, if the number of features is n and the sample size is m, then we have to run an algorithm on as many as $2^m \binom{2^n}{m}$ samples and test the learnability of 2^{2^n} concepts. This is doable when $n = 3$ (see Diettrich[7]) or 4, but soon becomes unaffordable when $n = 5$.

To reduce these computational costs, we can employ the following two techniques: First, we can exploit the fact that most of the learning algorithms are symmetric with respect to permutations and/or negations of input features. More precisely, if an algorithm learns a concept represented by a Boolean function $f(x_1, x_2, \ldots, x_i, \ldots, x_j, \ldots, x_n)$, then the same algorithm also learns the concepts represented by $f(x_1, x_2, \ldots, x_j, \ldots, x_i, \ldots, x_n)$, $f(x_1, x_2, \ldots, \bar{x}_i, \ldots, x_j, \ldots, x_n)$, and so on for all functions obtained by permuting and/or negating the features in f. These symmetry properties partition the space of concepts into equivalence classes such that it suffices to test one representative concept in each equivalence class to determine learnability for all concepts in the class. It turns out that for $n = 5$, the number of representative concepts one needs to consider is 1,228,158. Also, if the algorithm being tested is also symmetric with respect to complementing the target concept then this number if further reduced to 698,635. This is a considerable reduction since one has to consider more than 4 billion concepts in the exhaustive approach.

Second, we can measure the learnability of each concept statistically by running the learning algorithm on a *large* number of randomly chosen samples, rather than *all* the possible samples. The goal is to estimate the ratio of the samples on which the algorithm returns an ϵ-close hypothesis, and check that this is at least $1 - \delta$ in order to consider a concept learned.

The details of the above cost-reduction techniques are not central to this study. For a thorough discussion of these, we refer the interested reader to work by Almuallim.[3]

6.1 EXPERIMENTAL WORK

Using the above cost-reduction techniques, we experimentally measured the coverage figures for three algorithms: ID3,[13] FRINGE[11] and MDT, which is an exhaustive algorithm that finds a decision tree with fewest nodes consistent with the training sample. The coverage of these algorithms was measured for $n = 5$,

$\epsilon = 3/32$, $\delta = 0.1$ and $m = 8, 10, 12, 14$, and 16. Sampling in these experiments was done *without* replacement.

FRINGE is symmetric with respect to permuting and negating the features. Thus, the number of representative concepts we tested for this algorithm was 1,228,158. ID3 and MDT are in addition symmetric with respect to complementing the target concept as well, and thus the number of representative concepts for these algorithms was only 698,635 concepts.

Determining whether or not a concept is learned by an algorithm in the above setting was done in two passes:

- In the first pass, only 100 randomly chosen samples per concept were tested. Concepts for which the number of samples that resulted in ϵ-close hypothesis is less than 60 out of 100, were considered not learned and thus excluded.
- In the second pass, each algorithm was run on 10,000 randomly chosen samples for each of the remaining concepts. Three outcomes are then considered: (i) If the ratio of the samples on which the algorithm returned an ϵ-close hypothesis is less than 0.893, then the concept is considered unlearned. (ii) If the ratio is greater than 0.907, then the concept is considered learned. (iii) If the ratio is between 0.893 and 0.907, then we *hesitate* to make either decisions.

Due to the third outcome in the above process, the coverage figures are eventually given as a range $a \pm b$. The number of samples tested per concept and the margin ± 0.007 were chosen so that the probability of a wrong decision (considering an unlearned concept as learned and vice versa) becomes within 0.01 (see Almuallim[3] for details) that is, to achieve 99% level of significance.

The final results of our experiments are summarized in Table 1.

6.2 COVERAGE OF THE "BALLS" APPROACH

For comparison, let us compute the coverage obtained by the balls approach under the same conditions of our experiments. We look at two algorithms:

LARGE-BALL. Without loss of generality, assume that c_1 in Large-Ball is just the *nil* concept. This means that the algorithm guesses *negative* for all the examples not included in the sample. Let us define the *weight* of a concept as the number of positive examples in that concept. Then, the algorithm trivially learns all the concepts of weight 0 to 3 (i.e., $Ball(0, 3/32)$).

TABLE 1 Coverage figures for various algorithms.

Algorithm	Sample size				
	8	10	12	14	16
ID3	12 ± 0	332 ± 0	396 ± 0	$1,756 \pm 0$	$4,954 \pm 640$
FRINGE	12 ± 0	332 ± 0	396 ± 0	$1,756 \pm 0$	$5,284 \pm 970$
MDT	12 ± 0	12 ± 0	116 ± 40	496 ± 0	$3,694 \pm 0$
Large-Ball	5,489	5,489	5,489	41,449	41,449
Two-Large-Balls	10,978	10,978	10,978	82,898	82,898
Upper Bound	661,333	2,041,173	6,148,551	17,985,991	50,753,991

ALGORITHM Two-Large-Balls (*Sample*)
1. If $disagreement(Sample, c_1) < disagreement(Sample, \neg c_1)$, then $h = c_1$; else, if $disagreement(Sample, c_1) > disagreement(Sample, \neg c_1)$, then $h = \neg c_1$.
   Otherwise, arbitrarily set h to c_1 or $\neg c_1$.
2. Define the concept s as:
   $$s(X) = \begin{cases} 1 & \text{if } X \in Sample; \\ 0 & \text{otherwise.} \end{cases}$$
3. Define the concept p as:
   $$p(X) = \begin{cases} 1 & \text{if } X \in Sample \text{ and } X \text{ is a positive example;} \\ 0 & \text{otherwise.} \end{cases}$$
4. Return $(\neg x \wedge h) \vee p$.

FIGURE 5   The Two-Large-Balls Learning Algorithm.

Consider a concept c of weight 4. It should be obvious that any sample of c having one or more positive examples is mapped within ϵ from c. The only samples of c that are mapped to an ϵ-far hypothesis (the *nil* hypothesis) are those consisting of negative examples only. The probability of getting such a sample is just

$$\frac{\binom{32-4}{m}}{\binom{32}{m}}$$

which is less than 0.1 when $m \geq 14$. Therefore, the coverage is $\sum_{i=0}^{3} \binom{32}{i} = 5,489$ when $m = 8, 10$ or 12, and $\sum_{i=0}^{4} \binom{32}{i} = 41,449$ when $m = 14$ or 16.

THE TWO-LARGE-BALLS ALGORITHM. Consider the algorithm given in Figure 5. This algorithm is just a consistent version of the Two-Balls algorithm (Figure 1)—it modifies the final hypothesis so that there is no disagreement with the training sample.

Again, without loss of generality, assume that c_1 is the *nil* concept. This means that the algorithm classifies all the examples that are not in the training sample as positive if the majority of the examples in the sample are positive, or as negative otherwise (breaking ties arbitrarily). For those examples included in the training sample, the algorithm gives the same class as given in the sample.

If $m \geq 7$, then Two-Large-Balls behaves exactly like Large-Ball (with $c_1 = nil$) for all the concepts of weight up to 3 since, at the end of Step 1, h will definitely be the *nil* concept. If $m \geq 9$, then this can also be said about the concepts of weight 4.

Similarly, h will be the *true* concept with certainty if the target concept has weight 29 to 32 and $m \geq 7$, or if the target concept has weight 28 and $m \geq 9$. Thus, for these cases, Two-Large-Balls will again behave exactly as Large-Ball but with $c_1 = true$.

As a result, the coverage of Two-Large-Balls is twice as that of Large-Ball, that is, 10,978 when $m = 8, 10$, or 12 and 82,898 when $m = 14$ or 16.

6.3 DISCUSSION

Three important points can be seen in the coverage figures of Table 1:

1. MDT does not give better coverage than the heuristic algorithms ID3 and FRINGE.
2. The coverage of ID3 and FRINGE is disappointingly smaller than that of Large-Ball and Two-Large-Balls.
3. The coverage of all these algorithms is far below the upper bound of Theorem 1.

7. CONCLUSION AND FUTURE EXTENSIONS

We began this paper by suggesting that an important design criterion for learning algorithms should be the coverage of the algorithm. We presented the Multi-Balls algorithm and showed that it can achieve optimal coverage with a sample size that is within a constant factor of optimal. However, we then showed that a fairly trivial algorithm, Large-Ball, can also achieve very large coverage—larger than Multi-Balls in cases where ϵ is reasonably big. Experimental tests confirm that Large-Ball and a variation of Multi-Balls, have much better coverage than the popular ID3 algorithm and its relatives.

Why does Large-Ball strike us as trivial? Because it merely memorizes the training sample—it does not attempt to find any regularity in the data. Furthermore, the concepts it learns, while they are very numerous, are all located near a single concept. In short, the bias of Large-Ball is unlikely to be appropriate in real-world learning situations. This argument shows that coverage analysis alone is not sufficient to find a (practically) useful inductive bias.

This suggests that we combine coverage analysis with other methods for choosing inductive bias. For example, in a paper by Almuallim and Dietterich,[2] we described learning situations in which the MIN-FEATURES bias—the bias that prefers consistent concepts definable over fewer features—is appropriate. However, the MIN-FEATURES bias does not uniquely define a learning algorithm, because, given a training sample, there are typically many consistent hypotheses that have the same, minimum number of features. Hence, *within* the MIN-FEATURES bias, we could apply coverage analysis to design a learning algorithm that has the largest coverage among all algorithms that implement MIN-FEATURES.

In general, let $Pref(c_1, c_2)$ be a preference bias that prefers c_1 to c_2 in all cases where both concepts are consistent with the training sample. Let $Learns(L, c, m, \epsilon, \delta)$ be true if algorithm L can learn concept c from a sample of size m with accuracy and confidence parameters ϵ and δ. The *coverage within bias Pref* for L (with respect to m, ϵ, and δ), is the size of the set

$$C = \{c \mid Learns(L, c, m, \epsilon, \delta) \text{ and }$$
$$\forall c'\ Pref(c', c) \Rightarrow Learns(L, c', m, \epsilon, \delta)\} .$$

That is, a concept is "covered" only if all concepts preferred to it are also covered.

In conclusion, the results from this paper suggest that an important problem for future research is to design and analyze algorithms that have optimal coverage-within-bias for many of the popular biases. This will be particularly important for biases that are so weak that they do not have polynomial sample complexity.

ACKNOWLEDGMENTS

The authors gratefully acknowledge the support of the NSF under grant number IRI-86-57316. Thanks to Bella Bose for clarifying some issues on the Gilbert-Varshamov bound, and to Prasad Tadepalli for comments on an earlier draft of the paper. This work was conducted at Oregon State University while the first author was supported by a doctoral scholarship from King Fahd University of Petroleum and Minerals.

APPENDIX

LEMMA 3. (see Hoeffding[9])Let Y be a binomial random variable with t trails and p as the ratio of success. Then

$$Pr[Y \geq rt] = \sum_{i=\lceil rt \rceil}^{t} \binom{t}{i} p^i (1-p)^{t-i} \leq e^{-2(r-p)^2 t}$$

for any r such that $p < r \leq 1$, and

$$Pr[Y \leq rt] = \sum_{i=0}^{\lfloor rt \rfloor} \binom{t}{i} p^i (1-p)^{t-i} \leq e^{-2(p-r)^2 t}$$

for any r such that $0 \leq r < p$.

LEMMA 4. (see Chernoff[5]) Let Y be a binomial random variable with t trails and p as the ratio of success. Then

$$Pr[Y \geq rt] = \sum_{i=\lceil rt \rceil}^{t} \binom{t}{i} p^i (1-p)^{t-i} \leq e^{-\frac{(r-p)^2}{3p} t}$$

for any r such that $p < r \leq 1$, and

$$Pr[Y \leq rt] = \sum_{i=0}^{\lfloor rt \rfloor} \binom{t}{i} p^i (1-p)^{t-i} \leq e^{-\frac{(p-r)^2}{2p} t}$$

for any r such that $0 \leq r < p$.

LEMMA 5. For any $0 < \epsilon < 1/2$,

$$\sum_{i=0}^{\epsilon k} \binom{k}{i} \leq 2^{kH(\epsilon)}$$

where $H(\epsilon) = -\epsilon \log_2 \epsilon - (1 - \epsilon) \log_2(1 - \epsilon)$.

PROOF. For any integer r, it must be true that

$$2^{r(1-\epsilon)k} \sum_{i=0}^{\epsilon k} \binom{k}{i} \leq \sum_{i=0}^{\epsilon k} 2^{r(k-i)} \binom{k}{i}$$

$$\leq \sum_{i=0}^{k} 2^{r(k-i)} \binom{k}{i}$$

$$= \sum_{j=0}^{k} 2^{rj} \binom{k}{k-j}$$

$$= \sum_{j=0}^{k} 2^{rj} \binom{k}{j}$$

$$= (1 + 2^r)^k \, .$$

Thus

$$\sum_{i=0}^{\epsilon k} \binom{k}{i} \leq \frac{(1 + 2^r)^k}{2^{r(1-\epsilon)k}}$$

$$= \left[2^{-(1-\epsilon)r} + 2^{\epsilon r} \right]^k \, .$$

Now, letting $r = \log_2 (1 - \epsilon)/\epsilon$ (which is clearly positive for $0 < \epsilon < 1/2$), the last quantity is equal to

$$\left[2^{-(1-\epsilon) \log_2(1-\epsilon)+(1-\epsilon) \log_2 \epsilon} + 2^{\epsilon \log_2(1-\epsilon)-\epsilon \log_2 \epsilon} \right]^k$$

$$= \left[2^{-(1-\epsilon) \log_2(1-\epsilon)-\epsilon \log_2 \epsilon+\log_2 \epsilon} + 2^{-\epsilon \log_2 \epsilon-(1-\epsilon) \log_2(1-\epsilon)+\log_2(1-\epsilon)} \right]^k$$

$$= \left[2^{H(\epsilon)} \left\{ 2^{\log_2 \epsilon} + 2^{\log_2(1-\epsilon)} \right\} \right]^k$$

$$= 2^{kH(\epsilon)} [\epsilon + 1 - \epsilon]^k$$

$$= 2^{kH(\epsilon)} \, ,$$

which shows the lemma. $\square$

LEMMA 6. For any set S containing k distinct training examples and any concept $h \in 2^{U_n}$, the number of concepts that are consistent with the training examples in S and within ϵ of h is at most

$$\sum_{i=0}^{\epsilon 2^n} \binom{2^n - k}{i}.$$

PROOF. In how many ways can we construct a concept c such that c is consistent with S and within ϵ of h? For the k bits of c that correspond to the k examples of S, we have no choice but to follow the classification of these examples as given in S. For the remaining $2^n - k$ bits, we can afford to disagree with h on at most $\epsilon 2^n$ bits. Thus, there are at most

$$\sum_{i=0}^{\epsilon 2^n} \binom{2^n - k}{i}$$

ways to construct such c, and the lemma follows. $\square$

LEMMA 7. For any $0 < \epsilon < 1/2$ and any integer j such that $0 \le j \le \epsilon 2^n$, the quantity

$$2^k \binom{2^n - k}{j}$$

is monotonically increasing in k in the range $1 \le k < (1 - 2\epsilon)2^n$.

PROOF.

$$\frac{2^{k+1} \binom{2^n - k - 1}{j}}{2^k \binom{2^n - k}{j}}$$

$$= 2 \cdot \frac{(2^n - k - 1)(2^n - k - 2) \cdots (2^n - k - j)}{(2^n - k)(2^n - k - 1) \cdots (2^n - k - j + 1)}$$

$$= 2 \cdot \frac{2^n - k - j}{2^n - k}$$

$$= 2 \cdot \left(1 - \frac{j}{2^n - k}\right)$$

$$> 2 \cdot \left(1 - \frac{\epsilon 2^n}{2\epsilon 2^n}\right) \quad \text{(by substituting } j = \epsilon 2^n \text{ and } k = (1 - 2\epsilon)2^n)$$

$$= 1.$$

This shows the lemma. □

LEMMA 8. For any ϵ such that $\epsilon < 1/2$ and such that ϵk is an integer

$$\sum_{i=0}^{\epsilon k} \binom{k}{i} \leq \frac{1-\epsilon}{1-2\epsilon} \binom{k}{\epsilon k}.$$

PROOF. The left-hand side of the inequality is

$$\sum_{i=0}^{\epsilon k} \binom{k}{i} = \binom{k}{\epsilon k} + \binom{k}{\epsilon k - 1} + \cdots + \binom{k}{1} + \binom{k}{0}.$$

If we divide the second term by the first, the third by the second, the fourth by the third, and so on, we get

$$\frac{\epsilon k}{k - \epsilon k + 1}, \frac{\epsilon k - 1}{k - \epsilon k + 2}, \frac{\epsilon k - 2}{k - \epsilon k + 3}, \cdots \frac{2}{k-1}, \frac{1}{k}.$$

Note that these ratios are monotonically decreasing. Therefore, we can write

$$\sum_{i=0}^{\epsilon k} \binom{k}{i} \leq \binom{k}{\epsilon k}\left[1 + \left(\frac{\epsilon k}{k - \epsilon k + 1}\right) + \left(\frac{\epsilon k}{k - \epsilon k + 1}\right)^2 + \cdots \right.$$

$$\left. + \left(\frac{\epsilon k}{k - \epsilon k + 1}\right)^{\epsilon k - 1} + \left(\frac{\epsilon}{k - \epsilon k + 1}\right)^{\epsilon k}\right]$$

$$= \binom{k}{\epsilon k} \sum_{i=0}^{\epsilon k} \left[\frac{\epsilon k}{k - \epsilon k + 1}\right]^i$$

$$\leq \binom{k}{\epsilon k} \sum_{i=0}^{\infty} \left[\frac{\epsilon k}{k - \epsilon k + 1}\right]^i$$

$$\leq \binom{k}{\epsilon k} \sum_{i=0}^{\infty} \left[\frac{\epsilon}{1-\epsilon}\right]^i$$

$$= \frac{1-\epsilon}{1-2\epsilon} \binom{k}{\epsilon k}.$$

The last step follows by viewing $\sum_{i=0}^{\infty} [\epsilon/(1 - \epsilon)]^i$ as an infinite sum of a geometric series. This evaluates to $(1 - \epsilon)/(1 - 2\epsilon)$ provided that $\epsilon < 1/2$. □

LEMMA 9. Provided that $\epsilon < 1/4$ and $m < k/4$,

$$\sum_{i=0}^{\epsilon k} \binom{k-m}{i} \le 2e^{-\epsilon m} \binom{k}{\epsilon k}.$$

PROOF. First, note that

$$
\begin{aligned}
\frac{\binom{k-m}{\epsilon k}}{\binom{k}{\epsilon k}} &= \frac{(k-m)!}{(k-\epsilon k-m)!(\epsilon k)!} \cdot \frac{(k-\epsilon k)!(\epsilon k)!}{k!} \\
&= \frac{(k-\epsilon k)(k-\epsilon k-1)\cdots(k-\epsilon k-m+1)}{k(k-1)\cdots\cdots(k-m+1)} \\
&\le (1-\epsilon)^m \\
&\le e^{-\epsilon m}.
\end{aligned}
$$

Now, the quantity $\sum_{i=0}^{\epsilon k} \binom{k-m}{i}$ can be rewritten as

$$
\begin{aligned}
\sum_{i=0}^{(\frac{\epsilon k}{k-m})(k-m)} \binom{k-m}{i} &\le \frac{1-\frac{\epsilon k}{k-m}}{1-2\frac{\epsilon k}{k-m}} \binom{k-m}{\epsilon k} \qquad \text{(by Lemma 8)} \\
&= \frac{k-m-\epsilon k}{k-m-2\epsilon k} \binom{k-m}{\epsilon k} \\
&\le \frac{k-m-\epsilon k}{k-m-2\epsilon k} e^{-\epsilon m} \binom{k}{\epsilon k} \text{(by Eq. (10)).}
\end{aligned}
$$

The lemma follows by verifying that

$$\frac{k-m-\epsilon k}{k-m-2\epsilon k} \le 2$$

when $\epsilon < 1/4$ and $m < k/4$. This is equivalent to

$$k - m - \epsilon k \le 2k - 2m - 4\epsilon k$$

or

$$k \ge m + 3\epsilon k,$$

which is satisfied by the assumptions of the lemma. $\square$

LEMMA 10. There exists a value for α in the range $2\epsilon < \alpha < 1/2$ that satisfies Eq. (2), provided that $0 < \epsilon < 1/4$.

PROOF. First, note that $H(\alpha)$ is monotonically increasing in the range $0 < \alpha < 1/2$. Moreover, H is always strictly less than 1 except when $\alpha = 1/2$, where H becomes exactly 1.

Now, call the left-hand side of Eq. (2) $L(\alpha)$ and the right-hand side $R(\alpha)$. The reader can confirm the following:

- $L(2\epsilon) > 0$, since $2\epsilon < 1/2$ by assumption.
- $R(2\epsilon) = 0$, since $\beta(2\epsilon) = 1$.
- $L(1/2) = 0$, since $H(1/2) = 1$.
- $R(1/2) > 0$, since this is a product of positive quantities.

Therefore, $L(\alpha)$ and $R(\alpha)$ must intersect somewhere in the range $2\epsilon < \alpha < 1/2$ and the lemma holds. $\square$

REFERENCES

1. Almuallim, H., and T. G. Dietterich. "Learning with Many Irrelevant Features." *Proceedings of 9th National Conference on Artificial Intelligence* (AAAI-91) (1991): 547–552.
2. Almuallim, H. "Exploiting Symmetry Properties in the Evaluation of Inductive Learning Algorithms: An Empirical Domain-Independent Comparative Study." Technical Report, 1991-30-09, Department of Computer Science, Oregon State University, Corvallis, OR 97331-3202.
3. Almuallim, H. "Concept Coverage and Its Application to Two Learning Tasks." Ph.D. Thesis, Department of Computer Science, Oregon State University, Corvallis, Oregon. 1992.
4. Blumer, A., A. Ehrenfeucht, D. Haussler, and M. Warmuth. "Learnability and the Vapnik-Chervonenkis Dimension." Technical Report UCSC-CRL-87-20, Department of Computer and Information Sciences, University of California, Santa Cruz, Nov. 1987; *J. ACM* **36(4)** (1990): 929–965.
5. Chernoff, H. "A Measure of Asymptotic Efficiency for Tests of Hypothesis Based on the Sums of Observations." *Ann. Math. Stat.* **23** (1992): 493–509.
6. Haussler, D., and L. Pitt, eds. *Proceedings of the 1988 Workshop on Computational Learning Theory.* San Mateo, CA: Morgan Kaufmann, 1988.
7. Dietterich, T. G. "Limitations on Inductive Learning." In *Proceedings of the Sixth International Conference on Machine Learning*, 124–128. San Mateo, CA: Morgan Kaufmann, 1989.
8. Fulk, M. A., and J. Case, eds. *Proceedings of the Third Annual Workshop on Computational Learning Theory.* San Mateo, CA: Morgan Kaufmann, 1990.
9. Hoeffding, W. "Probability Inequalities for Sums of Bounded Random Variables." *J. Amer. Stat. Assoc.* **58** (1963): 13–30.
10. Natarajan, B. K. "On Learning Boolean Functions." In *Proceedings of the 19th Annual ACM Symposium on Theory of Computing*, 296–304. New York: ACM, 1987.
11. Pagallo, G., and D. Haussler. "Boolean Feature Discoverying Empirical Learning." *Machine Learning* **5(1)** (1990): 71–100.
12. Peterson, W. W., and E. J. Weldon. *Error Correcting Codes*, 86. Cambridge, MA: MIT Press, 1972.
13. Quinlan, J. R. "Induction of Decision Trees." *Machine Learning* **1(1)** (1986): 81–106.
14. Rivest, R. "Learning Decision Lists." *Machine Learning* **2(3)** (1987): 229–246.

15. Rivest, R., D. Haussler, and M. K. Warmuth, eds. *Proceedings of the Second Annual Workshop on Computational Learning Theory.* San Mateo, CA: Morgan Kaufmann, 1989.
16. Ross, S. *A First Course in Probability*, 111, 3rd ed. New York: Macmillan, 1988.
17. Valiant, L. G. "A Theory of the Learnable." *Comm. ACM* **27(11)** (1984): 1134–1142.
18. Valiant, L. G., and M. Warmuth, eds. *Proceedings of the Fourth Annual Workshop on Computational Learning Theory.* San Maateo, CA: Morgan Kaufmann, 1991.

Peter Cheeseman
Artificial Intelligence Research Branch, Mail Stop 269-2, NASA Ames Research Center, Moffett Field, CA 94035; e-mail: cheeseman@pluto.arc.nasa.gov

On Bayesian Model Selection

The task of inducing models (theories) from noisy data can be cast as the problem of finding the most probable model given the data. There is a straightforward Bayesian criterion for the relative probability of alternative models, although searching for the model with the maximum probability may be intractable. The Bayesian approach automatically gives the necessary tradeoff between model complexity and fit to the data, and so avoids the "overfitting" problem. The criterion for the most probable model does not depend on the language used to describe it, but it does depend on prior knowledge of the domain. In contrast, the syntactic measures of "bias" used in most machine learning research depends on the language used to describe the possible models. In addition, although finding the most probable domain model is often regarded as the goal of scientific investigation, in general, it is not the optimal means for making predictions.

1. INTRODUCTION

The fundamental question in statistical inference or machine learning is: "Given data and weak prior domain knowledge, what is the most probable model[1] of the given domain?" This question is at the opposite end of the knowledge spectrum from discovery-based learning and EBG (Explanation Based Generalization). The latter knowledge-intensive approaches pose the question "Given a strong theory of a domain, what are the interesting or important consequences of that theory?" This paper focuses on bottom-up model discovery, often called *inductive* or *empirical* learning, in which the basic problem is distinguishing the model from noise. At least since the time of William of Ockham (c. 1285–1349), people have known that one can always find a sufficiently complex "theory" to "explain" any outcome. For example, a set of N points can always be fitted exactly by a polynomial of order N. However, if there is error in the measurement of the points, as is usually the case, then the polynomial is partially fitting the inherent noise. Ockham's razor tells us not to "overfit" the data.

The solution to this well-known "overfitting" problem is to find a suitable tradeoff between the fit to data and the complexity of the model. Admonitions to keep the model "simple" are too vague for a computer. A model as complex as the data itself can fit the data exactly, but this has very little predictive value for new data. Conversely, models with very little structure do not predict the given data or new data very well. The basic problem then is to find the *appropriate* tradeoff. Some researchers in machine learning treat this as an empirical question,[16] whereas others avoid the problem entirely by investigating only deterministic domains with no noise in the data.

It is not widely known that the overfitting problem was solved in principle over 50 years ago. In 1939, Harold Jeffreys,[13] building on work going back to Bayes and Laplace, provided the theory and mathematical tools for finding the most probable model in noisy data, and successfully applied this method to many complex inductive problems. Bayesian theory explicitly trades model complexity, as determined by prior probabilities, against the (probabilistic) fit to the data. This tradeoff is a direct consequence of Bayes' theorem, requiring no additional assumptions. Bayesian theory was given a firmer foundation in a more recent proof by Cox.[9] This proof starts from some elementary principles that one would expect a rational agent to use in computing quantitative beliefs in propositions given incomplete or uncertain information. These principles lead directly to probability theory, and to Bayes' theorem in particular, as the logically sound way of making inferences under uncertainty. Any other method of inference necessarily violates these basic principles, and so is subject to undesirable behavior. For example, for many non-Bayesian inductive

---

[1]In this chapter, a model is any formal description (mathematical or logical) that assigns probabilities (including 1 or 0) to experimentally observable outcomes. Here, the term "model" can be interchanged with "assumptions," "hypothesis," "concept," or "theory"; the important property these terms share is the ability to make (probabilistic) predictions about possible observations.

methods in the literature, there is a dependence of the "best" model on the order in which the data is considered.

The Bayesian process of model discovery is outlined in the next two sections. This approach not only allows the comparison of models of the same complexity, such as alternative class descriptions with the same number of classes, but also allows the comparison of models with *different* complexity, such as class descriptions with different numbers of classes. In other words, Bayesian methods provide a solution to the problem of trading model complexity against the fit to the data from first principles, with no *ad hoc* assumptions.

2. BAYESIAN INFERENCE AND INFORMATION THEORY

Bayesian theory can be cast into an intuitively appealing form called the Minimum Message Length (MML) criterion[2] in which the most probable theory has the shortest encoding of the theory and data combined. By describing a new example as an instance of an existing abstraction, a shorter *total* message may result. The total message includes the information to encode the abstraction as well as the data given the abstraction. For example, an Identi-Kit picture that constructs a facial image from a small number of "standard" chins, mouths, eyes, etc., allows an enormous data compression compared to the original grey-scale image. In other words, if a set of face pictures has to be sent, then it requires much fewer bits to first transmit the "standard" features, then transmit each face as a combination of these features. If a new example (e.g., a dog's face) is sufficiently different from all existing abstractions, then there may be no encoding that is shorter than just describing the new example directly. In such cases MML indicates that the new example is very surprising relative to abstractions over the previous examples.

This is how the Bayesian (MML) method distinguishes signal from noise—if no encoding of the new example can be found that is shorter than a direct description of the example, then it is random as far as the existing theory is concerned. Such entirely new cases may eventually form the basis of new abstractions if there is some predictability among them. When a new example cannot be compactly described in terms of abstractions of previous data, it means that the previous data has very little predictive value for that example.

In more detail, assume the user has selected a set of discrete mutually exclusive and exhaustive hypotheses $\{H_0, H_1, \ldots, H_n\}$, and assigned *prior* probabilities $p(H_i|c)$, where c is the given general context. If the hypotheses (i.e., theories or models) are sufficiently concrete, they allow the calculation of the probability of

---

[2] For further information on the Minimum Message Length criterion, see Georgeff,[11] Rissanen,[17] and Wallace.[20]

observing any data set D—in other words, $p(D|H_i)$ is a consequence of the definition of H_i. This data probability is often referred to as the *likelihood*. Typically, you have some data D and want to know the (posterior) probability of a hypothesis given that data—i.e., $p(H_i|D)$. Often, you want the particular H_i that gives the maximum posterior probability, as defined by Bayes theorem:

$$p(H_i|D, c) = \frac{p(H_i|c)p(D|H_i, c)}{\sum_i p(H_i|c)p(D|H_i, c)}. \tag{1}$$

Taking the logarithm[3] of this expression turns products into sums, to give:

$$- \log p(H_i|D, c) = - \log p(H_i|c) + - \log p(D|H_i, c) + \text{constant}. \tag{2}$$

If you are only interested in the *relative* probability of different hypotheses, then the last (constant) term in Eq. (2) can be ignored. From information theory, $- \log p(\text{outcome})$ is the (theoretical) minimum message length to encode the particular outcome. This quantity can be thought of as a measure of "surprise" for the outcome. If the base of the log is two, then the message length is in bits, so surprising outcomes take many bits to encode.

The Minimum Message Length is the sum of two terms. The first term is the information to describe the hypothesis, which *increases* for more complex (less probable) hypotheses. The second term is the information required to encode the data given the hypothesis, and this term *decreases* for more complex hypotheses. From Eq. (2), it is clear that the maximum posterior probability criterion is equivalent to MML, which can be expressed symbolically as:

Total message length = Message length to describe the model
+ Message length to describe data given model.

It is the tradeoff between these two terms that gives a quantitative version of Ockham's razor. Although the user may have no interest in the likelihood term $p(D|H)$ *per se*, its evaluation is an essential part of finding the most probable model. Note that the Bayesian distinction between signal and noise is not absolute—the difference in message length between any two hypotheses is proportional to the relative probability of the hypotheses.[4] The next section outlines the practical application of this theory.[5]

---

[3] Since probabilities are ≤ 1, $- \log$ is used instead of log to ensure positive quantities.

[4] Specifically, if ΔM is the difference in message length between the two hypotheses (in bits), then their relative probability is given by $2^{\Delta M}$. It does not take may bits difference in message length before one hypothesis has negligible relative probability.

[5] For a more detailed description of the relationship between Bayesian probability and MML (see Wallace and Freeman[20]).

3. THE BAYESIAN INDUCTIVE PROCEDURE

The basic Bayesian approach to induction over a given a set of data is:

1. Define the model space. This is the space that you think contains the "true" model—finding this model is the goal of inductive learning.
2. Use domain knowledge to assign *prior probabilities* to the models and probability distributions on the values of model parameters.
3. Use Bayes' theorem to obtain the (possibly parameterized) posterior probability of models given the data. To compare models of different complexity, it is necessary to marginalize[6] all the parameters of the competing models. These marginalized probabilities are the probability of the particular model given the data; they give the relative probability of models, regardless of model complexity.
4. Choose a search algorithm to efficiently search the space of possible models for the (locally) maximum posterior probability model.
5. Stop the search when the most probable model is found, or when it is no longer worth searching to find more probable models. Typically, the stopping criterion will involve a tradeoff between model optimality and search time. For most interesting inductive search problems, this occurs because the goal of finding *the* optimal model is computationally too expensive because of local minima, so heuristic search methods are mandatory.

The following examples should make the above inductive steps clear.

3.1 EXAMPLE 1: CURVE FITTING

In curve fitting, the goal is to find a curve that summarizes all the actual information in a set of measurements. For example, modeling of variable-star luminosity as a function of time from a set of noisy measurements using a curve defined by an orthogonal series (e.g., Fourier series or Legendre polynomials). Such a curve describes the mean value of luminosity as a function of time, where the individual measurements deviate from the mean value mainly due to measurement error. This "noise" is often modeled by a Gaussian (normal) distribution, because the error is typically the result of many small independent errors. Choosing a particular series representation defines the model space (Step 1).

The fundamental problem is that the mean curve can always be made closer to the measurements by adding more coefficients to the series representation. If there are as many coefficients as there are measurements, an exact match can be found—clearly, at some point, the curve started fitting the noise. The basic question is: "How do you tell when the curve is fitting the noise?" Classical statistics has

[6]Marginalization is a procedure from probability theory in which variables are eliminated by summing over all the probabilities of the possible values of the variable to be eliminated.

provided many different answers to this question, the current most popular answer is some variety of cross validation.

In principle, the Bayesian answer to the question is simple. It takes a certain amount of information (measured in bits) to describe the coefficients of a particular curve—the more coefficients, the longer this model description.[12] On the other hand, with additional coefficients the information needed to describe the data decreases, because the curve can be chosen to be as close as possible to the measured values. Maximum likelihood methods (a classical statistical approach) are only concerned with fitting the data to the model (the likelihood term), and always favor adding more complexity to the model. The Bayesian approach, on the other hand, minimizes the *sum* of two competing quantities: the information required to describe the model and the information required to describe the data given the model (Step 3). The goal in Bayesian (or MML) search (Steps 4 and 5) is to find the model (or models) that require the minimum *total* information.

The Bayesian criterion has some important consequences. It does *not* state how much structure there is in the domain—only how much structure can be justified by the given data and the assumed model space. For example, in solar intensity measurements there may be a very weak periodic variation that is buried in the measurement noise. In this case, the Bayesian approach only says that the existence of such a weak periodic signal can be more compactly described by calling it noise. This conclusion may be reversed with more data. However, if the total message length is greatly reduced by assuming the model, this is overwhelming evidence that the modeled structure exists, even if it is not the whole story.

It is always possible that the model space does not contain the correct model, and in such cases the nearest approximation to the actual system in the given model space is found. For example, if the frequency of some signal is not constant, then modeling it with a Fourier series will find the nearest average frequency approximation. In other words, the discovery of some structure by asking one question does not mean that there is no more structure to be found by asking a better question. Deciding when the current model space is inadequate, and choosing a better one, is an area of active research.

3.2 EXAMPLE 2: UNSUPERVISED CLASSIFICATION

In unsupervised classification, a set of objects can be modeled by first describing a set of classes ("prototypes"), then describing the objects using these prototypical class descriptions.[7] Each description gives the probabilities of the observable features, assuming that the object belongs to the class. The class descriptions are

[7] Many view the problem of classification as partitioning the set objects according to some criterion and generating class descriptions from these partitions. This *conceptual clustering* approach[15] requires separate criteria for partitioning and for extraction class definitions. Since the end result of conceptual clustering is a set of class descriptions, the Bayesian approach has the advantage of working directly in class description space, and so avoids the need for *ad hoc* criteria.

chosen so that the information required to describe objects in the class is greatly reduced, because they are "close" to the class prototype. This information reduction arises because only the differences between the observed and expected (prototypical) values need to be described. It takes a certain amount of information (in bits) to describe a set of classes (i.e., the information required to state the probabilities of the features given an object belongs to the class). However, these probabilities reduce the information needed to describe the objects, by first describing each object's class membership, then describing how each object differs from the prototype.

The Bayesian approach involves finding the set of classes that minimizes the *total* information (class descriptions + objects given class descriptions). For a randomly chosen set of objects, it is very unlikely that class descriptions can be found for which the total information is less than that used to describe each object individually. Even though by chance some objects will tend to form "clusters," the clustering will be so weak that the information required to describe the class will not compensate for the reduced information required to describe the objects. This Bayesian/MML classification criterion has a long history.[7,8,19]

An important consequence of the Bayesian/MML approach to automatic (unsupervised) classification is that the assignment of objects to classes is necessarily probabilistic.[7,8] Such probabilistic class memberships capture the intuitions behind "family resemblances" described by philosophers, and the "fuzzy set" approach discussed by Cheeseman.[5] In the Bayesian approach, the probabilistic class assignments occur as a direct consequence of Bayes' theorem. A class description assigns probabilities to observable attributes, such as size (i.e., conditional probabilities of the observations given the object belongs to the class), so that when Bayes' theorem is used to invert this probability (i.e., the conditional probability of the class given the observations of the object), the resulting class probabilities are not unity. In other words, if you cannot say for certain what will be observed when you know that an object belongs to a particular class, then you cannot determine an object's class membership with certainty. In practice, if the classes are well separated, the probability of a particular object belonging to the nearest class is so close to unity that little is lost by assuming it belongs to that class.

4. DISCUSSION OF THE BAYESIAN PROCEDURE

Clearly, choosing the model space is a very important step—if an inappropriate choice is made, very poor results may be obtained. For example, using a polynomial representation to fit discontinuous curves is a poor choice. However, in a new domain, there is often little prior knowledge to guide the choice of model space. Fortunately, using a very general model will often turn up systematic effects even if that model is incorrect. These systematic effects are a statistical shadow of the real effects that are operating. Alternatively, you could choose the most general model possible, namely the set of Turing computable functions. This is the choice used in Kolomogorov/Chaitin complexity to establish basic asymptotic results.[4] However, this choice is impractical for inductive inference because the halting problem makes most functions in this space uncomputable in finite time.

Choice of model space is a way of introducing potentially important prior knowledge. If you choose the wrong model space, the Bayesian approach still gives the most probable answer in the space selected. For example, polynomials will fit the overall shape of a discontinuous curve, but will smooth out any discontinuities. If the curve is violently discontinuous, as it is for line spectra, the polynomials will probably not discover anything useful. Poor results are often a signal that a new model space should be tried. Typically, models include an explicit provision for "noise," and the part of the data that cannot be attributed to model is automatically attributed to "noise." One person's noise can be another person's signal.

A model definition essentially gives the probability of the possible outcomes (observables), assuming the model is true. These probabilities are defined by the model, although their value may be estimated from the data. For example, a class definition for "cat" assigns an expected body-weight distribution, expected number of ears, and so on, letting you calculate the probability that a given entity is in this class. The "cat" model does not give necessary and sufficient conditions for membership; it only assigns strong probabilities to particular observables if the entity is a cat.

The requirement that prior probabilities be assigned to the model space is often cited as an insurmountable difficulty, but the practical difficulties of assigning priors over a model space have been greatly exaggerated. Another objection is that prior probabilities introduces an undesirable degree of "subjectivity" into what should be an "objective" problem, as discussed by Cheeseman[6] and Berger.[1] These authors show that the subjective components of the so-called "objective" approach have been hidden, partly by calling them "assumptions." The view taken here is that in any real inductive problem there is always cogent prior knowledge, and failure to use it creates an illusion of "objectivity," while ignoring potentially important information. The Bayesian approach makes incorporation of prior information relatively easy and explicit.

Another advantage of assigning prior probabilities to models is that it does not limit the method to finite sets of theories. To prevent overfitting, machine learning and maximum likelihood approaches often arbitrarily limit the complexity of models they consider. However, this restriction is artificial and may exclude the best answer. For example, in polynomial curve fitting, the set of possible models is indefinitely extensible by adding more and more coefficients to the polynomial representation. Instead of arbitrarily placing an upper bound on the number of coefficients, the Bayesian method adds as many coefficients as are justified by the data. In this approach, it is the amount of data and its structure that limits the complexity of induced models. Note that in the case of polynomials, priors are required for the number of possible coefficients, as well as for parameters that describe the prior probability distribution for each coefficient.

Most interesting inductive problems involve search over a space of possible models, and finding the maximum posterior probability is the goal in such an optimization search. Standard numerical search algorithms, such as steepest decent, conjugate gradient methods, and the Newton-Raphson technique are possible choices of search procedures. Which algorithm is the best for a given problem depends on the amount of data and the properties of the search space. Unfortunately, most interesting inductive tasks have many local minima in a huge search space, making search for *the* optimal solution out of the question. In this case, it is appropriate to employ search methods that attempt to find good local minima, such as simulated annealing[14] or the EM algorithm.[10] There is no global method that is best for all problems or domains. In addition, much better models are usually found by extending the model space than by improving the search methods in a particular space.

5. OVERVIEW OF THE INDUCTIVE PROCESS

The Bayesian inductive process described in the previous section is only part of the whole inductive inference task. One way of viewing the whole task, including the evaluation of models, is shown schematically in Figure 1.

In this cycle, the domain expert first suggests a broad class of possible models, θ—which give a probability of observations x or $p(x|\theta)$. For example, $p(\text{weight}|\text{mean}, \text{variance})$, is the probability of a particular weight observation as a function of the model parameters "mean" and "variance." Model selection corresponds to Step 1 of the Bayesian procedure described earlier. The domain expert also provides prior probabilities over the possible models, $p(\theta)$—Step 2 of the procedure. For example, the expert may provide a prior distribution for the unknown parameters

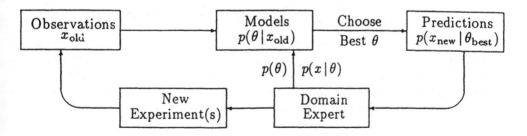

FIGURE 1   The hypothesize-and-test Cycle for Bayesian Induction.

"mean" and "variance." The product of $p(\theta)$ and $p(x|\theta)$ is proportional to the model probability given the data – $p(\theta|x)$. The goal of the Bayesian component of the inductive task is to maximize this posterior model probability, as shown on the top line of Figure 1. This search for the maximum posterior probability model corresponds to Steps 3 and 4 of the Bayesian procedure. In summary, Bayesian model discovery uses the (usually) weak prior information about which model is correct for a particular domain, $p(\theta)$, the data (x_{old}), and the probability of data given the (parametrized) model, $p(x|\theta)$, to define the posterior probability of the model given the data, $p(\theta|x)$. The Bayesian inductive procedure then searches for the maximum posterior model, or as close to it as the search procedure can find.

Although the domain expert plays a part in the overall process through model selection, prior probability assignment, and data selection, a more important role is in the evaluation of the current cycle (the bottom path illustrated in the figure). Typically, the domain experts pick the most likely model based on the current data, and use their domain knowledge to suggest new experiments that would generate more data relevent to supporting or denying the current most probable model. If the data fails to strongly confirm any particular model, this may suggest that the original model framework should be modified or expanded ($\theta \to \theta'$). The Bayesian analysis does not give direct information about this kind of model shift, but it does allow the relative evaluation of models within a given framework based on the given data.

In summary, Figure 1 suggests that modeling a domain from data is an iterative *process* that rarely succeeds in a one-shot pass. Each time around the cycle usually suggests new ways of understanding the domain, which in turn suggest new data to collect and analyze, and so on. There is no guarantee that the hypothesize and test cycle shown in Figure 1 will terminate, but the current most probable model makes the strongest predictions that are possible by any single model, given the available data and prior knowledge.

6. CONSEQUENCES OF THE BAYESIAN APPROACH

In the Bayesian approach, once the model space, priors, and data are chosen, all the degrees of freedom have been used up—there is only one answer to the question "what is the most probable model?" although finding it may be difficult. This approach has a number of consequences when applied to inductive learning and discovery that differ from much of the current practice, as described below.

6.1 LANGUAGE INVARIANCE

A basic principle in inductive learning is that the answer to the question "which model best fits the data?" should *not* depend on the language used to define the model, if the alternative languages are expressively equivalent.[8] For example, the discovery of similar three-dimensional shapes should depend on the intrinsic shapes of the objects and not on whether you use Cartesian or Polar coordinates to describe them, or on the particular viewpoint. However, the use of syntactic "biases," such as minimizing the number of rules in a particular rule syntax, violates this basic requirement. Consider the problem of finding the "best" model for a particular set of training examples using either a conjunctive normal form (CNF) or disjunctive normal form (DNF). A model in one language can be logically mapped into a corresponding model in the other language and *vice versa* (i.e., the languages are expressively equivalent). However, the "complexity" (bias) of the model description in the two languages can be very different. For example, describing a concept as "red or blue" is logically equivalent to describing the same concept as "not (yellow, green,··· , purple)," but these descriptions have very different syntactic complexities, and so could lead to different choices for the "best" model.

The Bayesian approach has the required property of language invariance. This is because the prior probabilities assigned to the components of one language are transformed correctly if the problem is mapped into another language. For example, a high prior probability on straight lines in a Cartesian representation is mapped into an equivalent high prior probability for straight lines in a polar representation, even though the representation of straight lines as equations in polar coordinates is more complex than for circles. Simple syntactic measures of complexity are generally not a good idea—the measure of model complexity (bias) should reflect the prior probabilities of the domain, not the language used to describe it. Note that there is a tendency for natural languages to evolve so that more probable cases are syntactically simpler. This means that syntactic measures of model complexity are often approximately correct.

6.2 MODELS AND SEARCH

The Bayesian approach splits the problem of finding the most probable model into two components: defining the *goal criterion* (e.g., MML) over possible models, and *searching* the space to find the best model under that criterion. This decomposition of the problem differs from that taken in most machine learning approaches. For example, by failing to specify their model space, some researchers implicitly question the need for explicit modeling or its appropriateness for a particular domain. For example, CNF forms, decision trees, and rule sets have all been used in machine learning, but there is usually no discussion as to why they provide suitable models (languages) for their intended applications.

---

[8] This principle was originally proposed in Wallace.[20]

Similarly, neural nets implicitly model domains, but the standard language of "hidden units," "connection weights," and activation functions is difficult to relate to prior knowledge of a particular domain. It is easy to find domains for which a standard language is totally inappropriate—not because it cannot represent the concept to be learned (although this can also be true), but because it requires a very complex description for common examples. For example, a one-dimensional vector of pixel values can represent any image, but it fails to compactly represent two-dimensional concepts, such as line, edge, and region. Because such two-dimensional concepts have very complex descriptions in a one-dimensional language, they are unlikely to be discovered by feeding a neural net raw one-dimensional pixel data.

Model space selection should reflect domain knowledge; it is not an arbitrary choice. For example, using a classification model to capture time-dependent behavior is not generally appropriate, whereas a Markov model or time series may be. Similarly, a Fourier series is appropriate for representing periodic models, whereas a polynomial series is not. Decision trees, such as those constructed by Quinlan's ID3,[16] implicitly assume the existence of a small number of highly relevant attributes instead of a large number of weakly informative attributes. Most of the literature on machine learning implicitly assumes that completely domain-independent model-free learning is possible, while Bayesian analysis denies this possibility.

7. WHY FIND THE BEST MODEL?

It might seem obvious that finding the most probable model is the goal of machine learning, but this assumption is questionable. The usual approaches base all predictions on the most probable model alone, as shown in Figure 1. However, if the goal is to make as accurate predictions as possible about future data, then taking a weighted average over possible models is optimal by the Bayesian criterion.[2,18] Basing all predictions on a single model is jumping to conclusions, and this is especially dangerous when alternative high-probability models make radically different predictions. Given the many hotly debated models of human evolution based on the current fossil evidence, it would be foolish to bet on one as correct. By taking a weighted average of the predictions from the different models, a more reliable prediction can be made (hedging your bets). This weighted prediction maps old data onto new via a weighted set of possible models, as shown in Figure 2.

If the goal is understanding of the domain, and the most probable model is far more likely than the alternatives, then finding this model and using it for prediction is reasonable. It also reduces the storage requirements and the computational effort of prediction. Although minimal computational effort may be important for humans, it is much less important for computers. Consequently, for *machine* learning, it is reasonable to keep all the high-probability models simultaneously and base

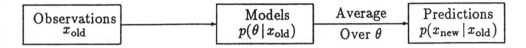

FIGURE 2   Optimal Model-Based Prediction.

predictions on a weighted combination. Retaining many alternatives lets new data change the relative weights of the competing models, instead of backtracking to previously abandoned alternatives, as required for the single best-model approach.

As shown in Figure 2, the use of a weighted combination of models to make predictions adds support to the philosophical approach called "instrumentalism."[3,18] In this approach, theories (models) have no intrinsic value, but are just convenient instruments for making predictions from previous experience. Bayesian theory confirms this view if predictive accuracy is the sole criterion. Even in this case, the prediction is influenced by the *class* of models considered. For example, in time series prediction, if the models are restricted to the first-order Markov assumption,[9] the results may differ from predictions extracted from models that depend on all the previous data.

However, this down-grading of the importance of theories ignores the fact that they are rarely generated in a vacuum, so that any theory generated must be evaluated in the context of existing theories in the same domain. For example, even if I discover a weak statistical correlation between rainfall in the Sahara and car accidents in Los Angeles, I will interpret this as a statistical fluke. In contrast, an identical correlation between rainfall and car accidents in Los Angeles would be interpreted as wet road conditions causing accidents. Why are two identical statistical results interpreted so differently? Clearly, in one case there is an obvious interpretation, whereas in the other case there is no plausible causal explanation.

This example suggests that in domains where there is already a strong theoretical framework (but one that is not strong enough to make specific predictions), the most probable model is *not* just dependent on the data but also on the prior probability of the model, as demanded by the Bayesian analysis. In practice, the "prior" probability of a model is more conveniently evaluated after the data has suggested strong candidates. If such a candidate can be plausibly "explained" in the context, this makes the model much more probable, but a failure to find a plausible explanation reduces the model's probability. If the statistical probability of a model from the data is overwhelming, but there is no plausible explanation within the current theoretical framework, this may indicate a problem with the basic theory. For instance, even though I have a very strong prior disbelief in ghosts, I would change my mind if presented with extremely strong evidence.

[9]The first-order Markov assumption is that the next value depends only on the present value, and not on the history.

---

8. CONCLUSIONS

In Figure 1 is shown the basic hypothesize-and-test cycle that governs human theory discovery. This cycle can be partially formalized so that the model evaluation part of the cycle (finding the most probable model) and predictions based on the model can be automated. The theory for model selection and probabilistic prediction is Bayesian inference, which also provides the necessary tradeoff between model complexity and the corresponding fit to the data. The problem of "overfitting" the data is well known, but many researchers seem unaware of the Bayesian solution. This solution can briefly be described as trying to find the model that minimizes the total message length (MML), where this is defined as the sum of the information needed to describe the model and the information to describe the data given the model.

The model with the minimum total message length is also the model with the maximum posterior probability, by Bayes theorem. Roughly speaking, a more complex model requires more bits to describe but, because this model can fit the data better, the message length to describe the data is correspondingly reduced. The basic question is whether the cost (in bits) of added model complexity is paid by improved prediction (also in bits). The goal is to find the model with the minimum *total* message length. For most interesting problems, achieving this goal is NP-hard, so heuristic methods are necessary that do not guarantee to find the MML model. Even when the best model is found, there is no guarantee that it is the "true" model, as more data may make a more complex model more probable.

The MML model is guaranteed to be the best *single* model predictor (in the given model space) for new data, but even better predictions can be made by taking an average of the predictions of all the models weighted by their respective posterior probability. This calls into question the utility of always selecting the best model and acting as if it is true, especially if prediction is the goal. Since a computer can entertain many hypotheses simultaneously, it is not good practice to always jump to conclusions unless the evidence overwhelmingly supports a particular hypothesis.

A major feature of the Bayesian/MML criterion for model selection is that it is independent of the language used to express the model (for expressively equivalent languages). This is possible because the prior probabilities over the model space are transformed accordingly if a different language is used. For example, a high prior probability of straight lines remains unaltered no matter what perspective is used. Methods that rely on syntactic "biases" do not have this language independence property. Although the Bayesian criterion for model selection is independent of the language, the average computational cost (mainly search) can differ considerably between languages. There is currently no theory on how to select a representation language to minimize computional cost.

An essential message of the Bayesian analysis is that inductive learning is inherently model dependent and that prior beliefs about the possible models affect the evaluation of the best (most probable) model. However, sufficient data will overcome prior beliefs if they do not fit the data well. Domain-independent or general (assumption-free) learning is not possible—every learning system implicitly or explicitly makes assumptions. The Bayesian approach makes the use of prior knowledge explicit so that these beliefs can be checked against the data. New models can be tried if the current models are found inadequate.

REFERENCES

1. Berger, J. O. "Statistical Analysis and the Illusion of Objectivity." *Am. Sci.* **76** (1988): 159–65.
2. Bretthorst, G. L. *Bayesian Spectrum Analysis and Parameter Estimation.* Berlin: Springer-Verlag, 1988.
3. Bridgman, P. W. *The Logic Of Modern Physics.* New York: Macmillan, 1927.
4. Chaitin, G. J. "On the Length of Programs for Computing Finite Sequences." *CACM* **13** (1966): 547–549.
5. Cheeseman, P. "Probabilistic Versus Fuzzy Reasoning." In *Uncertainty in Artificial Intelligence*, edited by L. N. Kanal and J. F. Lemme, 85–102. Amsterdam: Elsevier, 1986.
6. Cheeseman, P. "Inquiry into Computer Understanding + Discussions and Rebuttal." *Comp. Intel.* **4** (1988): 57–142.
7. Cheeseman P., D. Freeman, J. Kelly, M. Self, J. Stutz, and W. Taylor. "Autoclass: A Bayesian Classification System." In *Proceedings of the Fifth International Conference on Machine Learning*, 54–64. Ann Arbor, MI: Morgan Kaufmann, 1988.
8. Cheeseman, P., D. Freeman, J. Kelly, M. Self, J. Stutz, and W. Taylor. "Bayesian Classification." In *Proceedings of the Seventh National Conference on Artifical Intelligence*, 607–611. St. Paul, MN: Morgan Kaufmann, 1988.
9. Cox, R. T. *The Algebra of Probable Inference.* Baltimore: John Hopkins Press, 1961.
10. Dempster, A. P., N. M. Laird, and D. B. Rubin. "Maximum Likelihood from Incomplete Data via the EM Algorithm." *J. Roy. Stat. Soc., Series B* **39** (1977): 1–38.
11. Georgeff M. P., and C. S. Wallace. "A General Selection Criterion for Induction Inference." In *Proceedings of the European Conference on Artificial Intelligence*, 473–482. Amsterdam: Elsevier, 1984.
12. Gull, S. F. "Bayesian Inference and Maximum Entropy." In *Maximum Entropy and Bayesian Methods in Science and Engineering*, edited by G. J. Erickson and C. R. Smith, 53–74, Vol. 1. Dordrecht: Kluwer, 1988.
13. Jeffreys, H. *Theory of Probability.* Oxford: Clarendon Press, 1983.
14. Kirkpatrick, S., C. D. Gelatt, and M. P. Vecchi. "Optimization by Simulated Annealing." *Science* **220** (1983): 671–680.

15. Michalski, R., and R. Stepp. "Learning from Observation: Conceptual Clustering." In *Machine Learning: An AI Approach*, edited by R. S. Michalski, 331-=-363, Vol. 1. San Mateo, CA: Morgan Kaufmann, 1983.

16. Quinlan, J. R. "The Effect of Noise on Concept Learning." In *Machine Learning*, edited by R. S. Michalski, J. G. Carbonell, and T. M. Mitchell, 69–166, Vol. 2. San Mateo, CA: Morgan Kaufmann, 1986.

17. Rissanen, J. "Modeling by Shortest Data Description." *Automatica* **14** (1978): 445–471.

18. Self, M., and P. Cheeseman. "Bayesian Prediction for Artificial Intelligence." In *Proceedings of Workshop on Uncertainty in Artificial Intelligence*, 61–69, 1987.

19. Wallace, C. S., and D. M. Boulton. "An Information Measure for Classification." *Computer J.* **11** (1968): 185–195.

20. Wallace, C. S., and P. R. Freeman. "Estimation and Inference by Compact Encoding." *J. Roy. Stat. Soc., Series B* **49** (1987): 233–265.

Grace Wahba, Chong Gu, Yuedong Wang, and Richard Chappell
University of Wisconsin, Madison Statistics, Department TR 899, Madison, WI

Soft Classification, a.k.a. Risk Estimation, via Penalized Log Likelihood and Smoothing Spline Analysis of Variance

We discuss a class of methods for the problem of "soft" classification in supervised learning. In "hard" classification, it is assumed that any two examples with the same attribute vector will always be in the same class, (or have the same outcome) whereas in "soft" classification, two examples with the same attribute vector do not necessarily have the same outcome, but the *probability* of a particular outcome does depend on the attribute vector. In this chapter we will describe a family of methods that are well suited for the estimation of this *probability*. The method we describe will produce, for any value in a (reasonable) region of the attribute space, an estimate of the *probability* that the next example will be in class 1. Underlying these methods is an assumption that this probability varies in a smooth way (to be defined) as the predictor variables vary. The method combines results from *P*enalized log likelihood estimation, *S*moothing splines, and *A*nalysis of variance to get the PSA class of methods. In the process of describing PSA we discuss some issues concerning the computation of degrees of freedom for signal, which has wider ramifications for the minimization of generalization error in machine learning. As an illustration we apply the method to the Pima-Indian Diabetes data set in the UCI Repository, and compare the results to those of Smith et al.[35] who used the ADAP learning algorithm on this same data set to forecast the onset of diabetes mellitus. If the probabilities we obtain are thresholded to make a hard classification to compare with the hard classification of Smith et al.,[35] the results are very similar; however, the intermediate probabilities that we obtain provide useful and interpretable information on how the risk of diabetes varies with some of the risk factors.

1. INTRODUCTION TO "SOFT" CLASSIFICATION AND THE BIAS-VARIANCE TRADE-OFF

A typical problem in medical data analysis is the following: Records of attribute vectors as well as records of the outcome for each example (patient) for n examples are available as training data. Based on the training data, it is desired to predict the outcomes, for any new examples that may be presented in the future with only their attribute vectors. In this chapter we will consider only two outcomes (1 and 0), where 1 indicates that a particular medical condition of interest was (later) found to be true, and 0 indicates that it was found not to be true. As a concrete example, O'Sullivan, Yandell, and Raynor[31] (OYR) consider records from the Western Electric Health study which gave patient blood pressure and cholesterol level at the start of the study, and an indicator 1 or 0 indicating that the patient did or did not have a heart attack in the 19-year follow-up period. Assuming that the attribute pair of blood pressure and cholesterol level has been suitably scaled to a rectangle $\mathcal{T}$, the "hard" classification problem would be to partition $\mathcal{T}$, or, more precisely, some subregion of interest of it, into two non-overlapping regions, one labeled "1," and the other labeled "0." If a neural network (NN) is used for this task, the partition is generally not made explicit; however, if a new example is presented to the (trained) NN, the NN will produce a 1 or a 0 according to which region the attribute vector for the new example lies. In "soft" classification, the desired (trained) algorithm will produce not a 1 or 0 but a value p (usually strictly) between 1 and 0, which is an estimate of the probability that the new example is, or will be a "1." OYR is a prototype for the "soft" classification method that we will be describing. In order to do "soft" classification by the methods we will be discussing, we will be assuming that the desired probability varies "smoothly" with any continuous attribute (predictor variable). Categorical predictor variates will (later) be allowed, and if there are more than a few categories, some "smoothness" penalties on the categorical values will be required. We remark that to talk about probabilities we should carefully construct a "worldview" in which such probabilities make unambiguous sense, and we shall do that later.

It is well known that smoothness penalties and Bayes estimates are intimately related (see, for example, Kimeldorf and Wahba,[21,22] Wahba,[38,43] and Buntine and Weigend[3]). We will not discuss this further in the present chapter except to note that our philosophy with regard to the use of priors in Bayes estimates is to use them to generate reasonable penalty functionals with appropriate free parameters to generate a structure on which estimates can be "hung," (see Wahba,[43] Chapter 3; also Wahba[44]) and then use cross-validation (CV), generalized cross-validation (GCV), unbiased risk estimation (UBR), or some other *performance-oriented* method to choose the free (regularization) parameter(s) in the penalty functional to minimize some computable proxy for the generalization error (a.k.a. the bias-variance trade-off; see Geman, Bienenstock and Doursat.[8]) A person who completely believed the associated prior might use maximum likelihood to choose the free parameters, but maximum likelihood may not be robust against an unrealistic prior (that is, it may not do very well from the generalization point of view if the prior is not completely up to snuff); see Wahba.[40] Another proposal frequently put forward is to assign a hyperprior to the free parameters. However, except in particular cases where much is known *a priori*, there is no reason to believe that the use of hyperpriors will beat out a performance-oriented criterion which is a good proxy for the generalization error, assuming, of course, that low-generalization error is the true goal.

The "soft" classification structure that we will describe in this chapter historically begins with the penalized log likelihood risk estimate of O'Sullivan[29] and OYR, which was extended by Gu[11] in such a way that penalized log likelihood risk estimation could be combined with smoothing spline analysis of variance (SS-ANOVA) as described by Wahba and collaborators.[4,10,16,17,18,19,41,43] The SS-ANOVA allows a variety of interpretable structures for the possible relationships between the predictor variables and the outcome. Other recent work with the same goal but with technically different approaches than that described here include those by Gray[9] and Tibshirani and LeBlanc.[37] For an informative overview of this area from a statistician's point of view, see Ripley.[32]

Gu[13] has brought to the fore some rather subtle issues concerning the implementation of GCV and UBR in choosing (possibly multiple) smoothing parameters in the context of non-Gaussian (that is, nonquadratic) log likelihoods. Both of these estimates require the calculation of the *degrees of freedom for signal (df-signal)*. We claim that df-signal is going to be a key quantity in any method that does not seek to fit the data exactly, since it is intimately related to how close the model reproduces the training data. In this chapter we will review and discuss some of the results by Gu[13] and apply them to a *P*enalized log likelihood, *S*moothing spline, *A*nalysis of variance (PSA) model for "soft" classification. Our discussion concerning df-signal below is related to an intriguing proposal by Moody,[28] and has possible ramifications in other structures for machine learning, with respect to the calculation of df-signal in the general case where there is a nonquadratic optimization problem to be solved numerically. We will, in our example, use a UBR method (see

Craven and Wahba[6] and references cited there) modified for the binomial case and implemented by the self-voting algorithm by Gu,[13] where these subtle issues are discussed in some detail.

In Section 2 we review the penalized log likelihood estimate described in OYR and use that as a simple vehicle to describe the bias-variance trade-off (a.k.a. generalization error). In Section 3 we discuss the above-mentioned subtle issues in choosing the smoothing parameters and the key quantity df-signal. In Section 4 we describe the general PSA model, and discuss how to compute it. In Section 5 we apply several PSA models to the estimation of the risk of diabetes mellitus, from the Pima-Indian data set in the UCI Repository of Machine Learning Databases. We compare the best of the PSA models to the use of the ADAP NN classification algorithm as applied by Smith et al.[35] to the same Pima-Indian data set.

2. SOFT CLASSIFICATION AND PENALIZED LOG LIKELIHOOD RISK FACTOR ESTIMATION

First, to describe the "worldview" adopted in this chapter, let t be a vector of attributes, $t \in \Omega \in \mathcal{T}$, where Ω is some region of interest in attribute space $\mathcal{T}$. We imagine that the "world" consists of an arbitrarily large population of potential examples, whose attribute vectors are distributed in some way over Ω and, furthermore, considering all members of this "world" with attribute vector in a small neighborhood about t, the fraction of them that are 1's is $p(t)$. We are implicitly assuming that small neighborhoods about t can be defined.

Our training set is assumed to be a random sample of n examples from this population, whose classification is known, and our goal is to be able to estimate $p(t)$ for any $t \in \Omega$. In "soft" classification, we do not expect classification to be a "sure thing"; that is, we do not expect $p(t)$ to be 0 or 1 for large portions of Ω. Here is how we would use our estimate $\hat{p}(t)$ of $p(t)$—if our classification is the presence or absence of some medical condition, then, when a new patient (example) appears, with attribute vector t, we announce that their risk of getting the medical condition is approximately $\hat{p}(t)$; furthermore, (in some situations) we can also announce that if they change their attribute vector to t', then they can, approximately, change their risk to $\hat{p}(t')$. A "hard" classification can be done if desired, by thresholding at a fixed value of $\hat{p}$; however, in medical applications in particular, the "patient" would probably prefer knowing $\hat{p}$, rather than just which side of the threshold that they fell in. For example, the probability $\hat{p}$ may be useful in suggesting the urgency of treatment. The concept of a probability is also useful in large demographic studies, where one may wish to estimate what proportion of a population will later develop some medical condition.

We now review the penalized log likelihood risk estimate by O'Sullivan[29] and OYR. First, define the logit $f(t)$ by $f(t) = \log[p(t)/(1 - p(t))]$; the logit provides a convenient and commonly used means of transforming the unit interval into the real line; see, for example, McCullagh and Nelder.[27] In OYR, t is a vector containing two continuous variates, $t = (t_1, t_2)$, and the logit f is assumed to be "smooth" as a function of these continuous variates, in the sense of possessing square integrable second derivatives. More precisely OYR assume that f is in an appropriately defined collection of functions for which the thin plate spline "smoothness penalty" $J(f)$ defined by

$$ J(f) = \int_{-\infty}^{\infty} \int_{-\infty}^{\infty} \left(\frac{\partial^2 f}{\partial t_1^2} \right)^2 + 2 \left(\frac{\partial^2 f}{\partial t_1 \partial t_2} \right)^2 + \left(\frac{\partial^2 f}{\partial t_2^2} \right)^2 dt_1 dt_2 \qquad (2.1) $$

is well defined and finite. (See Wahba and Wendelberger,[45] or Wahba,[43] and references cited there for further information concerning this penalty functional.) The OYR estimate $p_\lambda(t)$ of $p(t)$ is then obtained, for any fixed nonnegative λ, as the minimizer, in the above-mentioned collection of functions, of

$$ - \log \text{ likelihood}\{\text{data}, f\} + \frac{n}{2} \lambda J(f) . \qquad (2.2) $$

The log likelihood is defined as follows: Let the training data be $\{y_i, t(i), i = 1, \ldots, n\}$ where y_i has the value 1 or 0 according to the classification of example i, and $t(i) = (t_1(i), t_2(i))$ be the attribute vector for example i. If the n examples are a random sample from our "world," then the likelihood of this data, given $p(\cdot)$, is (with some abuse of notation)

$$ \text{likelihood}\{y, p\} = \Pi_{i=1}^n p(t(i))^{y_i} (1 - p(t(i)))^{1 - y_i} , \qquad (2.3) $$

which is the product of n binomial likelihoods. By substituting in f and taking logs, we have

$$ - \log \text{ likelihood}\{y, f\} \equiv \mathcal{L}(y, f) = \sum_{i=1}^n \log(1 + e^{f(t(i))}) - y_i f(t(i)) . \qquad (2.4) $$

The minimizer, f_λ, of

$$ \mathcal{L}(y, f) + \frac{n}{2} \lambda J(f) \qquad (2.5) $$

will be taken as the estimate of $f(t)$. Assuming that the $t(i) = (t_1(i), t_2(i))$ do not fall on a straight line, f_λ is known to be in the n-dimensional space of functions with a representation

$$ f_\lambda(t) = d_0 + d_1 t_1 + d_2 t_2 + \sum_{i=1}^n c_i |t - t(i)|^2 \log |t - t(i)| , \qquad (2.6) $$

where $|t - t(i)|$ is the Euclidean distance between t and $t(i)$ and the $\{c_i\}$ satisfy the three conditions $0 = \sum_{i=1}^n c_i = \sum_{i=1}^n c_i t_1(i) = \sum_{i=1}^n c_i t_2(i)$; see Wahba,[43] Wahba and Wendelberger,[45] and references cited there. A Newton-Raphson iteration can be used to compute the coefficients. The likelihood function $\mathcal{L}$ will be maximized if $p(t(i))$ is 1 or 0 accordingly as y_i is 1 or 0; hence, the reader may be convinced that as $\lambda \to 0$, f_λ must tend to $+\infty$ or $-\infty$ at the data points. Thus, by letting λ be small, we can come close to fitting the data points, but it is fairly clear that unless the 1's and 0's are well segregated in attribute space, f_λ will be a very "wiggly" function and the generalization error (which we have not exactly defined yet) is likely to be large. For the moment, think of the generalization error as a failure of $\hat{p}(t) \equiv p_\lambda(t)$ to adequately approximate $p(t)$ according to some meaningful criterion. If λ is very large, it can be shown that f_λ will tend to a linear function of the components of t. Then, unless the true logit function is of this form, the generalization error can be expected to be large. We note that it is common in medical data analysis to fit a parametric model to the data in which the logit is linear in the attribute vector components.[27] The choice of λ here represents a trade-off between overfitting and underfitting the data, and this is the soft classification version of the bias-variance trade-off discussed by Geman et al.[8] In practice, it will generally be very important to obtain a good value of λ. Before proceeding, then, we should decide what we mean by a good value of λ. Given the family $p_\lambda, \lambda \geq 0$, we want to choose λ so that p_λ is close to the "true" but unknown p in some sense. Then, if a large number of new examples arrive with attribute vector in a neighborhood of t, $p_\lambda(t)$ will be a good estimate of the fraction of them that are 1's. "Closeness" can be defined in a number of reasonable ways, for example, as the norm of the difference between p_λ and p in some function space. In this chapter we will primarily use the Kullbach-Leibler distance $KL_\nu(p_\lambda, p)$, a commonly used criterion sometimes appearing in the NN literature under other names. (Note that it is not a real distance.) We will mention other criteria later. If $\nu(t)$ is some probability measure on $\mathcal{T}$, define $KL_\nu(p_\lambda, p)$ with respect to ν as

$$KL_\nu(p_\lambda, p) = - \int \left[p(t) \log \left(\frac{p_\lambda(t)}{p(t)} \right) + (1 - p(t)) \log \left(\frac{1 - p_\lambda(t)}{1 - p(t)} \right) \right] d\nu(t). \quad (2.7)$$

$KL_\nu(p_\lambda, p)$ as a measure of closeness of p_λ to p reflects the following "game": Nature chooses a new example with attribute t according to the probability distribution $\nu(t)$. Then the computer scientist–statistical data analyst (cs-sda) announces that the probability of a 1 for this example is $p_\lambda(t)$. Nature now chooses the outcome for this example as a 1 with probability $p(t)$ and 0 with probability $1 - p(t)$. If the example turns out to be a 1 the "loss" to the cs-sda is $- \log(p_\lambda(t)/p(t))$ and if the example is a 0, then the "loss" is $- \log(1 - p_\lambda(t)/1 - p(t))$. Thus the expected loss is

$$- \left[p(t) \log \left(\frac{p_\lambda(t)}{p(t)} \right) + (1 - p(t)) \log \left(\frac{1 - p_\lambda(t)}{1 - p(t)} \right) \right] \quad (2.8)$$

and averaging over the distribution ν of the t's gives Eq. (2.7). Note that the expected loss is minimized if $p_\lambda = p$. Since KL_ν is not computable from the data, it is necessary to develop a computable proxy for it. By a computable proxy is meant a function of λ that can be calculated from the training set which has the property that its minimizer is a good estimate of the minimizer of KL_ν. Note that to minimize KL_ν, it is only necessary to minimize

$$- \int [p(t)\log(p_\lambda(t)) + (1 - p(t))\log(1 - p_\lambda(t))]\, d\nu(t) \qquad (2.9)$$

over λ since Eq. (2.7) and Eq. (2.9) differ by something that does not depend on λ. Leaving-out-half cross validation ($1/2CV$) is one conceptually simple and generally defensible (albeit possibly wasteful) way of choosing λ to minimize a proxy for $KL_\nu(p_\lambda, p)$. The n examples are randomly (important!) divided in half and the first $n/2$ examples are used to compute p_λ for a series of trial values of λ. Recall that, since p_λ has a representation in terms of a set of basis functions, once the coefficients have been computed $p_\lambda(t)$ can be evaluated relatively cheaply for any attribute vector t in Ω. Then, the remaining $n/2$ examples are used to compute

$$\widehat{KL}_{1/2CV}(\lambda) = -\frac{2}{n}\sum_{i=n/2+1}^{n} [y_i \log p_\lambda(t(i)) + (1 - y_i)\log(1 - p_\lambda(t(i)))]$$

$$= -\frac{2}{n}\sum_{i=n/2+1}^{n} [y_i f_\lambda(t(i)) - \log(1 + e^{f_\lambda(t(i))})] \qquad (2.10)$$

for the trial values of λ. Since the expected value of y_i is $p(t(i))$, Eq. (2.10) is, for each λ, an unbiased estimate of Eq. (2.10) with $d\nu$ the sampling distribution of $t(n/2+1), \ldots, t(n)$. The parameter λ would then be chosen by minimizing Eq. (2.10) over the trial values. Note that it is inappropriate to just evaluate Eq. (2.10) using the same data that was used to obtain f_λ, as that would lead to overfitting the data. Variations on Eq. (2.10) are obtained by successively leaving out groups of data. A repeated leaving-out-one (or ordinary cross validation (OCV)) proxy for $\widehat{KL}_\nu(p_\lambda, p)$ would go as follows: Let $f_\lambda^{[k]}$ be the estimate of f_λ (i.e., the minimizer of Eq. (2.5) with the kth data point left out). Then the OCV proxy for KL_ν is

$$\widehat{KL}_{OCV}(\lambda) = -\frac{1}{n}\sum_{k=1}^{n} [y_k f_\lambda^{[k]}(t(k)) - \log(1 + e^{f_\lambda^{[k]}(t(i))})] \qquad (2.11)$$

and λ is chosen to minimize Eq. (2.11). Essentially this estimate was suggested by Cox and Chang[5] in the case of a single predictor variable with $J(f) = \int (f''(t))^2 dt$. In this case f_λ is a cubic spline, and Cox and Chang proposed a computational algorithm which used special computationally efficient methods available for polynomial splines. While OCV represents a relatively efficient use of the data, the computation required is likely to be expensive in general. In the next Section we will describe approximate GCV and UBR proxies for Eq. (2.9), which we have been able to compute in more complicated situations.

3. DF-SIGNAL AND THE GCV AND UBR ESTIMATES FOR λ

In order to understand the GCV and UBR estimates we will describe in the "soft" classification context, we will first describe their role in a simpler setup, namely, when $\mathcal{L}$ is quadratic in the data and the unknown. This is the situation where we have n examples $\{y_i, t(i), i = 1, 2, \ldots, n\}$, where our "world view" says that nature chooses t from the distribution ν, and then y_i is related to f by

$$y_i = f(t(i)) + \epsilon_i, \quad i = 1, \ldots, n, \tag{3.1}$$

where f is assumed to be smooth as before, and the ϵ_i are assumed to be independent, zero-mean Gaussian random variables with mean 0 and common, possibly unknown variance σ^2. It is desired to estimate f from this data. The estimate f_λ is the minimizer of

$$\sum_{i=1}^{n}(y_i - f(t(i)))^2 + n\lambda J(f). \tag{3.2}$$

These estimates are discussed by Wahba[43] and in many of the 13 pages of references cited there; see also Wahba.[44] There is a so-called smoother matrix $A(\lambda)$ defined by the property

$$\begin{pmatrix} f_\lambda(t(i)) \\ \vdots \\ f_\lambda(t(n)) \end{pmatrix} = A(\lambda)y. \tag{3.3}$$

Smoother matrices are symmetric nonnegative definite and have all their eigenvalues in the interval $[0, 1]$. By analogy with ordinary regression, the trace of $A(\lambda)$ is known as the degrees of freedom for signal (df-signal); see Wahba,[39] and Buja, Hastie, and Tibshirani.[2]

The OCV estimate of λ in this context was suggested by Wahba and Wold[46]; we review it here to show its relationship to GCV (generalized cross validation). Letting $f_\lambda^{[k]}$ be the minimizer of Eq. (3.1) with the kth data point left out, the OCV estimate of λ is the minimizer of $V_0(\lambda)$ defined by

$$V_0(\lambda) = \frac{1}{n}\sum_{k=1}^{n}(y_k - f_\lambda^{[k]}(t(k)))^2 \tag{3.4}$$

and the celebrated "leaving-out-one" lemma (a proof is in Wahba[43]) gives the identity

$$V_0(\lambda) \equiv \frac{1}{n}\sum_{k=1}^{n}\frac{(y_k - f_\lambda(t(k)))^2}{(1 - a_{kk}(\lambda))^2}, \tag{3.5}$$

where a_{kk} is the kkth entry of $A(\lambda)$. The GCV estimate of λ is the minimizer of $V(\lambda)$ defined by

$$V(\lambda) = \frac{1}{n} \sum_{k=1}^{n} \frac{(y_k - f_\lambda(t(k)))^2}{(1 - \frac{1}{n}\sum_{\ell=1}^{n} a_{\ell\ell}(\lambda))^2} \tag{3.6}$$

$$\equiv \frac{\frac{1}{n}\|(I - A(\lambda))y\|^2}{\left(\frac{1}{n}tr(I - A(\lambda))\right)^2}. \tag{3.7}$$

Both $V_0(\lambda)$ and $V(\lambda)$ are proxies for the criterion $R(\lambda)$ defined by

$$R(\lambda) = \int (f_\lambda(t) - f(t))^2 d\nu(t) \tag{3.8}$$

in the sense that the minimizers of $V(\lambda)$ and $V_0(\lambda)$ are good estimates of the minimizer of the expected value of $R(\lambda)$, with $V(\lambda)$ having superior theoretical properties under certain circumstances, and much superior computational properties. See Craven and Wahba[6] and Li.[23,24] The expected value here is taken over the random variables ϵ_i. The UBR estimate in this context was proposed by Craven and Wahba[6] based on Mallows celebrated C_p.[26] This estimate requires the knowledge of, or a good estimate of, σ^2 and is the minimizer of

$$U(\lambda) = \frac{1}{n}\|(I - A(\lambda))y\|^2 + 2\frac{\sigma^2}{n}trA(\lambda). \tag{3.9}$$

In what follows we will take the sampling distribution for the $\{t(i)\}$ as a proxy for $d\nu$ of Eq. (3.8). Then $U(\lambda)$ is a proxy for $R(\lambda)$ in that

$$EU(\lambda) = ER(\lambda) + \sigma^2. \tag{3.10}$$

We include a short proof since it is so simple: Let $f = (f(t(1)), \ldots, f(t(n)))'$, $\epsilon = (\epsilon_1, \ldots, \epsilon_n)'$, then

$$EU(\lambda) = E\frac{1}{n}\|(I - A(\lambda))(f + \epsilon)\|^2 + 2\frac{\sigma^2}{n}trA(\lambda)$$

$$= \frac{1}{n}\|(I - A(\lambda))f\|^2 + \frac{\sigma^2}{n}tr(I - A(\lambda))^2 + 2\frac{\sigma^2}{n}trA(\lambda)$$

$$= \frac{1}{n}\|(I - A(\lambda))f\| + \frac{\sigma^2}{n}trA^2(\lambda) + \sigma^2$$

$$= ER(\lambda) + \sigma^2.$$

We note that a crude version of the argument supporting the properties of $V(\lambda)$ as a proxy for $R(\lambda)$ in the case that σ^2 is not known goes as follows (see Craven and Wahba[6] for more details):

$$EV(\lambda) = \frac{[\frac{1}{n}\|(I - A(\lambda))f\|^2 + \sigma^2 tr(I - A(\lambda))]}{(1 - \frac{1}{n}trA(\lambda))^2}. \tag{3.11}$$

Assuming that $(1/n)\,tr\,A(\lambda)$ is small compared to 1 in the neighborhood of the optimal λ (a condition insuring that this is true is necessary for GCV to work well), then

$$EV(\lambda) \sim [\frac{1}{n}\|(I - A(\lambda))f\|^2 + \frac{1}{n}\sigma^2 tr(I - A(\lambda))^2][1 + \frac{2}{n}tr\,A(\lambda) + \dots]$$

$$\sim [\frac{1}{n}\|(I - A(\lambda)f\|^2 + \frac{\sigma^2}{n}tr\,A^2(\lambda) + \sigma^2][1 + o(1)]$$

$$\sim [ER(\lambda) + \sigma^2][(1 + o(1)].$$

Now, let us return to the case of soft classification, where $1/2 \sum_{i=1}^n (y_i - f(t(i)))^2$, which is a multiple of the negative Gaussian log likelihood, is replaced by a negative log likelihood of the general form

$$\mathcal{L}(y, f) = \sum_{i=1}^n [b(f(t(i))) - y_i f(t(i)))], \tag{3.12}$$

where $b(\cdot)$ is given. In the binomial case that we are discussing here $b(f) = \log(1 + e^f)$, but many likelihood functions can be represented this way. See McCullagh and Nelder.[27] For future use we note that in the binomial case it is easy to verify that

$$E(y_i) = p(t(i)) = \frac{e^{f(t(i))}}{1 + e^{f(t(i))}} = b'(f(t(i))), \tag{3.13}$$

$$\mathrm{Var}(y_i) = p(t(i))(1 - p(t(i))) = \frac{e^{f(t(i))}}{(1 + e^{f(t(i))})^2} = b''(f(t(i))); \tag{3.14}$$

however, these relations between the mean and variance of y_i and the first and second derivatives of b hold for any log likelihood of the form (3.12). Representing f either exactly by using a basis for the space of functions in Eq. (3.13), or approximately by suitable basis functions write

$$f \simeq \sum_{k=1}^N c_k B_k; \tag{3.15}$$

then we need to find $c = (c_1, \dots, c_N)'$ to minimize

$$I_\lambda(c) = \sum_{i=1}^n b\left(\sum_{k=1}^N c_k B_k(t(i)))\right) - y_i \left(\sum_{k=1}^N c_k B_k(t(i))\right) + \frac{n}{2}\lambda c'\Sigma c, \tag{3.16}$$

where Σ is the necessarily nonnegative definite matrix determined by $J(\sum_k c_k B_k) = c'\Sigma c$. Straightforward calculations show that the gradient ∇I_λ and the Hessian $\nabla^2 I_\lambda$ of I_λ are given by

$$
\nabla I_\lambda = \begin{pmatrix} \frac{\partial I_\lambda}{\partial c_1} \\ \vdots \\ \frac{\partial I_\lambda}{\partial c_N} \end{pmatrix} = X'(p_c - y) + n\lambda\Sigma c, \tag{3.17}
$$

$$
\{\nabla^2 I_\lambda\}_{jk} = \left\{ \frac{\partial^2 I_\lambda}{\partial c_j \partial c_k} \right\} = X'W_c X + n\lambda\Sigma, \tag{3.18}
$$

where X is the matrix with ijth entry $B_j(t(i))$, p_c is the vector with ith entry $p_c(t(i))$ given by $p_c(t(i)) = \frac{e^{f_c(t(i))}}{(1+e^{f_c(t(i))})}$ where $f_c(\cdot) = \sum_{k=1}^N c_k B_k(\cdot)$, and W_c is the diagonal matrix with iith entry $p_c(t(i))(1-p_c(t(i)))$; compare Eqs. (3.13) and (3.14). We next describe the Newton-Raphson iterate for c. Given the ℓth Newton-Raphson iterate $c^{(\ell)}$, a straightforward calculation shows that $c^{(\ell+1)}$ is given by

$$
c^{(\ell+1)} = c^{(\ell)} - (X'W_{c^{(\ell)}}X + n\lambda\Sigma)^{-1}(X'(y - p_{c^{(\ell)}}) + n\lambda\Sigma c^{(\ell)}) \tag{3.19}
$$

and another straightforward calculation shows that $c^{(\ell+1)}$ is the minimizer of

$$
I_\lambda^{(\ell)}(c) = \|z^{(\ell)} - W_{c^{(\ell)}}^{1/2} Xc\|^2 + n\lambda c'\Sigma c, \tag{3.20}
$$

where $z^{(\ell)}$, the pseudo-data, is given by

$$
z^{(\ell)} = W_{c^{(\ell)}}^{-1/2}(y - p_{c^{(\ell)}}) + W_{c^{(\ell)}}^{1/2} Xc^{(\ell)}. \tag{3.21}
$$

Next, we note that the "predicted" value $\hat{z}^{(\ell)} = W_{c^{(\ell)}}^{1/2} Xc$, where c is the minimizer of Eq. (3.20), is related to the pseudo-data $z^{(\ell)}$ by

$$
\hat{z}^{(\ell)} = A^{(\ell)}(\lambda)z^{(\ell)}, \tag{3.22}
$$

where $A^{(\ell)}(\lambda)$ is the smoother matrix given by

$$
A^{(\ell)}(\lambda) = W_{c^{(\ell)}}^{1/2} X(X'W_{c^{(\ell)}}X + n\lambda\Sigma)^{-1}X'W_{c^{(\ell)}}^{1/2}. \tag{3.23}
$$

Elsewhere,[1] it was proposed to obtain a GCV score for λ in Eq. (3.16) as follows: For fixed λ, iterate Eq. (3.19) to convergence. Define $V^{(\ell)}(\lambda)$ as

$$
V^{(\ell)}(\lambda) = \frac{\frac{1}{n}\|(I - A^{(\ell)}(\lambda))z^{(\ell)}\|^2}{\left(\frac{1}{n}tr(I - A^{(\ell)}(\lambda))\right)^2}. \tag{3.24}
$$

[1]See Wahba,[43] Section 9.2 in which the definition of λ there differs from the definition here by a factor of $n/2$. Please note the typographical error in (9.2.18) there where λ should be 2λ.

Letting L be the converged value of ℓ, compute

$$V^{(L)}(\lambda) = \frac{\frac{1}{n}\|(I - A^{(L)}(\lambda))z^{(L)}\|^2}{\left(\frac{1}{n}tr(I - A^{(L)}(\lambda))\right)^2} \sim \frac{\frac{1}{n}\|W_{c^{(L)}}^{-1/2}(y - p_{c^{(L)}})\|^2}{\left(\frac{1}{n}tr(I - A^{(L)}(\lambda))\right)^2} \qquad (3.25)$$

and minimize $V^{(L)}$ with respect to λ. Gu[13] found that the following algorithm for a GCV score for λ in this case was superior. Gu's algorithm goes as follows: Given a starting guess, from $c^{(\ell)}$, obtain $A^{(\ell)}(\lambda)$ and find $\lambda = \hat{\lambda}^{(\ell)}$ to minimize $V^{(\ell)}(\lambda)$; obtain $c^{(\ell+1)}$ by setting $\lambda = \hat{\lambda}^{(\ell)}$ in Eq. (3.19); iterate until convergence. We remark that the algorithms by Gu[13] and by Wahba,[43] and other algorithms, can be directly compared on simulated data by postulating a (synthetic) $p(\cdot)$ as "truth," generating attribute vectors $t(i), i = 1, \ldots, n$, and generating the y_i for these attribute vectors by a random mechanism which lets y_i be 1 with probability $p(t(i))$. One can then estimate p by various methods. Since the "true" p is known, an objective comparison can be made between the "true" p and the estimates, by computing the KL distance or other objective criterion. Gu made the comparison using the symmetrized KL distance $(= 1/2(KL(p, \hat{p}) + KL(\hat{p}, p)))$.

Now, considering the numerator in the right-hand side of Eq. (3.25), if we replace $\|W_{c^{(\ell)}}^{-1/2}(y - p_{c^{(\ell)}})\|^2$ by $\|W^{-1/2}(y - p_{c^{(\ell)}})\|^2$ where W is the diagonal matrix with iith entry $p(t(i))(1 - p(t(i)))$, we have a sum of squares of random variables involving $y_i/\sqrt{p(t(i))(1 - p(t(i)))}$ with variance $\sigma^2 = 1$. This suggests replacing the approximate GCV estimate V of Eq. (3.25) with the UBR estimate

$$U^{(\ell)}(\lambda) = \frac{1}{n}\|(I - A^{(\ell)}(\lambda))z^{(\ell)}\|^2 + 2\frac{\sigma^2}{n}tr A^{(\ell)}(\lambda) \qquad (3.26)$$

with $\sigma^2 = 1$. Gu[13] suggests using this unbiased risk estimate computed via the following algorithm: Given a starting guess, from $c^{(\ell)}$, obtain $A^{(\ell)}(\lambda)$ and find $\lambda = \hat{\lambda}^{(\ell)}$ to minimize $U^{(\ell)}(\lambda)$; obtain $c^{(\ell+1)}$ by setting $\lambda = \hat{\lambda}^{(\ell)}$ in Eq. (3.19); iterate until convergence. Monte Carlo studies by Gu[13] suggest that this estimate is better than the approximate GCV estimate computed in a similar manner, based on a comparison of the symmetrized KL distance. In the remainder of this chapter, we will be using Gu's algorithm for the unbiased risk estimate for λ. It would be nice to have a good understanding of the difference between the "iterate-to-convergence" algorithm and Gu's algorithm, both in the case of UBR and GCV. An argument in the UBR case is a little bit more transparent. To get a good estimate of λ from UBR in the Gaussian case, it is clear that it is necessary to have a reasonably good estimate of σ; furthermore, two λ's compared via $U(\lambda)$ on the basis of different values of σ^2 cannot be expected to be comparable. In the UBR estimate here, the variances for different y_i are in general different but, if $p(t(i))(1 - p(t(i)))$ were known, then the data would be rescaled by $\sqrt{p(t(i))(1 - p(t(i)))}$ so that the variances of the rescaled data would all be 1. The value of $\sqrt{p(t(i))(1 - p(t(i)))}$ is not known, but the rescaling is being done implicitly with an estimate of it. If the iteration is carried to convergence before $U(\lambda)$ is minimized, then the rescaling is being done with different estimates of the standard deviations for different λ's and the comparison of different λ's by looking at $U(\lambda)$ is not necessarily valid. In Gu's algorithm, at the ℓth iteration different λ's are being compared based on the *same* estimate diag $W_{c^{(\ell)}}$ of the variances of the y_i, so, at each iteration at least, $U(\lambda)$ for different λ's can be expected to be more directly comparable. O'Sullivan[30] has considered approximate UBR estimates in the case of penalized log-density and log-hazard estimates which involve the computation of an estimated degrees of freedom for signal. However, the estimate there does not include an implicit estimate of a variable variance.

Moody[28] discusses what may be considered a generalization of the unbiased risk estimate in situations which are in one sense more general than the setup we have been discussing. Since his generalization raises an important question for the case when the penalty functional or regularizer is not quadratic, we will discuss the relationship of his estimate to the UBR discussed here and note the issue. Moody assumes (in our notation) the model Eq. (3.1), where he assumes that ϵ_i are independently distributed with mean 0 and common variance σ^2, (either known or estimated) but not necessarily Gaussian. He considers more general methods of approximating f, he mentions "multilayer perceptrons and radial basis functions or other learning systems." Once the architecture is determined, he says that f will be estimated by f_ω determined by a set of weights ω_i (which play the role of our c), which in turn depend on a smoothing parameter λ. Moody next defines $\mathcal{E}(y,\omega)$ by

$$\mathcal{E}(y,\omega) = \frac{1}{n}\sum_{i=1}^{n} E(y_i, f_\omega(t(i))) \tag{3.27}$$

where $E(y_i, f_\omega(t(i))$ is an unspecified distance between y_i and $f_\omega(t(i)))$. For fixed λ, $f = f_{\lambda,\omega}$ is obtained by finding ω to minimize

$$I_\lambda(y,\omega) = \mathcal{E}(y,\omega) + \lambda S(\omega) \tag{3.28}$$

where $S(\omega)$ is a "smoothness" penalty on f_ω, not necessarily quadratic. Moody proposes that λ be estimated as the minimizer of $\mathcal{E}(y,w)+2\frac{\sigma^2}{n}\rho_{\text{eff}}(\lambda)$, where $\rho_{\text{eff}}(\lambda)$ is called by Moody "the effective number of parameters" and by us the "degrees of freedom for signal," and is the trace of what he calls the generalized influence matrix $G = \frac{1}{2}\tilde{X}H^{-1}\tilde{X}'$ where $\tilde{X}$ and H are the matrices defined by

$$\tilde{X}_{i\alpha} = \frac{\partial}{\partial y_i}\frac{\partial}{\partial \omega_\alpha}\mathcal{E}(\omega), \tag{3.29}$$

$$H_{\alpha\beta} = \frac{\partial}{\partial \omega_\alpha}\frac{\partial}{\partial \omega_\beta}I_\lambda(\omega). \tag{3.30}$$

It is easy to check that G plays the role of the influence matrix $A(\lambda)$ in the case that $\mathcal{E}$ and S are both quadratic with the y_i's treated as though they were independent with a common known variance, by considering

$$I_\lambda(y,\omega) = \|y - X\omega\|^2 + n\lambda\omega'\Sigma\omega, \qquad (3.31)$$

then $\tilde{X} = 2X, H = 2(X'X + n\lambda\Sigma)$, and $G = X(X'X + n\lambda\Sigma)^{-1}X'$, compare Eqs. (3.20), (3.22), and (3.23). If, however, $\partial/\partial y_i\partial/\partial\omega_\alpha\mathcal{E}(y,\omega)$ or $\partial/\partial\omega_\alpha\partial/\partial\omega_\beta S(\omega)$ depends on ω, then it is possible that *where* the derivatives are taken will make a difference, just as happens in the case considered earlier with $\mathcal{E}$ taken as $\mathcal{L}$.

We remark that in an entirely different context, varying df-signal can be related to varying the stopping criterion in an iterative fitting method (see Wahba[42]) we suspect that this phenomenon is fairly general in NN algorithms.

4. SMOOTHING SPLINE ANALYSIS OF VARIANCE (SS-ANOVA)

In the ANOVA approach to estimating a function of d variables, $f(t) = f(t_1,\ldots,t_d)$ is decomposed as

$$f(t) = \mu + \sum_\alpha f_\alpha(t_\alpha) + \sum_{\alpha<\beta} f_{\alpha\beta}(t_\alpha,t_\beta) + \sum_{\alpha<\beta<\gamma} f_{\alpha\beta\gamma}(t_\alpha,t_\beta,t_\gamma) + \cdots \qquad (4.1)$$

where the elements in the expansion are made unique in some manner or other, and, the expansion is truncated in some manner. See, for example, Stone,[36] Friedman,[7] and Buja, Hastie, and Tibshirani.[2] In the smoothing spline ANOVA context, with Gaussian data, the estimate $f_{\lambda,\theta}$ of f is obtained by finding $f_{\lambda,\theta}$ of the form of Eq. (4.1) in an appropriate function space (a reproducing kernel Hilbert space) to minimize

$$\frac{1}{n}\sum_{i=1}^n (y_i - f(t(i)))^2 + \lambda J_\theta(f) \qquad (4.2)$$

where

$$J_\theta(f) = \sum_{\alpha\in M} \theta_\alpha^{-1} J_\alpha(f_\alpha) + \sum_{\alpha,\beta\in M} \theta_{\alpha\beta}^{-1} J_{\alpha\beta}(f_{\alpha\beta}) + \cdots. \qquad (4.3)$$

The referenced function space has been constructed so that the mean μ, the main effects f_α, the two factor interactions $f_{\alpha\beta}$, and so forth are projections onto orthogonal subspaces whose elements satisfy certain side conditions. This generalizes the usual ANOVA decomposition familiar to veterans of some introductory statistics courses; see, for example, Hogg and Ledolter,[20] Chapter 6. Here M is the collection of indices for components with penalty functionals to be included in the model, the $J_\alpha, J_{\alpha\beta}$ and so forth are quadratic "smoothness" penalty functionals, and λ and the θ_β's satisfy an appropriate constraint for identifiability.

We will first describe what happens in this quadratic (Gaussian) context, then we will show how code in the quadratic case can be used as a subroutine in the computation of a PSA model for soft classification. References for the quadratic (SS-ANOVA) case are Gu,[10] Gu, Bates, Chen, and Wahba,[15] Chen, Gu, and Wahba,[4] Gu and Wahba,[16,17,18,19] and Wahba.[41,43]

First, for convenience, linearly relabel the, say, q terms included in the model in Eq. (4.3) so that β may stand for $\alpha, \alpha\beta, \alpha\beta\gamma$ and so forth, to obtain

$$J_\theta(f) = \sum_{\beta=1}^{q} \theta_\beta^{-1} J_\beta(f_\beta) \,. \tag{4.4}$$

It is known in the SS-ANOVA setup that the minimizer of Eq. (4.2) is in the n-dimensional space of functions with representation

$$f_{\lambda,\theta}(t) = \sum_{\nu=1}^{M} d_\nu \phi_\nu(t) + \sum_{i=1}^{n} c_i \sum_{\beta=1}^{q} \theta_\beta R_\beta(t, t(i)) \tag{4.5}$$

where the ϕ_ν span the null space of the penalty functional J_θ, the $R_\beta(\cdot, \cdot)$ are certain reproducing kernels associated with the corresponding terms J_β in the penalty functional and the $\{c_i\}$ satisfy the M conditions $\sum_{i=1}^{n} c_i \phi_\nu(t(i)) = 0, \nu = 1, \ldots, M$. Letting Σ_θ be the $n \times n$ matrix with ijth entry $\sum_{\beta=1}^{q} \theta_\beta R_\beta(t(i), t(j))$, the coefficients $d = (d_1, \ldots, d_M)'$ and $c = (c_1, \ldots, c_n)'$, are obtained by substituting Eq. (4.5) into Eq. (4.2) which then becomes[2]

$$\frac{1}{n}\|y - (\Sigma_\theta c + Td)\|^2 + \lambda c' \Sigma_\theta c \,. \tag{4.6}$$

The minimizing (c, d) satisfy

$$(\Sigma_\theta + n\lambda I)c + Td = y, \tag{4.7}$$

$$T'c = 0 \,. \tag{4.8}$$

See Wahba,[43] Chapter 10. The generic code RKPACK[10] can be used to compute $tr A(\lambda, \theta)$, where $A(\lambda, \theta)$ is the matrix which satisfies $\hat{y} = A(\lambda, \theta)y$, where $\hat{y} = (\Sigma_\theta c + Td)$, to determine $\lambda/\theta_\beta, \beta = 1, 2, \ldots, q$, by UBR or GCV, and to obtain c and d, given the ingredients y, Σ_θ and T, and in the case of UBR, σ^2.

We now return to the problem of soft classification, where we suppose that y_i is 1 or 0 with probability $p(t(i))$, and where $t = (t_1, \ldots, t_d)$. We suppose that $f(t) = \log[p(t)/(1 - p(t))]$ as before, but we will model f as

$$f(t) = \mu + \sum_{\alpha \in \mathcal{M}} f_\alpha(t_\alpha) + \sum_{\alpha,\beta \in \mathcal{M}} f_{\alpha\beta}(t_\alpha, t_\beta) + \cdots . \tag{4.9}$$

---

[2] By the properties of reproducing kernels, it can be shown in the setup discussed here that $J_\theta(\sum c_i \sum \theta_\beta R_\beta(\cdot, t(i))) = c'\Sigma_\theta c$.

Replacing Eq. (4.2) by

$$\mathcal{L}(y, f) + \frac{n}{2}\lambda J_\theta(f) \tag{4.10}$$

where $\mathcal{L}(y, f)$ is given by Eq. (2.4), it can be shown that the minimizer has a representation (4.5) for some (c, d), with the same M conditions on c. Furthermore, it is shown[3] that the Newton iterate $(c^{(\ell+1)}, d^{(\ell+1)})$ for the minimization of Eq. (4.10) with f as in Eq. (4.5), is the minimizer of

$$I_{\lambda,\theta}^{(\ell)}(c, d) = \|z^{(\ell)} - W_{(\ell)}^{1/2}(\Sigma_\theta c + Td)\|^2 + n\lambda c'\Sigma_\theta c, \tag{4.11}$$

where we are here and below writing the subscript (ℓ) as shorthand for the subscript $(c^{(\ell)}, d^{(\ell)})$, and $z^{(\ell)}$, the pseudo-data, is given by

$$z^{(\ell)} = W_{(\ell)}^{-1/2}(y - p_{(\ell)}) + W_{(\ell)}^{1/2}(\Sigma_\theta c^{(\ell)} + Td^{(\ell)}), \tag{4.12}$$

compare Eqs. (3.20) and (3.21). For fixed (ℓ), make the change of variables in Eq. (4.11):

$$\tilde{c} = W_{(\ell)}^{-1/2}c, \tag{4.13}$$

$$\tilde{\Sigma}_\theta = W_{(\ell)}^{1/2}\Sigma_\theta W_{(\ell)}^{1/2}, \tag{4.14}$$

$$\tilde{T} = W_{(\ell)}^{1/2}T, \tag{4.15}$$

$$\tilde{d} = d \tag{4.16}$$

to obtain

$$I_{\lambda,\theta}^{(\ell)}(\tilde{c}, \tilde{d}) = \|z^{(\ell)} - (\tilde{\Sigma}_\theta \tilde{c} + \tilde{T}\tilde{d})\|^2 + n\lambda \tilde{c}'\tilde{\Sigma}_\theta \tilde{c}. \tag{4.17}$$

Equation (4.17) is of the same form as Eq. (4.6). And $A^{(\ell)}(\lambda, \theta)$ is the matrix which satisfies $\hat{z}^{(\ell)} = A^{(\ell)}(\lambda, \theta)z^{(\ell)}$, where $\hat{z}^{(\ell)} = (\tilde{\Sigma}_\theta \tilde{c} + \tilde{T}\tilde{d})$. Given the ingredients $z^{(\ell)}, \tilde{\Sigma}_\theta$, and $\tilde{T}$, RKPACK can be called at the (ℓ)th step as a subroutine to obtain the UBR (or GCV) estimates of λ and θ_β, and (then) the Newton update with these updated values of the smoothing parameters. (RKPACK imposes conditions guaranteeing uniqueness; the solution only depends on the ratios λ/θ_β). Thus, the algorithm is: given a starting guess, from $c^{(\ell)}, d^{(\ell)}$, obtain $A^{(\ell)}(\lambda, \theta)$ and find $\lambda, \theta_\beta = \hat{\lambda}^{(\ell)}, \hat{\theta}_\beta^{(\ell)}$ to minimize $U^{(\ell)}(\lambda, \theta)$ given by Eq. (3.9) with $\sigma^2 = 1$, and $A^{(\ell)}(\lambda)$ replaced by $A^{(\ell)}(\lambda, \theta)$; obtain $c^{(\ell+1)}, d^{(\ell+1)}$ from Eq. (4.11) with $\lambda, \theta = \hat{\lambda}^{(\ell)}, \hat{\theta}^{(\ell)}$; iterate until convergence.

[3] See Gu.[11] Please note the following typographical errors in Gu[11]: $\tilde{c} = W_-^{-1/2}c$ and not $W_-^{1/2}c$, $\tilde{y} = W_-^{-1/2}(W_-\eta_- - u_-)$ and not $\tilde{y} = W_-^{1/2}(W_-\eta_- - u_-)$, also w_j in (2.6) should be w_{j_-}.

5. APPLICATION TO THE PIMA-INDIAN DATA SET

We have built and compared several PSA models for estimating $p(t)$ on the Pima Indians Diabetes Database which we retrieved from the UCI Repository of Machine Learning Databases and Domain Theories (ics.uci.edu: pub/machine-learning-databases) on October 7, 1992. This data set has also been analyzed using the ADAP learning algorithm by Smith et al.[35] so we will be able to compare some of our results with theirs. This database contains records of 768 instances, which were medical records from Pima-Indian women at least 21 years of age. Below is the list of eight attribute variables and the class variable (response):

```
1. Number of times pregnant
2. Plasma glucose concentration a 2 hours in an oral glucose tolerance test
3. Diastolic blood pressure (mm Hg)
4. Triceps skin fold thickness (mm)
5. 2-Hour serum insulin (mu U/ml)
6. Body mass index (weight in kg/(height in m)^2)
7. Diabetes pedigree function
8. Age (years)
9. Class variable (0 or 1)
```

The class variable was an indicator (1) for a positive test for diabetes between 1 and 5 years from the examination determining the other variables, or (0) a negative test for diabetes 5 or more years later. The repository index reports that there were 268 cases with "1" as their indicator and 500 with "0." It also reports that there are no missing attribute values; however, after some investigation into peculiar behavior of some of our results, box-plots of each set of attribute values revealed that there were 11 instances of 0 body mass index and 5 instances of 0 plasma glucose, both physical impossibilities(!). We have deleted those cases, leaving 752 instances for our experiments. Smith et al.[35] report that they used 576 randomly selected cases to train the ADAP algorithm, and then used the remaining 192 test cases as an evaluation set, to study the properties of the trained ADAP algorithm. Smith et al.[35] note that the ADAP algorithm is an "interactive associative learning model using the Hebbian learning rule," and give a brief description of the algorithm and further references. Given a new instance, the algorithm will output a score (real number) which is evidently intended to be larger if the new instance is more like the training 1's and smaller if the new instance is more like the training 0's. However, there is no suggestion that this score is intended to have the meaning of a probability. In any case, once a threshold level on this score is chosen, a "hard" classification (forecast) is made. The authors present a plot of specificity and sensitivity against a horizontal scale consisting of the ranks of the individuals in the the evaluation set ordered with respect to the output score. The sensitivity as a function of the rank is the fraction of true positives in the evaluation set with higher rank, and the specificity is the fraction of true negatives in the evaluation set with smaller rank. Thus, one can read the false positive and false negative rates off this plot as a function of threshold rank (or score, if it were provided). By inspection of the curves it appears that they cross at about rank 112 (out of 192). Smith et al.[35] report that the score at the crossing point is .448, and if this score is used to enforce a "hard" classification, then the rate of successful classification for both the true negatives and the true positives is 76%.

Our existing code is a big time and storage hog, as a result we found it necessary to be more modest than we would like in the data analysis. Thus, we decided to see how well we could do with fewer variables, and with a somewhat smaller training set. We randomly selected 500 instances out of 752 for the training set, and set aside the remaining 252 as the evaluation set. We used the glm function in S,[1] which implements the GLIM models of McCullagh and Nelder,[27] to fit several parametric models to the data in an effort to select a few of the most influential predictors. The GLIM model finds $f(t_1, \ldots, t_d)$ as a linear combination of simple parametric functions in the variables $(t_1, \ldots, t_d)$. The linear GLIM model would, for example, set $f(t_1, \ldots, t_d) = a + \sum_{\alpha=1}^{d} b_\alpha t_\alpha$, and find a and the b_α to minimize $\mathcal{L}(y, f)$ over f's of this form. The linear GLIM fit suggested that variables 1, 2, 6, and 7 were "significant" (assuming that you believed this model). Running one variable at a time through the linear GLIM model gave all relatively poor fits to the data compared to models with more than one variable, as measured by

$$\widehat{KL}_{\text{EVAL}} = -\frac{1}{252} \sum_{i=1}^{252} [y_i \hat{f}(t(i)) - \log(1 + e^{\hat{f}(t(i))})]. \tag{5.1}$$

where $\hat{f}$ is the GLIM model based on the training data and the sum in $\widehat{KL}_{\text{EVAL}}$ is over the evaluation data. Running the variables two at a time, the best pairwise variables according to $\widehat{KL}_{\text{EVAL}}$ were (2,6), (1,2), and (2,7), in that order, and the best of the three variable combinations was (1,2,6). For the application of PSA we decided to concentrate on a two variable model (2,6), a three-variable model (1,2,6), and a four-variable model (1,2,6,7). We considered variables 2, 6, and 7 as continuous variables, but we decided to consider variable 1 (number of pregnancies) as a categorical variable, with the four categories $C_1 = 0, C_2 = \{1, 2\}, C_3 = \{3, 4, 5\}$, and $C_4 = \{> 5\}$ We considered the following four models:

Model I : $f(t) = \mu + f_2(t_2) + f_6(t_6)$,
Model II : $f(t) = \mu + f_2(t_2) + f_6(t_6) + f_{2,6}(t_2, t_6)$,

$$\text{Model III} : f(t) = \mu + \sum_{k=1}^{3} \gamma_k I_k(t_1) + f_2(t_2) + f_6(t_6) + f_{2,6}(t_2, t_6),$$

where $I_k(t_k)$ is an indicator function which is 1 if variable 1 is from category C_k and 0 otherwise, and

$$\text{Model IV} : f(t) = \mu + \sum_{k=1}^{3} \gamma_k I_k(t_1) + f_2(t_2) + f_6(t_6) + f_7(t_7).$$

Models III and IV, which have linear combinations of a small number of unpenalized functions of known form (here indicator functions), are known as partial spline models; see Wahba.[43] From an algorithmic point of view, these functions are simply added to the set of functions spanning the null space of the penalty functional.[4]

Each of these models have a GLIM model as a special case, which is obtained by replacing $\mu + f_2(t_2) + f_6(t_6)$ by $\mu + a_2t_2 + a_6t_6$ or $\mu + f_2(t_2) + f_6(t_6) + f_{2,6}(t_2, t_6)$ by $\mu + a_2t_2 + a_6t_6 + a_{2,6}t_2t_6$, and similarly for t_7. The GLIM model would be fitted as a special case of the corresponding PSA model if all the λ/θ_β were estimated as ∞.

We can now compare all eight of these models by looking at their action on the 252 cases that have been left out. We estimate Eq. (2.9) for the PSA models by

$$\widehat{KL}_{\text{EVAL}} = -\frac{1}{252} \sum_{i=1}^{252} \left[y_i f_{\widehat{\lambda,\theta}}(t(i)) - \log \left(1 + e^{f_{\widehat{\lambda,\theta}}(t(i))} \right) \right], \qquad (5.2)$$

where again the sum is over the evaluation data and $f_{\widehat{\lambda,\theta}}$ is one of the four models that has been fit on the training set, using Gu's algorithm for the UBR to get $\widehat{\lambda}, \theta$. See Seaman and Hutchinson,[34] Wolpert,[47] and Schaffer[33] for closely related approaches to this *model selection* problem, and Gu[12] and Gu and Wahba[19] for philosophically different approaches to this problem, based on the sizes of each estimated component rather than a predictive criterion as we are using here. It would, of course, no longer be "fair" to compare the best of these models against the ADAP or other model on the basis of the *same* evaluation set, since the evaluation set has now been used to select the model. A "fair" comparison would be to take the model selected this way and compare it against a competitor on *another* evaluation set. In Table 1, we give $\widehat{KL}_{\text{EVAL}}$ the four PSA models and the four corresponding GLIM models.

The table identifies the PSA Model III as the "winner." We do not (as yet) have an objective criterion for saying which, if any of these eight models are significantly different (either in a statistical or a practical sense) from the "winner" although it appears that theoretical criteria relating to statistical significance can be developed. We remark that in PSA Model IV the main effect for variable 7, diabetes pedigree function, was quite small compared to the other main effects, and so PSA Models III and IV were very similar, in a practical sense. We will restrict further description to PSA Model III and its GLIM counterpart.

---

[4] The representations (4.5) for these models were constructed as by Gu and Wahba,[19] by rescaling t_2, t_6, and t_7 to the unit cube with the largest and smallest values mapped to 1 and 0, and using the reproducing kernels of Eqs. (4.1) and (4.2) of that paper as building blocks. The main effects penalty functionals were $J_\alpha(f_\alpha) = \int_0^1 (f''(t_\alpha))^2 dt_\alpha$, see Gu and Wahba[18,19] for further details. Gu and Wahba,[19] is available from `wahba@stat.wisc.edu` as UW-Madison Statistics Dept. TR 881.

TABLE 1 $\widehat{KL}_{\text{EVAL}}$

Model	PSA	GLIM
I	0.4929	0.5075
II	0.4861	0.5004
III	0.3925	0.5222
IV	0.4157	0.5226

The solid lines in Figure 1(a) and (b) give the main effects f_2 and f_6 for variables 2 (plasma glucose) and 6 (body mass index) for PSA Model III, and in Figure 1(c) gives the interaction term for PSA Model III. The dashed lines in Figure 1(a) and (b) give, for comparison, the corresponding GLIM main effects, which are, by construction, straight lines. And in Figure 1(d), we give the GLIM interaction term, which is of necessity bilinear in t_2 and t_6. The fitted interaction term in the PSA model actually was estimated as very close to bilinear in t_2 and t_6, and very small, possibly negligible. The interaction term in the GLIM model, which does not have a ready interpretation, was not statistically significant, according to the criteria of the GLIM code, which assumes that some GLIM model is true. In Figure 2, we give contour plots of $\hat{p}(t_1, t_2, t_6)$ corresponding to $t_1 = \{0\}, \{1, 2\}, \{3, 4\}$ and $\{> 5\}$ pregnancies, as a function of variables 2 and 6. Visually the plots for the first three categories do not appear much different, but being in the fourth category (≥ 5 pregnancies) appears to increase the risk at all levels of (t_2, t_6). On a logit scale, the difference between the fourth category and the average of the first three was about .92. In Figure 3, we give the same contour plots based on the GLIM model. In Figure 4, we give a plot of the body mass index *vs.* plasma glucose for the 500 member training set, 1's are plotted with a "*" and 0's are plotted with a "·." There are just two cases of body mass index above 55, so the models should not be taken too seriously much past 55. Methods for providing confidence statements for these model outputs which suggest the regions in which they can be trusted are discussed by Gu[14] and Gu and Wahba.[19]

Note that in this population all of the cases with body mass index less than about 23 did not later turn out to have diabetes. Similarly all of the cases with plasma glucose less than 78 did not later turn out to have diabetes. The greater flexibility in the PSA model allows the main effects to drop much more steeply than the GLIM model to accommodate this. Note also that the contribution of the main effects of body mass index in the PSA model is fairly flat along the range of about 30 to 40 body mass index, suggesting the desirability of keeping one's body mass index in that range.

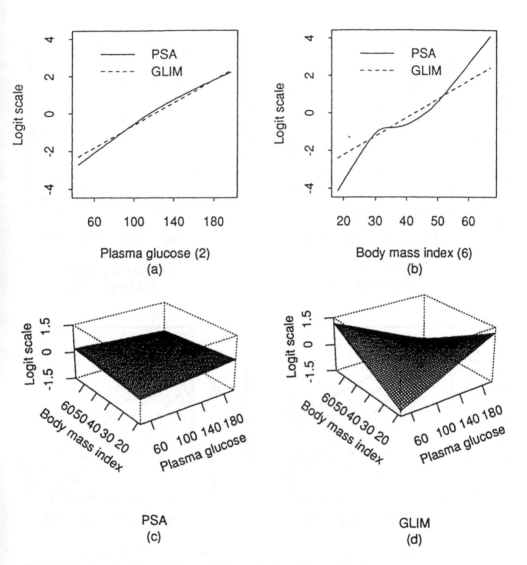

FIGURE 1   Logit main effects and interaction for PSA Model III and GLIM Model III.

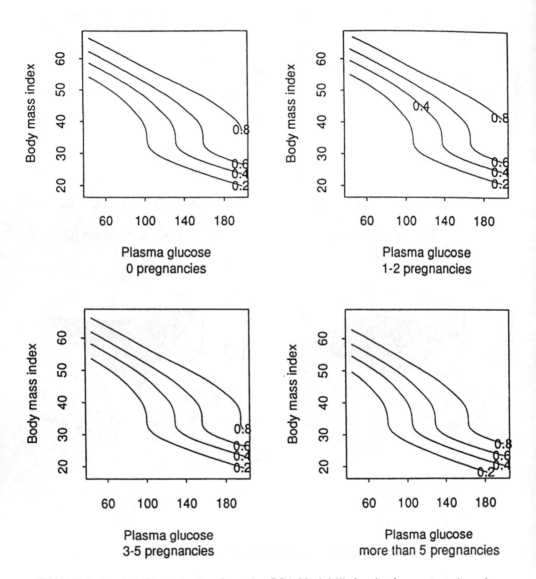

FIGURE 2   Probability estimates from the PSA Model III, for the four categories of variable 1, as a function of variables 2 and 6.

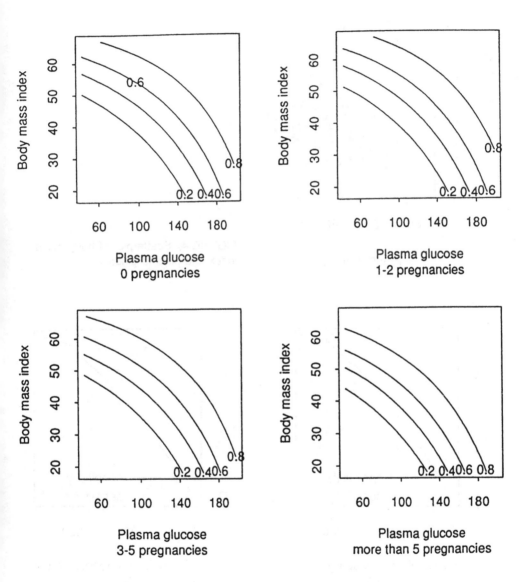

FIGURE 3 Probability estimates from the GLIM Model III, for the four categories of variable 1, as a function of variables 2 and 6.

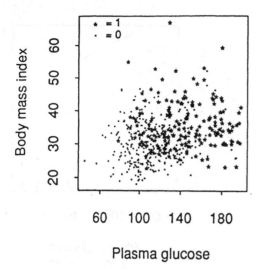

Plasma glucose

FIGURE 4   Scatterplot of body mass index and plasma glucose.

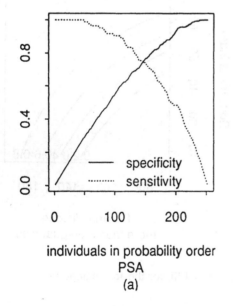

individuals in probability order
PSA
(a)

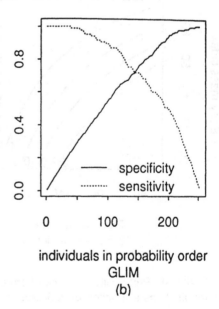

individuals in probability order
GLIM
(b)

FIGURE 5   Specificity and sensitivity for the PSA Model III and GLIM Model III.

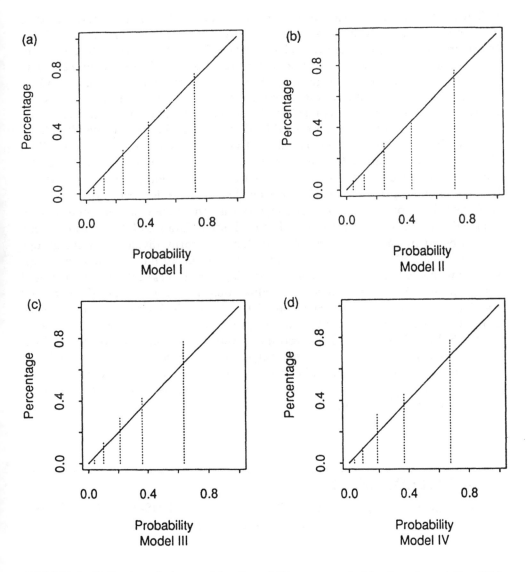

FIGURE 6 Estimates of observed fraction of 1's vs. expected fraction, by quintile, PSA models.

In order to compare this method with the analysis by Smith et al.,[35] we give in Figure 5(a) a plot of the sensitivity and the specificity scores for PSA and GLIM Model III, plotted against the ranked evaluation data, ranked according to the probability estimate. The sensitivity and specificity curves cross at about rank 146 (out of 252). If the score at this rank (which was about .74), were used to enforce a "hard" classification, then the rate of successful classification for both the true negatives and the true positives in the evaluation sample would be about 74%. In Figure 5(b), we give the same plots for the GLIM Model III. The successful classification rate at the crossover point for GLIM Model III was about 72%. The sensitivity vs. specificity curve given by Smith et al.[35] is visually very similar to Figure 5(a), the corresponding success rate for the ADAP analysis, using the crossover point as threshold was reported as 76%. This raises an interesting point about the ADAP algorithm. Its score report is evidently not claimed to be a probability; however, the Neyman-Pearson Lemma tells us that if we knew the correct probability, then any optimum threshold would depend on it. Thus, we might ask if the ADAP score (or other neural net scores) can be reasonably transformed into a probability by a monotone transformation. For more on this point, see Richard and Lippman.[25] In Figure 6, we provide a plot that suggests that the "score" $\hat{p}$ really does have a reasonable meaning as a probability. The evaluation data set of 252 cases has been rank ordered by their associated $\hat{p}$'s and then arbitrarily divided into five groups of size 50, 50, 50, 50, and 52 respectively. Returning to our "world view," suppose that the 252 probability estimates associated with these 252 cases represented "true" probabilities—say, the first 50 values were $p_1, \ldots, p_{50}$. Then the expected fraction of 1's in this group would be $1/50 \sum_{k=1}^{50} p_k$. In Figure 6(a)–(d) we have plotted the observed fractions for the five groups against their expected fractions, for the Spline Models I–IV. If the estimates were "on the money," they would fall on the solid line. We were somewhat surprised that Model I appears visually to be more "on the money" than Model III by this criterion, which of course is not exactly the same as the KL distance. At this point we have not made a study of the variability of these comparisons.

ACKNOWLEDGMENTS

Research supported by NSF Grants DMS-9121003 and DMS-9101730, NEI Grant ROI EY 099 46-01 and the Wisconsin Alumni Reserarch Foundation

REFERENCES

1. Becker, R., J. Chambers, and A. Wilks. *The New S Language*. Wadsworth, 1988.
2. Buja, A., T. Hastie, and R. Tibshirani. "Linear Smoothers and Additive Models" (with discussion). *Ann. Stat.* **17** (1989): 453–555.

3. Buntine, W., and A. Weigend. "Bayesian Back-Propagation." *Complex Systems* 5 (1991): 603–643.

4. Chen, Z., C. Gu, and G. Wahba. "Comments to 'Linear Smoothers and Additive Models,'" by Buja, Hastie, and Tibshirani *Ann. Stat.* 17 (1989): 515–521.

5. Cox, D., and Y. Chang. "Iterated State Space Algorithms and Cross Validation for Generalized Smoothing Splines." Technical Report 49, University of Illinois, Department of Statistics, 1990.

6. Craven, P., and G. Wahba. "Smoothing Noisy Data with Spline Functions: Estimating the Correct Degree of Smoothing by the Method of Generalized Cross-Validation." *Numer. Math.* 31 (1979): 377–403.

7. Friedman, J. "Multivariate Adaptive Regression Splines," (with discussion). *Ann. Stat.* 19 (1991): 1–141.

8. Geman, S., E. Bienenstock, and R. Doursat. "Neural Networks and the Bias/Variance Dilemma." *Neural Comp.* 4 (1992): 1–58.

9. Gray, R. "Flexible Methods for Analyzing Survival Data Using Splines with Applications to Breast Cancer Prognosis." *J. Amer. Stat. Assoc.* 87 (1992): 942–951.

10. Gu, C. "RKPACK and Its Applications: Fitting Smoothing Spline Models." In *Proceedings of the Statistical Computing Section*, 42–51. Alexandria, VA: American Statistical Association, 1989.

11. Gu, C. "Adaptive Spline Smoothing in Non-Gaussian Regression Models." *J. Amer. Stat. Assoc.* 85 (1990): 801–807.

12. Gu, C. "Diagnostics for Nonparametric Regression Models with Additive Terms." *J. Amer. Stat. Assoc.* 87 (1992): 1051–1057.

13. Gu, C. "Cross-Validating Non-Gaussian Data." *J. Comp. Graph. Stat.* 1 (1992): 169–179.

14. Gu, C. "Penalized Likelihood Regression: A Bayesian Analysis." *Stat. Sin.* 2 (1992): 255–264.

15. Gu, C., D. M. Bates, Z. Chen, and G. Wahba. "The Computation of GCV Functions Through Householder Tridiagonalization with Application to the Fitting of Interaction Spline Models." *SIAM J. Matrix Anal.* 10 (1989): 457–480.

16. Gu, C., and G. Wahba. "Comments to 'Multivariate Adaptive Regression Splines' by Friedman." *Ann. Stat.* 19 (1991): 115–123.

17. Gu, C., and G. Wahba. "Minimizing GCV/GML Scores with Multiple Smoothing Parameters via the Newton Method." *SIAM J. Sci. Stat. Comput.* 12 (1991): 383–398.

18. Gu, C., and G. Wahba. "Semiparametric ANOVA with Tensor Product Thin Plate Splines." *J. Roy. Stat. Soc. Ser. B* 55 (1993): 353–368.

19. Gu, C., and G. Wahba. "Smoothing Spline ANOVA With Component-Wise Bayesian 'Confidence Intervals.'" *J. Comp. Graph. Stat.* 2 (1993): 97–117.

20. Hogg, R., and J. Ledolter. *Engineering Statistics*. Macmillan, 1987.

21. Kimeldorf, G., and G. Wahba. "A Correspondence Between Bayesian Estimation of Stochastic Processes and Smoothing by Splines." *Ann. Math. Stat.* **41** (1970): 495–502.
22. Kimeldorf, G., and G. Wahba. "Some Results on Tchebycheffian Spline Functions." *J. Math. Anal. Appl.* **33** (1971): 82–95.
23. Li, K. C. "From Stein's Unbiased Risk Estimates to the Method of Generalized Cross-Validation." *Ann. Statist.* **13** (1985): 1352–1377.
24. Li, K. C. "Asymptotic Optimality of C_L and Generalized Cross Validation in Ridge Regression with Application to Spline Smoothing." *Ann. Stat.* **14** (1986): 1101–1112.
25. Richard, M., and R. Lippmann. "Neural Network Classifiers Estimate Bayesian *a posteriori* Probabilities." *Neural Comp.* **3** (1991): 461–483.
26. Mallows, C. "Some Comments on C_p." *Technometrics* **15** (1973): 661–675.
27. McCullagh, P., and J. Nelder. *Generalized Linear Models*, 2nd ed. Chapman and Hall, 1989.
28. Moody, J. "The Effective Number of Parameters: An Analysis of Generalization and Regularization in Nonlinear Learning Systems." In *Advances in Neural Information Processing Systems 4*, edited by J. Moody, S. Hanson, and R. Lippman. San Mateo: Kaufmann, 1991.
29. O'Sullivan, F. "The Analysis of Some Penalized Likelihood Estimation Schemes." Ph.D. Thesis, University of Wisconsin, Department of Statistics, 1983.
30. O'Sullivan, F. "Fast Computation of Fully Automated Log-Density and Log-Hazard Estimators." *SIAM J. Sci. Stat. Comp.* **9** (1988): 363–379.
31. O'Sullivan, F., B. Yandell, and W. Raynor. "Automatic Smoothing of Regression Functions in Generalized Linear Models." *J. Amer. Stat. Assoc.* **81** (1986): 96–103.
32. Ripley, B. "Neural Networks and Related Methods for Classification." *J. Roy. Statist. Soc.* **56** (1994): 3. Available by anonymous ftp from markov.stats.ox.ac.uk, 1992.
33. Schaffer, C. "Selecting a Classification Method by Cross-Validation." *Machine Learning* **13** (1993): 135–143.
34. Seaman, R., and M. Hutchinson. "Comparative Real Data Tests of Some Objective Aanalysis Methods by Withholding." *Aust. Met. Mag.* **33** (1985): 37–46.
35. Smith, J., J. Everhart, W. Dickson, W. Knowler, and R. Johannes. "Using the ADAP Learning Algorithm to Forecast the Onset of Diabetes Mellitus." In *Proceedings of the Symposium on Computer Applications and Medical Care*, 261–265. Washington, DC: IEEE Computer Society Press, 1988.
36. Stone, C. "Additive Regression and Other Nonparametric Models." *Ann. Stat.* **13** (1985): 689–705.
37. Tibshirani, R., and M. LeBlanc. "A Strategy for Binary Description and Classification." *J. Comp. Graph Stat.* **1** (1992): 3–20.

38. Wahba, G. "Improper Priors, Spline Smoothing and the Problem of Guarding Against Model Errors in Regression." *J. Roy. Stat. Soc. Ser. B* **40** (1978): 364–372.

39. Wahba, G. "Bayesian 'Confidence Intervals' for the Cross-Validated Smoothing Spline." *J. Roy. Stat. Soc. Ser. B* **45** (1983): 133–150.

40. Wahba, G. "A Comparison of GCV and GML for Choosing the Smoothing Parameter in the Generalized Spline Smoothing Problem." *Ann. Stat.* **13** (1985): 1378–1402.

41. Wahba, G. "Partial and Interaction Splines for the Semiparametric Estimation of Functions of Several Variables." In *Computer Science and Statistics: Proceedings of the 18th Symposium*, edited by T. Boardman, 75–80. Washington, DC: American Statistical Association, 1986.

42. Wahba, G. "Three Topics in Ill Posed Problems." In *Proceedings of the Alpine-U.S. Seminar on Inverse and Ill Posed Problems*, edited by H. Engl and C. Groetsch, 37–51. Orlando, FL: Academic Press, 1987.

43. Wahba, G. *Spline Models for Observational Data.* CBMS-NSF Regional Conference Series in Applied Mathematics, Vol. 59. Philadelphia, PA: SIAM, 1990.

44. Wahba, G. "Multivariate Function and Operator Estimation, Based on Smoothing Splines and Reproducing Kernels." In *Nonlinear Modeling and Forecasting*, edited by M. Casdagli and S. Eubank, 95–112. Santa Fe Institute Studies in the Sciences of Complexity, Proc. Vol. XII. Reading, MA: Addison-Wesley, 1992.

45. Wahba, G., and J. Wendelberger. "Some New Mathematical Methods for Variational Objective Analysis Using Splines and Cross-Validation." *Monthly Weather Rev.* **108** (1980): 1122–1145.

46. Wahba, G., and S. Wold. "A Completely Automatic French Curve." *Commun. Stat. Theory & Meth.* **4** (1975): 1–17.

47. Wolpert, D. "Stacked Generalization." *Neural Networks* **5** (1992): 241–259.

Leo Breiman
Statistics Department, University of California, Berkeley, CA

Current Research

Here is a summary of my current research projects, which are:

1. Optimizing Trees
2. Parallelizing Trees
3. Hypertrees
4. Hinges and Ramps
5. Stacked Predictors

1. OPTIMIZING TREES

Tree construction, as in CART, is a greedy algorithm.

A purity criterion function is specified and at each node the split selected is the one minimizing the sum of the impurities in the two children nodes. The procedure is to grow a large tree using this splitting algorithm and then to prune back and select the optimal pruned subtree using a test set or cross-validation.

This is a suboptimal procedure. It is not hard to construct examples of data where one can see that the one-step greedy algorithm produces a much less accurate tree than one could get knowing the structure of the data and selecting splits that produce a more optimal tree.

I know of two methods that have been proposed to get closer to optimality.

The first is to find the best k-step lookahead split at each node. For instance, the best two-step lookahead split would be obtained as follows: for each split of a node, search over all splits of its two children nodes and find the four grandchildren nodes that have minimum impurity. The sum of these four impurities is taken as the two-step criterion for the original split, and a search made for the split that minimizes this criterion.

Here is a quick computation of the number of operations involved. If there are N cases in the node and M variables, then the two-step lookahead has to examine $(MN)^2$ splits. If there are originally N cases in the data set, then the greedy algorithm has to examine about $MN \log(N)$ splits to grow the big tree. Two-step lookahead has to examine $(MN)^2 \log(N)$ splits. If $M = 100$ and $N = 1000$, the greedy algorithm examines about 10^6 splits. On a two-megaflop machine, this might take on the order of 10 cpu seconds. On the same machine, two-step lookahead would take about 12 days. One can do some clever things to cut computation time, but my estimate is that it could not be reduced more than an order of magnitude.

The other approach that has been taken is the use of annealing. Here one starts with a CART tree and then changes all of the splits by small amounts and uses annealing to try to find the global minimum of the resubstitution error rate. There have been a few papers on this approach. The most specific used an N-Cubed machine to try and optimize CART trees operating on three-class data with 2 variables and 300 cases. The analysis was confined to trees having five or fewer terminal nodes.

The results reported, averaged over twenty runs, was about a 5% decrease in test set misclassification error. The drawback is that even with so few variables, cases, and terminal nodes, each run took many hours on the massively parallel N-Cubed computer.

An approach that a recent Ph.D. student Nong Shang and I took follows from this idea: start with the root (top) node in the tree selected and consider varying the splits in this node, keeping constant all of the splits in the nodes below it. Find the split of the root node that minimizes the sum total impurity in the terminal nodes. Now do the same for each of the two children of the root node. Continue down the tree this way. When finished, start at the top again and work down. Continue until there are no more changes.

This seems like a good idea, but it needs changes to make it work more effectively. Part of the problem is that the iterative procedure is started on a many-node tree with the tree architecture already well developed. Our work showed that the process should be started much earlier, beginning with the two-node tree. A procedure was developed that keeps a list of the of the current best (lowest resubstitution error rate) m-node tree, $m = 1, 2, \ldots, M$ where M is large.

Start with m small and optimize this tree as above. Then use the CART procedure to expand this m-node tree to the largest tree possible. For each k, find the

k-node subtree of this big tree having minimum resubstitution error among all k-node subtrees. If any of these k-node subtrees has smaller resubstitution error than the current k-node tree on the list, then insert them into the list as the current best k-node tree. Continue until there is no more change.

This procedure has gone through extensive testing. For instance, on the data used in the annealing example cited above, it also produces a 5% reduction in the test set misclassification rate. On data designed to fool the greedy algorithm, it finds the right sequence of splits. The running times are modest. On the annealing example, the average running time for optimization was 114 cpu seconds on a Sparcstation 1+. On the three-class waveform data given in the CART book with 300 cases and 21 variables, the cpu time averaged about 600 seconds.

How much optimization helps is extremely data dependent. For instance, on the waveform data, there is no increase in accuracy. On other data sets, the increase is dramatic. The goal now is to get this written up.

2. PARALLELIZING TREES

Although CART runs fast on moderate-sized problems, it can be slow on massive problems. Furthermore, procedures to grow more accurate trees such as the above optimization or two-step lookahead slow it down considerably. Another current research project (joint with Phil Spector) is to develope a version of CART that runs in parallel on a net of SUN workstations.

Our approach is to let each workstation handle its own variables; i.e., if there are 100 variables in the data and 25 workstations in the net, then each workstation is assigned 4 of the variables. What we are learning is that for moderate-sized problems, the intermachine communication time is costly and there is little speed-up. As the size of the problem gets up to, say, 100,000 cases and 100 variables, there is a significant reduction in running time.

3. HYPERTREES

CART makes splits parallel to the coordinate axes, and fits the data in each node by a constant. There is a linear split option in CART but its a very ad hoc search procedure with largely untested performance. But even after a linear split, the data is fit by constants. It is pretty clear that improvement on many data sets could be obtained by both using linear combinations to split the data and then fitting via hyperplanes the data in the two children nodes. The problem was to find algorithms to do this that would run in reasonable amount of time. The latter requirement ruled out exhaustive search.

Recently, a fast iterative algorithm was found for fitting data by a hinge function. A hinge function consists of two hyperplanes continuously joined together at a hinge. (More description will be given below under 4.) This algorithm is being used to construct "hypertrees."

The idea is to fit the data in a node with a hinge function, then split the node on the hinge, leaving a fitted hyperplane in each of the two children nodes. I've written the code for a regression tree using this approach, and also allowing for the stepwise entry of variables (if desired) into the fitting of the hinge. Preliminary testing has been done, and as usual, the results are very data dependent.

It looks as though hypertrees rarely do worse than CART trees. Sometimes they give little improvement, sometimes a lot. It doesn't take much imagination to see that if the data has a strong local linear structure, hypertrees win big. But it is hard to tell *a priori* how much of a local linear structure is present.

Two things remain to do. First is more extensive testing. The second thing is that hypertrees can also be used in classification, but require some modification to the regression algorithm. This is on the list and hopefully will be completed this year.

4. RAMPS AND HINGES

Many current methods for fitting data in order to predict or classify have a common structure:

i.   A set of basis elements or "primitives" is defined.

ii. The data is used to select a subset of these basis elements.

iii. The predictor is formed as a linear combination of the selected basis elements.

For instance:

CART primitives:	Set of all indicator functions of hyper-rectangles
Neural Net Primitives:	Set of all sigmoids of linear functions
MARS primitives:	Set of all products of univariate linear splines

In recent work, we have used as primitives the set of hinge functions.

Hinge Function = two hyperplanes continuously joined at a "hinge."

More mathematically, if $\mathbf{x}$ is a vector of predictor variables, then a hinge function $h(\mathbf{x})$ is of the form

$$h(\mathbf{x}) = \min(\mathbf{b}_1\mathbf{x}, \mathbf{b}_2\mathbf{x})$$

where $\mathbf{b}_1\mathbf{x}, \mathbf{b}_2\mathbf{x}$ are two linear combinations of $\mathbf{x}$. We take one coordinate of $\mathbf{x}$ to be a constant so a linear combination of the x includes a constant term. Then the hinge is the set of all $\mathbf{x}$ such that $\mathbf{b}_1\mathbf{x} = \mathbf{b}_2\mathbf{x}$.

Now given data of the form $\{(y_n, \mathbf{x}_n), n = 1, \ldots, N\}$, this data is to be used to construct a predictor $f(\mathbf{x})$ of future y-values where $f(\mathbf{x})$ is to be a sum of hinge functions. Note that a sum of hinge functions provides a continuous piecewise linear fit.

There are two advantages of using hinge functions as primitives. One is a theorem that says that any sufficiently smooth function can be approximated arbitrarily closely by a sum of hinge functions. But this is what I called in my theory chapter a "comfort" theorem. Lots of sets of basis elements can be proved to have the same properties. What makes hinges desirable is that there is a lightning fast algorithm for fitting hinge functions.

The idea is simple: start with an arbitrary hinge. Then fit two least squares hyperplanes to the data on each side of the hinge. The intersection of these two hyperplanes forms a new hinge. Now repeat the procedure using the new hinge and keep iterating until convergence. In general, the algorithm converges rapidly to the right thing.[1]

The general strategy for fitting a sum of hinge functions is "greedy plus back-fitting." After a sum of $k - 1$ hinge functions is fitted to the data, the kth hinge function is fitted to the residuals. Then a sequence of backfitting cycles is carried out until convergence. In each cycle, the jth hinge is refitted to the residuals holding all other hinge functions fixed. This is done for $j = 1, \ldots, k$ in succession.

To give an idea of the speed of the hinge algorithm using greedy and backfitting, it was applied to noisy data with 1000 cases and 100 variables. A total of ten hinges were fit until the automatic stopping rule decided the fit was adequate. The process took 2.8 cpu minutes on an RS 6000 (12 megaflop). A version used for classification took 80 cpu seconds on 10-class data with 61 variables and 1000 cases with 10 hinge functions fitted.

Conceptually, there are two problems with hinge functions. In classification estimates of conditional probabilities are being constructed. Since these have a range in [0,1], hinge functions may not be a good set of primitives for approximating probabilities. The second issue is that hinge functions may not be a good set of primitives for fitting local bumps.

The thing to try to do, then, is to find a more suitable set of primitives that share the fast fitting property of hinges. This is being done in the current work on ramp functions joint with Jerome Friedman. A ramp function consists of three planes continuously joined together. One plane is horizontal at height zero. The second is also horizontal, but at height one. The third can be arbitrary. In three dimensions, one can visualize a ramp function as just that—a ramp leading from ground level to the second story.

Ramp functions have several advantages. One is that there is a rapid iterative algorithm for fitting ramps, similar to the hinge algorithm. The second is that a ramp function can be easily approximated by a sigmoid function of a linear function.

(To see this, note that all that has to be done is to approximate in the cross-section space.) After each ramp function is entered (using greedy), then it is approximated by a sigmoid, so the residuals for the next fit are based on the current sum of sigmoids.

The current status is that the programming is complete, but extensive testing needs to be done.

5. STACKED PREDICTORS

This grew out of a really nice idea published by David Wolpert. Suppose one has K predictors $f_l(\mathbf{x}), f_2(\mathbf{x}), \ldots, f_K(\mathbf{x})$ of y, all constructed from the data $\{(y_n, \mathbf{x}_n), n = 1, \ldots, , N\}$. For instance, here are two examples I have been working with:

i.   In linear regression with K variables, one regularizes by doing subset selection; i.e., $f_1(\mathbf{x})$ is the minimum residual-sum-of squares (RSS) linear regression based on a single variable, $f_2(\mathbf{x})$ is that minimum RSS linear regression based on two variables, etc. Thus, to get $f_1(\mathbf{x})$, look at all K single variable regressions and choose the one having minimum RSS.

ii. In CART trees, a big tree is grown. For each k, where $k = l, \ldots, K$, let $f_k(\mathbf{x})$ be the predictor obtained by finding that k terminal node subtree of the big tree which has minimum resubstitution error among all k terminal node subtrees.

The common approach in situations such as the above is to use the best of the K predictors, where the selection is based on a cross-validated estimate of the prediction error.

6. STACKED REGRESSION ALWAYS DOES BETTER!

But, there is a long history of statistical theory that says if you have a number of estimates, then the best single estimate is a combination of the individual estimate. This is like an aha! experience. Of course we should be able to do better by combining. The difficult question is how to combine. Let's restrict attention to linear combinations. That is, we want an estimate of the form

$$f(\mathbf{x}) = \sum_k c_k f_k(\mathbf{x}).$$

Now how can the coefficients $\{c_k\}$ be estimated? Generally, all of the predictors are trying to predict the same thing. If they are any good, they will have a large positive correlation. If we try to estimate the $\{c_k\}$ using ordinary least squares regression, the estimates will be all over the place, and $f(\mathbf{x})$ will be very noisy.

I tried to cut back the noise by using ridge regression. This gave an improvement, but no gold ring. Finally, a method was found that worked like gangbusters. To motivate it, note that one problem with linear least squares estimates of the $\{c_k\}$ is that $f(\mathbf{x})$ is all over the place. Even evaluated on the data $f(\mathbf{x}_n)$ may be much larger than the largest of the $f_k(\mathbf{x}_n)$, $k = l, \ldots, K$, or be less than the smallest. One way to settle down the combination estimate is to insist that for all n,

$$\min_k f_k(\mathbf{x}_n) \leq f(\mathbf{x}_n) \leq \max_k f_k(\mathbf{x}_n).$$

To satisfy these conditions, it is sufficient that the $\{c_k\}$ satisfy the constraints:

(CON) $\qquad\qquad c_k \geq 0, \quad k = l, \ldots, K \qquad \sum_k c_k = 1.$

So now the idea is to determine the $\{c_k\}$ by constrained least squares minimization.

There is one other piece needed to make this story complete. If the $\{c_k\}$ are determined by minimum RSS under the constraints (CON), then what is being minimized is the resubstitution error under (CON). What should be minimized is the cross-validation error under (CON). To do this, leave out the nth case $(y_n, \mathbf{x}_n)$ and recompute all of the predictors. Denote these leave-out-the-nth case predictors by $f_{k,(-n)}(\mathbf{x})$. Now take the $\{c_k\}$ to minimize

$$\sum_n (y_n - \sum_k c_k f_{k,(-n)}(\mathbf{x}_n))^2$$

under the constraints (CON).

I have carried out a systematic simulation of stacking in the linear case. It dramatically decreases the test set error in the subset selection situation as compared to choosing the best single predictor. A number of examples have also been run in stacking CART trees. The results show a systematic 5–10% decrease in test set's mean squared error over using the best single tree.

If the number of samples is large, computing the cross-validated predictors can become burdensome. But it seems that leaving out, say, a tenth of the data instead of leaving out one case does just as well and makes the computation feasible for large data sets. The results for the linear case are available in a technical report.[2]

REFERENCES

1. Breiman, L. "Hinging Hyperplanes for Noiseless Function Approximation, Regression and Classification." *IEEE Trans. Info. Theory* **39** (1993): 999–1013.
2. Breiman, L. "Stacking Regressions." Technical Report, Statistics Department, University of California at Berkeley, Berkeley, California, 1993.

Geoffrey E. Hinton† and Steven Nowlan‡
†Department of Computer Science, University of Toronto, Toronto, Canada M5S 1A4
‡Computational Neurobiology Laboratory, The Salk Institute, P. O. Box 85800, San Diego,
CA 92186-5800

Preface to "Simplifying Neural Networks by Soft Weight Sharing"

For a supervised neural network to generalize well, there must be less information in the weights than there is in the output vectors of the training cases. Researchers have considered many possible ways of limiting the information in the weights. One method is to divide the connections into subsets, and force the weights within a subset to be identical. If this "weight sharing" is based on an analysis of the natural symmetries of the task, it can be very effective.[2,3]

If we do not know, in advance, which connections should share weight values, we can still encourage solutions that share values by using an appropriate measure for the complexity of the weights. An obvious method is to use a finite number of discrete weight values. The number of bits required to specify each value is then $-\log p$ where p is the fraction of the weights having that value. In addition, some bits are required to specify the values used and their probability masses. If we use discrete weight values, it is obvious that the weights are simpler (i.e., take less bits to describe) if many of them are identical. Unfortunately, they are no simpler if they are **almost** identical so we have a nasty search space.

The search for simple weights would be much easier if we could find a measure of complexity that decreased smoothly as weight values became more similar to each other. The following chapter describes such a measure and shows that it leads to tractable searches and good generalization on two tasks. Instead of dividing the weights into discrete bins, the method models the distribution of weight values as a mixture of Gaussians. Each Gaussian acts like a "soft" bin. As the Gaussians adapt, the complexity term encourages the weights that fall under one Gaussian to form tighter clusters.

There is a weakness of the method described in the chapter that is not apparent for the two tasks we investigated but can be serious for harder tasks. If the network starts with small weights, it is easy for it to get very low complexity by collapsing all the weights to the same value and ignoring the task. To avoid this trivial local minimum it may be necessary to initially assign a lower importance to the complexity term, and then to increase the importance as the network learns to get the correct answers. Unfortunately this makes the whole method more *ad hoc.*

A second weakness is that our complexity measure, by focussing only on the probability density of a weight, implicitly assumes that all weights are encoded to the same accuracy. This is a very wasteful way to code weights when the output is more sensitive to some weights than to others. The method needs to be extended to allow different weights to have different accuracies. One method of doing this is described by Hinton and van Camp.[1]

Despite these weaknesses, the method does impressively well at predicting the sunspots time series, and it produces a simple network that is easy to interpret in terms of the dominant properties of the series.

Since we did this research, Ross Quinlan[4] has compared our method with decision trees. For the sunspots series, decision trees seem to be inferior to our method, even when the decision tree uses a linear model instead of a constant function at each leaf. Such "model trees" are described by Quinlan.[5] They are a hierarchical generalization of the TAR method by Tong and Lim mentioned in our chapter, and they work considerably better than TAR on the sunspots series. By combining model trees with another technique, Quinlan can obtain about the same performance as our method. He first builds a model tree, but then for each prediction he uses a few nearest neighbors to estimate a local correction term to add to the model's prediction. Of course, the predictions of our net could also be adjusted by using nearest neighbors to estimate a local correction term.

The real message of the following chapter is not that our scheme is the best way to keep the weights simple. We have shown that one relatively complicated method of keeping the weights simple can improve generalization dramatically, and we anticipate that other methods that work even better will soon be found.

REFERENCES

1. Hinton, G. E., and D. van Camp. "Keeping Neural Networks Simple by Minimizing the Description Length of the Weights." In *Proceedings of the Sixth Annual ACM Conference on Computational Learning Theory*, 5–13. Santa Cruz, 1993.
2. Lang, K. J., A. H. Waibel, and G. E. Hinton. "A Time-Delay Neural Network Architecture for Isolated Word Recognition." *Neural Networks* **3** (1990): 23–43.
3. le Cun, Y. "Generalization and Network Design Strategies." Technical Report No. CRG-TR-89-4, University of Toronto, 1989.
4. Quinlan, R. Personal communication, 1993.
5. Quinlan, R. "Learning with Continuous Classes." In *Proceedings 5th Australian Joint Conference on Artificial Intelligence*, 343–348. Singapore: World Scientific, 1992.

Steven J. Nowlan† and Geoffrey E. Hinton‡
†Computational Neurobiology Laboratory, The Salk Institute, P.O. Box 85800, San Diego, CA 92186-5800
‡Department of Computer Science, University of Toronto, Toronto, Canada M5S 1A4

Simplifying Neural Networks by Soft Weight Sharing

This paper originally appeared in Nowlan, S. J., and G. E. Hinton. "Simplifying Neural Networks by Soft Weight Sharing." *Neural Computation* 4 (1992): 173–193. Copyright © by MIT Press. Reprinted with permission.

One way of simplifying neural networks so they generalize better is to add an extra term to the error function that will penalize complexity. Simple versions of this approach include penalizing the sum of the squares of the weights, or penalizing the number of nonzero weights. We propose a more complicated penalty term in which the distribution of weight values is modeled as a mixture of multiple Gaussians. A set of weights is simple if the weights have high-probability densities under the mixture model. This can be achieved by clustering the weights into subsets with the weights in each cluster having very similar values. Since we do not know the appropriate means or variances of the clusters in advance, we allow the parameters of the mixture model to adapt at the same time as the network learns. Simulations on two different problems demonstrate that this complexity term is more effective than previous complexity terms.

INTRODUCTION

A major problem in training artificial neural networks is to ensure that they will generalize well to cases that they have not been trained on. Some recent theoretical results[3] have suggested that in order to guarantee good generalization, the amount of information required to directly specify the output vectors of all the training cases must be considerably larger than the number of independent weights in the network. In many practical problems there is only a small amount of labelled data available for training and this creates problems for any approach that uses a large, homogeneous network in order to avoid the detailed task analysis required to design a network with fewer independent weights and a specific architecture that is appropriate to the task. As a result, there has been much recent interest in techniques that can train large networks with relatively small amounts of labelled data and still provide good generalization performance.

One way to achieve this goal is to reduce the effective number of free parameters in the network. A number of authors (e.g., Rumelhart, Hinton, and Williams,[28] Ch. 8; Lang, Waibel, and Hinton;[14] and le Cun[15,16]) have proposed the idea of weight sharing, in which a single weight is shared among many connections in the network so that the number of adjustable weights in the network is much less than the number of connections. This approach is effective when the problem being addressed is quite well understood, so that it is possible to specify, in advance, which weights should be identical.[17]

Another approach is to use a network with too many weights, but to stop the training before overlearning on the training set has occurred.[21,32] In addition to the usual training and testing sets, a validation set is used. When performance on the validation set starts to *decrease*, the network is beginning to overfit the training set and training is stopped. Some experience with this technique has suggested that its effectiveness can be quite sensitive to the particular stopping criterion used.[32][1]

Yet another approach to the generalization problem attempts to remove excess weights from the network, either during or after training, to improve the generalization performance. Mozer and Smolensky[22] and le Cun et al.[18] have both proposed techniques in which a network is initially trained with an excess number of parameters and then a criterion is used to remove redundant parameters. The reduced network is then trained further. The cycle of reduction and retraining may be repeated more than once. The approach of Mozer and Smolensky estimates the relevance of individual *units* to network performance and removes redundant units and their weights. The method of le Cun et al. uses second-order gradient information to estimate the sensitivity of network performance to the removal of each weight, and removes the least critical weights.

---

[1] This approach is *not* the way in which cross validation is usually used in the statistics community. Usually a cross-validation set is used to determine the scale factor applied to a regularization term added to the optimization to prevent overfitting. We will see examples of this in the following sections.

An older and simpler approach to removing excess weights from a network is to add an extra term to the error function that penalizes complexity:

$$\text{cost} = \text{data-misfit} + \lambda\,\text{complexity}. \tag{1}$$

During learning, the network is trying to find a locally optimal trade-off between the data-misfit (the usual error term) and the complexity of the net. The relative importance of these two terms can be estimated by finding the value of λ that optimizes generalization to a validation set. Probably the simplest approximation to complexity is the sum of the squares of the weights, $\sum_i w_i^2$. Differentiating this complexity measure leads to simple *weight decay*[24] in which each weight decays towards zero at a rate that is proportional to its magnitude. This decay is countered by the gradient of the error term, so weights which are not critical to network performance, and hence always have small error gradients, decay away leaving only the weights necessary to solve the problem. At the end of learning, the magnitude of a weight is exactly proportional to its error derivative, which makes it particularly easy to interpret the weights (see, for example, Hinton[9]). Minimizing $\sum_i w_i^2$ is a well-known technique when fitting linear regression models that have too many degrees of freedom. One justification is that it minimizes the sensitivity of the output to noise in the input, since in a linear system the variance of the noise in the output is just the variance of the noise in the input multiplied by the squared weights.[13]

The use of a $\sum_i w_i^2$ penalty term can also be interpreted from a Bayesian perspective.[30] The "complexity" of a set of weights, $\lambda \sum_i w_i^2$, may be described as its negative-log probability density under a radially symmetric Gaussian prior distribution on the weights. The distribution is centered at the origin and has variance $1/\lambda$. For multilayer networks, it is hard to find a good theoretical justification for this prior, but Hinton[10] justifies it empirically by showing that it greatly improves generalization on a very difficult task. More recently, MacKay[19] has shown that even better generalization can be achieved by using different values of λ for the weights in different layers.

A MORE COMPLEX MEASURE OF NETWORK COMPLEXITY

One potential drawback of $\sum_i w_i^2$ as a penalty term is that it can favor two weak interactions over one strong one. For example, if a unit receives input from two units that are highly correlated with each other, its behavior will be similar if the two connections have weights of w and 0 or weights of $w/2$ and $w/2$. The penalty term favors the latter because

$$\left(\frac{w}{2}\right)^2 + \left(\frac{w}{2}\right)^2 < w^2 + 0^2. \tag{2}$$

If we wish to drive small weights towards zero without forcing large weights away from the values they need to model the data, we can use a prior which is a mixture of a narrow (n) and a broad (b) Gaussian, both centered at zero.

$$p(w) = \pi_n \frac{1}{\sqrt{2\pi}\sigma_n} e^{-w^2/2\sigma_n^2} + \pi_b \frac{1}{\sqrt{2\pi}\sigma_b} e^{-w^2/2\sigma_b^2} \tag{3}$$

where π_n and π_b are the mixing proportions of the two Gaussians and are therefore constrained to sum to 1.

Assuming that the weight values were generated from a Gaussian mixture, the conditional probability that a particular weight, w_i, was generated by a particular Gaussian, j, is called the *responsibility*[2] of that Gaussian for the weight and is given by:

$$r_j(w_i) = \frac{\pi_j p_j(w_i)}{\sum_k \pi_k p_k(w_i)} \tag{4}$$

where $p_j(w_i)$ is the probability density of w_i under Gaussian j.

When the mixing proportions of the two Gaussians are comparable, the narrow Gaussian gets most of the responsibility for a small weight. Adopting the Bayesian perspective, the cost of a weight under the narrow Gaussian is proportional to $w^2/2\sigma_n^2$. As long as σ_n is quite small there will be strong pressure to reduce the magnitude of small weights even further. Conversely, the broad Gaussian takes most of the responsibility for large weight values, so there is much less pressure to reduce them. In the limiting case when the broad Gaussian becomes a uniform distribution, there is almost no pressure to reduce very large weights because they are almost certainly generated by the uniform distribution. A complexity term very similar to this limiting case has been used successfully by Weigend et al.[32] to improve generalization for a time series prediction task.[3]

There is an alternative justification for using a complexity term that is a mixture of a uniform distribution and a narrow, zero-mean Gaussian. The negative-log probability is approximately constant for large weights but smoothly approaches a much lower value as the weight approaches zero. So the complexity cost is a smoothed version of the obvious discrete cost function that has a value of zero for weights that are zero and a value of 1 for all other weights. This smoothed cost function is suitable for gradient descent learning, whereas the discrete one is not.

---

[2] This is more commonly referred to as the *posterior probability* of Gaussian j given weight w_i.

[3] See Nowlan[23] for a precise description of the relationship between mixture models and the model used by Weigend et al.[32]

ADAPTIVE GAUSSIAN MIXTURES AND SOFT WEIGHT SHARING

A mixture of a narrow, zero-mean Gaussian with a broad Gaussian or a uniform allows us to favor networks with many near-zero weights, and this improves generalization on many tasks, particularly those in which there is some natural measure of locality that determines which units need to interact and which do not. But practical experience with hand-coded weight constraints has also shown that great improvements can be achieved by constraining particular subsets of the weights to share the same value.[14,16] Mixtures of zero-mean Gaussians and uniforms cannot implement this type of symmetry constraint. If, however, we use multiple Gaussians and allow their means and variances to adapt as the network learns, we can implement a "soft" version of weight sharing in which the learning algorithm decides for itself which weights should be tied together. (We may also allow the mixing proportions to adapt so that we are not assuming all sets of tied weights are the same size.)

If we know, in advance, that two connections should probably have the same weight, we can introduce a complexity penalty proportional to the squared difference between the two weights. But if we do not know which pairs of weights should be the same it is harder to see how to favor solutions in which the weights are divided into subsets and the weights within a subset are nearly identical. We now show that a mixture of Gaussian models can achieve just this effect. The basic idea is that a Gaussian that takes responsibility for a subset of the weights will squeeze those weights together since it can then have a lower variance and, hence, assign a higher probability density to each weight. If the Gaussians all start with high variance, the initial division of weights into subsets will be very soft. As the variances shrink and the network learns, the decisions about how to group the weights into subsets are influenced by the task the network is learning to perform.

AN UPDATE ALGORITHM

To make the intuitive ideas of the previous section a bit more concrete, we may define a cost function of the general form given in Eq. (1):

$$C = \frac{K}{\sigma_y^2} \sum_c \frac{1}{2}(y_c - d_c)^2 - \sum_i \log\left(\sum_j \pi_j p_j(w_i)\right) \tag{5}$$

where σ_y^2 is the variance of the squared error and each $p_j(w_i)$ is a Gaussian density with mean μ_j and standard deviation σ_j. We will optimize this function using some form of gradient descent to adjust the w_i as well as the mixture parameters π_j, μ_j, and σ_j, and σ_y.[4]

The partial derivative of C with respect to each weight is the sum of the usual squared error derivative and a term due to the complexity cost for the weight:

$$\frac{\partial C}{\partial w_i} = \frac{K}{\sigma_y^2} \sum_c (y_c - d_c) \frac{\partial y_c}{\partial w_i} - \sum_j r_j(w_i) \frac{(\mu_j - w_i)}{\sigma_j^2}. \tag{6}$$

The derivative of the complexity cost term is simply a weighted sum of the difference between the weight value and the center of each of the Gaussians. The weighting factors are the *responsibility* measures defined in Eq. (4) and, if over time, a single Gaussian claims most of the responsibility for a particular weight the effect of the complexity cost term is simply to pull the weight towards the center of the responsible Gaussian. The strength of this force is inversely proportional to the variance of the Gaussian. Notice also that since the derivative of the complexity cost term is actually a sum of forces exerted by each Gaussian, the *net* force exerted on the weight can be very small even when the forces exerted by some of the individual Gaussians are quite large (i.e., a weight placed midway between two Gaussians of equal variance has zero net force acting on it). This allows one to set up initial conditions in which each Gaussian accounts quite well for at least some of the weights in the network, but the overall force on any weight due to the complexity term is negligible so the weights are initially driven primarily by the derivative of the data-misfit term.

The partial derivatives of C with respect to the means and variances of the Gaussians in the mixture have similar forms:

$$\frac{\partial C}{\partial \mu_j} = \sum_i r_j(w_i) \frac{(\mu_j - w_i)}{\sigma_j^2}, \tag{7}$$

$$\frac{\partial C}{\partial \sigma_j} = -\sum_i r_j(w_i) \frac{(\mu_j - w_i)^2 - \sigma_j^2}{\sigma_j^3}. \tag{8}$$

The derivative for μ_j simply drives μ_j toward the *weighted* average of the set of weights Gaussian j is responsible for. Similarly the derivative for σ_j drives it toward the weighted average of the squared deviations of these weights about μ_j. The derivation of the partial of C with respect to the mixing proportions is slightly less straightforward since we must worry about maintaining the constraint that the

---

[4] $1/\sigma_y^2$ may be thought of as playing the same role as λ in Eq. (1) in determining a trade-off between the misfit and complexity costs. σ_y is reestimated as learning proceeds so this trade-off is not constant. K is a factor that adjusts for the effective number of degrees of freedom (or number of well determined parameters) in a problem. For the simulations described here its value was close to 1.0 and was determined by cross validation.

mixing proportions must sum to 1. Appropriate use of a Lagrange multiplier and a bit of algebraic manipulation leads to the simple expression:

$$\frac{\partial C}{\partial \pi_j} = \sum_i \left(1 - \frac{r_j(w_i)}{\pi_j} \right). \tag{9}$$

Once again the result is intuitive; π_j is moved towards the average responsibility of Gaussian j for all of the weights.

The partial derivatives of C with respect to each of the mixture parameters are simple enough that, for fixed values of the responsibilities, the exact minimizer can be found analytically with ease (for example, the minimizer for Eq. (9) is simply $\mu_j = \sum_i r_j(w_i) w_i \sum_i r_j(w_i)$). This suggests that one could proceed by simply recomputing the $r_j(w_i)$ after each weight update and setting the μ_j, σ_j, and π_j to their exact minimizers given the current $r_j(w_i)$. In fact the process of recomputing the $r_j(w_i)$ and then setting all the parameters to their analytic minimizers corresponds to one iteration of the *EM* algorithm applied to mixture estimation.[5] This is a sensible and quite efficient algorithm to use for estimating the mixture parameters when we are dealing with a *stationary* data distribution. However, in the case we are considering, it is clear that the "data" we are modeling, the set of w_i, does not have a stationary distribution. In order to avoid stability problems, it is very important that the rate of change of our mixture parameters be tied to the rate of change of the weights themselves. For this reason, we choose to update all of the parameters $(w_i, \mu_j, \sigma_j, \pi_j)$ *simultaneously* using a conjugate gradient descent procedure.[5]

Before considering applications of the method outlined above, we need to consider briefly the issue of initializing all of our parameters appropriately. It is well known that maximum likelihood methods for fitting mixtures can be very sensitive to poor initial conditions.[20] For example, if one component of a mixture initially has little responsibility for any of the weights in the network, its mixing proportion is driven rapidly toward zero and it is very difficult to recover from this situation. Fortunately, in the case of a network we usually know the initial weight distribution and so we can initialize the mixture appropriately. Commonly, we initialize the network weights so they are uniformly distributed over an interval $[-W, W]$. In this case we may initialize the means of the Gaussians so they are spaced evenly over the interval $[-W, W]$, and set all of the variances equal to the spacing between adjacent means and the mixing proportions equal to each other. This ensures that each component in the mixture initially has the same total responsibility over the entire set of weights,[6] and also produces sufficient counterbalance between the forces from each Gaussian so most of the weights in the network initially receive

---

[5] Any method of gradient descent could be used in the parameter update; however, the conjugate gradient technique is quite fast and avoids the need to tune optimization parameters such as step size or momentum rate.

[6] There is a minor edge effect for the two most extreme components.

very little net force from the complexity measure. This initialization procedure is used for all of the simulations discussed in this chapter.

There are two additional tricks used in the simulations discussed in this chapter. The variances of the mixture components, σ_j^2, must, of course, be restricted to be positive and in addition if the variance of any component is allowed to approach 0 too closely, the likelihood may become unbounded. To maintain the positivity constraint and at the same time make it difficult for the variance of a component to approach 0, we define the variance of the components in terms of a set of auxiliary variables:

$$\sigma_j^2 = e^{\gamma_j} \tag{10}$$

where the value of γ_j is unrestricted. The gradient descent is performed on the set of γ_j rather than directly on the σ_j. In a similar fashion, the mixing proportions π_j must also be positive and in addition sum to one. Rather than deal with the Lagrange multipliers required to enforce these constraints as in Eq. (9), we define the mixing proportions in terms of a second set of unconstrained auxiliary variables:

$$\pi_k = \frac{e^{\kappa_k}}{\sum_j e^{\kappa_j}} . \tag{11}$$

The gradient of the cost function with respect to the κ_j has a particularly simple form: $\partial C / \partial \kappa_j = \sum_i (\kappa_j - r_j(w_i))$.[7]

RESULTS ON A TOY PROBLEM

In this section, we report on some simulations that compare the generalization performance of networks trained using the cost criterion given in Eq. (5) to networks trained in three other ways:

- No cost term to penalize complexity.
- No explicit complexity cost term, but use of a validation set to terminate learning.
- The complexity cost term used by Weigend et al.[32][8]

The problem chosen for this comparison was a 20 input, one output shift detection network (see Figure 1).

[7] The use of the γ_j and κ_j may be thought of simply as a technique for getting a better conditioned optimization problem.
[8] With a fixed value of λ chosen by cross validation.

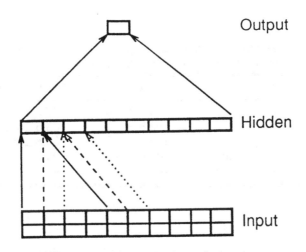

Output

Hidden

Input

FIGURE 1   Shift detection network used for generalization simulations. The output unit
was connected to a bias unit and the ten hidden units. Each hidden unit was connected
to the bias unit and to eight input units, four from the first block of ten inputs and the
corresponding four in the second block of ten inputs. The solid, dashed, and dotted
lines show the group of input units connected to the first, second, and third hidden
units, respectively.

The network had 20 input units, 10 hidden units, and a single output unit
and contained 101 weights. The first ten input units in this network were given a
random binary pattern, and the second group of ten input units were given the
same pattern circularly shifted by 1 bit left or right. The desired output of the
network was +1 for a left shift and −1 for a right shift.

A data set of 2400 patterns was created by randomly generating a ten-bit string,
and choosing with equal probability to shift the string left or right.[9] The data set
was divided into 100 training cases, 1000 validation cases, and 1300 test cases. The
training set was deliberately chosen to be very small ($< 5\%$ of possible patterns) to
explore the region in which complexity penalties should have the largest impact.

Networks were trained with a conjugate gradient technique and, except for the
networks trained using cross validation, training was stopped as soon as 100% correct performance was achieved on the training set. For the networks trained with
cross validation, training was stopped when three consecutive weight updates[10]

---

[9] Since there are only 2048 distinct cases, this set of 2400 did contain some duplicates.

[10] A weight update refers to the final weight change accepted at the end of a single line search.

TABLE 1 Summary of generalization performance of five different
training techniques on the shift detection problem.

Method	Train % Correct	Test % Correct
Back Prop.	100.0 ± 0.0	67.3 ± 5.7
Cross Valid.	98.8 ± 1.1	83.5 ± 5.1
Weight Decay	100.0 ± 0.0	89.8 ± 3.0
Soft-share - 5 Comp.	100.0 ± 0.0	95.6 ± 2.7
Soft-share - 10 Comp.	100.0 ± 0.0	97.1 ± 2.1

produced an increase in the error on the validation set and the weights were then
reset to the weights that achieved the lowest error on the validation set before
testing for generalization. For the technique described by Weigend et al.,[32] $\lambda =
5.0 \times 10^{-5}$ in all simulations.[11] Simulations using Eq. (5) were performed with
Gaussian mixtures containing five and ten components. Each component had its
own mean (μ_j), variance (σ_j), and mixing proportion (π_j). The parameters of the
mixture distribution were continuously reestimated as the weights were changed as
was the normalizing factor for the squared error (σ_y).

Ten simulations were performed with each method, starting from ten different
initial weight sets (i.e., each method used the same ten initial weight configurations).
The simulation results are summarized in Table 1.

The first column indicates the method used in training the network, while the
second and third columns present the performance on the training and test sets
respectively (plus or minus one standard deviation).

All three methods that employ some form of weight modeling performed significantly better on the test set than networks trained using backpropagation without a
complexity penalty ($p \gg 0.9999$).[12] The network trained with cross validation also
performs better than a network trained without a complexity penalty ($p > 0.995$).
The two soft-sharing models perform better than cross validation ($p > 0.995$ for the
five-component mixture, $p > 0.999$ for the ten-component mixture). The evidence
that Weigend et al.'s form of weight decay is superior to cross validation on this
problem is very weak ($p < 0.9$). Finally, the two soft-sharing models are significantly
better than the weight decay model ($p > 0.999$ for the ten-component mixture and

[11]This value was selected by performing simulations with λ ranging between 1.0×10^{-1} and
1.0×10^{-9} and choosing the value of λ that gave the best performance on the cross-validation set.
[12]All statistical comparisons are based on a t-test with 19 degrees of freedom. The variable p
denotes the probability of rejecting the hypothesis that the two samples being compared have the
same mean value.

$p > 0.99$ for the five-component mixture). The difference in the performance of the five- and ten-component mixtures is not significant.

A typical set of weights learned by the soft-sharing model with a ten-component mixture is shown in Figure 2 and the final mixture density is shown in Figure 3.

The weight model in this case contains four primary weight clusters: large magnitude positive and negative weights and small magnitude positive and negative weights. These four distinct classes may also be seen clearly in the weight diagram (Figure 2).

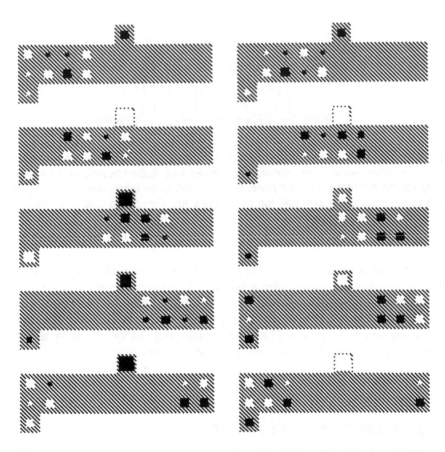

FIGURE 2 A diagram of weights discovered for the shift problem by a model that employed a ten-component mixture for the complexity cost. Black squares are negative weights and white values are positive, with the size of the square proportional to the magnitude of the weight. Weights are shown for all ten hidden units. The bottom row of each block represents the bias, the next two rows are the weights from the 20 input units, and the top row is the weight to the output unit.

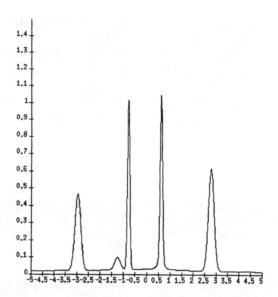

FIGURE 3   Final mixture probability density for the set of weights shown in Figure 2. Five of the components in the mixture can be seen as distinct bumps in the probability density. Of the remaining five components, two have been eliminated by having their mixing proportions go to zero and the other three are very broad and form the baseline offset of the density function.

What is perhaps most interesting about the mixture probability density shown in Figure 3 is that it *does not* have a significant component with mean 0. The classical assumption that the network contains a large number of inessential weights which can be eliminated to improve generalization is not appropriate for this problem and network architecture. This may explain why the weight decay model used by Weigend et al.[32] performs relatively poorly in this situation.

RESULTS ON A REAL PROBLEM

The second task chosen to evaluate the effectiveness of the cost criterion of Eq. (5) was the prediction of the yearly sunspot average from the averages of previous years. This task has been well studied as a time series prediction benchmark in the statistics literature[25,26] and has also been investigated by Weigend et al.[32] using a cost criterion similar to the one discussed in Section 2.

The network architecture used was identical to the one used in the study by Weigend et al.: The network had 12 input units that represented the yearly average from the preceding 12 years, 8 hidden units, and a single output unit that represented the prediction for the average number of sunspots in the current year. Yearly sunspot data from 1700 to 1920 was used to train the network to perform this one-step prediction task, and the evaluation of the network was based on data from 1921 to 1955.[13] The evaluation of prediction performance used the *average relative variance* (*arv*) measure discussed by Weigend et al.[32]:

$$arv(S) = \frac{\sum_{k \in S}(\text{target}_k - \text{prediction}_k)^2}{\sum_{k \in S}(\text{target}_k - \text{mean}_S)^2} \tag{12}$$

where S is a set of target values and mean$_S$ is the average of those target values.

Simulations were performed using the same conjugate gradient method used in the previous section. Complexity measures based on Gaussian mixtures with three and eight components were used and ten simulations were performed with each (using the same training data but different initial weight configurations). The results of these simulations are summarized in Table 2 along with the best result obtained by Weigend et al.[32] (*WRH*), the bilinear autoregression model of Tong and Lim[31] (*TAR*)[14], and the multilayer RBF network of He and Lapedes[8] (*RBF*). All figures represent the *arv* on the test set. For the mixture complexity models, this is the *average* over the ten simulations, plus or minus one standard deviation.

TABLE 2 Summary of average relative variance of five different models on the one-step sunspot prediction problem.

Method	Test arv
TAR	0.097
RBF	0.092
WRH	0.086
Soft-share - 3 Comp.	0.077 ± 0.0029
Soft-share - 8 Comp.	0.072 ± 0.0022

[13] The authors wish to thank Andreas Weigend for providing his version of this data.
[14] This was the model favored by Priestly[26] in a recent evaluation of classical statistical approaches to this task.

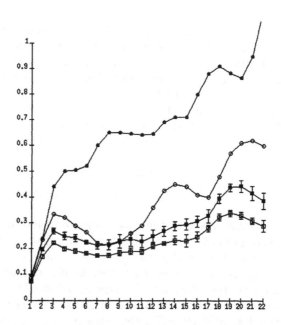

FIGURE 4 Average relative I-times iterated prediction variance versus number of prediction iterations for the sunspot time series from 1921 to 1955. Closed circles represent the TAR model, open circles the WRH model, closed squares the three-component complexity model, and open squares the eight-component complexity model. One standard deviation error bars are shown for the three- and eight-component complexity models.

Since the results for the models other than the mixture complexity trained networks are based on a single simulation, it is difficult to assign statistical significance to the differences shown in Table 2. We may note, however, that the difference between the three- and eight-component mixture complexity models is significant ($p > 0.95$) and the differences between the eight-component model and the other models are much larger.

Weigend et al. point out that for time series prediction tasks, such as the sunspot task, a much more interesting measure of performance is the ability of the model to predict more than one time step into the future. One way to approach the multistep prediction problem is to use *iterated single-step prediction*. In this method, the predicted output is fed back as input for the next prediction and all other input units have their values shifted back one unit. Thus the input typically consists of a combination of actual and predicted values.

We can define the predicted value for time t, obtained after I iterations to be $\hat{x}_{t,I}$. The prediction error will depend not only on I but also on the time $(t - I)$ when the iteration was started. In order to account for both effects, Weigend et al.

suggested the *average relative I-times iterated prediction variance* as a performance measure for iterated prediction:

$$\frac{1}{\hat{\sigma}^2} \frac{1}{M} \sum_{m=1}^{M} (x_{m+I} - \hat{x}_{m+I,I})^2 \tag{13}$$

where M is the number of different start times for iterated prediction and $\hat{\sigma}$ is the estimated standard deviation of the set of target values. In Figure 4 we plot this measure (computed over the test set from 1921 to 1955) as a function of the number of prediction iterations for the simulations using the three- and eight-component complexity measures, the Tong and Lim model (TAR), and the model from Weigend et al. that produced the lowest single step arv (WRH). The plots for the three- and eight-component complexity models are the averages over ten simulations with the error bars indicating the plus or minus one standard deviation intervals. Once again, the differences between the three- and eight-component models are significant for all numbers of iterations.

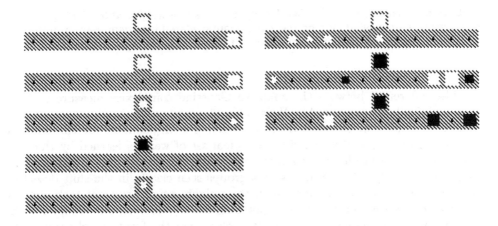

FIGURE 5   A diagram of weights discovered for the sunspot prediction problem by a model which employed an eight-component mixture for the complexity cost. Weights are shown for all eight hidden units. For each unit, the weights coming from the 12 inputs are shown in a row with the single weight to the output immediately above the row. The biases of the hidden units, which are not shown, were, with one exception, small negative numbers very close in value to most of the other weights in the network. The first three units in the left column all represent the simple rule that the number of sunspots depends on the number in the previous year. The last two units in this column compute a simple moving average. The three units on the right represent more interesting rules. The first captures the 11-year cycle, the second recognizes when a peak has just passed, and third appears to prevent the prediction from rising too soon if a peak happened nine years ago and the recent activity is low.

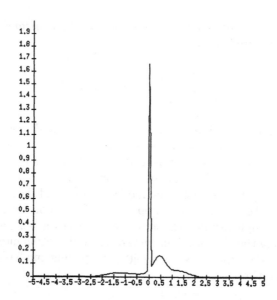

FIGURE 6   Final mixture probability density for the set of weights shown in Figure 5. The density is dominated by a narrow component centered very near zero, with the remaining components blending into a skewed distribution with a peak around 0.5.

The differences between the adaptive Gaussian complexity measure and the fixed complexity measure used by Weigend et al. are not as dramatic on the sunspot task as they were in the shift detection task. The explanation for this may be seen in Figures 5 and 6 which show a typical set of weights learned by the soft-sharing model with eight mixture components and the corresponding final mixture probability density. The distinct weight groups seen clearly in the shift detection task (Figure 2) are not as apparent in the weights for the sunspot task and the final weight distribution for the sunspot task is very smeared out except for one very strong sharp component near 0. It is clear that the fixed model assumed by Weigend et al. is much more appropriate for the sunspot prediction task than it was for the shift detection task.

A MINIMUM DESCRIPTION LENGTH PERSPECTIVE

A number of authors have suggested that when attempting to approximate an unknown function with some parametric approximation scheme (such as a network), the proper measure to optimize combines an estimate of the cost of the misfit with an estimate of the cost of describing the parametric approximation.[1,2,27] Such a measure is often referred to as a minimum description length criterion (MDL), and typically has the general form

$$MDL = \sum_{\text{messages}} -\log p(\text{message}) + \sum_{\text{parameters}} -\log p(\text{parameter}).$$

For a supervised network, the parameters are the weights and the messages are the desired outputs. If we assume that the output errors are Gaussian and that the weights are encoded using a mixture of Gaussians probability model, the description length is approximated by Eq. (5).

The expression in Eq. (5) does not include the cost of encoding the means and variances of the mixture components or the mixing proportions of the mixture density. Since the mixture usually contains a small number of components (fewer than ten usually) and there are only three parameters associated with each component, the cost of encoding these parameters is negligible compared to the cost of encoding the weights in most networks of interest.[15] In addition, since the *number* of components in the distribution does not change during the optimization, if the component parameters are all encoded with the same fixed precision, the cost of the mixture parameters is simply a constant offset which is ignored in the optimization.[16]

There is one important aspect of estimating the cost of describing a weight that we have ignored. We have assumed that the cost of a weight is the negative logarithm of a probability *density* function evaluated at the weight value, but this ignores the accuracy with which the weight must be described. We are really interested in the probability *mass* of a particular small interval of values for the parameter, and this means that we should integrate our density function over this interval to estimate the cost of each weight. We have implicitly assumed that this integration region has the same (infinitesimal) width for every weight, and so the probability of a weight is simply proportional to the density function. This ignores the fact that most networks are generally much more sensitive to small changes in some weight values than others, so some weights need to be encoded more accurately than others.[17]

The sensitivity of a network to a small change in a weight is determined by the curvature of the error surface. One could evaluate the curvature by computing the Hessian and make the width of the integration region for each weight inversely proportional to the curvature along each weight dimension. To be perfectly accurate, one would need to integrate the joint probability density function for all of the weights over a region determined by the complete Hessian (since the directions of maximum curvature are often not perfectly aligned with the weight axes).

---

[15] In order to provide enough data to fit the mixture density, one should have an order of magnitude more weights than components in the mixture.

[16] This ignores the possibility of not encoding the parameters of components whose mixing proportions approach 0.

[17] Other things being equal, we should prefer networks in which the outputs are less sensitive to the precise weight values, since then the weight values can be encoded imprecisely without causing large output errors.

This process would be computationally very costly, and an adequate approximation might be obtainable by using a diagonal approximation to the Hessian and treating each weight independently (as advocated by le Cun et al.[18]). We see no reason why our method of estimating the probability density should not be combined with a method for estimating the integration interval. For small intervals, this is particularly easy since the probability mass is approximately the width of the interval times the height of the density function so these two terms are additive in the log probability domain.

A BAYESIAN PERSPECTIVE

As a number of authors have recently pointed out,[4,19,23,33] Eq. (1) can be derived from the principles of Bayesian inference. If we have a set of models $M_1, M_2, \ldots$ which are competing to account for the data, Bayesian inference is concerned with how we should update our belief in the relative plausibility of each of these models in light of the data D. If $P(M_i|D)$ is the plausibility of model M_i given we have observed D, Bayes rule states:

$$P(M_i|D) = \frac{P(D|M_i)P(M_i)}{P(D)} \tag{14}$$

where $P(D|M_i)$ is a measure of how well model i predicts the data and $P(M_i)$ is our belief in the plausibility of model i before we have seen any data. Here $P(D)$ is simply a normalizing factor to ensure our beliefs add up to one. If we are only interested in comparing alternate models, $P(D)$ can be ignored and in the log domain Eq. (1) becomes Eq. (14) with the data-misfit cost equal to $\log P(D|M_i)$ and the complexity cost equal to $\log P(M_i)$. If we are only considering a single network architecture, $P(M_i)$ becomes a prior distribution over the set of possible weights.[18]

What Eq. (14) highlights is that in the Bayesian framework our complexity cost should be *independent* of our data. This is certainly true when the complexity is the sum of the squares of the weights, and also holds for models such as the one used by Weigend et al. However, the mixture densities discussed in this chapter are clearly not independent of the data and cannot be regarded as classical Bayesian priors.

The complexity cost we are using corresponds more closely to a Bayesian *hyperprior*[11,6]: We have specified a particular family of distributions from which the prior will be drawn but have left the parameters of the prior (π_j, μ_j, σ_j) undetermined. Members of this family of distributions have the common feature of favoring sets

---

[18]Much more interesting results are obtained when we apply this framework to making choices among many architectures; see MacKay for some elegant examples.[19]

of weights in which the weights *in a set* are clustered about a small number of values.[19] When using a hyper-prior, we can deal with the *hyper-parameters* either by marginalizing over them (in effect, integrating them out),[4] or by allowing the data (i.e., the weights) to determine their values *a posteriori*.[20] We have used this second approach which is advocated by Gull and Skilling,[6] who have shown that the use of such flexible hyper-priors can lead to considerable improvement in the quality of image reconstructions[7,29] compared to the use of more classical priors.

The trick of optimizing γ_j rather than σ_j (discussed at the end of Section 4) may also be justified within the Bayesian framework. In order to estimate our hyper-parameters, we should properly specify prior distributions for each. If these priors are uninformative,[21] then the estimated values of the hyper-parameters are determined entirely by the data. A parameter like σ_j is known as a *scale* parameter (it affects the width of the distribution) while parameters like μ_j are known as *location* parameters (they affect the position of the distribution). (See Jeffreys for further discussion.[12]) An uninformative prior for a location parameter is uniform in the parameter, but an uninformative prior for a scale parameter is uniform in the log of the parameter (i.e., uniform in γ_j rather than σ_j; see Gull[6]). It is more consistent from this perspective to treat γ_j and μ_j similarly, rather than σ_j and μ_j.

SUMMARY

The simulations we have described provide evidence that the use of a more sophisticated model for the distribution of weights in a network can lead to better generalization performance than a simpler form of weight decay, or techniques that control the learning time. The better generalization performance comes at the cost of greater complexity in the optimization of the weights. The effectiveness of the technique is likely to be somewhat problem dependent, but one advantage offered by the more sophisticated model is its ability to automatically adapt the model of the weight distribution to individual problems.

[19] The locations of these clusters will generally be different for different sets of weights.

[20] In principle, both approaches will lead to the same posterior distribution over the weights and the same ultimate choice of weights for the network. The difference lies in whether we are searching over a joint space of weights and hyper-parameters or using prior analytic simplifications to reduce the search to some manifold in weight space alone.

[21] A prior which contains no initial bias except for a possible range constraint.

ACKNOWLEDGEMENTS

This research was funded by grants from the Ontario Information Technology Research Center, the Canadian Natural Science and Engineering Research Council, the Howard Hughes Medical Institute. Hinton is the Noranda fellow of the Canadian Institute for Advanced Research.

REFERENCES

1. Akaike, H. "Information Theory and an Extension of the Maximum Likelihood Principle." In *Proceedings 2nd International Symposium on Information Theory*, edited by B. N. Petrov and F. Csaki, 267–281. Budapest, Hungary: Akademia Kiado, 1973.
2. Barron, A. R., and R. L. Barron. "Statistical Learning Networks: A Unifying View." In *1988 Symposium on the Interface: Statistics and Computing Science*. Symposium held April 21–23, 1988 in Reston, Virginia.
3. Baum, E. B., and D. Haussle. "What Size Net Gives Valid Generalization?" *Neural Comp.* **1** (1989): 151–160.
4. Buntine, W. L., and A. S. Weigend. "Bayesian Back-Propagation." Submitted for publication, 1991.
5. Dempster, A. P., N. M. Laird, and D. B. Rubin. "Maximum Likelihood from Incomplete Data via the EM algorithm." *Proc. Roy. Stat. Soc.* **B-39** (1977): 1–38.
6. Gull, S. F. "Bayesian Inductive Inference and Maximum Entropy." In *Maximum Entropy and Bayesian Methods in Science and Engineering*, edited by G. J. Erickson and C. R. Smith. Kluwer, 1988.
7. Gull, S. F. "Developments in Maximum Entropy Data Analysis." In *Maximum Entropy and Bayesian Methods (8th Workshop)*, edited by J. Skilling. Kluwer, 1989.
8. He, X., and A. Lapedes. "Nonlinear Modelling and Prediction by Successive Approximation Using Radial Basis Functions." Technical Report No. LA-UR-91-1375, Los Alamos National Laboratory, Los Alamos, NM, 1991.
9. Hinton, G. E. "Learning Distributed Representations of Concepts." In *Proceedings of the Eighth Annual Conference of the Cognitive Science Society*, 1–12. Hillsdale, NJ: Erlbaum, 1986.
10. Hinton, G. E. "Learning Translation Invariant Recognition in a Massively Parallel Network." In *Proceedings of the Conference on Parallel Architectures and Languages Europe*. Eindhoven, 1987.

11. Jaynes, E. T. "Bayesian Methods: General Background." In *Maximum Entropy and Bayesian Methods in Applied Statistics*, edited by J. H. Justice, 1–25. Cambridge, MA: Cambridge University Press, 1986.
12. Jeffreys, H. *Theory of Probability*. Oxford: Oxford University Press, 1939.
13. Kohonen, T. *Associative Memory: A System-Theoretical Approach*. Berlin: Springer-Verlag, 1977.
14. Lang, K. J., A. H. Waibel, and G. E. Hinton. "A Time-Delay Neural Network Architecture for Isolated Word Recognition." *Neural Networks* **3** (1990): 23–43.
15. le Cun, Y. "Modèles Connexionnistes de l'Apprentissage." Université Pierre et Marie Curie, Paris, France, 1987.
16. le Cun, Y. "Generalization and Network Design Strategies." Technical Report No. CRG-TR-89-4, University of Toronto, 1989.
17. Le Cun, Y., B. Boser, J. S. Denker, D. Henderson, R. E. Howard, W. Hubbard, and L. D. Jackel. "Handwritten Digit Recognition with a Back-Propagation Network." In *Advances in Neural Information Processing Systems 2*, edited by D. S. Touretzky, 396–404. San Mateo, C: Morgan Kauffman, 1990.
18. Le Cun, Y., J. Denker, S. Solla, R. E. Howard, and L. D. Jackel. "Optimal Brain Damage." In *Advances in Neural Information Processing Systems II*, edited by D. S. Touretzky. San Mateo, CA: Morgan Kauffman, 1990.
19. MacKay, David J. C. "Bayesian Modelling and Neural Networks." Ph.D. Thesis, Computation and Neural Systems, California Institute of Technology, Pasadena, CA, 1991.
20. McLachlan, G. J., and K. E. Basford. *Mixture Models: Inference and Applications to Clustering*, edited by McLachlan and Basford, Ch. 1 and 2, 1–60. Marcel Dekker, 1988.
21. Morgan, N., and H. Bourlard. "Generalization and Parameter Estimation in Feedforward Nets: Some Experiments." Technical Report No. TR-89-017, International Computer Science Institute, Berkeley, CA, 1989.
22. Mozer, M. C., and P. Smolensky. "Using Relevance to Reduce Network Size Automatically." *Connection Sci.* **1(1)** (1989): 3–16.
23. Nowlan, S. J. "Soft Competitive Adaptation: Neural Network Learning Algorithms based on Fitting Statistical Mixtures." Ph.D. Thesis, School of Computer Science, Carnegie-Mellon University, Pittsburgh, PA, 1991.
24. Plaut, D. C., S. J. Nowlan, and G. E. Hinton. "Experiments on Learning by Back-Propagation." Technical Report No. CMU-CS-86-126, Carnegie-Mellon University, Pittsburgh PA 15213, 1986.
25. Priestley, M. B. *Spectral Analysis and Time Series*. New York: Academic Press, 1991.
26. Priestley, M. B. *Nonlinear and Nonstationary Time Series Analysis*. New York: Academic Press, 1991.
27. Rissanen, J. "Modelling by Shortest Data Description." *Automatica* **14** (1978): 465–471.

28. Rumelhart, D. E., J. L. McClelland, and the PDP research group. In *Parallel Distributed Processing: Explorations in the Microstructure of Cognition*, Vols. I and II. Cambridge, MA: MIT Press, 1986.
29. Skilling, J. "Classic Maximum Entropy." In *Maximum Entropy and Bayesian Methods (8th Workshop)*. Kluwer, 1989.
30. Szeliski, R. Personal communication, 1985.
31. Tong, H., and K. S. Lim. "Threshold Autoregression, Limit Cycles, and Cyclical Data." *J. Roy. Stat. Soc. B* **42** (1980).
32. Weigend, A. S., B. A. Huberman, and D. E. Rumelhart. "Predicting the Future: A Connectionist Approach." *Intl. J. Neural Sys.* **1** (1990).
33. Weigend, A. S., D. E. Rumelhart, and B. A. Huberman. "Generalization by Weight-Elimination with Application to Forecasting." In *Advances in Neural Information Processing Systems 3*, edited by R. P. Lippmann, J. E. Moody, and D. S. Touretzky, 875–882. San Mateo, CA: Morgan Kaufmann, 1991.

Thomas G. Dietterich[†] and Ghulum Bakiri[††]
[†]Department of Computer Science, Oregon State University, Corvallis, OR 97331-3202
[††]Department of Computer Science, University of Bahrain, Isa Town, Bahrain

Error-Correcting Output Codes: A General Method for Improving Multiclass Inductive Learning Programs

Multiclass learning problems involve finding a definition for an unknown function $f(\mathbf{x})$ whose range is a discrete set containing $k > 2$ values (i.e., k "classes"). The definition is acquired by studying large collections of training examples of the form $\langle \mathbf{x}_i, f(\mathbf{x}_i) \rangle$. Existing approaches to this problem include (a) direct application of multiclass algorithms such as the decision-tree algorithms ID3 and CART, (b) application of binary concept learning algorithms to learn individual binary functions for each of the k classes, and (c) application of binary concept learning algorithms with distributed output codes such as those employed by Sejnowski and Rosenberg in the NETtalk system.[20] This chapter compares these three approaches to a new technique in which BCH error-correcting codes are employed as a distributed output representation. We show that these output representations improve the performance of ID3 on the NETtalk task and of backpropagation on an isolated-letter speech-recognition task. These results demonstrate that error-correcting output codes provide a general-purpose method for improving the performance of inductive learning programs on multiclass problems.

The Mathematics of Generalization, Ed. David Wolpert, SFI Studies in
the Sciences of Complexity, Proc. Vol XX, Addison-Wesley, 1995   **395**

1. INTRODUCTION

The task of learning from examples is to find an approximate definition for an unknown function $f(\mathbf{x})$ given training examples of the form $\langle \mathbf{x}_i, f(\mathbf{x}_i) \rangle$. For cases in which f takes only the values $\{0, 1\}$—binary functions—there are many algorithms available. For example, the decision-tree methods, such as ID3[13,15] and CART,[3] can construct trees whose leaves are labeled with binary values. Most artificial neural network algorithms, such as the perceptron algorithm[18] and the error backpropagation (BP) algorithm,[19] are best suited to learning binary functions. Theoretical studies of learning have focused almost entirely on learning binary functions.[7,8,17,22]

In many real-world learning tasks, however, the unknown function f takes on values from a discrete set of "classes": $\{c_1, \ldots, c_k\}$. For example, in medical diagnosis, the function might map a description of a patient to one of k possible diseases. In digit recognition, the function maps each hand-printed digit to one of $k = 10$ classes.

Decision-tree algorithms can be easily generalized to handle these "multiclass" learning tasks. Each leaf of the decision tree can be labeled with one of the k classes, and internal nodes can be selected to discriminate among these classes. We will call this the *direct multiclass* approach.

Connectionist algorithms are more difficult to apply to multiclass problems, however. The standard approach is to learn k individual binary functions $f_1, \ldots, f_k$, one for each class. To assign a new case, $\mathbf{x}$ to one of these classes, each of the f_i is evaluated on $\mathbf{x}$, and $\mathbf{x}$ is assigned the class j of the function f_j that returns the highest activation.[12] We will call this the *one-per-class* approach, since one binary function is learned for each class.

Finally, a third approach is to employ a *distributed output code*. Each class is assigned a unique binary string of length n; we will refer to these as "codewords." Then n binary functions are learned, one for each bit position in these binary strings. These binary functions are usually chosen to be meaningful, and often independent, properties in the domain. For example, in the NETtalk system,[20] a 26-bit distributed code was used to represent phonemes and stresses. The individual binary functions (bit positions) in this code corresponded to properties of phonemes and stresses, such as "voiced," "labial," and "stop." By representing enough distinctive properties of phonemes and stresses, each phoneme/stress combination can have a unique codeword.

For distributed output codes, training is accomplished as follows. For an example from class i, the desired outputs of the n binary functions are specified by the codeword for class i. With artificial neural networks, these n functions can be implemented by the n output units of a single network. With decision trees, n separate decision trees are learned, one for each bit position in the output code.

New values of $\mathbf{x}$ are classified by evaluating each of the n binary functions to generate an n-bit string s. This string is then compared to each of the k codewords, and **x** is assigned to the class whose codeword is closest, according to some distance measure, to the generated string s.

This review of methods for handling multiclass problems raises several interesting questions. First, how do the methods compare in terms of their ability to classify unseen examples correctly? Second, are some of the methods more difficult to train than others (i.e., do they require more training examples to achieve the same level of performance)? Third, are there principled methods for designing good distributed output codes?

To answer these questions, this chapter begins with a study in which the decision-tree algorithm ID3 is applied to the NETtalk task[20] using three different techniques: the direct multiclass approach, the one-per-class approach, and the distributed output code approach. The results show that the multiclass and distributed output code approaches generalize much better than the one-per-class approach.

It is helpful to visualize the output code of a learning system as a matrix whose rows are the classes and whose columns are the n binary functions corresponding to the bit positions in the codewords. In the one-per-class approach, there are k rows and k columns, and the matrix has 1's only on the diagonal. In the distributed output code approach, there are k rows and n columns, and the rows of the matrix give the codewords for the classes.

From this perspective, the two methods are closely related. A codeword is assigned to each class, and new examples are classified by decoding to the nearest of the codewords. This perspective suggests that a better distributed output code could be designed using error-correcting code methods. Good error-correcting codes choose the individual code words so that they are well separated in Hamming distance. The potential benefit of such error correction is that the system could recover from errors made in learning the individual binary functions. If the minimum Hamming distance between any two codewords is d, then $\lfloor (d-1)/2 \rfloor$ errors can be corrected.

The "code" corresponding to the one-per-class approach has a minimum Hamming distance of 2, so it cannot correct any errors. Similarly, many distributed output codes have small Hamming distances, because the columns correspond to meaningful orthogonal properties of the domain. In the Sejnowski-Rosenberg code, for example, the minimum Hamming distance is 1, because there are phonemes that differ only in whether they are voiced or unvoiced. These observations suggest that error-correcting output codes (ECC) could be very beneficial.

On the other hand, unlike either the one-per-class or distributed-output-code approaches, the individual bit positions of error-correcting codes will not be meaningful in the domain. They will constitute arbitrary disjunctions of the original k classes. If these functions are difficult to learn, then they may negate the benefit of the error correction.

We investigate this approach by employing BCH error-correcting codes.[2,9] The results show that while the individual binary functions are indeed more difficult to learn, the generalization performance of the system is improved.

Furthermore, as the length of the code n is increased, additional performance improvements are obtained.

Following this, we replicate these results on the ISOLET isolated-letter speech recognition task[4] using a variation on the back-propagation algorithm. Our error-correcting codes give the best performance attained so far by any method on this task. This shows that the method is domain-independent and algorithm-independent.

2. A COMPARISON OF THREE MULTICLASS METHODS ON THE NETTALK TASK

2.1 THE NETTALK TASK

In Sejnowski and Rosenberg's NETtalk system,[20] the task is to map from English words (i.e., strings of letters) into strings of phonemes and stresses. For example, f("lollypop") = ("lal-ipap", ">1<>0>2<"). Where "lal-ipap" is a string of phonemes, and ">1<>0>2<" is a string of stress symbols. There are 54 phonemes and 6 stresses in the NETtalk formulation of this task. Note that the phonemes and stresses are aligned with the letters of the original word.

As defined, f is a very complex discrete mapping with a very large range. Sejnowski and Rosenberg reformulated f to be a mapping g from a seven-letter window to a phoneme/stress pair representing the pronunciation of the letter at the center of the window. For example, the word "lollypop" would be converted into eight separate seven-letter windows:

```
g("___loll") = ("l", ">")
g("__lolly") = ("a", "1")
g("_lollyp") = ("l", "<")
g("lollypo") = ("-", ">")
g("ollypop") = ("i", "0")
g("llypop_") = ("p", ">")
g("lypop__") = ("a", "2")
g("ypop___") = ("p", "<")
```

The function g is applied to each of these eight windows, and then the results are concatenated to obtain the phoneme and stress strings. This mapping function g now has a range of 324 possible phoneme/stress pairs. This is the task that we shall consider in this chapter.

2.2 THE DATA SET

Sejnowski and Rosenberg provided us with a dictionary of 20,003 words and their corresponding phoneme and stress strings. From this dictionary we drew at random (and without replacement), a training set of 1000 words, and a testing set of 1000 words. It turns out that of the 324 possible phoneme/stress pairs, only 126 appear in the training set, because many phoneme/stress combinations make no sense (e.g., consonants rarely receive stresses). Hence, in all of the experiments in this chapter, the number of output classes is only 126.

2.3 INPUT AND OUTPUT REPRESENTATIONS

In all of the experiments in this chapter, the input representation scheme introduced by Sejnowski and Rosenberg for the seven-letter windows is employed. In this scheme, the window is represented as the concatenation of seven 29-bit strings. Each 29-bit string represents a letter (one bit for each letter, period, comma, and blank) and, hence, only one bit is set to 1 in each 29-bit string. This produces a string of 203 bits (i.e., 203 binary features) for each window. Experiments by Shavlik, Mooney, and Towell[21] showed that this representation was better than treating each letter in the window as a single feature with 29 possible values. Of course, many other input representations could be used. Indeed, in most applications of machine learning, high performance is obtained by engineering the input representation to incorporate prior knowledge about the task. However, an important goal for machine learning research is to reduce the need to perform this kind of "representation engineering." In this chapter, we show that general techniques for changing the *output* representation can also improve performance.

The representation of the output classes varies, of course, from one multiclass approach to another. For the direct multiclass method, the output class is represented by a single variable that can take on 126 possible values (one for each phoneme/stress pair that appears in the training data). For the one-per-class approach, the output class is represented by 126 binary variables, one for each class.

For the distributed-output-code approach, we employ the code developed by Sejnowski and Rosenberg. We used the Hamming distance between two bit-strings to measure distance. Ties were broken in favor of the phoneme/stress pair that appeared more frequently in the training data. In a paper by Dietterich, Hild, and Bakiri,[5] we called this "observed decoding."

2.4 THE ID3 LEARNING ALGORITHM

ID3 is a simple decision-tree learning algorithm developed by Ross Quinlan.[13,15] In our implementation, we did not employ windowing, CHI-square forward pruning,[14] or any kind of reverse pruning.[16] Experiments we have reported show that these pruning methods do not improve performance.[6] We did apply one simple kind of forward pruning to handle inconsistencies in the training data: If at some point in the tree-growing process all training examples agreed on the values of all features— and yet disagreed on the class—then growth of the tree was terminated in a leaf and the class having the most training examples was chosen as the label for that leaf (ties were broken arbitrarily for multiclass ID3; ties were broken in favor of class 0 for binary ID3).

In the direct multiclass approach, ID3 is applied once to produce a decision tree whose leaves are labeled with one of the 126 phoneme/stress classes. In the one-per-class approach, ID3 is applied 126 times to learn a separate decision tree for each class. When learning class i, all training examples in other classes are considered to be "negative examples" for this class. When the 126 trees are applied to classify examples from the test set, ties are broken in favor of the more frequently occurring phoneme/stress pair (as observed in the training set). In particular, if none of the trees classifies a test case as positive, then the most frequently occurring phoneme/stress pair is guessed. In the distributed output code approach, ID3 is applied 26 times, once for each bit-position in the output code.

2.5 RESULTS

In Table 1 we show the percent correct (over the 1000-word test set) for words, letters, phonemes, and stresses. A word is classified correctly if each letter in the word is correct. A letter is correct if the phoneme and stress assigned to that letter are both correct. For the one-per-class and distributed-output-code methods, the phoneme is correct if all bits coding for the phoneme are correct (after mapping to the nearest legal codeword and breaking ties by frequency). Similarly, the stress is correct if all bits coding for the stress are correct.

There are several things to note. First, the direct multiclass and distributed output codes performed equally well. Indeed, the statistical test for the difference

TABLE 1 Comparison of three multiclass methods.

| METHOD | LEVEL OF AGGREGATION (% CORRECT) | | | | TREE STATISTICS | | |
	Word	Letter	Phoneme	Stress	N	Avg. Leaves	Avg. Depth
Direct Multiclass	13.5	70.8	81.1	78.3	1	2652.0	73.0
One-per-class	8.7	66.7	76.4	74.5	126	34.9	10.5
Distributed	12.5	69.6	81.3	7.2	26	270.0	29.3

TABLE 2 Performance of error-correcting output codes.

| BCH CODE | | | | LEVEL OF AGGREGATION | | | | TREE STATISTICS | | |
| Phoneme | | Stress | | % Correct (1000-Word Test Set) | | | | | Avg. | Avg. |
n	d	n	d	Word	Letter	Phon.	Stress	N	Leaves	Depth
10	3	9	3	13.3	69.8	80.3	80.6	19	677.4	51.9
14	5	11	5	14.4	70.9	82.3	80.3	25	684.7	53.1
21	7	13	7	17.2	72.2	83.9	80.4	34	681.4	53.9
26	11	13	11	17.5	72.3	84.2	80.4	39	700.5	56.4
31	15	30	15	19.9	73.8	84.8	81.5	61	667.8	52.7
62	31	30	15	20.6	74.1	85.4	81.6	77	669.9	53.3
127	63	30	15	20.8	74.4	85.7	81.6	157	661.6	54.8

of two proportions cannot distinguish them. Second, the one-per-class method performed markedly worse, and all differences in the table between this method and the others are significant at or below the .01 level.

3. ERROR-CORRECTING CODES

The satisfactory performance of distributed output codes prompted us to explore the utility of good error-correcting codes. We applied BCH methods[11] to design error-correcting codes of varying lengths. These methods guarantee that the rows of the code (i.e., the codewords) will be separated from each other by some minimum Hamming distance d.

In Table 2 we show the results of training ID3 with distributed error-correcting output codes of varying lengths. Phonemes and stresses were encoded separately, although this turns out to be unimportant. Columns headed n show the length of the code, and columns headed d show the Hamming distance between any two code words.

The first thing to note is that the performance of even the simplest (19-bit) BCH code is superior to the 26-bit Sejnowski-Rosenberg code at the letter and word levels. Better still, performance improves monotonically as the length (and error-correcting power) of the code increases. The long codes perform much better than either the direct multiclass or Sejnowski-Rosenberg approaches at all levels of aggregation (e.g., 74.4% correct at the letter level versus 70.8% for direct multiclass).

Not surprisingly, the individual bits of these error-correcting codes are much more difficult to learn than the bits in the one-per-class approach or the Sejnowski-Rosenberg distributed code. Specifically, the average number of leaves in each tree in the error-correcting codes is roughly 665, whereas the one-per-class trees had only 35 leaves and the Sejnowski-Rosenberg trees had 270 leaves. Clearly, distributed output codes do not produce results that are easy to understand!

The fact that performance continues to improve as the code gets longer suggests that we could obtain arbitrarily good performance if we used arbitrarily long codes. Indeed, this follows from information theory under the assumption that the errors in the various bit positions are independent. However, because each of the bits is learned using the same body of training examples, it is clear that the errors are not independent. We have measured the correlation coefficients between the errors committed in each pair of bit positions for our BCH codes. All coefficients are positive, and many of them are larger than 0.30. Hence, there must come a point of diminishing returns where further increases in code length will not improve performance. An open problem is to predict where this breakeven point will occur.

3.1 ERROR-CORRECTING CODES AND SMALL TRAINING SETS

Given that the individual binary functions require much larger decision trees for the error-correcting codes than for the other methods, it is important to ask whether error-correcting codes can work well with smaller sample sizes. It is well established that small training samples cannot support very complex hypotheses.

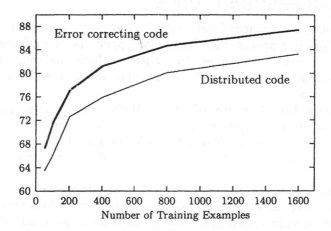

FIGURE 1 Learning curves showing "% phonemes correct" for the distributed output code and for the 93-bit error-correcting code (63 phoneme bits with $d = 31$, 30 stress bits with $d = 15$).

To address this question, In Figure 1 we show learning curves for the distributed output code and for the 93-bit error-correcting code (63 phoneme bits, 30 stress bits). At all sample sizes, the performance of the error-correcting configuration is better than the Sejnowski-Rosenberg distributed code. Hence, even for small samples, error-correcting codes can be recommended.

4. REPLICATION IN ISOLATED-LETTER SPEECH RECOGNITION

To test whether error-correcting output codes provide a general method for boosting the performance of inductive learning algorithms, we applied them in a second domain and with a different learning algorithm. Specifically, we studied the domain of isolated-letter speech recognition and the back-propagation learning algorithm.

In the isolated-letter speech-recognition task, the "name" of a single letter is spoken by an unknown speaker and the task is to assign this to one of 26 classes corresponding to the letters of the alphabet. Ron Cole gave us access to his ISOLET database of 7,797 training examples of spoken letters.[4] The database was recorded from 150 speakers balanced for sex and representing many different accents and dialects. Each speaker spoke each of the 26 letters twice (except for a few cases). The database is subdivided into five parts (named ISOLET1, ISOLET2, etc.) of 30 speakers each.

Cole's group has developed a set of 617 features describing each example. Each feature has been scaled to fall in the range $[-1, +1]$. We employed the opt implementation of back-propagation with conjugate gradient optimization in all of our experiments.[1]

In our experiments, we compared the one-per-class approach to a 30-bit ($d = 15$) BCH code and a 62-bit ($d = 31$) BCH code. In each case, we used a standard three-layer network (one input layer, one hidden layer, and one output layer). In the one-per-class method, test examples are assigned to the class whose output unit gives the highest activation. In the error-correcting code case, test examples are assigned to the class whose output codeword is the closest to the activation vector produced by the network as measured by the following distance metric: $\sum_i |\mathrm{act}_i - \mathrm{code}_i|$.

One advantage of conjugate-gradient optimization is that, unlike back-propagation with momentum, it does not require the user to specify a learning rate or a momentum parameter. There are, however, three parameters that must be specified by the user: (a) the starting random-number seed (used to initialize the artificial neural network), (b) the number of hidden units, and (c) the total-summed squared error (TSS) at which training should be halted (this avoids over training).

TABLE 3 Parameter values selected via cross validation.

Configuration	# Hidden Units	Best TSS
one-per-class	78	10.50
30-bit ECC	156	142.66
62-bit ECC	156	161.76

TABLE 4 Performance in the isolated letter domain.

Configuration	Actual TSS	% Correct	% Error
one-per-class	10.51	95.83	4.17
30-bit ECC	142.26	96.73	3.27
62-bit ECC	161.85	95.96	4.04

To determine good values for these parameters, we followed the "cross-validation" training methodology advocated by Lang, Waibel, and Hinton.[10] The training data were broken into three sets:

- The **training set** consisting of the 3,120 letters spoken by 60 speakers. (These are the examples in Cole's files ISOLET1 and ISOLET2.)
- The **cross-validation set** consisting of 3,118 letters spoken by 60 additional speakers. (These are the examples in files ISOLET3 and ISOLET4.)
- The **test set** consisting of 1,559 letters spoken by 30 additional speakers. (These are the examples in file ISOLET5.)

The idea is to vary the parameters while training on the training set and testing on the cross-validation set. The parameter values giving the best performance on the cross-validation set are then used to train a network using the *union* of the training and cross-validation sets, and this network is then tested against the test set. We varied the number of hidden units between 35 and 182, and, for each number of hidden units, we tried four different random seeds. In Table 3 we show the parameter values that were found by cross validation to give the best results.

In Table 4 we show the results of training each configuration on the combined training and cross-validation sets and testing on the test set. Both error-correcting configurations perform better than the one-per-class configuration. The results are not statistically significant (according to the test for the difference of two proportions), but this could be fixed by using a larger test set. The results are very definitely significant from a practical standpoint: The error rate has been reduced by more than 20%. This is the best known error rate for this task.

5. CONCLUSIONS

The experiments in this chapter demonstrate that error-correcting output codes provide an excellent method for applying binary learning algorithms to multiclass learning problems. In particular, error-correcting output codes outperform the direct multiclass method, the one-per-class method, and a domain-specific distributed output code (the Sejnowski-Rosenberg code for the NETtalk domain). Furthermore, the error-correcting output codes improve performance in two very different domains and with two quite different learning algorithms.

We have investigated many other issues concerning error-correcting output codes, but, due to lack of space, these could not be included in this chapter. Briefly, we have demonstrated that codes generated at random can act as excellent error-correcting codes. Experiments have also been conducted that show that training multiple neural networks and combining their outputs by "voting" does not yield as much improvement as error-correcting codes.

ACKNOWLEDGMENTS

The authors thank Terry Sejnowski for making available the NETtalk dictionary and Ron Cole and Mark Fanty for making available the ISOLET database. The authors also thank National Science Foundation for its support under grants IRI-86-57316 and CCR-87-16748. Ghulum Bakiri was supported by Bahrain University.

REFERENCES

1. Barnard, E., and R. A. Cole. "A Neural-Net Training Program Based on Conjugate-Gradient Optimization." Report No. CSE 89-014, Beaverton, OR, 1989.
2. Bose, R. C., and D. K. Ray-Chaudhuri. "On a Class of Error-Correcting Binary Group Codes." *Inf. Cntl.* **3** (1960): 68–79.

3. Breiman, L., J. H. Friedman, R. A. Olshen, and C. J. Stone. *Classification and Regression Trees.* Monterey, CA: Wadsworth and Brooks, 1984.

4. Cole, R., Y. Muthusamy, and M. Fanty. "The ISOLET Spoken Letter Database." Report No. CSE 90-004, Oregon Graduate Institute, Beaverton, OR, 1990.

5. Dietterich, T. G., H. Hild, and G. Bakiri. "A Comparative Study of ID3 and Backpropagation for English Text-to-Speech Mapping." In *Seventh International Conference on Machine Learning,* 24–31. Austin, TX: Morgan Kaufmann, 1990.

6. Dietterich, T. G., H. Hild, and G. Bakiri. "A Comparison of ID3 and Backpropagation for English Text-to-Speech Mapping." Report No. 90-30-4, Oregon State University, Corvallis, OR, 1990.

7. Fulk, M. A., and J. Case, eds. *COLT '90: Proceedings of the Third Annual Workshop on Computational Learning Theory.* Rochester, NY: Morgan Kaufmann, 1990.

8. Haussler, D., and L. Pitt, eds. *COLT '88: Proceedings of the Second Annual Workshop on Computational Learning Theory.* Cambridge, MA: Morgan Kaufmann, 1988.

9. Hocquenghem, A. "Codes Corecteurs d'Erreurs." *Chiffres* **2** (1959): 147–156.

10. Lang, K. J., A. H. Waibel, and G. E. Hinton. "A Time-Delay Neural Network Architecture for Isolated Word Recognition." *Neural Nets.* **3** (1990): 33–43.

11. Lin, S., and D. J. Costello, Jr. *Error Control Coding: Fundamentals and Applications.* Englewood Cliffs: Prentice-Hall, 1983.

12. Nilsson, N. J. *Learning Machines.* New York: McGraw Hill, 1965.

13. Quinlan, J. R. "Learning Efficient Classification Procedures and Their Application to Chess Endgames." In *Machine Learning,* edited by R. S. Michalski, J. Carbonell, and T. M. Mitchell, 463–482. Vol. I. Palo Alto, CA: Tioga Press, 1983.

14. Quinlan, J. R. "The Effect of Noise on Concept Learning." In *Machine Learning,* edited by R. S. Michalski, J. Carbonell, and T. M. Mitchell, 149–166. Vol. II. Palo Alto, CA: Tioga Press, 1986.

15. Quinlan, J. R. "Induction of Decision Trees." *Mach. Learn.* **1(1)** (1986): 81–106.

16. Quinlan, J. R. "Simplifying Decision Trees." *Int. J. Man-Mach. Stud.* **27** (1987): 221–234.

17. Rivest, R. L., D. Haussler, and M. K. Warmuth, eds. *COLT '89: Proceedings of the Second Annual Workshop on Computational Learning Theory.* Santa Cruz, CA: Morgan Kaufmann, 1989.

18. Rosenblatt, F. "The Perceptron." *Psych. Rev.* **65(6)** (1958): 386–408.

19. Rumelhart, D. E., G. E. Hinton, and R. J. Williams. "Learning Internal Representations by Error Propagation." In *Parallel Distributed Processing,* edited by D. E. Rumelhart and J. L. McClelland, 318–362. Vol. I. Cambridge, MA: MIT Press, 1986.

20. Sejnowski, T. J., and C. R. Rosenberg. "Parallel Networks that Learn to Pronouce English Text." *Complex Systems* **1** (1987): 145–168.
21. Shavlik, J. W., R. J. Mooney, and G. G. Towell. "Symbolic and Neural Learning Algorithms: An Experimental Comparison." *Mach. Learn.* **6** (1990): 111–144.
22. Valiant, L. G. "A Theory of the Learnable." *CACM* **27** (1984): 1134–1142.

John S. Denker and Christopher C. J. Burges
AT&T Bell Laboratories, Holmdel, NJ 07733

Image Segmentation and Recognition

We have constructed a system for recognizing multicharacter images. This is a nontrivial extension of our previous work on single-character images.

We show how a highly structured neural network can be combined with a dynamic programming lattice to construct a system that learns to recognize multicharacter images.

It is somewhat surprising that a very good single-character recognizer does not, in general, form a good basis for a multicharacter recognizer. The correct solution depends on three key ideas: (1) a method for normalizing probabilities correctly, to preserve information on the quality of the segmentation; (2) a method for giving credit for multiple segmentations that assign the same interpretation to the image; and (3) a method that combines recognition and segmentation into a single adaptive process, trained to maximize the score of the right answer.

We also discuss improved ways of analyzing recognizer performance.

A major part of this work is devoted to giving our methods a good theoretical footing. In particular, we do *not* start by asserting that maximum likelihood is obviously the right thing to do. Instead, the problem is formalized in terms of a probability measure; the learning algorithm must then be arranged to make this probability conform to the customer's needs.

This formulation can be applied to other segmentation problems such as speech recognition.

Our recognizer using these principles works noticeably better than the previous state of the art.

1. OVERVIEW

We believe that the future of supervised learning will include attacking progressively more complex problems, using correspondingly more complex learning machines. In order to engineer such complex machines, we hope to have systematic procedures for combining a number of modules, each of which handles parts of the task. We illustrate this with a case study: we have combined a neural network (known to be good for recognizing single characters) with a dynamic programming lattice (known to be good for dealing with sequences) to produce a multicharacter recognizer. The principles involved should be quite broadly applicable.

We begin with a terse overview. Many of the buzzwords used here will be defined more clearly in the following sections.

The task at hand is to build an image recognizer. The input to the recognizer is a multicharacter image, such as the ZIP Code shown in Figure 1. The desired output of our recognizer is the best interpretation of the image (e.g., "35133") along with a good estimate of the probability that that interpretation is correct.

As mentioned in the abstract, we have constructed a character recognizer that incorporates three key ideas, as discussed in the following three subsections.

FIGURE 1   A typical image.

1.1 NORMALIZATION OF PROBABILITIES

Our first design for a multicharacter recognizer (MCR) naturally used our trusty single-character recognizer (SCR) as a building block. We have since discovered that this is not quite optimal. Specifically, the normalization that is appropriate for an SCR throws away information which is needed for proper segmentation.

The correct strategy has two steps:

1. Compute a score for each possible multicharacter path (where a "path" specifies the interpretation *and* segmentation) by combining the scores of the individual characters (and whatever other information is available).
2. Normalize these scores by dividing by the sum over all paths.

We present below a firm theoretical support for this scheme. In particular, it would be clearly nonoptimal to perform the required operations in the reverse order:

1'. Normalize on a per character basis.

2'. Form a multicharacter score by multiplying the normalized per character scores.

1.2 MULTIPLE SEGMENTATIONS

The normalized score for a path gives the joint probability of a given interpretation *and* segmentation. But at the end of the day, the customer cares about the correct intepretation; the segmentation problem is just a step along the way. Therefore, we compute the score for a given interpretation by summing over all paths that give that interpretation. This, too, has a firm theoretical foundation. The required sum can be performed very efficiently using the "forward"[1] algorithm.

The difficult step is that we must find the *best* interpretation, i.e., the interpretation that maximizes the aforementioned sum over paths. The Viterbi algorithm[17,9] can efficiently find the best scoring path, but it is harder to find the best *sum* directly in closed form. If we could assume that each sum was dominated by its largest term, Viterbi would be exact. In practice, we find that Viterbi, while not always exact, is good enough to identify the one or two interpretations that are *candidates* for having the best sum. We can then run the forward algorithm to check these candidates, i.e., to compute their exact score.

1.3 LEARNING

For a single-character recognizer, we use a neural net to produce a score that reflects the conditional probability of an interpretation, *given* the segmentation. For a multicharacter recognizer, we need more detailed information, namely the joint probability of interpretation *and* segmentation. Since the training process determines what the network will actually do, we have designed a training process that takes the needs of the MCR into account.

Specifically, the block diagram of the MCR consists of a large number of SCRs (each of which evaluates the score of a segment) plus a lattice (that finds the best segmentation, i.e., the best way of combining segments). This idea[15,2,10,11,8] has been around for some time. Obviously "backprop" can be used to train the SCR, but it has not been 100% obvious how to choose the training targets that backprop (supposedly) needs. The new scheme is to treat the whole MCR (lattice and all) as one huge network and perform backprop (or a generalization thereof) on the whole thing.

We use the Baum-Welch algorithm[1] (also known as forward-backward or E-M) to calculate derivatives. This tells us how sensitive the output is to each adjustable parameter in the MCR. Baum-Welch is to forward as backprop is to fprop.

In essence, during training, each parameter is adjusted in a direction that will increase the score of the correct answer. (Since the answers are normalized, this automatically decreases the score of wrong answers.)

This training scheme has two advantages:

i. As mentioned above, it teaches the neural net to produce information about the quality of the segmentation in addition to information about what the interpretation would be if we were given the correct segmentation.

ii. It teaches the network to avoid *near mistakes,* not just *mistakes.* That is: under the old training scheme, the highest scoring path was identified. If it was incorrect, the training process would lower its score. As soon as the wrong path's score dropped below the right path's score, no further training could be applied, unless the wrong path had the same segmentation as the right path. Fortunately for us, in most cases the segmentations *were* the same, so the old training scheme worked quite well in practice. In those cases where the segmentations were different, the training could well produce bad segmentations with scores only infinitesimally lower than the good segmentation—leading to poor robustness. The new training scheme does not have this vulnerability. Since it involves a sum over paths, not a max over paths, it pushes all good paths up and pushes all bad paths down, and keeps pushing regardless of what the best path happens to be.

And finally, the bottom line, we have implemented these ideas and the resulting recognizer just plain works better, as will be discussed later.

2. THE DYNAMIC PROGRAMMING APPROACH TO SEGMENTATION

For some time now we have had a good single-character recognizer (SCR). If all images were naturally divided into segments containing one character apiece, the multicharacter recognizer (MCR) would be straightforward. But, in general, real images contain (a) fragmented characters consisting of multiple disconnected strokes, and (b) characters that nestle, touch, and overlap. In Figure 1 is exhibited both of these problems: the horizontal stroke that is supposed to form the top of the "5" is disconnected from the body of the "5" and is connected to the following "1" instead.

In Figure 2 are several touching characters, and one character nestled under the overhang of another.

The strategy may be summarized as follows:

- Divide the image into cells of manageable size.
- Combine the cells in various ways to form "segments.
- Form a sequence of segments (called a "segmentation") that accounts for the whole image.
- Assign scores to various possible segmentations.
- Find the interpretation of the highest scoring segmentation(s).

One significant advantage of this strategy is that the system does not have to generate the correct segmentation initially: it merely has to create a set of hypothetical segmentations which *contains* the correct segmentation(s).[7,6] Previous approaches to this problem envisioned a "pipeline" architecture in which the segmentation process was completed before the single-character recognition process began; in contrast we envision a system in which the segmenter calls a (modified) single-character recognizer as a subroutine.

The "second generation" segmenter is described more fully elsewhere[7,6] but will be briefly described here. It starts by cutting the image into cells. The cuts are not necessarily vertical. While the "first generation" segmenter[14] makes extensive use of "connected component analysis" (CCA), the second-generation system does not; we are working on a segmenter that unifies the cut generation and CCA.

One or more cells are combined to make a segment. The objective is to create segments that contain exactly one character. The second-generation segmenter uses cells in a left-to-right order and only makes segments out of *consecutive* cells; there are one or two examples (out of several thousand) in the training set where this restriction is nonoptimal, because one character overhangs another.

Since the number of ways that cells can be combined to make a segment grows combinatorially with the number of cells, we cannot afford to chop the image into ultra-small cells. Therefore, the preprocessor embodies heuristics that will identify

FIGURE 2   Touching and nestling.

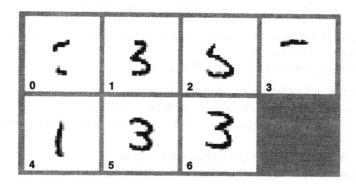

FIGURE 3   The cells.

a set of "good" cuts, including some "obvious" cuts. The cells between good cuts are generally much more than one pixel wide. The set of good cuts will generally contain more cuts than needed.

The image in Figure 1 was divided into the seven cells shown in Figure 3.

Segments are sometimes called boxes, in analogy with the "boxes" and "glue" used by the typesetting language TEX.

A segmentation is defined to be a sequence of boxes. Ideally the boxes would abut one another left to right, but in practice we must allow for "glue" between boxes. Positive glue means a sliver of image between the two boxes is skipped, negative glue means two boxes overlap, zero glue means they abut.

In Figure 4 we show the segments constructed for the example image; in the corner of each image is a list of numbers designating the cells from which the segment was constructed.

In Figure 5 we show three of the many possible segmentations of our example image. Interestingly, the (2) segment appears in the third slot of the first segmentation, and the second slot of the second segmentation. Thus the segmentation algorithm cannot simply keep track of which segments are used; it must keep track of which segment appears in which slot of the answer.

It is advantageous[6,7] to formulate the segmentation problem as a "best path through graph" problem, as we will now summarize.

We will distinguish between the input space (i.e., image pixels) and the output space (i.e., character interpretations). Borrowing some of the methods used[17] for segmenting speech signals, we will speak of the output space in terms of "time slots." In the previous examples there were five time slots—because we are expecting a five-digit answer.

At this level of description we have a two-dimensional array of nodes, indexed by (segment-ID, time-slot), where segment-ID is an index into the set of allowed segments for this image.

This can be drawn as shown in Figure 6, where each • is a node in the graph. Note there is a special start-node before the first time and to the left of the leftmost segment, and a special end-node after the last time and to the right of the rightmost segment. The arcs constituting all valid segmentations are also shown.

For reasons that will become clear in a moment, we subdivide the time axis, as shown in Figure 7; the label m is for morning and e is for evening. In this graph, there are two kinds of arcs: glue arcs and recognition arcs. All arcs are directed; we think of them as left to right, although one could just as well solve the problem right to left instead. Recognition arcs (rec-arcs) are daytime arcs, from a morning to an evening. Glue-arcs are nighttime arcs, from an evening to the next morning.

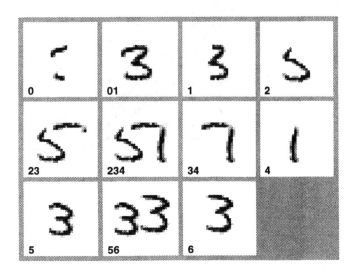

FIGURE 4   The segments.

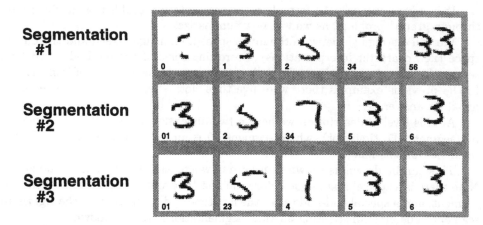

Segmentation #1

Segmentation #2

Segmentation #3

FIGURE 5   Three segmentations.

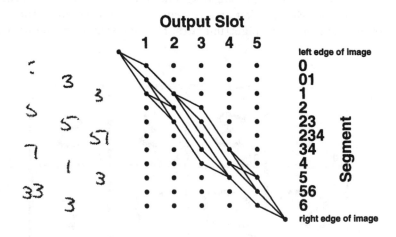

FIGURE 6   Basic Lattice.

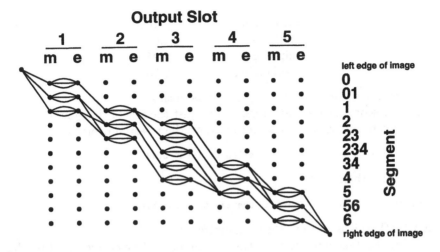

FIGURE 7   Expanded Lattice.

2.1 GLUE ARCS

A given evening node will generally have several glue arcs leaving it. For example, a (1 2) node might originate the following arcs:

(1 2) $\longrightarrow$ (3)	"abut"	
(1 2) $\longrightarrow$ (3 4)	"abut"	
(1 2) $\longrightarrow$ (4)	"skip"	
(1 2) $\longrightarrow$ (2 3)	"overlap"	

The parameters of our implementation are currently set to allow only abutment, and possibly skips connected to the start and end nodes.

The number of arcs can be reduced by using the information about obvious cuts; no segment can straddle an obvious cut. This is technically a statement about node building, but it has a big impact on arc building, because we do careful pruning of nodes and arcs. We designate "live nodes" as ones that are descendents of the start node and (simultaneously) ancestors of the end node. Arcs to or from a nonlive node are deleted from the graph, to prevent unnecessary computation.

We experimented with permitting skips and overlaps within the image, but it didn't help. Too many new possibilities, not many GOOD new possibilities. The score of such arcs should be determined by our knowledge about the probability of images with and without arcs of a given type. For instance, we might assign scores as follows:

type of arc	score
abutting	1.0
skipping	0.95
overlapping	0.90

where scores have the same meaning as probabilities.

The glue-score can depend on more than just the size of the glue. For instance, we may want to discourage total overlap more than partial overlap:

    (2)   --> (2 3)   (total overlap of first box)

is much scarier than

    (1 2) --> (2 3)   (partial overlap of first box)

Explain WHY this is scarier. Show examples!!!!!!!

We can also take into consideration the number of *pixels* traversed by the glue (and in the boxes at the head and tail of the glue) rather than just using the id numbers of the cells involved.

2.2 RECOGNITION ARCS

We assign each rec-arc four attributes: origin-node, destination-node, interpretation, and score. Note that we can have more than one arc connecting a given pair of nodes; indeed, recognition arcs come in groups of N, where N is the size of the recognition alphabet ($N = 10$ for digit recognition). For clarity, only $N = 3$ recognition arcs per group are drawn in Figure 7.

As shown in the figure, recognition arcs connect the morning and evening nodes of a given segment. The score of each rec-arc is determined by calling the single-character recognizer on the corresponding segment, i.e., box of pixels. There is only one call to the SCR per row of the lattice.

2.3 PATHS, SEGMENTATIONS, INTERPRETATIONS

We will consider paths connecting the start-node to the end-node; the path will be a sequence of arcs of the form [G R G R G R G R G R G] where G stands for glue-arc and R stands for rec-arc.

In this formalism the segmentation corresponding to a given path is independent of the rec-arcs and consists of the list of segments visited by the glue arcs in the path.

Analogously, an "interpretation" is a string of characters, e.g., "07733." For a five-digit string, there will be 10^5 possible interpretations per segmentation. For each path, the segmentation is formed by concatenating the interpretations of the rec-arcs making up the path.

The score of a path is formed by multiplying the scores of the constituent glue-arcs and rec-arcs. Presently all glue-arcs have a score of 1.0; the rec-arc scores are computed by a neural network as discussed in Section 4.

To summarize this subsection, if you take a path and ignore the rec-arcs, you get a segmentation consisting of a list of glue-arcs; conversely, if you take a path and leave out the glue-arcs, you get get an interpretation, that consists of a list of rec-arcs.

2.4 RUNNER-UP INFORMATION

There are a number of situations in which it is useful to know the scores for more than one interpretation of the image. If nothing else, it is useful during training to know how the bad-guy scores are changing relative to the good-guy scores. (A training scheme that increases the good-guy scores is a failure if it increases bad-guy scores by the same amount.)

The famous Viterbi algorithm[17,9] operates in terms of paths. It will find the best scoring path and tell you its score. There exist algorithms that will find the K best paths through the lattice, but as likely as not, many of these paths will have the same interpretation as the best one (i.e., different segmentations with the same interpretation). It could be quite inefficient to enumerate all these paths and then search for one that differs in interpretation. Therefore, we now present an algorithm that directly finds the best path *that differs in interpretation* from the very best path.

We will do this using a lattice that has twice as many nodes as the lattice in Figure 7. One part of the new lattice will be called the "gold plane" and will operate as previously described to identify the "gold path," i.e., the very best interpretation of the image. The other part of the new lattice will be called the silver plane and will be used to compute the "silver path," i.e., the best path that differs *in interpretation* from the gold path.

The nodes in the new lattice are identified by (segment-ID, time-slot, m/e, g/s) where m/e specifies morning versus evening, and g/s specifies gold versus silver.

The left part of Figure 8 is a close-up view of part of the gold plane. The ordinary Viterbi algorithm has been used to identify the gold path, which is shown with a heavy line. The arcs connecting the gold and silver planes are disabled while the gold path is being computed, and are not shown in this part of the figure.

After the gold path has been identified, the system is ready to compute the silver path. Each evening node in the silver plane has $2N - 1$ incoming arcs: N from the corresponding morning node in the silver plane, and $N - 1$ from the corresponding morning node in the gold plane. The missing arc is the one whose interpretation matches the gold path's interpretation for that output time slot. This is portrayed in the right part of Figure 8. The column headed "2G" represents time slot 2

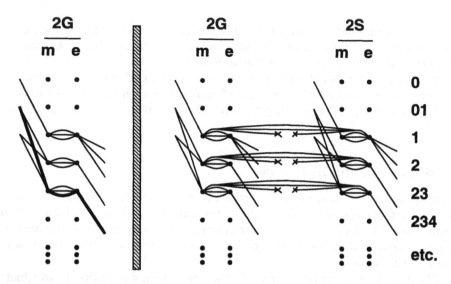

FIGURE 8   Runner-Up Lattice.

in the gold plane, and the column headed "2S" is the corresponding part of the silver plane.

Glue-arcs stay in their plane.

There is only one start node, and it is in the gold plane. The end node of the gold path is in the gold plane, and the end node of the silver path is defined to be in the silver plane. Any path that reaches the silver plane must have differed (in interpretation!) from the gold path in at least one place, because the only arcs that go from one plane to the other are restricted to differ in the required way. It is, of course, possible that the silver path differs from the gold path in more than one slot.

The computation for the silver path is of the same order of complexity as for the gold path—which means it is negligible compared to the nonrecurring cost of computing the rec-arc scores. Also note that during the computation of the runnerup, no nodes in the gold plane need be updated.

3. ASSIGNING PROBABILITIES TO COMPETING INTERPRETATIONS

We assign a number called an "R-score" to each possible outcome of the segmentation/recognition process. In general, R will be a function of the interpretation C, the segmentation S, and the image I.

In the first part of this section we will be intentionally vague about the meaning of the R-score, in order to maximize the generality of some theoretical results. But to fix ideas, it won't hurt if you visualize $R(C, S, I)$ as equaling the probability $P(C, S, I)$, i.e., the joint probability that the data set contains a data point where character string C and segmentation S are attached to image I. For starters, we will explicitly require that R be nonnegative.

As discussed in the previous section, since we are building upon our single-character recognizer (SCR), one way of constructing the R-score of a segmentation would be to multiply the r-scores of the constituent segments, where the r-score (with a little r) is computed by the SCR. Fancier factorizations of R will be considered below.

Let us assume for the moment that the SCR produces properly normalized r-scores. That is, for each single character, the r-scores add up to unity. This will guarantee that R-scores of a given segmentation will also be normalized. Therefore, we don't need to worry about the case where one interpretation has an R-score of .99, while some other interpretation has an R-score of .95 (which would be a problem since the alternatives would add up to more than unity).

On the other hand, there is no satisfactory way the SCR can be normalized so that scores of *different* segmentations can be compared. In particular all the segments in the second and third segmentations shown in Figure 5 would receive good r-scores, so both segmentations would be assigned good R-scores.

It absolutely will not suffice to normalize these R-scores by dividing by the total of all R-scores. This is infeasible to compute and gives absurd results anyway.

Therefore, we must retract the assumption that the r-scores are normalized (i.e., per character normalization) and find a new way to normalize the R-scores (i.e., per image normalization).

To formalize the task: we want to find the most probable character string, and estimate its probability, given the image.

Let

I	$:=$	the "image"
C	$:=$	the "interpretation" (a string of characters)
C_t	$:=$	the character in the tth position of C
S	$:=$	the "segmentation" (a string of segments)
S_t	$:=$	the segment in the tth position of S

The single-character recognizer returns an r-score which depends on S_t, C_t, and I. Note: it is more common to think of the recognizer as returning a vector of scores, depending on S_t and I, but it is formally equivalent and more convenient for present purposes to pick out the score for each C_t separately. For example, if we are considering the image in Figure 1 and its segment (01) as shown in Figure 4, then $r(\text{"3,"}$ seg, img$)$ will be large while $r(\text{"1,"}$ seg, img$)$ will be small.

Using a set of SCRs (plus whatever else you can think of) let's assume we can assign a score to a given segmentation and interpretation; call it $R(C, S, I)$. Arrange it to be positive by construction. As will be discussed later, we want to train the network so that R will be more or less a probability, but we don't want to write it as $P(C, S, I)$ just yet.

Remark: there is an axiomatic definition of probability measure. Specifically, a *measure* is a mapping from sets to numbers; it must be positive, and the measure of the union of disjoint sets must equal the sum of the measures of the ingredients. A *probability measure* has the further property that it is bounded above; typically we arrange this bound to be unity. Anything that satisfies the axioms can justly be called "a" probability; we will be using several such probabilities, and must be careful not to call any of them "the" probability without qualification.

John Bridle[4,5] suggested computing the quantity

$$Q(C|S, I) := \frac{R(C, S, I)}{\sum_{c'} R(C', S, I)} \tag{1}$$

where the sum runs over all possible interpretations. This is called "softmax" normalization.

It is easy to verify that $\sum_C Q(C|S, I) = 1$, and that $Q(C|S, I)$ is positive. Therefore, Q satisfies the axioms of probability, and the placement of the "given" bar in Eq. (1) is consistent with the usual notation of conditional probabilities. For a single-character recognizer where the segmentation S can be considered a "given," this scheme (with an elaboration to be discussed below) is an excellent way of doing the normalization. For suitable functions R, this gives a good estimate of the actual probability of the interpretation C being correct.

Many people were surprised to discover that softmax scores cannot serve as the basis of a multicharacter recognizer (MCR). To see this, compute instead

$$Q(C, S|I) := \frac{R(C, S, I)}{\sum_{C'S'} R(C', S', I)} \tag{2}$$

and

$$Q(C|I) := \sum_S R(C, S|I). \tag{3}$$

The last formula is very important because the probability $P(C|I)$ (or the expected cost based thereon) is what the customer cares about! In particular he often wants to know the max and argmax (w.r.t. C) of $P(C|I)$ for each image I. If we are to have any hope that Q will be a good estimate of P, it must be normalized correctly. $Q(C|I)$ cannot be computed by summing the softmax result $Q(C|S, I)$ over S; the given-bar is on the wrong side of S.

To make this problem clear, consider (again) the competing segmentations in Figure 5. Segmentations #2 and #3 are the serious contenders. Most humans prefer segmentation #3 (with interpretation "35133") over segmentation #2 (with interpretation "35733"). Segment (34) which appears in segmentation #2 is a perfectly good "7," so the preference must be based on the judgement that segment (2) is not a good "5."

Humans can make this judgement, but an MCR based on a softmax SCR will have trouble. The problem is that the conditional probability is nearly 100% that *if* the pixels in segment (2) represent a digit, *then* it must be a "5." Since softmax provides just such a conditional probability, it is doomed.

The MCR needs the *joint* probability that the segmentation is correct *and* the interpretation is correct. Softmax throws away the segmentation-quality information prematurely.

To quantify this, let us factor $R(C, S, I) = T(C, S, I) U(S)$, where $T(C, S, I)$ is constructed to carry very little information about the quality of the proposed segmentation S—that information being carried instead by $U(S)$. To fix ideas, imagine $T(C, S, I)$ to be the conditional probability $P(C, I|S)$, while $U(S)$ is the "marginal" $P(S)$. Plugging in, we get:

$$Q(C|S, I) := \frac{T(C, S, I)U(S)}{\sum_{c'} T(C', S, I)U(S)}. \tag{4}$$

Alas, the factor $U(S)$ drops out of this expression. In contrast, the S-dependence does not drop out of Eq. (2), since it has an S in the numerator and an S' in the denominator.

As foreshadowed above, let us assume (with modest loss of generality) that the recognition results are context independent, so we can factor R as

$$R(C, S, I) = \prod_t r(C_t, S_t, I) \tag{5}$$

where the product runs over all output slots t (e.g., 1 through 5, for ZIP Codes). The output slots are called t to suggest "time" in analogy to speech segmentation models.

Additional factors expressing the "quality of segmentation" (e.g., the score of glue arcs as discussed in Section 2) have been omitted from this expression for clarity; restoring them to this and subsequent expressions is routine.

We assume r is positive. We will try to arrange it to be an increasing function of the probability that the (individual) character c goes with (individual) segment s.

Plugging in, we conclude that the customer may want to know the max and argmax (w.r.t. C, for a given I) of the quantity

$$Q(C|I) = \frac{\sum_{S''} \prod_t r(C_t, s_t'', I)}{\sum_{C'} \sum_{S'} \prod_t r(C', S_t', I)}. \tag{6}$$

It is useful to consider the equivalent expression

$$Q(C|I) = \frac{\sum_{C''} \sum_{S'} \prod_t r(C_t, S_t'', I)\delta_{C''}^C}{\sum_{c'} \sum_{S'} \prod_t r(C_t', S_t', I)} \tag{7}$$

where δ is the Kronecker delta. This form makes it clear that both numerator and denominator are "lattice sums"; i.e., the sums run over all paths in the lattice. The summand is slightly different for the numerator and denominator. Such sums can be computed very efficiently via the "forward" algorithm.

The forward algorithm is very similar to the Viterbi algorithm. They have the same computational complexity; the former computes a sum where the latter computes a max.

The first step in computing $\max Q(C|I)$ is to compute the denominator using the forward algorithm, that does exactly what we want. The denominator is independent of C.

As for the numerator, we need to compute the sum *and* maximize over all interpretations C. For any particular C, it is fast (using the forward algorithm) to compute the sum. It is also fast (using Viterbi) to compute the max (over all C) of the summand. Viterbi will not compute the max of the whole sum, which is what we would like.

To paraphrase Abraham Lincoln, we can easily maximize one term over all paths, and we can easily evaluate all the terms for some of the paths, but we can't so easily maximize all the terms over all the paths.

We could just approximate the sum by its largest term, in which case Viterbi would give us the answer. A better approximation is to use Viterbi to find the interpretation that contributes the largest term to the sum, and then use forward to evaluate *all* the terms with that interpretation. This gives the exact score $Q1$ for that interpretation; the only question is whether it is the genuine maximum. Since Q is manifestly normalized, if $Q1$ is greater than one-half, we know there can be no other possibility.

If $Q1$ does not "use up" enough of the probability measure, we can use second-interpretation Viterbi to identify another contender, and then use forward to compute its score. In principle this process could continue indefinitely. Let $Q*$ denote the sum of the scores of the already checked interpretations; if at any point $Q*$ differs from unity by less than the best score already found, then we have certainly identified the winner (and precisely computed its score).

This technique is known in the speech recognition community, but is rarely used, apparently because the number of candidates gets out of hand. In our case, though, one or two candidates usually suffice to "use up" essentially 100% of the probability, at which point we know our method has found the exact answer.

There exist other schemes for finding and/or approximating the max of the sum, but we won't discuss them here.

3.1 COOPERATING SEGMENTATIONS

To reiterate, our MCR assigns scores to interpretations. The score is a ratio: the denominator is a sum over all paths, and the numerator is a restricted sum, restricted to paths having the given interpretation.

We now explain why it is advantageous to evaluate the restricted sum in the numerator, as opposed to making the Viterbi approximation that the sum is equal to its largest term. Consider the two segmentations [(01) (23) (4) (5) (6)] and [(1) (23) (4) (5) (6)]. Since we see from Figure 4 that segment (01) and segment (1) are both perfectly good "3"s, both segmentations will receive high scores; let's assume their scores are essentially equal. Both paths will contribute to the sum in the denominator. If only one of them contributed to the numerator, the score would be reduced by a factor of two—a very significant amount.

---

4. TRAINING

In the previous section, we showed that for any r-score function, the Q function derived from it would be normalized like a probability, so it could be considered "a" probability in the abstract, formal sense. Our task is not complete, however, until we show how to create an r-score function that leads to "the" probability, or at least to "a" probability that meets the customer's needs.

We work within the framework of supervised learning. That is, for each image I^α in the training set, there is a "desired" interpretation $C*^\alpha$. A straightforward learning principle is to say that we want to increase the expected score of the right answer, $\langle Q(C*^\alpha \,|\, I) \rangle$, where the expectation involves averaging over all elements α in the training set. Because Q is normalized, this principle automatically implies that the score of undesired interpretations should decrease.

Consider the lattice for our example image and its 11 segments. The lattice calculates one final output, the Q score for the desired answer. This number is a function of 110 numbers that are input to the lattice, namely the scores of the recognition arcs (ten per segment). It is useful to calculate the sensitivity of the output to each of these inputs. This can be done using the Baum-Welch algorithm (also called the forward-backward algorithm). Specifically, Baum-Welch will give 110 partial derivatives, one for each input. The meaning of each one is clear: if the derivative is x, it means that if the score of that rec-arc went up by one unit, the final Q score would go up by x units (to first order).

The situation for our example is shown in Figure 9. A white box indicates a positive derivative, while a black box denotes a negative derivative. We can see that increasing the response of "3" detector neuron to segment (01) will improve the overall score. It also makes sense that it would hurt the overall score if the (2) segment were recognized as a "3" or a "5"—because that would increase the R-score of paths that give the wrong answer. The strong negative derivative on the "7" response to segment (34) will be discussed later.

In our recognizer, the r-scores come from a feedforward neural network called LeNet.[12] The ten output signals of LeNet are exponentiated to give the r-scores.

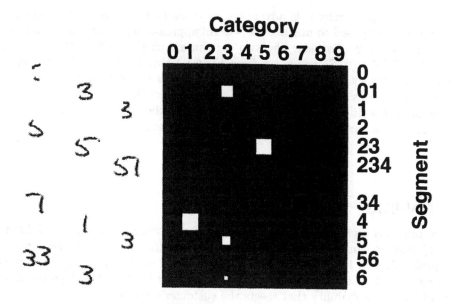

FIGURE 9   Derivatives w.r.t. Recognition Arcs.

In the example, the whole MCR consists of the lattice plus 11 copies of LeNet (one for each segment). All these networks share the same set of weights.

Now, we can consider the whole MCR as a complicated neural network. We can backpropagate through the whole shebang. The objective function, as we said, is the average Q score of the desired answers. By the chain rule, the gradient in LeNet's weight space is where the · indicates a sum over all 110 r-scores, and where W is the vector of weights, i.e., the adjustable parameters of the neural network. The first factor is computed using Baum-Welch, while the second factor is computed using standard backprop. To put it another way, the derivatives calculated using Baum-Welch (and displayed in Figure 9) are used as initial deltas at the top layer of LeNet. Backprop does not require targets; all it requires is initial deltas.

Taking a small step in weight space in the direction of the gradient is guaranteed to increase the objective function.

The large black blob in column "7" and row (34) of Figure 9 means that LeNet will be trained *not* to recognize such a segment as a "7." Penalizing this image could be a problem, since the segment really looks like a "7," but fortunately this is not the only image in the training set. We rely on other images to generate a countervailing training signal. In the end, images like this will be recognized as "7"s, but the score (i.e., the estimated probability) will be less than 100%, which seems reasonable.

4.1 REFINEMENTS

The learning principle "increase Q" was suggested as being reasonable; it was not claimed to be necessary or sufficient. Here are several refinements to the basic picture:

- In particular, $\log Q$ is at least as attractive an objective function as Q. In fact, we use $\log Q$. Because the log function is steeper near zero, this causes low scoring patterns to be emphasized, relatively speaking.
  We take the logarithm of the r-score, too; that is, we set the initial deltas to $\partial \log Q / \partial \log r$. (Remember that $\log r$ is the actual output of LeNet.) Observations indicate that with this choice, the deltas are reasonably uniform in size from pattern to pattern.

- If all r-scores are multiplied by a constant, the Q value is unchanged. Therefore, an additional principle is required to promote a reasonable scale for r. For starters, we check for saturated output units on LeNet; a special initial delta is applied if necessary to bring them out of saturation.

- The basic formulation above does not allow the winning Q score (for any particular image) to be less than 10^{-5} (or more generally, N^{-T}, where N is the number of characters in the alphabet, and T is the length of the interpretation). The minimum occurs when all N^T interpretations are equally likely. We need to question the assumption that there are only 10^5 interpretations for a ZIP Code image. In particular, the image may not be a ZIP Code at all. The solution is to add a junk category, making $N' = N + 1$ categories.
  When adding a junk category, some care is required, lest the whole formalism fall down. In particular, there might be a very large number of ways of segmenting the image such that LeNet is absolutely sure, correctly, that the resulting characters are junk. This is not a problem in the numerator, since we just constrain the search to find the highest scoring nonjunk path. The pitfall is that high-scoring junk paths could contribute to the denominator, lowering the score of the right answer even when there is not anything actually wrong. This pitfall can be avoided. It is important to realize that the R-score of a path involves not just the probability of the interpretation, but also the probability of the segmentation.
  In the present implementation, each segment effectively has a junk unit with output level fixed at -0.8, in units where the activation level of ordinary units' normal range is -1.0 to $+1.0$, and extreme range is -1.7 to $+1.7$.
  We now return to the issue of the multiplicative scale factor of the r-scores (which is equivalent to an additive scale on the LeNet output levels). In cases where LeNet is sure of the classification, the junk unit nibbles away a tiny amount of the probability measure. The only way the Q score can improve is for the magnitude of the LeNet outputs to increase. If unchecked, this would cause a dreadful divergence of the weights. We counter this tendency by applying a tiny amount of "state decay," which is implemented as an additional small, arbitrary, negative contribution to the initial deltas. It tells the neurons, if you

haven't got anything better to do, drift downwards. Eventually they drift down until the junk unit starts to make a significant difference, at which point the delta from Baum-Welch is big enough to balance the state decay.

The presence of the junk unit now imparts an absolute scale to the output neurons. Neurons can be high or low with respect to the junk neuron, which is more meaningful than being high or low with respect to other unmoored neurons.

5. EXPERIMENTAL RESULTS

This work used some 11,000 images of ZIP Codes. Approximately 7,000 were chosen at random for the training set, 3,000 for the test set, and 1,000 were reserved for the validation set. About 6% of the images correspond to nine-digit ZIP Codes, and the rest to five-digit ZIP Codes. They were digitized in black and white at 212 dots per inch. The data was collected by SUNY Buffalo, and is referred to as the "hwb" set by them. All images were lifted from live mail at a mail sorting center, and had been rejected (as unclassifiable) by the "MLOCR" (Multi-Line OCR) machines currently used by the U.S. Postal Service.

In most anticipated applications of our recognizer, the cost of punting (i.e., rejecting an image as unclassifiable) is small compared to the cost of a substitution error. To make this clear, consider the following "value model": the recognizer is offered a sack of mail. For every percent that it classifies correctly, the customer will pay $1.00 but, for every percent that it classifies incorrectly, there is a penalty of $10.00. There is no cost for rejecting an item since the customer can send the item through the existing system and be no worse off than he is now.

The recognizer bases its punting decision on the Q-score; if the score exceeds a threshold, the item is accepted; otherwise, it is rejected. The optimal value of the threshold can be determined by varying the threshold and using the value model, as is done in Figure 10. The bottom axis indicates what percentage of the mail is accepted. The upper curve is the value measured on the test set using a $10.00 penalty, and the lower curve is the same but using a $20.00 penalty.

Obviously, at 0% acceptance, the value is zero—no payment, no penalty. For small values of acceptance, the value rises with unit slope, because the recognizer can choose to accept only the images that it is sure of. As we move toward the right side of the graph, at some point the recognizer must begin accepting lower scoring images and taking the attendant risk. The 100% acceptance point is grossly suboptimal.

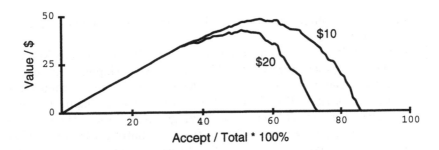

FIGURE 10   Recognizer value curves.

We find it highly advantageous to think of recognizer performance in these terms, as opposed to (for instance) considering the total number of errors made. It is all too easy to make a design change that is superficially desirable because it reduces the total number of errors, yet unfortunately increases the score of some of the remaining errors (and/or decreases the score of some of the correct items) in such a way that the peak of the value curve is actually lowered.

5.1 COMPARISON WITH OUR PREVIOUS RECOGNIZERS

To make a reliable comparison between the old and new ways of building an MCR, we use an idea advocated by Leon Bottou,[3] which we call the *cross-error matrix*. It emphasizes the distinctions between networks which is a big win over the low-tech procedure of evaluating each network separately and comparing the scores. All our recognizers perform so well that a validation set of "only" 1000 zips seems frighteningly small; we have to do the statistics carefully lest we get fooled by fluctuations.

Suppose we have two result-files "v2" and "vj"—the scores for two different MCRs on the validation set. Only the network training procedure is different; the network architecture, preprocessor, and everything else are the same. If you know that vj has a punting rate of 37.9% (at 60% correct) while v2 has a punting rate of 38.6% (also at 60% correct), it is clear that vj is better than v2, but you ought to wonder whether the difference is statistically significant.

But consider the following table:

		vj			
		Ri	Pu	Wr	Ign
	Ri	575	26	0	601
v2	Pu	25	353	1	379
	Wr	0	7	13	20
	Ign	600	386	14	1000

where the row and column headers are abbreviations for the words Right, Punt, Wrong, and Ignore.

The diagonal is easy to understand: there are 575 cases where *both* networks got the image right (above threshold), 353 cases where both punted it, and 13 cases where both got it badly wrong. There are 1000 cases total, so the counts are easily converted to percentages. The rightmost column tells what happens if we ignore vj; for instance, there are 601 images that v2 got right (above threshold) and 20 images that it got badly wrong. Similarly, from the bottom row we see that there are 14 images that vj got badly wrong. The 14 versus 20 ratio is more informative than the 386 versus 379 ratio, but still not overwhelming.

The most interesting information comes from the 7 and the 1 just off the diagonal. That means that if we exclude the 13 cases that *both* networks got wrong, vj makes only 1 mistake that v2 didn't make, while v2 makes 7 mistakes that vj didn't. Most people consider 1 versus 7 rather more convincing than 14 versus 20.

In the example above, we knew all along that vj was better than v2. The latter uses the network that has been our state-of-the-art standard network for almost a year, but it was not particularly matched to the segmenter (candidate cut generator) and alignment lattice we were using. In contrast, vj had been trained (using Baum-Welch and all the latest tricks) with the new segmenter and lattice. A comparison that is more fair (but in some ways harder to interpret) is v1 versus vj; v1 lets the standard net use its favorite segmenter and lattice.

		Ri	Pu	Wr	Ign
	Ri	574	26	0	600
v1	Pu	26	353	3	382
	Wr	0	7	1	18
	Ign	600	386	14	1000

In this case the advantage is less extreme, but this constitutes reasonably good evidence that the new MCR (built according to the principles presented in this chapter) really is better than the old system.

The experiments presented here are preliminary; we expect substantial improvements in the future.

6. REMARKS

It is worth calling attention to certain ideas that were *not* used in deriving the results of this chapter. In particular, we have not used the fashionable, but hard to justify, "principle" of maximum likelihood as our starting point. Rather, our training uses the principle that we want the score of the right answer to increase, on average.

Recall that "likelihood" is a technical term referring to the probability of the training data given the model. In contrast, the derivation presented here revolves around the probability of correct classification, given the training data; this is *not* a likelihood, maximal or otherwise. The mathematics of hidden Markov models[16,17] (HMMs)—which we have not invoked—tends to focus attention on likelihoods, since the HMM is clearly a data-generating model rather than a classifier per se.

We are quite aware that HMM theory can be used to motivate the construction of a lattice-based classifier that is similar in many respects to our design, but we feel that the derivation presented here is simpler and easier to justify, and makes more clear what probabilities are conditioned on what.

7. CONCLUSIONS

We have constructed a multicharacter recognizer by combining neural networks (to evaluate individual segments) with a lattice (to handle the segmentation problem). We have used measure theory to understand how multiple segmentations contribute to the final score—an estimate of the probability of correct segmentation. We discovered that normal single-character recognizers do not supply the information needed for segmentation, but the neural network can be retrained to provide this information. The multicharacter recognizer is trained as a single adaptive system: error-correction information is propagated through the lattice and into the networks.

Preliminary experiments have demonstrated the advantages of this approach, and we have adopted it as the basis of further work.

ACKNOWLEDGMENTS

This report is a snapshot of a long-running collaboration. We are especially indebted to Esther Levin, Yann leCun, Leon Bottou, Craig Nohl, and Yoshua Bengio.

REFERENCES

Cited References

1. Baum, L. E. "An Inequality and Associated Maximization Technique in Statistical Estimation for Probabilistic Functions of a Markof Process." *Inequalities* **3** (1972): 1–8.
2. Bengio, Y., R. deMori, G. Flammia, and R. Kompe. "Global Optimization of a Neural Network—Hidden Markov Model Hybrid." *Proc. of EuroSpeech 91* (1991)
3. Bottou, . Private communication.
4. Bridle, J. S. "Probabilistic Interpretation of Feedforward Classification Network Outputs, with Relationships to Statistical Pattern Recognition." In *Neuro-computing: Algorithms, Architectures and Applications*, edited by F. Fogelman and J. Hérault. Berlin: Springer-Verlag, 1989.
5. Bridle, J. S. "Training Stochastic Model Recognition Algorithms as Networks Can Lead to Maximum Mutual Information Estimation of Parameters." In *Advances in Neural Information Processing 2*, edited by David Touretzky. Palo Alto, CA: Morgan Kaufman, 1990.
6. Burges, C. J. C., O. Matan, J. Bromley, C. E. Stenard. "Rapid Segmentation and Classification of Handwritten Postal Delivery Addresses Using Neural Network Technology." Interim Report, Task Order Number 104230-90-C-2456, USPS Reference Library, Washington, D.C., August 1991.
7. Burges, C. J. C., O. Matan, Y. leCun, J. S. Denker, L. D. Jackel, C. E. Stenard, C. R. Nohl, and J. I. Ben. "Shortest Path Segmentation: A Method for Training a Neural Network to Recognize Character Strings." *IJCNN Conference Proceedings* **3** (1992): 165–172.
8. Driancourt, X., L. Bottou, and P. Gallinari. "Learning Vector Quantization, Multilayer Perceptron and Dynamic Programming: Comparison and Cooperation." In *Proceedings of the IJCNN 91*, 815–819, 1991.
9. Forney, G. D., Jr. "The Viterbi Algorithm." *Proc. IEEE* **61** (1978): 268–278.
10. Franzini, M., K. Lee, and A. Waibel. "Connectionist Viterbi Training: A New Hybrid Method for Continuous Speech Recognition." *Proceedings of ICASSP 90*, 1990.
11. Haffner, P., M. Franzini, and A. Waibel. "Integrating Time-Alignment and Neural Networks for High-Performance Continuous Speech Recognition." *Proceedings of ICASSP 91*, 1991.
12. le Cun, Y., B. Boser, J. S. Denker, D. Henderson, R. E. Howard, W. Hubbard, and L. D. Jackel. "Handwritten Digit Recognition with a Back-Propagation Network." In *Advances in Neural Information Processing 2*, edited by David Touretzky, 396–404. Palo Alto, CA: Morgan Kaufman, 1990.
13. Levinson, S. E. "Structural Methods in Automatic Speech Recognition." *Proc. IEEE* **73(11)** (November 1985).

14. Matan, O., J. Bromley, C. J. C. Burges, J. S. Denker, L. D. Jackel, Y. leCun, E. P. D. Pednault, W. D. Satterfield, C. E. Stenard, and T. J. Thompson. "Reading Handwritten Digits: A ZIP Code Recognition System." *IEEE Computer* **25(7)** (1992): 59–63 .
15. Matan, O., C. J. C. Burges, Y. LeCun, and J. S. Denker. "Multi-Digit Recognition Using a Space Displacement Neural Network." *Neural Information Processing Systems*, edited by J. M. Moody, S. J. Hanson, and R. P. Lippman, **4**. Palo Alto, CA: Morgan Kaufmann, 1992.
16. Pednault, E. P. D. "A Hidden Markov Model for Resolving Segmentation and Interpretation Ambiguities in Unconstrained Handwriting Recognition." Technical Memorandum 11352-090929-01TM, Bell Labs, 1992.
17. Rabiner, L. R. "A Tutorial on Hidden Markov Models and Selected Applications in Speech Recognition." *Proc. IEEE* **77(2)** (1989): 257–286.

Additional Scribblings

1. Baum, and Eagon. "An Inequality with Applications to Statistical Prediction for Functions of Markov Processes and to a Model of Ecology." *Bull. Amer. Math. Soc.* **73** (1963): 360.
2. Baum et al. "A maximization technique occurring in the statistical analysis of probabilistic functions of Markov chains." *Ann. Math. Stat.* **41** (1970): 164.
3. Baum. "An Inequality and Associated Maximization Technique in Statistical Estimation of Probabilistic Functions of Markov Processes." *Inequalities* **3** (1972): 1–8.

Tutorials

1. Levinson, Rabiner, and Sondhi. "An Introduction to the Application of the Theory of Probabilistic Functions of a Markov Process to Automatic Speech Recognition." *Bell Sys. Tech. J.* **64(4)** (1983): 1035.
2. Rabiner, L. R., and Juang. "An Introduction to Hidden Markov Models." *IEEE ASSP* **3(1)** (1986): 4–16
3. Rabiner, L. R., and B. H. Huang, "An Introduction to Hidden Markov Models." *IEEE ASSP* (1986): 4–16.

HM Contributions

1. Bose, C. , and S. Kuo, "Connected and Degraded Text Recognition Using a Hidden Markov Model." Technical Manuscript No.11224-911001-12TM.
2. Agazzi, O., S. Kuo, E. Levin, and R. Pieraccini. "Connected and Degraded Text Recognition Using Planar Hidden Markov Models." To appear in Proceedings of ICASSP 93.
3. Lecolinet, E., and J. Moreau. "A New System for Automatic Segmentation and Recognition of Unconstrained Handwritten ZIP Codes." Paper presented at the 6th Scandinavian Conference on Image Analysis, Finland, June 1989. (I have a copy but can't find it.)

4. Lecolinet, E., and J. P. Crettez. "A Graphme-Based Segmentation Technique for Cursive Script Recognition." Paper presented at ICDAR 91 First International Conference on Document Analysis and Recognition, Saint-Malo, France. (I just received a copy of this.)

Index